4th International Symposium on Transportation Studies in Developing Countries

IOP Conference Series: Earth and Environmental Science
Volume 1294

Bandung, Indonesia
11 - 12 November 2023

ISBN: 978-1-7138-9120-8
ISSN: 1755-1307

Printed from e-media with permission by:

Curran Associates, Inc.
57 Morehouse Lane
Red Hook, NY 12571

CURRAN ASSOCIATES INC.
proceedings.com

Some format issues inherent in the e-media version may also appear in this print version.

Institute of Physics
Dirac House, Temple Back
Bristol BS1 6BE UK

Phone: 44 1 17 929 7481
Fax: 44 1 17 920 0979

techtracking@iop.org

Additional copies of this publication are available from:

Curran Associates, Inc.
57 Morehouse Lane
Red Hook, NY 12571 USA
Phone: 845-758-0400
Fax: 845-758-2633
Email: curran@proceedings.com
Web: www.proceedings.com

TABLE OF CONTENTS

IOP Conf. Series: Earth and Environmental Science 1294 (2024) 011001 doi:10.1088/1755-1315/1294/1/011001

4th International Symposium on Transportation Studies in Developing Countries
11th – 12th November 2023

Institut Teknologi Bandung

Forum Studi Transportasi antar Perguruan Tinggi (FSTPT), the preeminent transportation academic forum in Indonesia, consistently organised the International Symposium on Transportation Studies in Developing Countries (ISTSDC), a conference where international scholars and researchers convene to discuss transportation-related issues. Institut Teknologi Bandung (ITB) hosted the 4th ISTSDC from 11th to 12th November 2023.

The 4th ISTSDC aims to bring together leading international researchers, developers, and users of advanced technologies from transportation engineering disciplines to discuss theoretical and practical issues in the field.

This conference covers various topics, namely :

- Intelligent Transport System
- Land Use and Transport Interactions
- Social and Environmental Impact
- Transport Safety, Security, and Climate Change Adaptation
- Design and Analysis for Transport Infrastructure
- Traffic Engineering and Management
- Public Transport
- Transport Economics and Finance
- Freight and Logistics
- Transport Policy, Regulation, and Management
- Geotechnics and Pavement Technology
- Marine Transport and Inland Waterways
- Transport Infrastructure and Management
- Railway Transport
- Air Transport

The conference has received a total of 68 submissions of academic papers, out of which 44 articles have been deemed suitable for presentation and inclusion in the conference proceedings. The contributors represent several academic institutions in Indonesia, Japan, Australia, and Hungary, which can be listed as follows:

- Budapest University of Technology and Economics (Hungary)
- Kochi University of Technology (Japan)
- University of New South Wales (Australia)
- Institut Teknologi Bandung (Indonesia)
- Universitas Gajah Mada (Indonesia)
- Universitas Indonesia (Indonesia)
- Universitas Islam Sultan Agung ITB (Indonesia)
- Politeknik Transportasi Darat Bali (Indonesia)
- Universitas Khairun (Indonesia)
- Universitas Sumatera Utara (Indonesia)
- Politeknik Negeri Bali (Indonesia)
- Universitas Syiah Kuala (Indonesia)
- Universitas Jember (Indonesia)
- Universitas Sebelas Maret (Indonesia)
- Petra Christian University (Indonesia)
- Universitas Diponegoro (Indonesia)
- Universitas Katolik Parahyangan (Indonesia)
- Institut Teknologi Nasional (Indonesia)
- Kementerian Perhubungan Republik Indonesia
- Universitas Jenderal Achmad Yani (Indonesia)
- Universitas Islam Riau (Indonesia)

Additionally, the conference invited distinguished researchers to deliver keynote speeches, including:

- Dr. Chandra Balijepalli (Institute for Transport Study University of Leeds, UK)

- Prof. Hiroaki Nishiuchi (School of Systems Engineering, Kochi University of Technology, Japan)
- Prof. Tri Basuki Joewono (Parahyangan Catholic University)
- Prof. Raphael Grzebieta (Transport and Road Safety Research, University of New South Wales, AUS, Department of Forensic Medicine, Monash University, AUS)

In addition to ITB, the conference was also co-hosted by other universities in Bandung, namely :

- Universitas Katolik Parahyangan (Unpar)
- Universitas Jenderal Ahmad Yani (Unjani)
- Institut Teknologi Nasional (ITENAS)
- Universitas Logistik & Bisnis Internasional (ULBI)
- Universitas Pendidikan Indonesia (UPI)
- Universitas Kristen Maranatha (UKM)
- Politeknik Negeri Bandung (POLBAN)
- Universitas Islam Bandung (Unisba)
- Universitas Langlangbuana (UNLA)
- Universitas Sangga Buana YPKP (USBYPKP)
- Universitas Wiyana Mukti (Unwim)

Those universities are then working together to ensure a successful conference in Bandung. The committees are then listed as follows:

Steering Committee :

- Chairman of FSTPT
- Advisory board of FTSPT
- Dean of Faculty of Civil and Environmental Engineering, ITB
- Dean of Faculty of School of Architecture, Planning, and Policy Development, ITB
- Chairman of Transportation Research Group, ITB

IOP Conf. Series: Earth and Environmental Science 1294 (2024) 011001 doi:10.1088/1755-1315/1294/1/011001

Scientific Committee:

- Prof. Bagus Hario Setiadji (UNDIP, Indonesia)
- Prawira Fajarindra Belgiawan, Ph.D (ITB, Indonesia)
- Prof. Yusak Susilo (University of Natural Resources and Life Sciences, Vienna, Austria)
- Dr. Ari K. M. Tarigan (University of Stavanger (UiS), Norway)
- Dr. Anugrah Ilahi (United Arab Emirates University, United Arab Emirates)
- Dr. H.R. Pasindu (University of Moratuwa, Srilanka)
- Febri Zukhruf, Ph.D (ITB, Indonesia)
- Prof. Muhammad Zudhy Irawan (Universitas Gadjah Mada, Indonesia)
- Mr. Berlian Kushari (UII, Indonesia)

Organizing Committee:

- R. Sony Sulaksono Wibowo, Ph.D. (ITB)
- Dr. Eng. Widyarini Weningtyas (ITB)
- Mr. Andrean MaulanaITENAS)
- Prof. Ina Helena Agustina (UNISBA)
- Dr. Aine Kusumawati (ITB)
- Dr. Harmein Rahman (ITB)
- Dr. Russ Bona Frazila (ITB)
- Dr. Sofyan Triana (ITENAS)
- Mr. Tilaka Wasanta (UNPAR)
- Dr. Eri Susanto Hariadi (ITB)
- Mr. Barkah Wahyu (ITENAS)
- Mr. Teguh Tuhu Prasetyo (ULBI)
- Dr. An An Anisarida (UNWIM)
- Mr. Pradhana W Nariendra (ULBI)
- Mr. Andreas Danu Prakoso (ITB)
- Mr. Angga Marditama Sultan Sufanir (POLBAN)
- Dr. Ir. R. Didin Kusdian (USBYPKP)

- Taufiq Nugroho, Ph.D. (ITB)
- Dr. Nurlela Kumala Dewi (ULBI)
- Mr. Ignatius Sudarsono,(UNLA)
- Dr. Juang Akbardin (UPI)
- Mr. Ronni Ibnu Sabil (UNJANI)
- Ms. Dizka Triarsani Ramadhanti (ITB)

Moreover, the success of the 4th ISTSDC can be attributed to the assistance provided by various sponsors, specifically:

- ASTRA Infra
- Jasamarga
- Angkasa Pura I
- DSI
- Pelindo

The documentation of the conference can be seen in the figures below.

IOP Conf. Series: Earth and Environmental Science 1294 (2024) 011001 doi:10.1088/1755-1315/1294/1/011001

Figure 1 Keynote Speakers in 4th ISTSDC

Figure 2 Keynote Speech by Prof. Hiroaki Nishiuchi

Figure 3 Keynote Speech by Prof. Tri Basuki Joewono

IOP Conf. Series: Earth and Environmental Science 1294 (2024) 011001 doi:10.1088/1755-1315/1294/1/011001

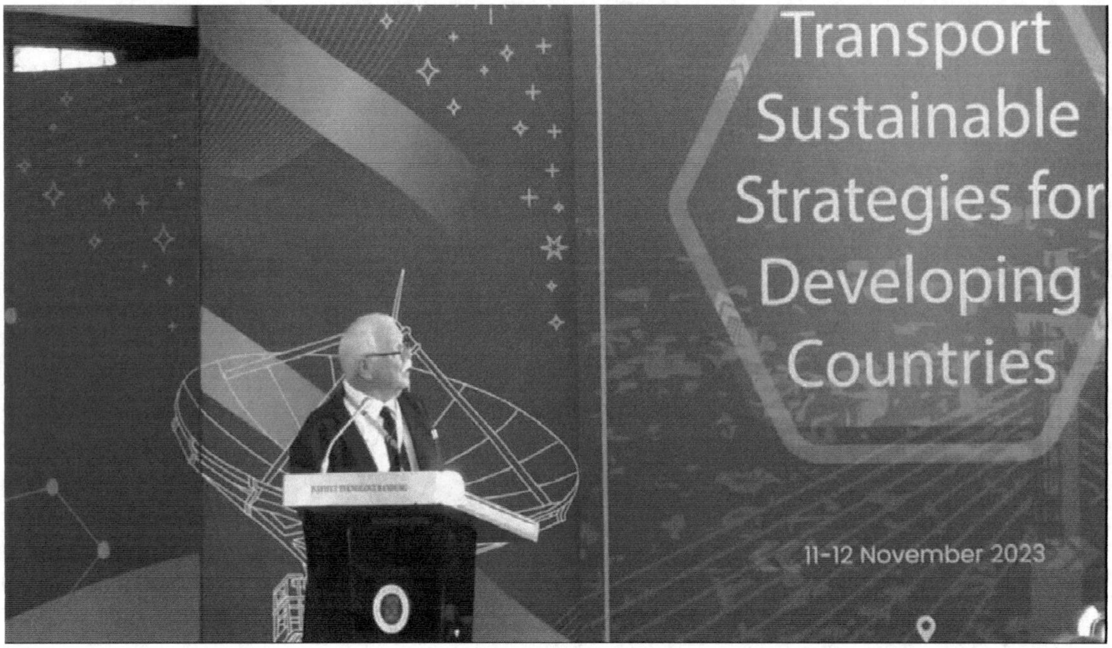

Figure 4 Keynote Speech by Prof. Raphael Grzebieta

Figure 5 Keynote Speech by Dr. Chandra Balijepalli

Figure 6 Presentation of Parallel Session

IOP Conf. Series: Earth and Environmental Science 1294 (2024) 011002 doi:10.1088/1755-1315/1294/1/011002

Peer Review Statement

All papers published in this volume have been reviewed through processes administered by the Editors. Reviews were conducted by expert referees to the professional and scientific standards expected of a proceedings journal published by IOP Publishing.

- **Type of peer review:** Single Anonymous
- **Conference submission management system:** Morressier
- **Number of submissions received:** 68
- **Number of submissions sent for review:** 65
- **Number of submissions accepted:** 44
- **Acceptance Rate (Submissions Accepted / Submissions Received × 100):** 64.7
- **Average number of reviews per paper:** 1.44
- **Total number of reviewers involved:** 29
- **Contact person for queries:**
  Name: Prof. Bagus Hario Setiadji
  Email: bhsetiadji@ft.undip.ac.id
  Affiliation: Universitas Diponegoro

IOP Conf. Series: Earth and Environmental Science 1294 (2024) 012001 doi:10.1088/1755-1315/1294/1/012001

Examining urban development trends and characteristics: future of public transportation in greater bandung

I P Suharjo[1], **A A R Leksono**[1], **M F Aththaariq**[1], **U A Adnan**[1], **Evangelista**[1], and **G D Priadi**[1]

[1]Department of Civil Engineering, Bandung Institute of Technology, Bandung, 40132, Indonesia

E-mail: iqbalprasetyo22@gmail.com

Abstract: This article investigates the urban development trends and characteristics of Greater Bandung, Indonesia, with the goal of projecting the future urban landscape and proposing government actions. The study analyzes the existing situation of urban development in the region by utilizing government regulations, statutes, and planning documents, including designs. The findings provide a complete knowledge of how Greater Bandung's urban landscape is anticipated to grow and highlight significant areas that require attention. The analysis identifies significant inadequacies in transportation infrastructure and underlines the need for government intervention to remedy these deficiencies by reviewing existing policies and plans. Transportation modelling was employed to give validation and sense in numerical sense of travel volumes in the roads of Bandung, helping to form a concise recommendation. The article will be useful for policymakers, urban planners, and stakeholders interested in the future development of Public Transportation in Greater Bandung. These recommendations encompass transit hubs and corridors suggestions to accommodate future growth of the city and its surrounding areas. The result of the study analysis shows that the development of the Greater Bandung Area can be done by prioritizing 7 (seven) main points of interest which are translated into 9 (nine) nodes. These nodes can be connected into 2 (two) main corridors as the backbone and 21 feeder corridors.

1. Introduction

Urban development is an essential aspect in modern societies, shaping the physical, social, and economic landscape of cities. In the context of Indonesia, rapid urbanization has posed significant challenges in managing urban growth and ensuring sustainable development.

The development of major cities is clearly evident with the establishment of metropolitan areas. Metropolitan areas have been defined in [1] on Spatial Planning as self-contained urban areas or cores that are functionally connected to the surrounding areas. The fundamental reason for establishing metropolitan areas is the orientation towards achieving improvement in economic aspects, welfare, modernity, and sustainability. Bandung Raya is one of the metropolitan regions characterized by its central growth, along with Bodebek Karpur Metropolitan and Greater Cirebon Metropolitan [2].

Greater Bandung consists of 6 regencies and cities across the basin in West Java Province. With an area of 343.087 Ha, it is estimated that the population of Bandung Metropolitan Area will reach 12 million inhabitants in 2030 [3]. Greater Bandung covers the area of Bandung City, Cimahi City, Padalarang Regency, West Bandung Regency, Bandung Regency, and Sumedang Regency. The region as a whole has been enjoying economic growth to this day and has been targeted to be among one of the world-class tourism destinations in Indonesia by the government.

The arrangement regarding Greater Bandung is contained in the Presidential Regulation No. 45 of 2018 [4] which discusses national strategic areas. There are several main issues in its development into a metropolitan area, one of which is about transportation, namely the movement to and from the main cities (Bandung City and Cimahi City). To support the acceleration of these issues, the Bandung Basin Area Management Agency (Badan Pengelola Kawasan Cekungan Bandung) was formed through West Java Governor Regulation No. 86 of 2020 [5] with the tasks and functions of planning, managing, developing, cross-regional public services, and controlling space utilization in the Greater Bandung.

The Greater Bandung has experienced considerable urban sprawl in recent years, resulting in the formation of Greater Bandung as a metropolitan area encompassing several surrounding cities and regencies. The amount of movement and growth that has occurred without being balanced by adjustments to its infrastructure has led to traffic saturation[6]. This is in accordance with the characteristics of big cities in Indonesia in the transportation sector which often occurs is congestion [7].

Understanding the trends and characteristics of urban development in Greater Bandung is crucial for effective planning and decision-making processes. It provides insights into the current state of the urban landscape, identifies potential flaws, and guides interventions to shape a more sustainable and livable future for the region. While some studies have explored specific aspects of urban development in Bandung, there remains a gap in research that comprehensively examines the urban development trends and characteristics across Greater Bandung as a whole.

The purpose of this research paper is to address this gap by conducting literature study of the urban development trends and characteristics of areas within Greater Bandung, Indonesia. The paper aims to provide an understanding of the future urban landscape of the region based on government laws, statutes, and planning documents, including blueprints. By examining these official documents, the research seeks to identify the existing flaws in the urban development trajectory of Greater Bandung and propose potential interventions that the government could undertake.

Moreover, this research places special emphasis on the importance of urban transit and transportation infrastructure development as a key aspect of urban planning. Efficient and sustainable transportation networks are essential for addressing the challenges posed by rapid urbanization, such as traffic congestion, air pollution, and inadequate mobility options. By analyzing existing policies and plans, this study aims to shed light on potential shortcomings in transportation infrastructure and provide recommendations for government intervention in this critical area.

2. Literature Study

2.1 Bandung Metropolitan Area

Subroto [8] stated that the built-up area in the Greater Bandung is currently concentrated in Bandung City with its total area reaching a total proportion of 90.66%, followed by Cimahi City with 88.35%. Limited land in both urban areas causes sporadic and uncontrolled expansion of built-up area growth to the surrounding area and is referred to as urban sprawl.

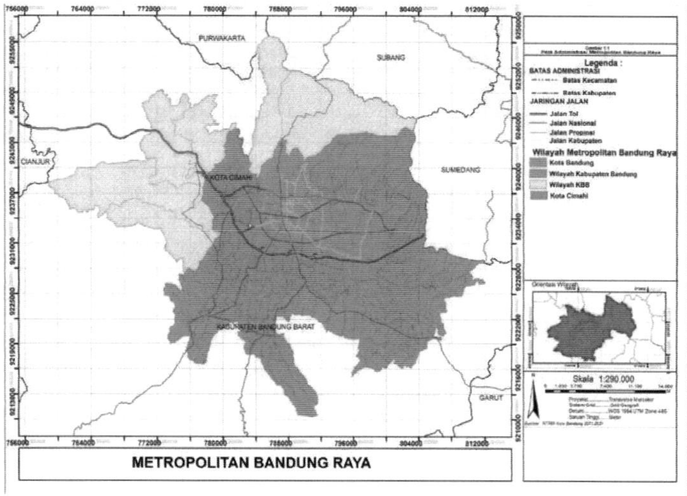

Figure 1. Greater Bandung Area [9].

Bandung City and Cimahi City became the core area surrounded by 15 urban areas spread across West Bandung Regency, Bandung Regency, and Sumedang Regency and finally formed a metropolitan area. The selection of the development of these 15 areas is based on the potential of the population density and the potential movement of new trip generation and trip attraction which exists in each urban area.

The direction of the 15 urban areas development is based on Presidential Regulation No. 45 of 2018 on Urban Area Spatial Plan of Greater Bandung [4], namely: development relies on local strengths and potentials by holding the potential of the area fundamentally, including community empowerment, development of tourist activity centers, development of supporting facilities and infrastructure, and integration between nature conservation areas, cultivation, and tourism. Therefore, transportation is essential to actualize an integrated urban system between core urban areas from and to the surrounding area. The zoning of the area is shown in the following table:

Table 1. Zoning of Urban Areas in Greater Bandung.

No	West Bandung Residence	Bandung Residence	Sumedang Residence
1	Padalarang-Ngamprah	Soreang-Kutawaringin	
2	Cipatat	Margahayu-Margaasih	
3	Batujajar	Majalaya-Ciparay	
4	Cihampelas	Baleendah-Dayeuhkolot-Bojongsoang	Jatinangor-Tanjungsari
5	Lembang	Perkotaan Banjaran	
6	Cipendeuy-Cikalong Wetan	Cicalengka	
7	Cililin	Cileunyi-Rancaekek	

From each urban area development based on regional regulations regarding the Greater Bandung Spatial and Regional Plan (RTRW), the main points of activity are obtained in the table below:

Table 2. Activity Centres in Greater Bandung.

No	Legal Basis	Region	Activity Center Area	Description
1	Regional Regulation No. 5 of 2022 on Bandung City Spatial and Regional Plan (RTRW) 2022-2042 [10]	Bandung City	1. Western Side City Center Area Development Center (PPK) (SWK Bojonagara, Cibeunying, Tegalega, Karees) 2. East Gedebage Area Development Center (PPK) (SWK Arcamanik, Ujung Berung, Kordon, Gedebage)	• Bojongnagara- Airport and strategic industrial area (Aerobiopolis) • Cibeunying- Protection of heritage and culinary buildings • Tegalega-Creative Industry (Mediapolis) • Karees- Integrated creative industry (Karyapolis) • Arcamanik – Sports Coaching Center • Ujung Berung – Cultural Arts Development Center • Kordon – Art Products and Services • Gedebage – Integrated education and industrial center
2	Regional Regulation No. 27 of 2016 on Bandung Residence Spatial and Regional Plan (RTRW) 2016-2036 [11]	Bandung Residence	1. Soreang-Kutawaringin-Ketapang 2. Margahayu-Margaasih 3. Cileunyi-Rancaekek 4. Cicalengka 5. Majalaya-Ciparay 6. Baleendah-Dayeuhkolot-Bojongsoang 7. Banjaran	• Government, trade, and residential areas • Rubber and Food Industry • Residential, agricultural and conservation areas • Industrial Area • Residential and education areas • Agricultural Industry and Residential
3	Regional Regulation No. 2 of 2012 on West Bandung Residence Spatial and Regional Plan (RTRW) 2009-2029 [12]	West Bandung Residence	Development Areas 1. WP Padalarang 2. WP Lembang 3. WP Cikalong Wetan 4. WP Cililin	• Padalarang – Residential, trade centers and services • Lembang – Tourism areas, conservation and residential area • Cikalong Wetan – Agroindustry • Cililin – Farm products cultivation and conservation

Table 3 Activity Centres in Greater Bandung (continued).

No	Legal Basis	Region	Activity Center Area	Description
4	Regional Regulation No. 4 of 2013 on Cimahi City Spatial and Regional Plan (RTRW) 2012-2032 [13]	Cimahi City	Area Development Centers (Pusat Pengembangan Kawasan) 1. PPK Baros 2. PPK Alun-Alun	• SWK A – Tourism Area • SWK B – Higher Education Center • SWK C – Goods and Services Trade Center area development • SWK D – Creative Industry development • SWK E – Military Complex and Heritage Area
5	Regional Regulation No. 4 of 2018 on Sumedang Residence Spatial and Regional Plan (RTRW) 2022-2042 [14]	Sumedang Residence	Jatinangor - Tanjungsari	Higher Education Center and Residential.

From the activity area center that has been described in RTRW for each regional area, further corrections are made to existing conditions and developments which became the center of movement of the new trip generation and trip attraction with tourism potential, commercial activities, residential areas, industry, thus "Point of Interest" is obtained. The intended movements and pulls are used to express current and forecast future movements.

2.2 Corridor and Route Selection Method

An indication of the need for transportation facilities is the increase in population, regional development, growth of activity centers, increase in urbanization, increase in income, the desire to travel will increase. The use of private vehicles is certainly not a solution as the main means of transportation for Greater Bandung. In current condition of Bandung City, the number of private vehicle ownership almost equals the population in Bandung City. Increased traffic volume induced by this current trend will overpopulate the road and cause frequent delays in travel time across the road networks, bringing financial losses and environmental impacts to a concerning level. Therefore, the development of public transport facilities and infrastructure should be a major focus for the municipal and central government. Furthermore, the development will provide increased integration and connectivity of each region of Greater Bandung.

Tamin [7] explains the basic concept of the four step modelling namely:

a. Trip Generation

This is a step that estimates the number of movements originating and attracted from a zone or land use. With output results in the form of generation/attraction from one zone to another.

b. Trip Distribution

IOP Conf. Series: Earth and Environmental Science 1294 (2024) 012001 doi:10.1088/1755-1315/1294/1/012001

This is the step that estimates the distribution of movements that go to and leave the zone. The distribution of movements is represented by a desire line or an Origin-Destination matrix. So that the output at this stage can see the distribution of movements from the origin zone to the destination zone.

c. Moda Split

It is the step that determines the proportion of travellers in the form of people and goods who use certain modes of transport.

d. Trip Assignment

Trip Assignment is the last step. In selecting a route there are several reasons or determining factors such as travel time, time value, and travel costs.

In planning a route or transport corridor there are several theories used such as:

2.2.1 Trunk and Feeder

This concept is a service pattern with routes that only operate on the main corridor (trunk) and are equipped with the smaller feeder routes. Feeders aim to connect the main corridor with the initial and final destinations of passengers. Therefore, passengers need to transfer between the two routes (ITDP, 2016). The Trunk - Feeder concept is a system that is considered more effective and efficient. Operating the mode parallel to the road will not overload the road network.

2.2.2 Direct Service

Direct Service is a concept or system that can minimize the distribution of passengers from the feeder to the station or bus stop so that it will save passenger travel time.

This system needs to be put in place to improve the performance of public transport. With the concept of feeders and separate stops, passengers must transfer from one service to another service at the final terminal. As it cannot cut in the middle of the line, to save travel time, it is necessary to increase the range of areas covered by public transport, and serve the increase in passengers, it is necessary to implement direct service.

Figure 2. Direct Service Route in Jakarta (ITDP, 2016).

3. Methodology

The study is conducted with steps illustrated by the flow chart in Figure 2. Broadly speaking, there are two main works that will be carried out in this study, namely:

1. Spatial analysis and regional planning
   The initial stage of this research is to collect data through various media, books, scientific publications, the use of government regulations, articles of associations, and planning documents including design, as well as direct observation in the field. We also conducted undisclosed interview with anonymous experts and local transportation stakeholders to further understand the current and future urban and transportation planning trends in Greater Bandung
2. Transport modeling analysis
   Using the data from spatial analysis, we have collected points of interest based on possible trip generation trends on current and future land-use. Trip distribution and modal split were done using PTV VISSUM software to project the volume of trips. Lastly, we examined the segments, their respective trips, and accessibility to find possible corridors to be recommended.

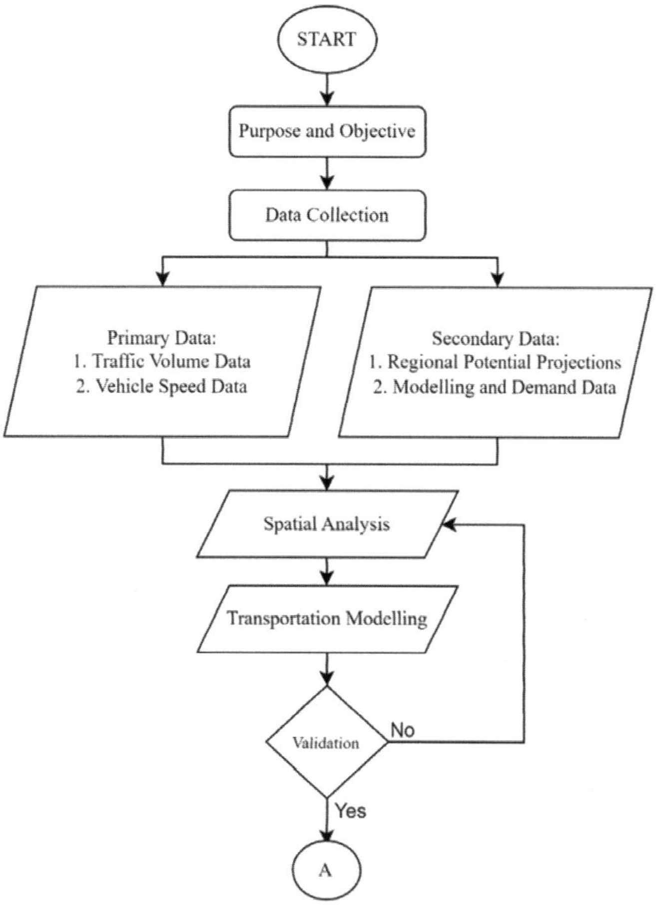

Figure 3. Research Framework.

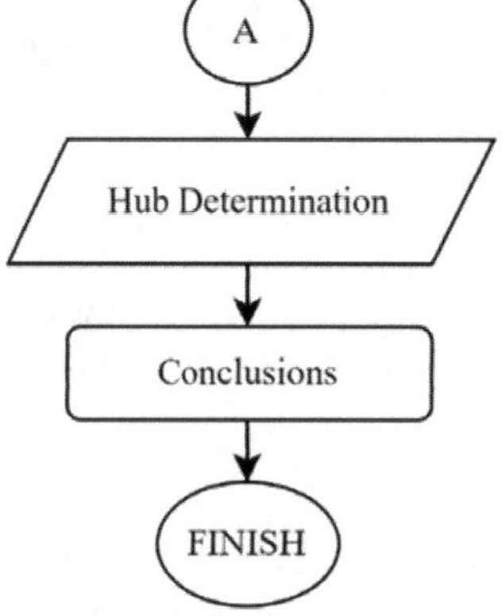

Figure 4. Research Framework (continued).

4. Analysis Results

4.1. Future Demand in Greater Bandung

Based on the analysis study conducted, Greater Bandung has several locations that attract attention. The points are translated into 7 (seven) points of interest.

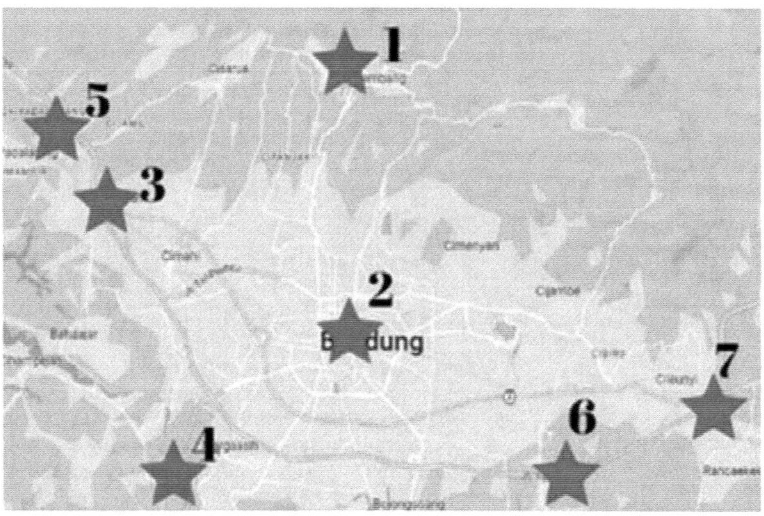

Figure 5. Points of Interest (Google Maps, 2023).

The points were chosen as a result of an analysis conducted based on the potential movement and development of the area in the Greater Bandung Area which is explained in the table as follows:

Table 4. Area Descriptions.

No	Area	Description
1	Lembang	• Development of a world-class agro-tourism area. potential of more than 10,000 visitors at 3 (three) tourist attractions alone (Lembang Park Zoo, Floating market, and Great Asia Africa). • Generally a tourism area which may generate trips with all its attractions. • Plans to be its own city in the future.
2 and 3	Bandung and Cimahi City	Greater Bandung's core regions (Existing travel demands and volumes are already very high)
4	Soreang	• Agglomeration of 5 areas (Soreang, Kutawaringin, Ketapang, Margahayu, Margaasih) which plans to build a CBD area (business centre) with apartments, hotels, business commerce centre inside. • Textile industrial complex. • Capital of Bandung Residence. Potential development of new settlements due to easy access to the toll road to and from the core city (Bandung)
5 and 6	Padalarang (KBB) and Gedebage (Bandung City)	• High-Speed Railway stops in Padalarang Station and new urban area development. • Ease of Access and its impact on further sprawl. There are 2 stations available as access hubs, so the potential for urban sprawl is even greater, which comes not from the Bandung area alone, but also residents from Jakarta. The chance for potential growth is utilised by the private sector to cooperate in developing the potential of the city area, one of which is a residential area, industry, religious tourism, artificial entertainment tourism. (Radarbandung.id, 2022).
7	Rancaekek and Jatinangor	Fulfilment of movement service facilities from the Eastern part of Greater Bandung (starting point of the Cisumdawu Toll Road, meeting point of 4 national routes, the final end of the Purbaleunyi Toll Road) and the development of the Rancaekek Industrial Area, as well as the center of higher education in Jatinangor.

The first area is in the West Bandung Regency, namely Lembang, which is known to be a tourist area that has a very high attraction both on weekends and weekdays. It is necessary to develop transportation service infrastructure to reduce congestion, other considerations of this area are the geographical conditions in the form of hillside plateaus and small roads. The second area is the core cities, namely Bandung City and Cimahi City. The third area is in the Soreang area, the capital of Bandung Regency, which has the potential to develop the Central Business District area. The existing conditions in this area are dominated by residential areas that shift towards the core area. The fourth area is where the Jakarta-Bandung high-speed train station is located, namely in Padalarang, West Bandung Regency, and the Tegalluar area adjacent to Gedebage, potential for the emergence of the new trip attraction and generation. The last area is in the Rancaekek area in Bandung City which will develop into a Rancaekek industrial area. In Sumedang regency, there is the Jatinangor area which is an educational area.

4.2. Traffic Demand and Modelling

The transportation modelling was conducted using Origins-Destinations Matrix input data that was applied to the existing road network model with the whole Greater Bandung Area as coverage area, resulting in the largest private vehicle flows in the southern region and the largest VCR >0.8 in the core area of Bandung City.

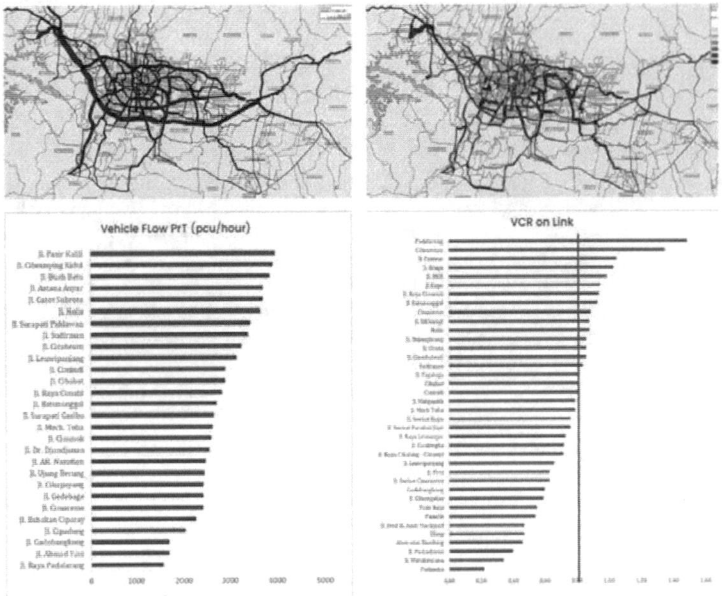

Figure 6. Vehicle Flow in Greater Bandung.

The West - East movement is the highest PrT demand on the Purbaleunyi toll road and Soekarno Hatta Road with a value of > 4,000 pcu/hour. While the North - South link movement is the largest vehicle flow in the Pasirkaliki area with 3,968 pcu/hour and the highest VCR values are in the Padalarang and Cibeureum zones with values of 1.49 and 1.35. Desire line public transport movements for Greater Bandung, all movements towards the city core.

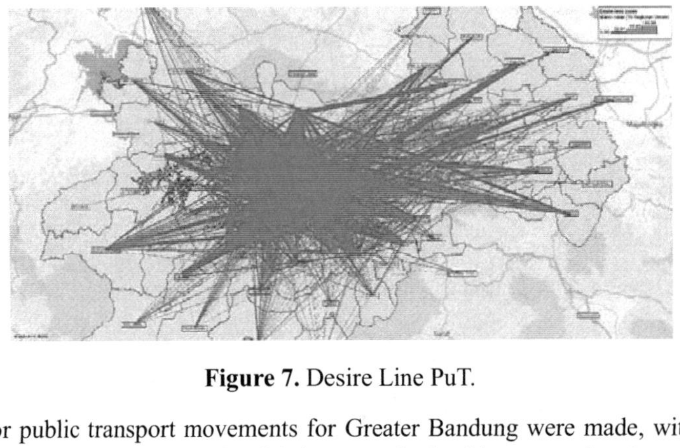

Figure 7. Desire Line PuT.

Desire line for public transport movements for Greater Bandung were made, with all movements coming towards the city core. Model validation was obtained by comparing the modelled data with available traffic counting data (primary data from surveys at: Gedebage, Lembang, Padalarang, Pusdai, and Batununggal. From the model validation results, the PrT (private vehicles) calibration was then carried out with "Demand Matrix Correction", it was found that the goodness of fit of the model expressed in the R-square parameter was 0.6231 which indicated that the model was valid in representing current traffic conditions.

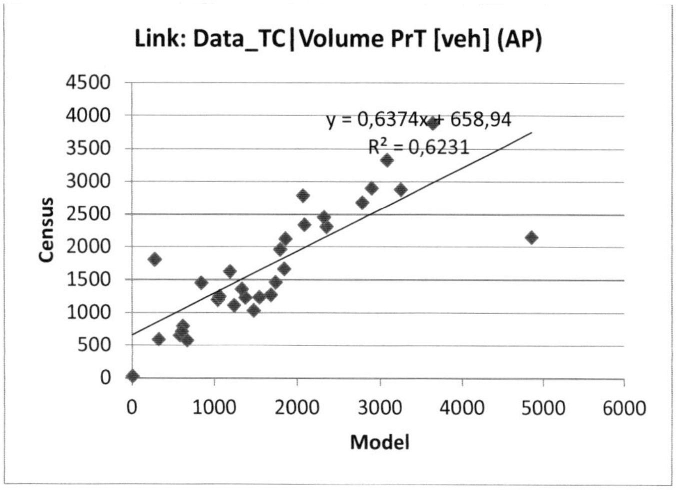

Figure 8. Model Validation.

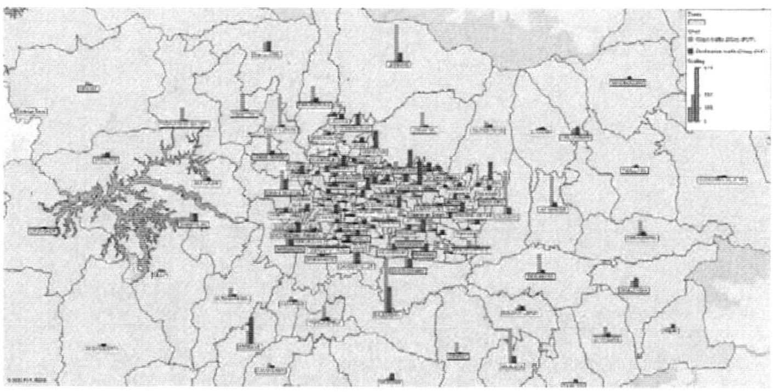

Figure 9. Trip Zone Value for Public Transport (PuT).

The largest public transport generation values are in the Baleendah, Cileunyi, Margahayu, Lembang, Majalaya, Bojongsoang, Jatinangor, Cimahi, and Dayeuhkolot zones. While the largest public transport attraction values are in the Babakan Tarogong, Soreang, Baleendah, Antapani, and Cigadung zones.

To conclude the process, we obtained several points of interest from the results of spatial analysis, which will generate massive trips in the future due to their land use. They were then validated by the transport modelling process, yielding results satisfying the standard for public transit demand. The modelling also shows urgency in the development of alternative modes of travel with VCR values being as high as 1.49.

4.3. Hubs and Corridors Placement

At this stage, the points of interest will then be defined as hubs. Designation of the hubs is carried out to create a flow of relationships between the areas to be served. The placement of these hubs and their subsequent corridors depend on the factors such as the future trips between zones, existing modes of travel, road segment dimensions, land ownerships, and existing boundaries. Based on the results of the hub point analysis, 2 (two) categories were obtained based on the orientation of the main service corridor, namely vertical (north-south) and horizontal (east-west). The following are the results of the area development analysis that produced the main corridors and their supporters:

Table 5. Hubs Oriented on The Horizontal Movements.

No	Hub	Code	Route Descriptions
1	Padalarang Station	1A 1B 1C	St. Padalarang-Cimahi St. Padalarang-Ngamprah St. Padalarang-KBP
2	Alun-Alun Cimahi	2A 2B 2C	Alun-Alun Cimahi-Term. Leuwipanjang Alun-Alun Cimahi-Cililin Alun-Alun Cimahi-Lembang
3	Leuwipanjang Terminal	3A 3B 3C 8A	Term. Leuwipanjang-Gedebage Term. Leuwipanjang-Soreang Term. Leuwipanjang-Soreang Alun-alun Kota Bandung - Term. Leuwipanjang
4	Gedebage	4A 4B 4C 4D	Gedebage-Cibiru Gedebage-Majalaya Gedebage-Term.Cicaheum Gedebage-Stasiun Tegalluar
5	Cibiru	5A 5B 5C 5D	Cibiru-Univ. Padjajaran Cibiru-Majalaya Cibiru-Rancaekek Cibiru-Term.Cicaheum

Table 6. Hubs Oriented on Vertical (North-South) Movements.

No	Hub	Code	Route Descriptions
6	Babakan Siliwangi	6A 6B 6C	Babakan Siliwangi-Padjajaran Babakan Siliwangi-Lembang Babakan Siliwangi-Term. Cicaheum
7	Padjajaran	7A 7B	Padjajaran - Alun-Alun Kota Bandung Padjajaran – Bandara Husein
8	Alun-alun Kota Bandung	8A 8B	Alun-Alun – Term. Leuwipanjang Alun-Alun – Kiara Artha Park
9	Soreang	9A	Soreang – Alun-Alun Ciwidey

Based on the results of the analysis of the development of the area, the results of the overall recapitulation of the corridors formed include:

1. Main Corridors (2 Corridors)
   a. North-South(Babakan Siliwangi-Soreang) : 26,4 km
   b. East-West (Padalarang-Jatinangor) : 40,5 km
2. Feeder (21 Corridors) : Total 272,9 km

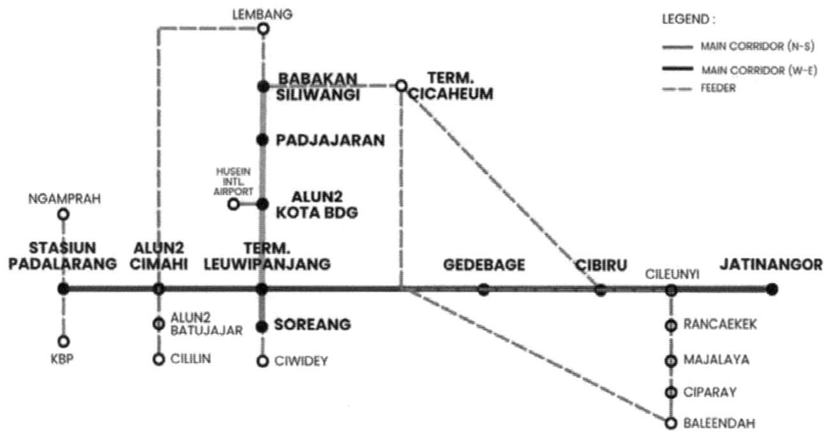

Figure 10. Sketch of Future Transit Hubs and Corridors in Greater Bandung.

5. Conclusion and Discussion

Greater Bandung as a National Strategic Area (Kawasan Strategis Nasional) with its regional characteristics, existing road network, and distribution of built-up areas requires transport solutions that accommodate existing movements. Based on the analysis of the study, the development of the area can be done by prioritising 7 (seven) main points of interest which are translated into 9 (nine) hub points. The result is that there are 2 (two) main corridors as the backbone and 21 feeder corridors.

Acknowledgments
Many thanks to Ir. R. Sony Sulaksono Wibowo, M.T, Ph.D dan Dr. Eng., Ir. Febri Zukhruf, S.T, M.T as Lecturers and Andrean Maulana, S.T, M.T as the Class Assistant in the Group Field Project Class in Transportation Engineering Research Group of Department of Civil Engineering, Bandung Institute of Technology.

References

[1] Presiden Republik Indonesia 2007 *Undang-Undang No 26 Tahun 2007 tentang Penataan Ruang*
[2] Pemerintah Provinsi Jawa Barat 2014 *Regional Development Plan In West Java Province*
[3] Dinas Perhubungan Jawa Barat 2022 *Pengembangan Layanan Angkutan Penumpang Umum di Bandung Raya*
[4] Presiden Republik Indonesia 2018 *Peraturan Presiden Republik Indonesia Nomor 45 Tahun 2018 tentang Rencana Tata Ruang Kawasan Perkotaan Cekungan Bandung*
[5] Gubernur Jawa Barat 2020 *Peraturan Gubernur Jawa Barat Nomor 86 Tahun 2020 tentang Badan Pengelola Kawasan Perkotaan Cekungan Bandung*
[6] Saputra, I. dan Mulyanisa, Z. M. 2018 Probabilitas Peralihan Moda Pengguna Kendaraan Pribadi (Mobil) ke Monorel Bandung Raya *Jurnal Wilayah dan Kota* volume 5 (1), 1-2.
[7] Tamin, O. Z. 2000 *Perencanaan dan Pemodelan Transportasi Edisi Kedua*

IOP Conf. Series: Earth and Environmental Science 1294 (2024) 012001 doi:10.1088/1755-1315/1294/1/012001

[8] Pikiranrakyat 2022 *Kota Bandung dan Cimahi Padat, Badan Pengelola Kawasan Perkotaan Cekungan Bandung Kaji 15 Kawasan Perkotaan*

[9] Nugraha, A. 2018 Kajian Remote Parking Untuk Melayani Komuter di Metropolitan Bandung Raya (Studi Kasus : Jalan Terusan Buah Batu dan Jalan Mochamad Toha. *Repository Universitas Pasundan*, hal 6-7.

[10] Wali Kota Bandung 2022 *Peraturan Daerah Kota Bandung Nomor 5 Tahun 2022 tentang Rencana Tata Ruang Wilayah Kota Bandung Tahun 2022-2042*

[11] Bupati Bandung. 2016. *Peraturan Daerah Kabupaten Bandung Nomor 27 Tahun 2016 tentang Rencana Tata Ruang Wilayah Kabupaten Bandung Tahun 2016-2036*. Bandung.

[12] Bupati Bandung Barat 2012 *Peraturan Daerah Kabupaten Bandung Barat Nomor 2 Tahun 2012 tentang Rencana Tata Ruang Wilayah Kabupaten Bandung Barat Tahun 2009-2029*

[13] Wali Kota Cimahi 2013 *Peraturan Daerah Kota Cimahi Nomor 4 Tahun 2013 tentang Rencana Tata Ruang Wilayah Kota Cimahi Tahun 2012-2032*

[14] Bupati Sumedang 2018 *Peraturan Daerah Kabupaten Sumedang Nomor 4 Tahun 2018 tentang Rencana Tata Ruang Wilayah Kabupaten Sumedang Tahun 2018-2038*

[15] The Institute for Transportation and Development Policy 2016 *The BRT Standard*

[16] The Institute for Transportation and Development Policy 2016 *Implementasi Transportasi Publik Rendah Karbon – Direct Service Report 2012*

IOP Conf. Series: Earth and Environmental Science 1294 (2024) 012002 doi:10.1088/1755-1315/1294/1/012002

Walkability assessment and pedestrian way planning on Mercedes Benz Road Corridor, Cicadas, Bogor Regency

N N Hayati[1], G Sulistyaningtyas[1], Daud Nawir[2], S Z Ishak[3, 4]

[1] Urban and Regional Planning Study Program, Faculty of Engineering, Universitas Jember (UNEJ), Jember-68121, East Java, Indonesia

[3] Civil Engineering Study Program, Faculty of Engineering, Universitas Borneo Tarakan (UBT), Tarakan City-77115, North Borneo, Indonesia

[3] Malaysia Institute of Transport (MITRANS), Universiti Teknologi MARA (UiTM), 40450 Shah Alam, Selangor, Malaysia

[4] School of Civil Engineering, College of Engineering, Universiti Teknologi MARA (UiTM), 40450 Shah Alam, Selangor, Malaysia

Corresponding author: nunung.nuring@unej.ac.id

Abstract. A pedestrian way is a space for pedestrians to walk safely. Walkability is a conceptual term for measuring connectivity and quality levels of various pedestrian ways. Walkability assessment can be measured by assessing the infrastructure available for pedestrians. The Mercedes Benz Road, Cicadas, Bogor Regency is the location used as the research object. It is used as a research location because the land use is dominated by trade and services as many as 116 units and industry as many as 26 units. A large enough demand is not balanced with the procurement of pedestrian facilities that are adequate. This study aims to determine the level of walkability and the need for pedestrian walkways. The method used is the Global Walkability Index and the Greenshields model. The results of this study found that all segments are slightly walkable. In segment 1, 1.5 m and 2 m wide sidewalks were designed. In segment 2, 1.5 m and 1.7 m wide sidewalks were intended, and in segment 3, 2 m wide sidewalks were designed on each side.

1. Introduction

Walking is a component of the transportation system that can connect various activity functions in urban areas [1]. Walking is an essential component of an urban transportation system, as it combines modes of transportation and destinations [2], [3]. The provision of pedestrian paths can provide benefits such as creating a better life and providing convenient transportation alternatives [4], [5]. Due to the broad scope of the concept of walkability, it can be recognized as a plan related to the living environment and urban life [6]. A walkable city is an urban planning supported by transportation that can trigger its people to walk to carry out daily activities [7]. The functions and benefits of pedestrian paths based on the Minister of Public Works Regulation Number 03/PRT/M/2014 [8] concerning guidelines for planning, providing, and utilizing pedestrian network infrastructure and facilities in urban areas, namely to facilitate the movement of pedestrians from one place to another by ensuring aspects of pedestrian safety and comfort.

In Indonesia, pedestrian ways are still being developed because many pedestrian ways still have problems. The condition of pedestrian paths that change functions to become places for street vendors to sell and illegal parking lots are some examples of the issues that continue to occur [9]. To create a pedestrian-friendly environment, good planning is needed by considering the characteristics of pedestrians and the area with the concept of walkability that is friendly to pedestrians [10], [11].

The research was conducted on the Mercedes Benz Road, Cicadas, Bogor Regency, along 2.05 Km. Based on RTRW 2016-2036 of Bogor Regency [12], the Mercedes Benz Road is part of the Tlajung Udik - Wanaherang - Cileungsi road section, which is developed into a primary collector road IV [12]. Mercedes Benz Road has land use as a mixed-use area with a dominance of trade and services of as many as 116 units and an industry of 26 units. This creates a large enough pull so that walking is one of the quite attractive modes [13]–[15]. There are no pedestrian ways at the research location, and many side obstacles make pedestrians need to go down the road when walking [16]. The government has made a policy to provide connected, quality, and equitable public works infrastructure, and there is also a strategy to add quality pedestrian paths (RENSTRA PUPR Bogor Regency, 2018-2023 [17]).

Based on current conditions, it is necessary to evaluate the pedestrian walkways to identify the level of safety and comfort of pedestrians on Jalan Mercedes Benz, Cicadas, and Bogor Regency. The method used is the Greenshields model analysis to determine the need for the width of the pedestrian ways at the research location [16]. The Global Walkability Index (GWI) method is also used, which can measure the level of the walkability of a city [7]. The results of this study are the level of walkability and the need for pedestrian walkways in each segment.

2. Method

2.1 Study Sites
The research location in this study is on Mercedes Benz Road, Cicadas, Gunungputri District, Bogor Regency, with a total road length of 2.05 Km. The research location is divided into three segments, and each segment is divided into two observation points: observation point A for the west and observation point B for the east (Figure 1). The division of these segments was based on road intersections.

2.2 Data Collecting
The data collection method in this study was carried out in two ways, namely by primary and secondary data collection methods. Primary data collection methods in this study were direct observation in the field and questionnaires [19]. The pedestrian observation activities are divided into several times, on weekends and weekdays. The pedestrian survey was divided into a survey of the number of pedestrians and a survey of pedestrian speed. Questionnaires were also distributed to respondents, namely pedestrians at the research location. The number of respondents is divided into each segment determined, that is, 32 respondents in each segment. In comparison, secondary data is in various books, journals, theses, and state documents. Some of the documents used in this research are documents related to spatial planning and pedestrian way procurement regulations.

2.3 Research Variables
In the study, two objectives were divided into three variables. Then, each variable was split into sub-variables and indicators. The variables of this study can be seen in Table 1.

IOP Conf. Series: Earth and Environmental Science 1294 (2024) 012002 doi:10.1088/1755-1315/1294/1/012002

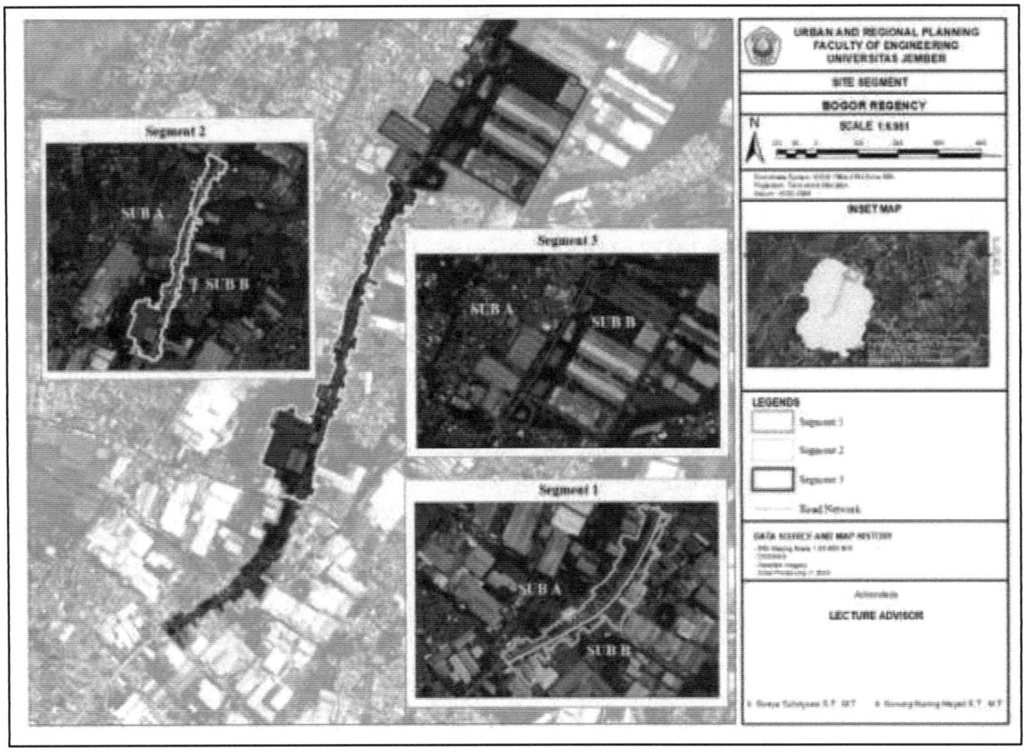

Figure 1. Map of the Research Location.

Table 1. Research Variables.

No.	Purpose	Variable	Sub Variable	Indicator
1.	Knowing the characteristics of pedestrian ways on Mercedes Benz, Gunungputri, Bogor Regency	Characteristics of pedestrian way	Pedestrian ways geometry	- Width - Pavement type - Side obstacles
			Pedestrian Way support facilities	- Green lane - Lighting - Seating - Safety fence - trash cans - ramps and disabled lanes - markings, signs, and information boards - bus stops
		Greenshields	The Relation between Volume, Speed, and Density	- Number of pedestrians - Pedestrian speed - Pedestrian density
2.	It was knowing the level of walkability by measuring the level of walkability on Mercedes Benz Street, Gunungputri District, Bogor Regency.	*Walkability*	*Walkability Index*	- Walkway mode conflict - Pedestrian path availability - Crosswalk availability - Safety of crosswalks - Motorist behavior - Availability of supporting facilities - Supporting infrastructure for people with disabilities - Constraints or barriers - Security and crime

2.4 Data Collecting
This research uses three analyses: descriptive analysis, Greenshields model analysis, and Global Walkability Index. The descriptive analysis research method carried out in this study explains the data obtained in the form of narrative descriptions [20]. The characteristics of the relation between volume, speed, and density can be studied with a mathematical relationship between the three parameters, namely speed, flow, and density of pedestrian traffic on a road section. The derivation of a model capable of expressing or representing the relation between these three variables is the Greenshields linear model, where the mathematical relationship between speed and density is assumed to be linear [21]. The result of the Greenshields analysis is the need for the width of the pedestrian path [22]. In addition, the Global Walkability Index analysis is used to determine whether an area is pedestrian-friendly or not [14]. Walkability assessment aims to find out how friendly an area is for walking and review the level of pedestrian services [23].

3. Result and Discussion

3.1 Overview of Mercedes Benz Road, Cicadas, Bogor Regency
It is known that trade and services and industrial land uses dominate segment 1. In addition, there are also residential and educational land uses. In segment 2, it can be seen that trade and service land uses dominate it; there are also industrial, worship, health, and office land uses. Segment 3 is dominated by industrial land use, but residential land use, trade, and services exist. Thus, it can be seen that the entire research corridor is dominated by industrial land use and also trade and services.

The research location has not been reached by mass urban public transportation services in the form of commuters, so the local community can only rely on mass transportation in the form of city transportation (angkot) and private vehicles. Based on BPS Gunung Putri in figures 2021 [18], Cicadas only has four public transportation routes with high occupancy of public transportation passengers.

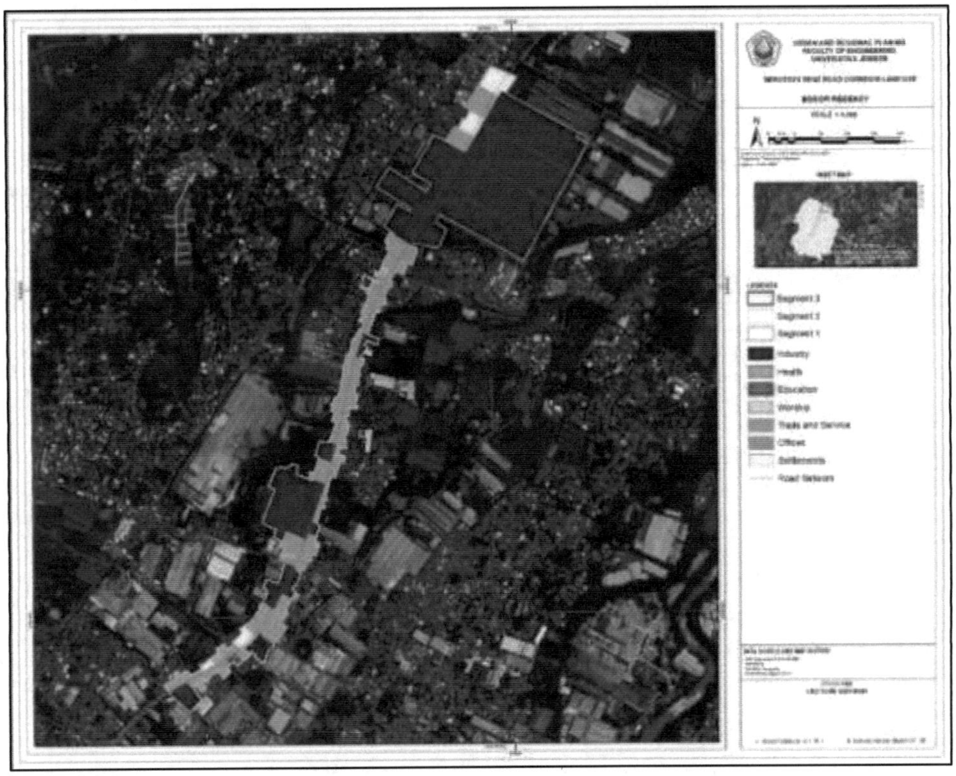

Figure 2. Research Corridor Land Use Map.

3.2 *Pedestrian Walkway Characteristics*

The characteristics of the pedestrian way are categorized by each segment, with length, width, and side obstructions discussed (Table 2).

Table 2. Characteristics of Pedestrian Way.

Segment	Section	Pedestrian Way Characteristics		
		Length	Side Road With	Side Obstacles
Segment 1	A	550 m	1,5 m	- Street Vendor
	B	550 m	2 m	- Flowerpot - Parking lot
Segment 2	A	900 m	1,5 m	- Street Vendor - Parking lot
	B	900 m	1,7 m	- Street Vendor - Parking lot
Segment 3	A	600 m	2 m	- Street Vendor - Tree
	B	600 m	2 m	- Street Vendor - Tree - Signage

3.3 *Effective Width of Pedestrian Way Needs*

Pedestrian volume and speed data were obtained based on observations in the Mercedes Benz Road corridor four days a week, divided into two weekdays and two weekends. Surveys were conducted in each direction with a time interval of 15 minutes in the morning (07.00 – 08.00 WIB), afternoon (12.00 – 13.00 WIB), afternoon (15.00-16.00 WIB), and evening (19.00 – 20.00 WIB). In addition, observations were also made of pedestrian travel time along 10 meters in both directions to determine pedestrian speed.

3.3.1 *Effective Width Requirement of Segment 1 Pedestrian Walkway.* Speed data and pedestrian volume in segment one were taken from the Jalan Pancasila IV intersection to the Jalan Pancasila V intersection. Segment 1 pedestrian volume is calculated at 15-minute intervals (Figure 3). Next, calculate the pedestrian speed in segment one based on observations of 10 pedestrians passing along 10 m (Table 3).

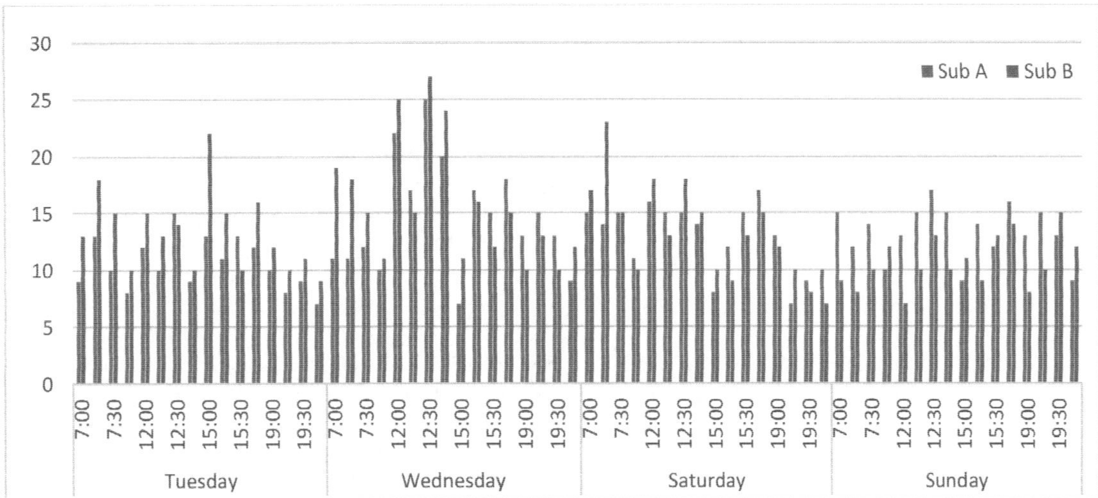

Figure 3. Pedestrian Volume Segment 1.

Table 3. Pedestrian Speed Segment 1.

Hour	Pedestrian Travel Time (10 m)										Average Travel Time
	1	2	3	1	2	6	1	2	9	1	
SUB A											
7:00	8,84	8,04	9,12	8,75	7,6	8,9	6,4	7,83	8,55	7,65	8,168
7:15	7,93	8,45	8,97	7,13	6,97	8,65	7,43	6,82	7,35	7,14	7,684
7:30	8,2	7,8	7,25	7,35	8,12	8,2	8,62	8,92	9,1	8,72	8,228
8:00	8,12	7,7	7,32	7,65	7,8	7,12	7,26	7,34	8,53	8,14	7,698
12:00	9,45	8,05	8,79	6,89	6,8	8,35	8,63	9,72	9,65	8,64	8,497
12:15	9,57	8,62	8,94	8,87	8,98	8,65	9,42	9,26	7,27	8,7	8,828
12:30	9,1	8,97	8,23	7	6,58	6,62	7,15	8,46	8,18	8,23	7,852
13:00	7,23	6,5	6,35	6,52	7,65	7,4	7,35	8,79	8,75	8,71	7,525
15:00	7,96	7,85	9,35	9,67	8,36	8,75	9,02	7,8	9,23	8,3	8,629
15:15	9,65	8,23	8,76	7,62	7,09	8,37	8,5	8,62	6,75	6,7	8,029
15:30	9,23	8,95	8,35	8,42	8,46	8,62	7,37	8,62	8,12	7,35	8,349
16:00	8,32	8,36	8,94	8,32	7,4	7,35	8,32	7,62	7,56	7,35	7,954
19:00	9,88	9,8	6,8	8,05	8,41	7,78	6,39	9,34	6,39	7,62	8,046
19:15	8,92	6,71	8,9	8,97	7,93	7,17	7,7	7,72	8,24	8,75	8,101
19:30	7,56	7,62	8,61	7,83	7,46	7,25	8,12	8,42	7,64	9,21	7,972
20:00	8,64	8,42	8,6	7,25	7,34	7,5	9,23	8,79	8,54	7,62	8,193
SUB B											
7:00	11,4	12,47	12	9,82	10	8,75	12,6	8,73	8,8	6,5	10,107
7:15	10,65	9,75	9,89	9,7	11,05	10,73	7,62	8,41	8,13	8,2	9,413
7:30	8,73	7,64	7,35	8,12	8,46	8,92	9,32	7,65	7,36	7,98	8,153
8:00	6,94	8,72	9,65	7,62	7,94	7,36	8,62	8,34	8,56	8,46	8,221
12:00	9,74	8,86	9,82	6,79	11,5	7,6	11,21	9,85	7,65	7,9	9,092
12:15	7,62	7,34	7,5	6,89	7,65	8,64	8,35	8,46	7,64	7,92	7,801
12:30	8,61	8,35	8,67	8,25	8,63	12,69	9,62	9,08	9,98	9	9,288
13:00	8,24	7,7	7,65	7,95	8,65	8,25	8,56	9,64	7,82	8,94	8,34
15:00	10,8	11,72	9,8	6,52	7,98	8,36	8,75	8,68	7,25	7,43	8,729
15:15	7,63	7,89	8,17	8,65	7,62	10,2	8,36	8,75	8,11	9,57	8,495
15:30	8,65	8,35	8,45	7,95	8,92	9,23	9,24	8,15	8,25	8,36	8,555
16:00	9,25	7,85	7,65	7,62	7,35	8,6	8,61	8,35	7,64	7,6	8,052
19:00	8,52	7,61	9,5	8,55	6,85	9,81	10,49	9,5	8,75	8,6	8,818
19:15	8,15	8,1	10,45	9,63	8	8,84	9,56	9,16	9,52	8,73	9,014
19:30	8,25	8,64	8,32	8,3	8,64	7,52	7,56	7,64	9,1	8,93	8,29
20:00	7,16	8,35	9,2	8,34	8,64	7,71	7,65	7,62	8,32	8,5	8,149

3.3.2 *Calculating Maximum Volume.* The Greenshields linear model equation calculates the maximum volume of pedestrians passing through segment 1 (Table 4).

Table 4. Analysis of Volume, Speed, and Density Relationships Greenshields Method Segment 1.

Hour	Volume (X.Y)	Speed (Y)	Density (X)	X^2
7:00	2	65,66	0,03046	0,00093
7:15	2	70,19	0,02850	0,00081
7:30	2	73,26	0,02730	0,00075
8:00	1	75,38	0,01327	0,00018
12:00	2	68,22	0,02932	0,00086
12:15	2	66,24	0,03019	0,00091
12:30	2	76,66	0,02609	0,00068
13:00	2	75,64	0,02644	0,00070
15:00	1	69,13	0,01447	0,00021
15:15	2	72,62	0,02754	0,00076
15:30	2	70,99	0,02817	0,00079
16:00	2	74,97	0,02668	0,00071

Hour	Volume (X.Y)	Speed (Y)	Density (X)	X^2
19:00	2	71,16	0,02811	0,00079
19:15	2	70,11	0,02853	0,00081
19:30	2	73,79	0,02710	0,00073
20:00	1	73,43	0,00000	0,00000
Total	29	1147,46	0,39215	0,01062

To find a linear regression analysis, an analysis was carried out based on the results of the previous tabulation by taking the total value of each existing variable.

$$\boldsymbol{B} = \frac{16.(29)-(0{,}40577).(1147{,}46)}{16.(0{,}01081)-(0{,}40577)^2} = \frac{464-465{,}602}{0{,}17294-0{,}16465} = \frac{-1{,}60281}{0{,}00829} = \mathbf{-193{,}3397} \quad (1)$$

$$\boldsymbol{A} = \frac{(0{,}01081).(1147{,}46)-(0{,}40577).(29)}{16.(0{,}01081)-(0{,}40577)^2} = \frac{12{,}40244-11{,}76726}{0{,}17297-0{,}16465} = \frac{0{,}63519}{0{,}00829} = \mathbf{76{,}61956} \quad (2)$$

So, Sff = A = 76,61956

$$\boldsymbol{Dj} = \frac{-A}{B} = \frac{-76{,}61956}{-193{,}3397} = \mathbf{0{,}3962} \quad (3)$$

The Sff and Dj values are known with the above calculations to determine the Vm value. The measure of the Vm value can be seen in the equation (4).

$$\boldsymbol{Vm} = \frac{Dj \times Sff}{4} = \frac{0{,}256 \times 77{,}942}{4} = \mathbf{8\ ped/min} \quad (4)$$

3.3.3 *Relation between Speed (S), Density (D), and Volume (V) Segment 1.* The relationship between volume and density can be known in equation (5) and presented in Figure 4 (a).

$$\boldsymbol{V = 77{,}94284.D - 0{,}3962.D^2} \quad (5)$$

The relationship between speed and density can be known in equation (6) and presented in Figure 4 (b).

$$\boldsymbol{S = 77{,}94284 - 0{,}3962\,.D} \quad (6)$$

The relationship between volume and speed can be known in equation (7) and presented in Figure 4 (c).

$$\boldsymbol{V = 0{,}256.S - 0{,}3962.S^2} \quad (7)$$

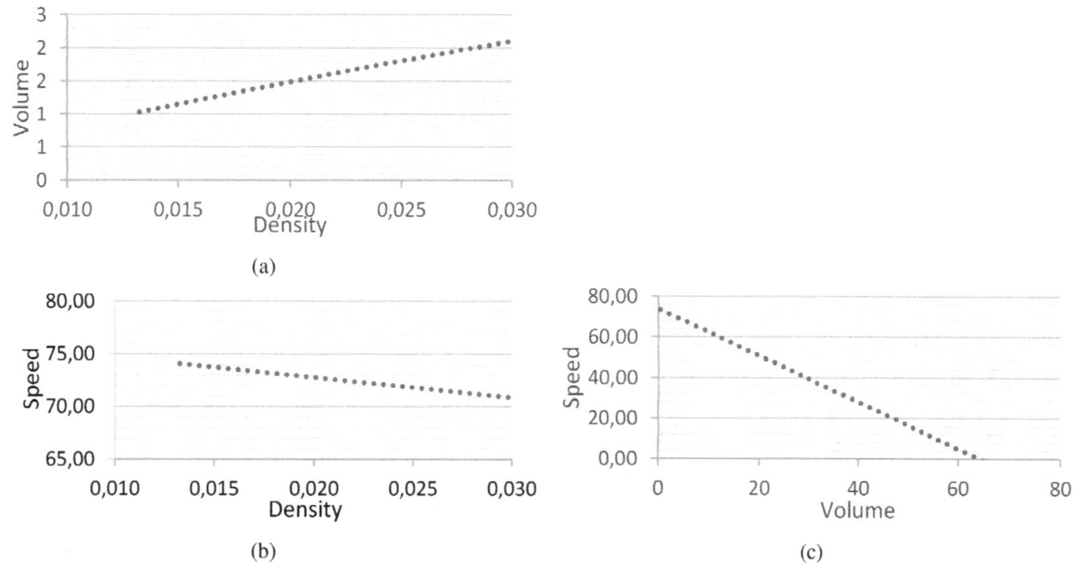

Figure 4. Graph of Speed, Density, and Volume Relationship of Greenshields Model Segment 1.

3.4 Effective Sidewalk Width Dimension

Based on the calculation of the maximum pedestrian flow in Segment 1, it is known that the expected pedestrian value is two people. Furthermore, the effective sidewalk width dimensions can be calculated in equation (8).

$$W = \left(\frac{P}{35}\right) + n = \left(\frac{8}{35}\right) + 1 = 1,23\ meter \quad (8)$$

Based on the calculations (8), it can be seen that the width of the pedestrian walkway required in segment 1 is 1.23 meters. However, because the research corridor is included in the industrial area, the width of the pedestrian path in sub-A is 1.5 m wide, and sub-B is 2 m wide to adjust to the existing conditions of the width of the existing sidewalk.

3.4.1 *Effective Width Requirement of Segment 2 Pedestrian Walkway.* Speed and pedestrian volume data in segment two were taken at the intersection of Jalan Pancasila V to the intersection of Jalan Utama Perumahan Villa Asri I. Segment 2 pedestrian volume was calculated at 15-minute intervals (Figure 5). Next, calculate the pedestrian speed in segment two based on observations of 10 pedestrians passing along 10 m (Table 5).

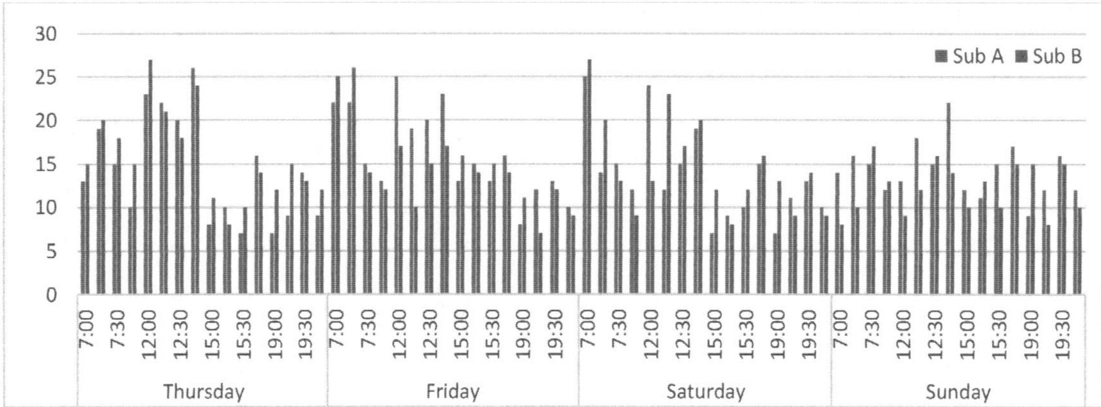

Figure 5. Pedestrian Volume Segment 2.

Table 5. Pedestrian Speed Segment 2.

Hour	Pedestrian Travel Time (10 m)										Average Travel Time
	1	2	3	4	5	6	7	8	9	10	
SUB A											
7:00	6,7	7,06	8,33	5,97	7,48	6,7	6,77	7,44	12	11,5	8,00
7:15	9,25	6,98	8,3	5,24	9,56	9,5	7,59	1,42	7,9	8,34	7,41
7:30	8,25	8,34	8,51	8,35	8,45	7,65	7,24	7,7	7,65	7,34	7,95
8:00	7,82	7,35	7,64	8,64	6,34	7,46	8,52	8,61	7,34	8,32	7,80
12:00	7,94	7,86	6,75	9,66	8,62	8,75	7,41	8,48	7,27	8,23	8,10
12:15	7,36	7,8	7,89	6,75	9,09	9	8,38	7,42	8,4	9,2	8,13
12:30	8,43	8,5	8,64	9,37	8,54	8,32	8,41	9,2	8,97	8,52	8,69
13:00	8,27	8,35	7,45	8,65	7,83	9,2	9,23	8,31	9,23	8,65	8,52
15:00	8,3	7,55	8,23	8,12	7,25	7,36	8,21	8,1	7,94	9,85	8,09
15:15	8,64	9,63	12,06	8,37	8,14	8,67	7,95	9,12	8,64	7,93	8,92
15:30	9,14	8,34	8,46	8,51	8,23	8,31	8,64	6,98	7,32	7,45	8,14
16:00	8,37	8,45	8,14	8,2	7,89	7,64	8,64	8,73	8,61	9,61	8,43
19:00	8,11	7,62	6,87	8,61	8,73	8,69	9,24	7,74	8,65	6,82	8,11
19:15	8,38	8,94	8,4	8	7,64	7,66	6,47	8,43	8,31	8,75	8,10
19:30	8,64	8,58	7,34	7,82	8,345	8,46	8,51	8,94	8,62	9,38	8,46
20:00	7,25	7,65	8,45	8,36	8,45	8,34	8,62	8,52	8,35	9,12	8,31

Hour	Pedestrian Travel Time (10 m)										Average Travel Time
	1	2	3	4	5	6	7	8	9	10	
SUB B											
7:00	8,48	8,5	7,05	8,5	7,74	7,7	10,8	6,65	8,25	8,3	8,197
7:15	11,1	8,12	8,74	8,7	9,82	11,6	9,21	9,2	9,63	7,03	9,315
7:30	10,23	9,45	8,45	8,52	8,64	8,45	8,94	7,98	7,46	7,62	8,574
8:00	7,65	8,64	8,76	8,62	8,43	8,35	8,64	9,2	7,42	7,98	8,369
12:00	8,35	7,5	7,12	8,61	9,2	8,62	7,64	7,67	7,71	7,64	8,006
12:15	4,25	4,5	7,17	8,87	11,42	15	7,4	5,7	5,6	8,6	7,851
12:30	7,84	7,62	7,45	8,35	8,46	9,12	7,63	7,84	8,36	8,42	8,109
13:00	9,31	8,35	8,75	8,63	8,27	8,65	9,23	8,36	7,54	8,64	8,573
15:00	6,19	8,91	7,24	7,52	8,55	8,12	8	7,35	7,4	8,75	7,803
15:15	7,65	7,92	9,53	9,72	9,15	10,82	9,72	8,35	7,38	8	8,824
15:30	8,68	7,56	8,36	7,65	7,71	9,1	7,65	6,54	6,56	7,32	7,713
16:00	8,54	8,62	7,73	7,64	8,62	7,85	7,58	7,65	7,56	7,45	7,924
19:00	7,43	7,75	8,67	8,92	8,14	8,19	7,64	7,51	8,34	7,68	8,027
19:15	8,75	7,38	7,14	8,36	7,9	7,85	8,72	6,8	9,87	8,62	8,139
19:30	7,86	7,94	8,65	8,45	7,38	7,43	8,64	8,72	9,1	7,62	8,179
20:00	7,61	6,89	7,82	7,38	7,62	8,62	8,34	8,56	8,45	8,61	7,99

3.4.2 *Calculating Maximum Volume.* The Greenshields linear model equation was used to calculate the maximum volume of pedestrians passing through segment 2 (Table 6).

Table 6. Analysis of Volume, Speed, and Density Relationships Greenshields Method Segment 2.

Hour	Volume (X.Y)	Speed (Y)	Density (X)	X^2
7:00	2	74,11	0,027	0,00073
7:15	3	71,76	0,042	0,00175
7:30	2	72,63	0,028	0,00076
8:00	2	74,20	0,027	0,00073
12:00	2	74,52	0,027	0,00072
12:15	2	75,09	0,027	0,00071
12:30	2	71,43	0,028	0,00078
13:00	3	70,22	0,043	0,00183
15:00	1	75,50	0,013	0,00018
15:15	2	67,65	0,030	0,00087
15:30	2	75,71	0,026	0,00070
16:00	2	73,39	0,027	0,00074
19:00	1	74,37	0,013	0,00018
19:15	1	73,91	0,014	0,00018
19:30	2	72,10	0,028	0,00077
20:00	1	73,62	0,014	0,00018
Total	30	1170,20	0,412	0,01181

To find a linear regression analysis, an analysis was carried out based on the results of the previous tabulation by taking the total value of each existing variable.

$$\boldsymbol{B} = \frac{16.(30)-(0{,}412).(1170{,}2)}{16.(0{,}01181)-(0{,}412)^2} = \frac{480-482{,}42576}{0{,}18893-0{,}16996} = \frac{-2{,}42576}{0{,}01897} = \mathbf{-127{,}8892} \tag{9}$$

$$\boldsymbol{A} = \frac{(0{,}01181).(1170{,}2)-(0{,}412).(30)}{16.(0{,}01181)-(0{,}412)^2} = \frac{13{,}81757-12{,}36783}{0{,}18893-0{,}16996} = \frac{1{,}44974}{0{,}01897} = \mathbf{76{,}43245} \tag{10}$$

So, Sff = A = 76,43245

$$\boldsymbol{Dj} = \frac{-A}{B} = \frac{-76{,}43245}{-127{,}8892} = \mathbf{0{,}597646} \tag{11}$$

The Sff and Dj values are known with the above calculations to determine the Vm value. The measure of the Vm value can be seen in the equation (12).

$$\boldsymbol{Vm = \frac{Dj \times Sff}{4} = \frac{0{,}597646 \times 76{,}43245}{4} = 11\ ped/min} \tag{12}$$

3.4.3 *Relation between Speed (S), Density (D), and Volume (V) Segment 2.* The relationship between volume and density can be known in equation (13) and presented in Figure 6 (a).

$$\boldsymbol{V = 76{,}43245.D - 127{,}89.D^2} \tag{13}$$

The relationship between speed and density can be known in equation (14) and presented in Figure 6 (b).

$$\boldsymbol{S = 76{,}43245 - 127{,}89.D} \tag{14}$$

The relationship between volume and speed can be known in equation (15) and presented in Figure 6 (c).

$$\boldsymbol{V = 0{,}597646.S - 127{,}89.S^2} \tag{15}$$

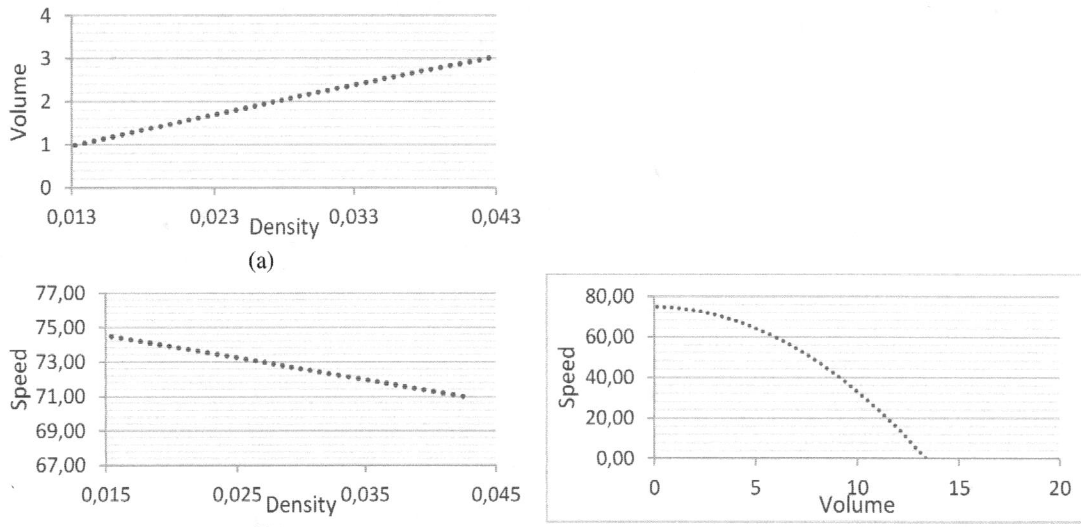

Figure 6. Graph of Speed, Density, and Volume Relationship of Greenshields Model Segment 2.

3.4.4 *Effective Sidewalk Width Dimension Segment 2.* Based on the calculation of the maximum pedestrian flow in segment 2, it is known that the expected pedestrian value is two people. Furthermore, the effective sidewalk width dimensions can be calculated in equation (16).

$$\boldsymbol{W = \left(\frac{P}{35}\right) + n = \left(\frac{11}{35}\right) + 1 = 1{,}31\ meter} \tag{16}$$

Based on the calculations (16), it can be seen that the width of the pedestrian path required in segment 2 is 1.31 meters. However, because the research corridor is included in the industrial area, the width of the pedestrian path in sub-A is 1.5 m wide, and sub-B is 1.7 m wide to adjust to the existing conditions of the existing sidewalk width.

3.4.5 *Effective Width Requirements of Segment 3 Pedestrian Walkaway.* Data on pedestrian speed and volume in segment three were taken from the intersection of Jalan Utama Perumahan Villa Asri I to the intersection of Jalan Utama Perumahan Taman Putri. Segment 1 pedestrian volume is calculated at 15-

minute intervals (Figure 7). Next, calculate the pedestrian speed in segment three based on observations of 10 pedestrians passing along 10 m (Table 7).

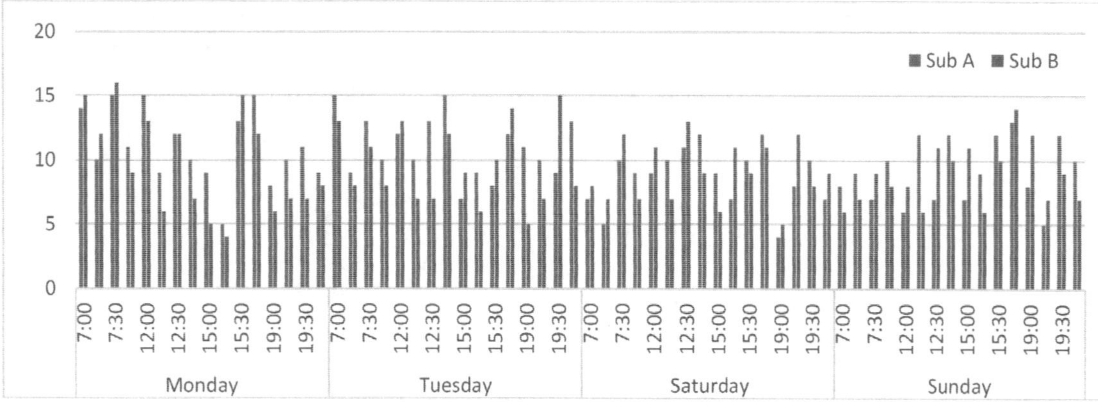

Figure 7. Pedestrian Volume Segment 3

Table 7. Pedestrian Speed Segment 3

Hour	Pedestrian Travel Time (10 m)										Average Travel Time
	1	2	3	1	2	6	1	2	9	1	
SUB A											
7:00	10,25	9,78	9,54	9,2	9,32	10,5	10,61	8,73	8,94	9,12	9,599
7:15	8,35	7,76	8,35	9,55	7,82	8,77	8,31	7,65	7,86	7,78	8,22
7:30	8,23	8,45	8,65	8,12	8,35	8,47	9,8	8,13	9,2	8,64	8,604
8:00	7,16	7,64	8,43	8,3	6,98	7,86	7,92	8,56	8,49	8,54	7,988
12:00	9,39	8,64	8,72	9,36	9,4	10,31	9,67	9,72	8,73	8,38	9,232
12:15	8,3	7,24	7,83	7,26	7,73	7,62	8,2	7,36	8,3	6,21	7,605
12:30	9,13	8,61	7,45	8,52	8,62	8,35	7,64	7,98	8,26	8,34	8,29
13:00	6,94	7,62	8,21	7,86	7,9	6,55	8,61	8,36	7,45	7,43	7,693
15:00	7,62	7,6	8,12	9,37	8,55	7,45	6,98	8,43	7,65	7,73	7,95
15:15	8,22	7,68	7,73	6,62	7,37	8,62	8	8,91	7,62	8,4	7,917
15:30	6,64	7,35	7,46	7,68	7,62	8,43	7,6	7,68	8,12	8,97	7,755
16:00	8	8,12	8,2	8,46	8,36	8,52	8,36	7,86	7,64	7,62	8,114
19:00	6,63	7,48	7,94	7,95	7,93	8,14	8,12	7,23	8,74	7,27	7,743
19:15	8,67	7,41	8,64	8,12	6,37	7,58	9,66	8,26	7,06	7,43	7,92
19:30	7,64	7,72	7,61	8,34	8,42	8,39	6,94	7,34	8,42	8,6	7,942
20:00	7,4	7,37	8,65	8,62	8,34	6,98	8,48	8,46	7,62	7,64	7,956
SUB B											
7:00	8,84	8,65	7,39	8,95	8,71	8,64	9,67	7,7	8,72	9,83	8,71
7:15	8,35	6,94	7,5	7,68	8,93	8,87	8,36	8,32	8,45	7,35	8,075
7:30	8,43	8,45	7,67	8,34	8,61	8,62	8,45	8,67	8,71	9,12	8,507
8:00	8,22	8,15	8,43	9,17	7,91	8,13	7,8	7,43	7,72	7,64	8,06
12:00	9,11	8,65	9,73	9,7	8,32	7,12	8,67	8,82	7,34	7,96	8,542
12:15	8,73	8,62	8,64	8,37	7,82	7,73	8	8,18	7,35	7,62	8,106
12:30	8,67	8,89	8,84	9,61	7,34	7,62	7,75	8,63	9,71	8,34	8,54
13:00	8,2	8,19	8,34	7,82	7,68	9,18	8,92	7,68	7,49	7,62	8,112
15:00	8,37	10,11	8,64	8,93	8,72	7,65	9,73	7,68	9,15	9,22	8,82
15:15	9,34	9,62	7,35	7,82	8,67	8,38	8,84	9,75	10,2	8,39	8,836
15:30	7,91	7,88	9,3	7,75	8,34	8,29	7,82	7,46	7,62	7,83	8,02
16:00	8,29	9,1	7,92	8,2	8,19	8,62	7,86	8,5	9,34	8,43	8,445
19:00	8,62	8,97	7,62	8,43	9,76	9,7	8,61	8,98	9,53	9,75	8,997
19:15	9,86	9,8	7,31	8,68	7,43	7,35	8,97	9,65	9,69	8,7	8,744
19:30	7,86	7,86	7,35	7,4	8,46	9,11	8,72	8,82	8,56	8,62	8,276
20:00	8,34	8,86	7	7,36	7,34	8,64	8,62	9,43	9,63	9,42	8,464

3.4.6 *Calculating Maximum Volume.* The Greenshields linear model equation was used to calculate the maximum volume of pedestrians passing through segment 3 (Table 8).

Table 8. Analysis of Volume, Speed, and Density Relationships Greenshields Method Segment 3.

Hour	Volume (X.Y)	Speed (Y)	Density (X)	X^2
7:00	2	65,542	0,03052	0,00093
7:15	1	73,642	0,01358	0,00018
7:30	2	70,130	0,02852	0,00081
8:00	1	74,776	0,01337	0,00018
12:00	2	67,514	0,02962	0,00088
12:15	1	76,380	0,01309	0,00017
12:30	2	71,301	0,02805	0,00079
13:00	1	75,925	0,01317	0,00017
15:00	1	71,556	0,01398	0,00020
15:15	1	71,629	0,01396	0,00019
15:30	1	76,070	0,01315	0,00017
16:00	2	72,468	0,02760	0,00076
19:00	1	71,685	0,01395	0,00019
19:15	1	72,012	0,01389	0,00019
19:30	1	73,992	0,01352	0,00018
20:00	1	73,082	0,01368	0,00019
JUMLAH	21	1157,703	0,29364	0,00620

To find a linear regression analysis, an analysis is carried out based on the results of the previous tabulation by taking the total value of each existing variable.

$$B = \frac{16.(21)-(0{,}29364).(1157{,}703)}{16.(0{,}00620)-(029364)^2} = \frac{336-339{,}94503}{0{,}09918-0{,}08622} = \frac{-3{,}94503}{0{,}01296} = -304{,}3969 \quad (17)$$

$$A = \frac{(0{,}0062).(1157{,}703)-(0{,}29364).(21)}{16.(0{,}00620)-(0{,}29364)^2} = \frac{7{,}17654-6{,}16639}{0{,}09918-0{,}08622} = \frac{1{,}01015}{0{,}01296} = 77{,}94284 \quad (18)$$

So, Sff = A = 77,94284

$$Dj = \frac{-A}{B} = \frac{-77{,}94284}{-304{,}3969} = 0{,}256 \quad (19)$$

The Sff and Dj values are known with the above calculations to determine the Vm value. The measure of the Vm value can be seen in equation (20).

$$Vm = \frac{Dj \times Sff}{4} = \frac{0{,}256 \times 77{,}942}{4} = 5 \text{ pedestrian/min} \quad (20)$$

3.4.7 *Relation between Speed (S), Density (D), and Volume (V) Segment 3.* The relationship between volume and density can be known in equation (21) and presented in Figure 8 (a).

$$V = 77{,}942.D - 304{,}46.D^2 \quad (21)$$

The relationship between speed and density can be known in equation (22) and presented in Figure 8 (b).

$$S = 77{,}942 - 304{,}46.D \quad (22)$$

The relationship between volume and speed can be known in equation (23) and presented in Figure 8 (c).

$$V = 0{,}256.S - 304{,}46.S^2 \quad (23)$$

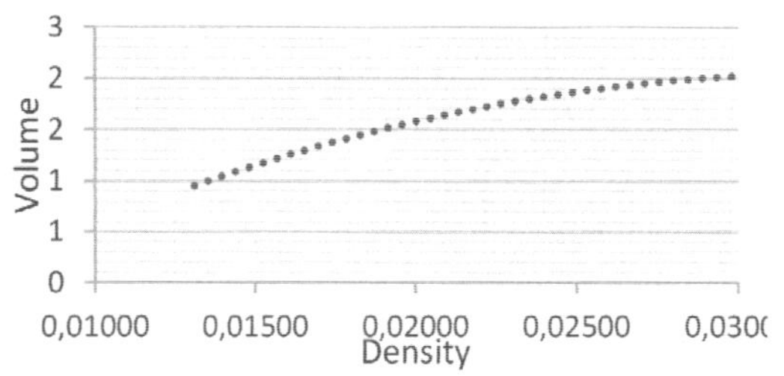

(a)

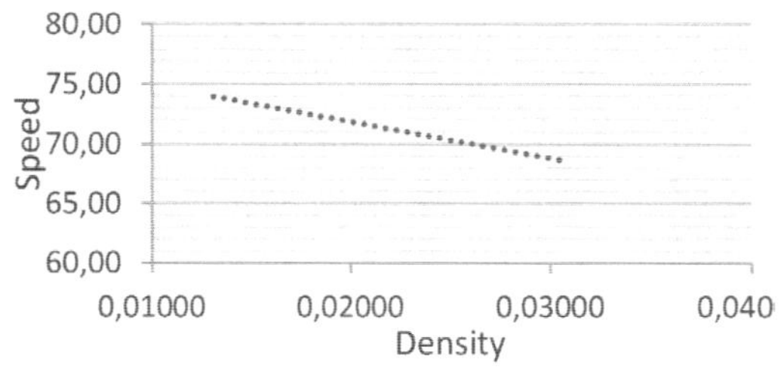

(b)

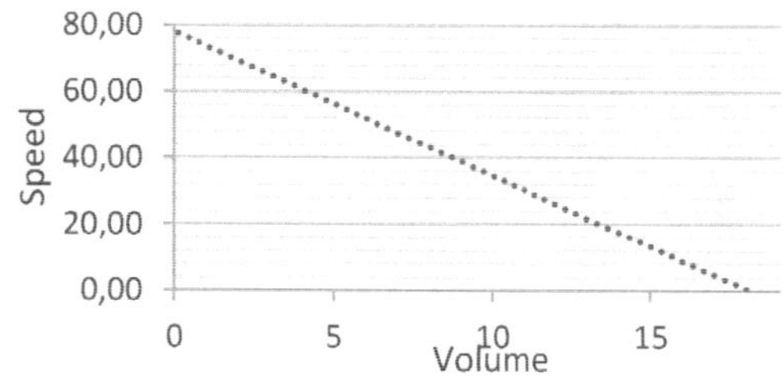

(c)

Figure 8. Graph of Speed and Density Relationship of Greenshields Model Segment 3.

3.4.8 *Effective Sidewalk Width Dimension Segment 3.* Based on the calculation of the maximum pedestrian flow in segment 3, it is known that the expected pedestrian value is one person. Furthermore, the measure of the effective sidewalk width dimensions can be seen in the equation below.

$$\boldsymbol{W} = \left(\frac{P}{35}\right) + \boldsymbol{n} = \left(\frac{5}{35}\right) + \mathbf{1} = \mathbf{1,14}\ \boldsymbol{meter} \tag{24}$$

Based on the calculations (24), it can be seen that the width of the pedestrian path required in segment 3 is 1.14 meters. However, because the research corridor is included in the industrial area, the width of the pedestrian path in sub-A is 2 m wide, and sub-B is 2 m wide to adjust to the existing conditions of the width of the existing sidewalk.

3.5 Walkability Level Analysis

3.5.1 *Walkability Level Analysis Segment 1.* In segment 1, with a length of 527.3 m, school, and industrial land are used to attract people to walk.

Table 9. Calculation of Segment 1 Walkability Level.

No	Variable	Weight	Average Scoring	Weight × Average Score
1.	Pedestrian conflicts	15	3,2	47,3
2.	Pedestrian path availability	25	1,3	33,6
3.	Availability of crosswalks	10	2,2	22,2
4.	Safety of crosswalks	10	2,8	27,5
5.	Motorist behavior	5	2,7	13,3
6.	Support facilities	10	2,8	28,1
7.	Disability support infrastructure	10	1,2	12,2
8.	Obstacles or barriers	10	2,2	21,9
9.	Safety from crime	5	3,1	15,3
	Walkability Index			44,3

Based on the results of Table 9, it can be seen that the walkability value in Segment 1 is 44,3 and is included in the Walk at your own risk category because it has a value in the range of 0 - 49 [24].

3.5.2 *Walkability Level Analysis Segment 2.* In segment two, industrial land uses, schools, trade, and services attract people to walk.

Table 10. Segment 2 Walkability Rate Calculation.

No	Variable	Weight	Average Scoring	Weight × Average Score
1.	Pedestrian conflicts	15	3,2	48,3
2.	Pedestrian path availability	25	3,2	48,3
3.	Availability of crosswalks	10	1,5	38,3
4.	Safety of crosswalks	10	2,0	20,0
5.	Motorist behavior	5	2,5	25,0
6.	Support facilities	10	2,3	11,7
7.	Disability support infrastructure	10	2,6	25,9
8.	Obstacles or barriers	10	1,3	13,4
9.	Safety from crime	5	2,1	21,3
	Walkability Index			43,6

In Segment 2, it is known that the walkability index value is 43,6 (Table 10), which means that this segment is included in the Walk at your own risk category.

3.5.3 *Walkability Level Analysis Segment 3. Segment 3, with a length of 646.7 m, has industrial and residential land uses that attract people to walk.*

Table 11. Segment 3 Walkability Rate Calculation.

No	Variable	Weight	Average Scoring	Weight × Average Score
1.	Pedestrian conflicts	15	4,0	60,0
2.	Pedestrian path availability	25	1,1	27,3
3.	Availability of crosswalks	10	1,1	10,9
4.	Safety of crosswalks	10	3,3	33,1
5.	Motorist behavior	5	3,5	17,3
6.	Support facilities	10	2,3	23,1
7.	Disability support infrastructure	10	1,1	10,6
8.	Obstacles or barriers	10	2,6	26,3
9.	Safety from crime	5	4,5	22,7
	Walkability Index			46,3

Based on the calculation results above (Table 11), it is known that segment 3 has a walkability level value of 46,3. Thus, segment three is included in the Walk at Your Risk classification because the walkability index value is 0 - 49.

4. Conclusions

Providing pedestrian ways is a form of fulfillment of rights to pedestrians so that users can feel safe and comfortable. In addition, a walkable pedestrian path can attract the surrounding community to walk instead of using private vehicles. Based on the green shield model calculation, the width requirement of the pedestrian path in Segment 1 is 1.23 m, segment 2 is 1.31 m, and Segment 3 is 1.14 m. The effective width obtained does not match the needs of the pedestrian path in Segment 1. The effective width obtained does not follow the pedestrian ways minimum requirements. It is not following Permen PU No. 03/PRT/M/2014 [8], so it is adjusted to the existing conditions. For walkability assessment, segment 1 has a walkability index value of 44,3, segment 2 has a walkability index value of 43,6, and Segment 3 has a walkability index value of 46,3. All segments are classified as "walk at your own risk", so improvements to the pedestrian facilities are recommended.

Based on the research results, the Bogor Regency government can use the study results to provide pedestrian ways with a minimum width of 2 m in spatial documents. The government can accommodate it in the infrastructure network plan section in the Gunungputri District RDTR document and the Bogor Regency RTRW.

5. References

[1] Paradita N B 2021 Pengembangan jalur pejalan kaki dengan konsep walkability pada kawasan dengan pola penggunaan lahan campuran studi kasus: Jalan Z.A. Pagar Alam (Sumatera: Institut Teknologi Sumatera)
[2] Knapskog M, Hagen O H, Tennøy A and Rynning M K 2019 Exploring ways of measuring walkability Transp. Res. Procedia 41 264–282
[3] Hillnhutter H 2016 Pedestrian access to public transport (Stavanger: University of Stavanger)
[4] Shaaban K 2019 Assessing sidewalk and corridor walkability in developing countries Sustain 11 1¬19
[5] Singh R 2016 Factors affecting walkability of neighborhoods Procedia - Soc. Behav. Sci 216 643–654
[6] Shields R, Gomes da Silva E J, Lima e Lima T and Osorio N 2023 Walkability: a review of trends J. Urban 16 19–41
[7] Lestari A R 2019 Kajian konsep walkable city di Kota Pekanbaru (Pekanbaru: Universitas Islam Riau)
[8] Umum M P 2014 Permen PU Nomor 03/PRT/M/2014
[9] Pratiwi I D, Agustin I W and Waloeya B S 2016 Evaluasi kinerja jalur pejalan kaki di jalan merdeka kota malang Plan. Urban Reg. Environ 5 1–8
[10] Saeid S and El Sayed M 2023 Improving a pedestrian-friendly environment in low income Egyptian districts MSA Eng. J. 2 1147–1172
[11] Gori S, Nigro M and Petrelli M 2014 Walkability indicators for pedestrian-friendly design Transp. Res. Rec. 2464 38–45
[12] Kabupaten Bogor P 2016 Rencana Tata Ruang Wilayah Rtrwk Bogor pp. 1–70.
[13] Lam T M et al. 2022 Development of an objectively measured walkability index for the Netherlands Int. J. Behav. Nutr. Phys. Act. 19 1–16
[14] Frank L D et al. 2010 The development of a walkability index: Application to the neighborhood quality of life study Br. J. Sports Med. 44 924–933
[15] Barbosa J P D A S, Guerra P H, Santos C de O, Nunes A P de O B, Turrell G and Florindo A A 2019 Walkability, overweight, and obesity in adults: A systematic review of observational studies Int. J. Environ. Res. Public Health 16
[16] Zulrehansyah M 2021 Analisis Volume, Kecepatan, dan Kepadatan Lalu Lintas dengan Metode Greenshields dan Greenberg (Pekanbaru: Universitas Islam Riau)
[17] Bogor D P K 2018 RENSTRA PUPR Kabupaten Bogor pp. 248–253.
[18] Semiawan P 2010 Metode penelitian kualitatif (Jakarta)
[19] Yusuf A M 2017 Metode penelitian kuantitatif, kualitatif, dan penelitian gabungan (Jakarta) 4
[20] Mantik V G, Timboeleng J A and Jefferson L 2015 Perencanaan kebutuhan pedestrian pada Ruas Jalan Suprapto Kota Manado Tekno 13
[21] Radjawane L E 2021 Study of Pedestrians proportion to roadside friction index IOP Conf. Ser. Earth Environ. Sci. 841
[22] Endarwati M C, Setyawan A and Marison O 2018 Penilaian walkability score index pada pusat pelayanan dalam menuju Kota Malang berkelanjutan (assessment of walkability score index at services center towards sustainable Malang City) Semin. Nas. Kota Berkelanjutan 1 174
[24] Leather J, Fabian H, Gota S and Mejia A 2011 Walkability and Pedestrian Facilities in Asian Cities State and Issues (Mandaluyong, Philippines)

IOP Conf. Series: Earth and Environmental Science 1294 (2024) 012003 doi:10.1088/1755-1315/1294/1/012003

Road performance analysis due to Sayung Industrial Park development

R D Sirait[1], R J Sumabrata and N T Anumasta

Department of Civil Engineering, Faculty of Engineering, Universitas Indonesia, Depok, West Java, 16424, Indonesia

[1] ssirait.rachel@gmail.com

Abstract. This study aims to analyze the impact of Sayung Industrial Park development to the surrounding road network. The study merely relies on available secondary data, including recent transport study in Semarang Raya, trip rate, and previous research related on industrial area occupancy growth. Road performance in the terms of degree of saturation and travel speed are analyzed using a simplified four-step model using CUBE software. The results show that in 2023, assuming 50% occupancy of the industrial park, the average travel speed decreases by 8.5% and the average degree of saturation increases by 18.7%. While in 2030, assuming 100% occupancy, the road performance is worsened, the average travel speed decreases by 46.4% and the average degree of saturation increases by 45.5%. A new additional lane in each direction on Semarang–Demak main corridor is simulated to anticipate the full operational of Sayung Industrial Park in the future. The result shows an improvement on road performance.

1. Introduction

Sayung Industrial Park is an industrial area in Demak Regency managed by JatengLand. The site is conveniently situated next to Semarang–Demak national road, as well as the Semarang–Demak toll road. In addition, there are several infrastructure facilities in the vicinity that support distribution operations for industries. These include Tanjung Emas Port, Ahmad Yani Airport, and Tawang Station [1].

The road network is crucial for industrial activities. Industrial activities process raw materials into goods production [2]. The demand for goods by consumers leads to increased trips in industrial areas. The trips are carried out for the distribution of raw materials and marketing of products to consumers.

The activities taking place in Sayung Industrial Park are predicted to generate trips that affect the performance of nearby roads. To determine the impact and anticipation, it's necessary to conduct an analysis of road performance. The CUBE program is used to analyze the traffic model.

2. Literature study

2.1. Degree of saturation (V/C ratio)

The ratio of traffic flow to capacity is the main factor in determining the performance level of a road segment. The degree of saturation is a tool for examining traffic patterns in terms of speed. The degree of saturation plays a crucial role in determining whether there are issues with the capacity of a particular road segment [3].

2.2. *Travel speed*
According to Indonesia Highway Capacity Manual (IHCM) 1997, travel speed refers to the average speed of light vehicles on a road segment, expressed in km/h. Travel speed is the primary measure of road segment performance because it is easy to measure. In addition, the value of speed plays an essential role in economic analysis related to road user costs [3].

2.3. *Route selection*
The weighted amount or function of time, distance, and cost are used as the generalized cost as shown in equation (1). The equation is used in the CUBE model of the Transportation Laboratory of the Faculty of Engineering, Universitas Indonesia, in making toll road traffic models or other sources from the CUBE manual.

$$COST_k = TIME + (A \times DISTANCE) + (TOLL_k / VOT_k) \quad (1)$$

$COST_k$	=	Total cost of type k vehicles to cross the segment in unit time (minutes)
$TIME$	=	Time taken by the vehicle to cross the segment (minutes)
A	=	Coefficient of type k vehicle operating costs (minutes/km)
$DISTANCE$	=	Segment length (km)
$TOLL_k$	=	Toll fee for type k vehicles (Rp); on a closed system toll road, the rate is converted in advance according to the length of the segment (Rp/km × the length of the section in km)
VOT_k	=	Time value of vehicle type k (Rp/minute)

2.4. *Capacity restrained*
According to the capacity restrained assignment approach, if the traffic on a particular stretch of road increases but the capacity remains constant, then the time taken to travel on it will also increase. The volume-delay function used during traffic assignment in this study adopts the BPR formula. The equation used is as follows:

$$t_c = t_0 \times (1 + \alpha \times (V/C)^{\beta}) \quad (2)$$

t_c	=	Vehicle travel time in traffic flow conditions of V (minutes)
t_0	=	Vehicle travel time in free flow conditions (minutes)
V	=	Traffic volume (pcu/h)
C	=	Road capacity (pcu/h)
α, β	=	Coefficient

3. Methodology
The study solely relies on available secondary data, including recent transport study in Semarang Raya, trip rate, and previous research related on industrial area growth. The modeling process involves a simplified four-step model, which includes trip generation, trip distribution, and trip assignment—mode choice is not part of this research model since the secondary trip data used is already in vehicle trip unit. The processing of data for road capacity and free flow speed follows the guidelines set out in IHCM 1997.

To determine the number of trips generated by industrial park activities, a trip rate calculation is used along with several assumptions. Afterwards, a set of origin-destination matrices (ODM) is developed using the Furness method. The results of the ODM iteration are then used as input into the CUBE program for the assignment stage. These results are then analyzed to determine the impact of industrial park development.

4. Data interpretation and analysis

4.1. Road/link data

Geometric data consists of main roads outside and within the industrial park. The geometry of the roads outside the area is obtained from the Transportation Laboratory of the Faculty of Engineering, while the roads inside the area are measured with the help of Google Earth. The calculation of free flow capacity and speed refers to the IHCM 1997. Urban roads have four adjustment factors: carriageway width, directional split, side friction, and city size. For freeways, adjustments are only made to the carriageway width.

Table 1. Road type, widths, capacity, and free flow speed (outside the industrial park).

Segment	Road type	Eff. width (m)	Lane width (m)	Shoulder width (m)	Capacity (pcu/h)	Free flow speed (km/h)
Semarang–Demak Toll Road	4/2D	14.4	3.60	3	8,832	89
Semarang–Demak Road	4/2D	14	3.50	0	6,072	52
Onggorawe Road	2/2UD	4	2.00	0	645	24
Manageng Road	2/2UD	4.5	2.25	0	1,058	27
Ki Godek Road	2/2UD	4.5	2.25	0	1,058	27
Guntur Road	2/2UD	4.5	2.25	0	1,058	27

Source: Authors' own elaboration based on Google Street View

Table 2. Typical road type, widths, capacity, and free flow speed plan (inside the industrial park).

Road type	Eff. width (m)	Lane width (m)	Curb width (m)	Capacity (pcu/h)	Free flow speed (km/h)
6/2D	24	4	4	10,051	58
6/2D	19.5	3.25	4	8,933	52
4/2UD	12	3	4	2,794	46
4/2D	12	3	4	5,950	49

Source: Authors' own elaboration based on Google Street View and Sayung Industrial Park Masterplan

4.2. Traffic analysis zone

There are 48 traffic analysis zones in total, comprising 41 internal zones (zone 1-41) and 7 external zones (zone 42-48). In the scenario without the industrial park development, there are four internal existing zones (zone 38-41), and in the scenario where the industrial park exists/developed, the number of zones increases to 48. The region's zones are numbered from 1 to 37.

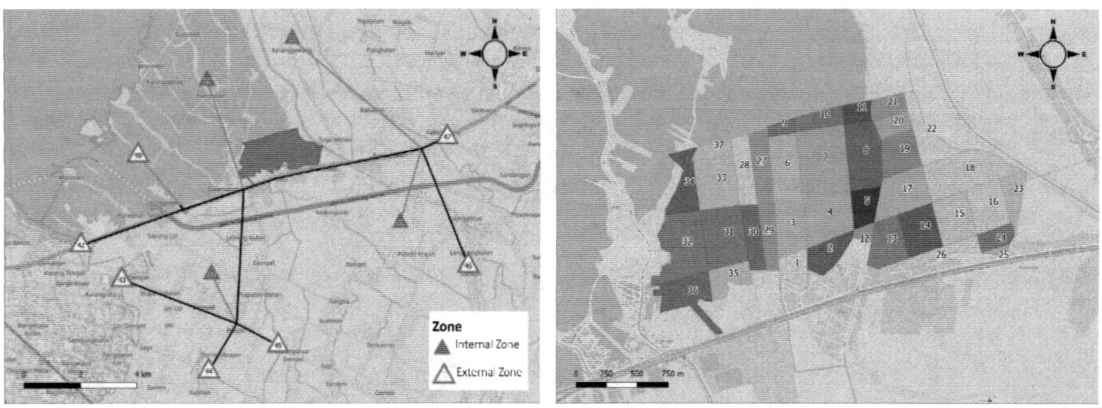

Figure 1. Traffic zone.

4.3. *Trip generation estimation*

4.3.1. Industrial area occupancy level. The occupancy level assumption is obtained from interpolating the Kendal Industrial Park occupancy level. The pandemic condition affects the level of occupancy in industrial areas. Between 2018 and 2022, the typical company investment was between 1–3 ha. However, this amount increased to 10 ha in 2023 [4]. The occupancy rate in 2016 was 3% [5], in 2020 was 30% [6], in 2021 was 47% [7], and in 2023 was 80% [4]. According to the estimates, the occupancy rate for all upcoming constructions in 2023 is predicted to be 50%. The occupancy rate for phase 1 is 80%, and for phase 2 is 50%. It is assumed that the area will be fully occupied by the year 2030.

4.3.2. Vehicle composition. Vehicle composition data is used to calculate the number of vehicles in passenger car units (pcu). In general, the composition of vehicles within the area is assumed to be the same as that outside the area because other industries have been established in the vicinity. In certain zones, the composition of vehicles needs to be adjusted according to their function.

4.3.3. Industrial park trip generation. To obtain the building floor area, the land area in the industrial park is multiplied by 0.5. The number of trips generated by each zone is determined by multiplying the trip rate by the number of units measured based on its land function. This results in trips being measured in units of trips per hour for each zone [8].

Table 3. Trip rate.

No.	Function	ITE category	Unit of measure	Trips per unit
1.	Industrial	Industrial park	1,000 SF GFA	0.40
2.	Open green space	Public park	acres	0.11
3.	Residential	Single-family detached housing	*dwelling units*	0.99
4.	Clinic	Clinic	1,000 SF GFA	3.28
5.	Management office	General office building	1,000 SF GFA	1.15
6.	Commercial	Business park	1,000 SF GFA	0.42
7.	Warehousing	Warehousing	1,000 SF GFA	0.19
8.	Water treatment plant	Industrial	1,000 SF GFA	0.40
9.	wastewater treatment plant	Industrial	1,000 SF GFA	0.40
10.	Mosque	Mosque	1,000 SF GFA	4.22
11.	Cargo terminal	Truck terminal	1,000 SF GFA	1.72
12.	Sport center	Health/fitness club	1,000 SF GFA	3.45

Source: Institute of Transportation Engineers, 2017

4.4. *Trip distribution assumptions*

There are four ODM prepared for each scenarios: the 2023 scenario without industrial park (2023-TKI), the 2023 scenario with industrial park (2023-KI), the 2030 scenario without industrial park (2030-TKI), and the 2030 scenario with industrial park (2030-KI). The transportation information for trips outside the industrial zone is sourced from the Transportation Laboratory. For industrial zones, it is only expected that some particular zones (zone 1, 2, 5, 8, 35, and 36) those will generate trips (produce and attract) with other zones within the industrial zone itself. The distribution of trips in the industrial park refers to the proportion of trips generated by each zone in the area. Distribution of trips outside the industrial park refers to the proportion of trips from zone 40.

The available study on Tegineneng Industrial Park was used as a reference to estimate the increase in the number of trips resulting from the development of the industrial park. Before the area was developed, there were 186,876 vehicles per day on trips. After the development, the number increased to 279,441 vehicles per day [9]. Based on the data, it can be inferred that 33.1% of trips originate from industrial areas. This value is used for the distribution of trips.

4.5. Assignment

The assignment process is carried out with generalized cost. To calculate travel time, the capacity restrained method with equation (2) is used during the traffic assignment. The values of coefficients α and β used in this method are 0.15 and 4.0, respectively. To assign traffic on the CUBE, the coefficients in the table below are utilized.

Table 4. Coefficient (A).

Vehicle type	A (min./km)
Motorcycle (MCY)	0.20
Passenger car (CAR)	0.30
Light goods vehicle (LGV)	0.30
Medium goods vehicle (MGV)	0.35
Heavy goods vehicle (HGV)	0.40

Source: Transport Lab. Engineering Faculty of Universitas Indonesia, 2022

4.6. Validation

Validation of travel time is conducted by comparing the model results and actual travel time (based on Google Maps) on two road segments, segment 1 is Semarang - Demak Road and segment 2 is Manageng Road and Ki Godek Road.

Table 5. Travel time validation.

Road segment	Travel time (min.)		% Error
	Model	Actual*	
Segment 1	31.1	29	7.21%
Segment 2	14.1	14	0.71%
	Average		3.96%

*) Actual travel time are based on Google Maps

In addition, the validation process includes data on the number of vehicles observed along Semarang–Demak Road. According to the survey, the number of vehicles recorded was 2,404 veh./h. However, the CUBE results showed 2,830 veh./h. Based on the percentage of modeling data errors compared to the actual data, which is 17.7%.

4.7. Trip distribution

4.7.1. Year 2023. In general, zones 41 and 43 seem to generate the least amount of trips. The thickest desire line indicates the highest number of trips, linking zones 42 with 47 and 42 with 48. The most number of vehicles originating from the industrial area go to these three zones.

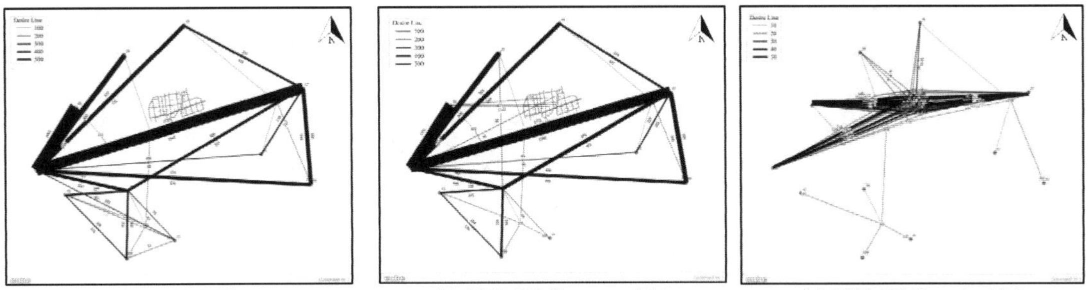

(a) 2023-TKI (b) 2023-KI (from outside) (c) 2023-KI (from inside)

Figure 2. Desire line (2023).

4.7.2. Year 2030. The distribution of trips between zones in 2030 is more spread out than in 2023. The majority of trips originating from the industrial park development are headed toward zones 42, 47, and 48. The road connecting these three zones is called Semarang–Demak Road. As a result, the industrial park has the greatest impact on this road.

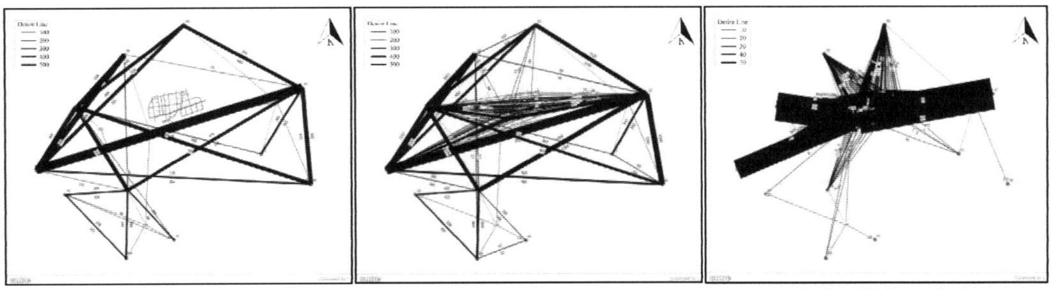

(a) 2030-TKI (b) 2030-KI (from outside) (c) 2030-KI (from inside)

Figure 3. Desire line (2030).

4.8. Road performance

By 2030, travel speed on multiple roads near the entrance of Sayung Industrial Park will decrease. By 2023, the development of industrial park will cause a decrease in travel speed on Semarang–Demak Road and Onggorawe Road. This decrease in speed will extend to all road segments by 2030 due to continued development.

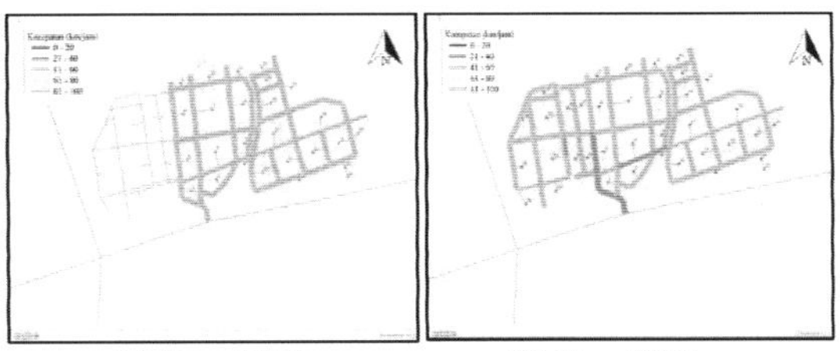

(a) Year 2023 (b) Year 2030

Figure 4. Travel time inside the industrial park.

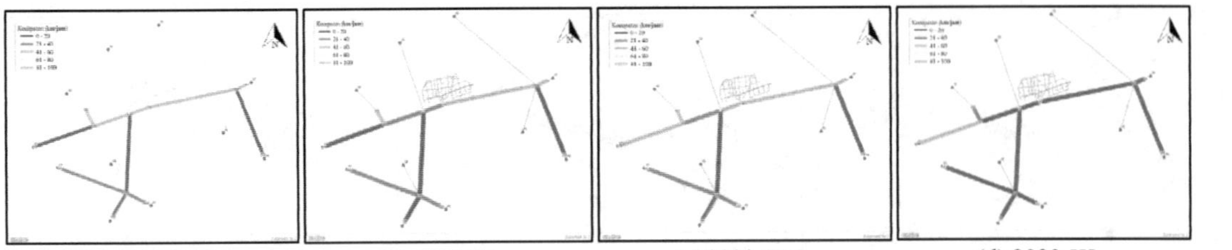

(a) 2023-TKI (b) 2023-KI (c) 2030-TKI (d) 2030-KI

Figure 5. Travel speed outside the industrial park.

The degree of saturation on roads in industrial zones is increasing, especially near the entrances to these areas. The degree of saturation of roads located outside the industrial park increases significantly in each road segment. The most significant changes are observed on the toll road and Semarang – Demak Road.

The road performance in the area has declined from 2023 to 2030, as indicated by the decrease in travel speed from 47.39 km/h to 45.47 km/h. The degree of saturation has also decreased from 0.05 to 0.28. Despite this, drivers still have some freedom in determining the speed of their vehicle based on the degree of saturation value.

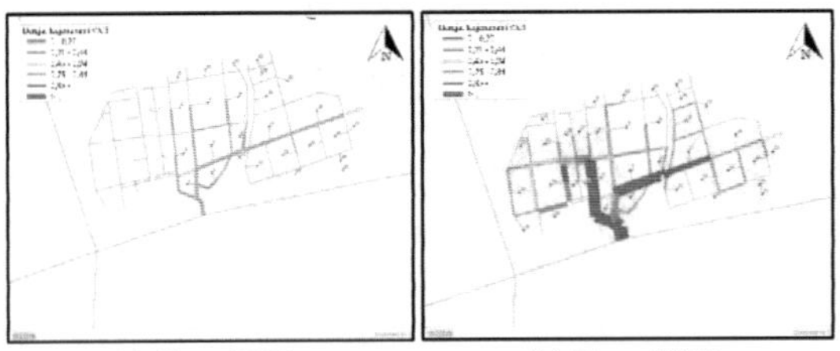

(a) Year 2023 (b) Year 2030

Figure 6. Degree of saturation inside the industrial park.

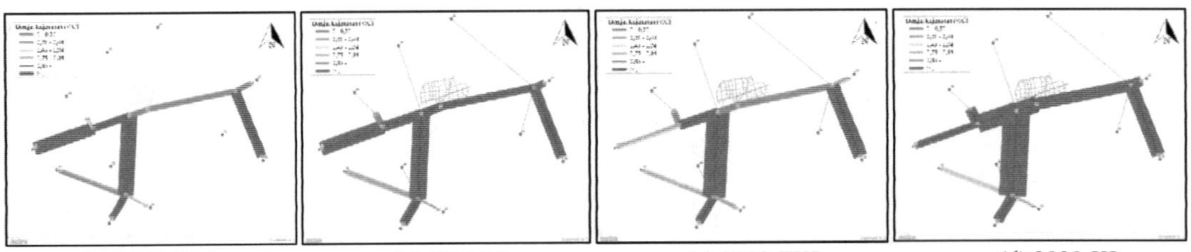

(a) 2023-TKI (b) 2023-KI (c) 2030-TKI (d) 2030-KI

Figure 7. Degree of saturation outside the industrial park.

Table 6. Road performance.

		2023			2030		
		TKI	KI	Change	TKI	KI	Change
Inside the industrial park	Travel speed	-	47.4 km/h	-	-	45.5 km/h	-
	V/C ratio	-	0.05	-	-	0.28	-
Outside the industrial park (excludes toll road)	Travel speed	27.6 km/h	24.4 km/h	-11.6%	29.5 km/h	17.5 km/h	-40.7%
	V/C ratio	1.30	1.41	+8.6%	1,35	1,91	+41.7%
Toll road	Travel speed	85.2 km/h	80.5 km/h	-5.5%	80.1 km/h	38.4 km/h	-52.1%
	V/C ratio	0.69	0.89	+28.7%	0.90	1.71	+90.4%

In 2023, when the industrial park is developed and 50% of the area is occupied, the average travel speed will change by 8.54% and the degree of saturation will increase by 18.68%. In 2030, the impact of industrial park will be greater when the occupancy rate is at 100%. At full occupancy, there will be a 46.39% increase in the degree of saturation and the average travel speed will be 65.54% lower.

4.9. Do-something scenario (DS)

In the year 2030, the performance of roads is expected to decrease due to development and traffic growth. Therefore, it is necessary to change the road geometry to improve road performance.

- Lane addition on the Semarang–Demak Toll Road from 4/2D to 6/2D
- Lane addition on Semarang–Demak Road from 4/2D to 6/2D
- Road widening on Onggorawe and Guntur roads to 3.5 m

Table 7. Road type, capacity, and free flow speed improvement.

Road Segment	Before			After (DS)		
	Road type	Capacity (pcu/h)	Free flow speed (km/h)	Road type	Capacity (pcu/h)	Free flow speed (km/h)
Semarang–Demak Toll Road	4/2D	8,832	89	6/2D	13,248	92
Semarang–Demak Road	4/2D	6,072	53	6/2D	9,163	54
Onggorawe Road	2/2UD	645	24	2/2UD	2,581	38
Guntur Road	2/2UD	1,058	27	2/2UD	2,581	38

By increasing the road capacity according to the improvement scenario (DS) as proposed above, the average performance on roads outside the area and toll roads increases. The travel speed on the toll road has gone up to 73.1 km/h, and the degree of saturation has decreased to 1.14. On roads outside the area, the travel speed increases to 34.2 km/h, and the degree of saturation decreases to 0.99.

Table 8. Road performance outside the industrial park (do-something scenario).

Category	Travel speed (km/h)	V/C ratio
Non-toll road	34	0.99
Toll road	73	1.14

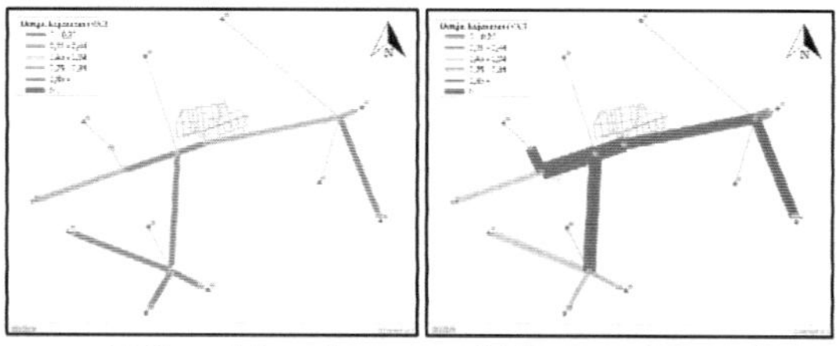

(a) Travel Speed (b) Degree of Saturation

Figure 8. Road performance (do-something scenario).

In addition to increasing the capacity of the road, other alternative solution could be proposed to minimize the use of private vehicles by providing public transportation to limit the movement of private vehicles. The presence of adequate and reliable public transportation services could reduce the dependence on private vehicles. In Central Java, the government has been developing bus rapid transit (BRT) system, but currently the service has not yet reach Sayung Regency [10]. The nearest Trans Semarang BRT stop is at Genuk Market, which is 8.2 km away, while the nearest Trans Jateng BRT stop is at Godong Terminal, which is 30.8 km from the industrial area. Therefore, additional BRT routes are needed by considering passenger needs. Moreover, another solution that could be done is to limit the movement of private vehicles. Private vehicle movement can be limited during peak hours. Further analysis with related data is needed to determine the detailed calculations of these two solutions.

5. Conclusion

- The development of Sayung Industrial Park with an assumption of 50% occupancy in 2023 will result in an average decrease in speed of 8.54% and an average increase in degree of saturation of 18.68%.
- The development of the Sayung Industrial Park with the assumption of 100% occupancy in 2030 will result in an average decrease in speed of 46.39% and an average increase in the degree of saturation of 66.06%.
- Road performance inside the industrial area has decreased. Vehicle speed decreased from 47.39 km/h to 45.47 km/h, while the road degree of saturation increased from 0.05 to 0.28.
- The do-something scenario can improve road performance by 2030.

References

[1] Dinas Penanaman Modal dan Pelayanan Terpadu Satu Pintu Provinsi Jawa Tengah 2019 *Perencanaan Penyediaan Infrastruktur Pendukung Kawasan Industri di Jawa Tengah Semarang*

[2] Rahayu A M C, Raharjo E P, Dwipayana A D and Suraharta I M 2022 *Identifikasi Faktor Pengungkit Bangkitan dan Tarikan Perjalanan Kegiatan Industri (Studi Kasus Kawasan Industri di Kabupaten Bekasi)* (Jurnal Teknologi Transportasi dan Logistik vol 3) chapter 1 pp 23-28.

[3] Direktorat Jenderal Bina Marga 1997 *Manual Kapasitas Jalan Indonesia* (Jakarta: Bina Marga)

[4] Putri M N 2023 *Mantap! Kawasan Industri Kendal Terisi 80 Persen, 90 Investor Tanam Modal* (Bisnis.solopos.com) https://bisnis.solopos.com/mantap-kawasan-industri-kendal-terisi-80-persen-90-investor-tanam-modal-1665018

[5] Kementerian Perindustrian Republik Indonesia 2016 *KIK Pacu Kontribusi Jateng untuk Ekonomi Nasional* (Kemenperin.go.id.) https://www.kendalindustrialpark.co.id/post/index/94/kik-pacu-kontribusi-jateng-untuk-ekonomi-nasional

[6] Perdana A P 2021 *Kawasan Ekonomi Kendal Bakal Serap Puluhan Ribu Tenaga Kerja* (Kompas.id.) https://www.kompas.id/baca/ekonomi/2021/09/07/tor-kek-kek-kendal-bakal-serap-10-000-tenaga-kerja

[7] Kementerian Perindustrian Republik Indonesia 2021 *Kemenperin: Penjualan Naik, Kawasan Industri Jadi Magnet Investasi* (Kemenperin.go.id.) https://www.kemenperin.go.id/artikel/22630/Kemenperin:-Penjualan-Naik,-Kawasan-Industri-Jadi-Magnet-Investasi

[8] Institute of Transportation Engineers 2017 *Trip Generation Manual 10th Edition*

[9] Octavia A 2022 *Dampak Rencana Pengembangan Kawasan Industri Strategis di Kecamatan Tegineneng terhadap Pola Pergerakan di Kabupaten Pesawaran* (Bekasi: Politeknik Transportasi Darat Indonesia – STTD)

[10] Rahayu A M C, Raharjo E P, Dwipayana A D, and Suraharta I M 2022 *Integrasi antar Transportasi Umum di Kota Semarang* (Jurnal Pengembangan Kota vol 10) chapter 1 pp 36-46.

IOP Conf. Series: Earth and Environmental Science 1294 (2024) 012004 doi:10.1088/1755-1315/1294/1/012004

Analysis of road performance due to trip generation from the development of Batang Integrated Industrial Estate

R D Antoro[1], R. J Sumabrata and N T Anumasta

Department of Civil Engineering, Faculty of Engineering, Universitas Indonesia, Depok, West Java, 16424, Indonesia

[1] rahmdwiantoro@gmail.com

Abstract. The development of the Batang Integrated Industrial Estate (KIT Batang) is predicted to affect the performance of the surrounding roads. This study aims to analyze roads performance around industrial estate by referring to the Indonesian Highway Capacity Manual 1997 using a "modified four-step model" (overriding the mode choice stage). The result shows that in 2024, the presence of KIT Batang will not have a significant impact on the surrounding roads network. However in 2029, there will be a significant changes, indicated by worsening road performance—in this case referring to travel speed, service level, and volume-capacity ratio—where toll roads have shown an average level of service F. The do-something scenario is carried out by widening the toll and arterial roads in 2029. Analysis of simulation results shows improved performance on toll roads with an average service level from F to C and on arterial roads from C to B.

1. Introduction

In general, KIT Batang has an area of 4,300 ha. KIT Batang itself is designed to have three clusters, where currently the construction has entered phase 1 in cluster 1 covering an area of 450 ha, where the area being built is allocated as textile and battery industrial area. The supporting infrastructure at the Batang KIT phase 1 itself is targeted to operate in the first half of 2024 which includes gas, electricity, raw water and other supporting infrastructures, so that tenants can start operating in 2024 [1].

With the presence of the KIT Batang development, it will certainly have an impact on current and future traffic conditions. Therefore, further studies and research are needed regarding the performance of roads due to the development of industrial area infrastructure.

2. Literature Study

2.1. The Four-Step Model of Transportation Planning

The most popular and frequently used concept of transportation planning is the "four-step model of transportation planning". This model consists of Trip Generation-Trip Distribution-Mode Choice-and Route Choice. The trip distribution stage produces the origin-destination (OD) matrix, which can be obtained using the Furness method as in equation (1) [2].

$$T_{id} = t_{id}E_i \text{ and } T_{id} = t_{id}E_d \tag{1}$$

T_{id} = Total future movement from origin zone i to destination zone d
t_{id} = Total present movement from origin zone i to destination zone d
E_i, E_d = The growth rate of origin zone i and destination zone d

At the route choice stage, the route is selected based on the distribution of trips by assuming that each driver chooses a route that minimizes their travel costs where the following generalized cost equation is used.

$$COST_k = TIME + (A \times DISTANCE) + (TOLL_k / VOT_k) \quad (2)$$

$COST_k$ = Total cost of type k vehicles to traverse the segment in unit time (minutes)
$TIME$ = Vehicle travel time to cross the segment (minutes)
A = Weight, coefficient of type k vehicle operating costs (minutes/km)
$DISTANCE$ = Section length (km)
$TOLL_k$ = Type k vehicle toll fees (Rp); on a closed system toll road the rate is converted first according to the length of the section (Rp/km x the length of the section in km = Rp)
VOT_k = Type k vehicle time value (Rp/minute)

The capacity restrained (CR) method is used in route choice. The route is chosen by considering the capacity (section) as the limit or impedance of the vehicle's trip when passing through a road section. The volume-delay function equation from the BPR formula is used as in equation (3).

$$t_c = t_0 \times (1 + \alpha \times (V/C)^{\beta}) \quad (3)$$

t_c = Vehicle travel time in traffic flow conditions of V (minutes)
t_0 = Vehicle travel time in free flow conditions (minutes)
V = Traffic volume (pcu/hour)
C = Road section capacity (pcu/hour)
α, β = Coefficient

2.2. Road Performance
The capacity (C) and free flow speed (FV) are measured to obtain road performance. Capacity is the maximum traffic flow through a point on the path that can be withheld per unit hour under certain conditions. Free flow speed (FV) is the speed at zero vehicle flow rate or the speed that the driver will choose if driving without being influenced by other motorized vehicles. When the number of traffic flows (Q) is known, the degree of saturation value can be determined by comparing the value of traffic flow to its capacity so that the level of service of a road section can be identified [3].

3. Research Methodology
This study only relies on the secondary data obtained form related authorized agencies. The modeling process carried out using a simplified four-step model in transportation planning by eliminating the mode choice stage due to secondary data obtained is already in vehicle trip units. The Semarang Raya traffic model from the FT UI Transportation Laboratory study in 2022 is used as the basis for the trips number and distribution outside the industrial area for further details on trips within the industrial estate. Trip generation within the industrial estate is determined using trip rates by Institute of Transportation Engineers based on the land use area. The OD matrix is determined by an analogy approach using the Furness method, which will be input into the cube software during the route selection process. The results of the running process on the cube software will be used to analyze road performance due to the industrial area. Road performance analysis is carried out in 2024 and 2029.

4. Data Calculation and Interpretation

4.1. Road Geometry Data

The geometry data of the roads are obtained from the FT UI Transportation Laboratory. The calculation of road's capacity and free flow speed refers to the Indonesian Highway Capacity Manual 1997. The calculation for urban roads have four adjustment factors consisting of carriageway width, directional split, side friction, and city size. For freeways, only the adjustment factor for carriageway width is used.

Table 1. Road Geometry Data

No	Road Segment	Road Category	Road Type	Width (m)			Capacity[a] (pcu/h)	Free Flow Speed (km/h)
				Lane	Shoulder	Total		
1	Jalan Tol Semarang – Batang (Main Road)	Toll	4/2 D	3.6	3	20.4	4,655.20	88.80
2	Jalan Tol Semarang – Batang (Ramp/Toll Gate)	Toll	2/2 D	4	3	14	2,438.00	92.00
3	Jalan Kawasan Industri Batang	Industrial Estate	4/2 D	3.5	0	14	2,853.84	49.12
			2/2 UD	3.5	0	7	1,213.07	36.31
4	Jalan Raya Pantura	Primary Arterial	4/2 UD	3.5	0	14	2,594.40	45.06
			2/2 UD	3.5	0	7	1,213.07	36.31
5	Jalan Gringsing – Kedawung	Local	2/2 UD	2.5	0	5	679.32	28.10

[a] calculated per direction

Source : FT UI Transportation Laboratory (2022)

4.2. Traffic Analysis Zone

The internal zones used in the Batang KIT are 23 zones inside the industrial area (TAZ 1-23) and four zones outside the industrial area (TAZ 24-27). In addition, there are five external zones (TAZ 28-32) at each end of the road under study.

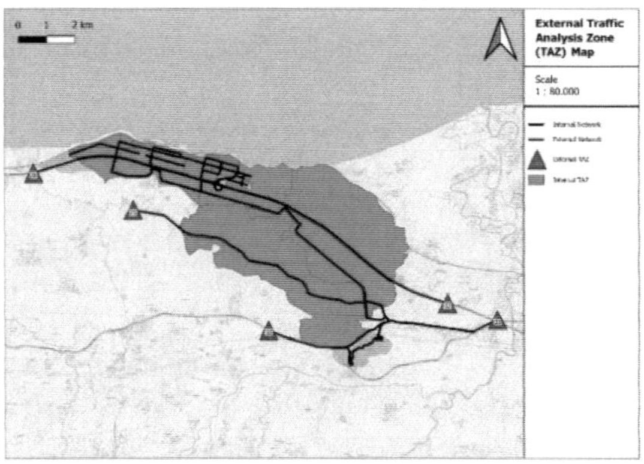

Figure 1. Traffic Analysis Zone (External).

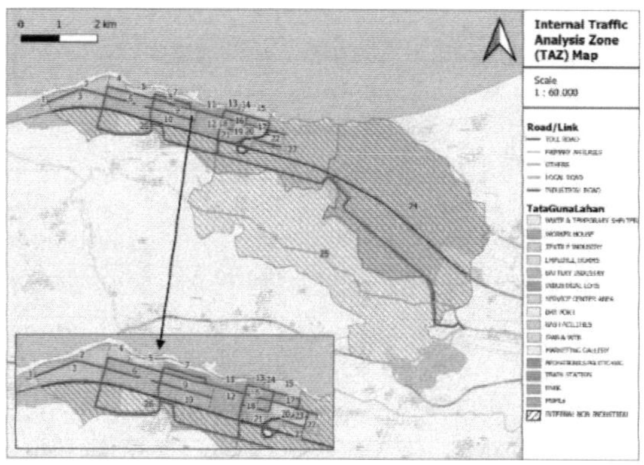

Figure 2. Traffic Analysis Zone (Internal).

4.3. Trip Generation Calculation

The calculation of trip generation stage uses the following assumptions.

4.3.1. Trip Generation Rates. The following is the trip rates for each type of land use. This value is used as a reference for calculating trip generation by using a trip rate based on known area units [4].

Table 2. Trip Rates for Each Type of Land Use

No.	Function	ITE Category	Unit of Measure	Trips per Unit
1	Textile and battery Industry	General Light Industrial	1000 SF GFA	0.63
2	WTP, Temporary shelter, gas facilities	Utilities	1000 SF GFA	2.27
3	Employee horms	Multi-Family Housing (Mid Rise)	*Dwelling Units*	0.44
4	Service center area	Clinic, police station, fire station, mosque (average)	1000 SF GFA	2.05
5	Marketing gallery/office	General Office Building	1000 SF GFA	1.15
6	Industrial lots, MSMEs	Industrial park	1000 SF GFA	0.40
7	Dry port, train station	Intermodal truck terminal	1000 SF GFA	1.72
8	Mechatronics polytechnic	University/college	1000 SF GFA	1.17
9	Open green space	Public park	Acres	0.11

Source : Institute of Transportation Engineers (2017)

4.3.2. Building Floor Coefficient (KLB). KLB is used as a reference for building floor area in each land use. By using six building references with industrial designation, i.e cellulose factory in Way Kanan [5], cellulose acetate and acetate ahydride mill in Karanganyar [6], PT Komatsu Remanufacturing Asia in North Jakarta [7], garment factory in Karanganyar [8], PT TACI in Cikarang [9], and PT Banten Energy International in Serang [10], the average of KLB 0.5 is obtained.

4.3.3. Industrial Area Occupancy Rate. It is necessary to assume occupancy rate of industrial estates in year 2024 and 2029 by using eight industrial estate references, i.e Jababeka Cikarang Industrial Estate [11], MM2100 Bekasi Industrial Estate [12], Bukit Semarang Baru Industrial Estate, Terboyo Industrial Estate Semarang, Candi Industrial Estate Semarang [13], JIEP Pulogadung Industrial Estate [14], Medan Industrial Estate [15], Bekasi EJIP Industrial Estate [16]. With reference to the year KIT Batang was founded in 2023, the assumption that the industrial area is occupied after 1 year of operation (2024) is 3% and after 6 years of operation (2029) is 18% by using interpolation.

4.3.4. Vehicle Composition. Five types of vehicles were used in this study, i.e motorcycles (MCY), cars (CAR), light goods vehicles (LGV), medium goods vehicles (MGV), and heavy goods vehicles (HGV). By referring to references, the composition of vehicles produced for each type of land use is obtained. Land use itself is divided into seven types, including industrial, residential, institutional, medical, recreational, ports/terminals, and office.

4.3.5. Trip Generation based on Vehicle Type. Referring to all the assumptions used, trip generation values are obtained in 2024 and 2029 for each type of vehicle across all land use types.

Table 3. Trip Generation based on Vehicle Type

TAZ	Land Use	Number of Trips [vehicle/hour]									
		2024 (Occupancy 3%)					2029 (Occupancy 18%)				
		MCY	CAR	LGV	MGV	HGV	MCY	CAR	LGV	MGV	HGV
1	Industrial	1.01	4.18	1.12	2.73	5.81	6.08	25.05	6.70	16.37	34.85
2	Industrial	2.25	9.27	2.48	6.06	12.90	13.49	55.63	14.87	36.35	77.38
3	Industrial	2.81	11.59	3.10	7.57	16.12	16.87	69.53	18.59	45.44	96.73
4	Residential	4.00	1.47	0.24	0.18	0.00	23.99	8.82	1.41	1.06	0.00
5	Medical	11.85	5.82	1.81	0.60	0.00	71.09	34.94	10.84	3.61	0.00
6	Industrial	3.09	12.75	3.41	8.33	17.73	18.55	76.49	20.45	49.98	106.40
7	Industrial	0.47	1.93	0.52	1.26	2.69	2.81	11.59	3.10	7.57	16.12

TAZ	Land Use	Number of Trips [vehicle/hour]									
		2024 (Occupancy 3%)					2029 (Occupancy 18%)				
		MCY	CAR	LGV	MGV	HGV	MCY	CAR	LGV	MGV	HGV
8	Industrial	0.47	1.93	0.52	1.26	2.69	2.81	11.59	3.10	7.57	16.12
9	Industrial	1.55	6.37	1.70	4.17	8.87	9.28	38.24	10.22	24.99	53.20
10	Industrial	1.55	6.37	1.70	4.17	8.87	9.28	38.24	10.22	24.99	53.20
11	Institutional	14.15	3.82	0.57	0.57	0.00	84.91	22.95	3.44	3.44	0.00
12	Recreational	0.14	0.11	0.01	0.00	0.00	0.86	0.66	0.05	0.00	0.00
13	Port/ Terminal	1.53	6.33	1.69	4.14	8.80	9.21	37.97	10.15	24.81	52.82
14	Industrial	0.76	3.13	0.84	2.05	4.36	4.56	18.79	5.02	12.28	26.14
15	Port/ Terminal	5.18	21.36	5.71	13.96	29.71	31.08	128.14	34.26	83.74	178.25
16	Industrial	0.56	2.30	0.61	1.50	3.20	3.35	13.80	3.69	9.02	19.19
17	Industrial	0.56	2.30	0.61	1.50	3.20	3.35	13.80	3.69	9.02	19.19
18	Industrial	0.27	1.10	0.30	0.72	1.54	1.61	6.62	1.77	4.33	9.21
19	Recreational	0.03	0.02	0.00	0.00	0.00	0.16	0.12	0.01	0.00	0.00
20	Office	3.38	3.01	0.75	0.38	0.00	20.30	18.05	4.51	2.26	0.00
21	Industrial	0.67	2.76	0.74	1.80	3.84	4.02	16.56	4.43	10.82	23.03
22	Industrial	1.01	4.18	1.12	2.73	5.81	6.08	25.05	6.70	16.37	34.85
23	Residential	2.00	0.74	0.12	0.09	0.00	12.00	4.41	0.71	0.53	0.00
Total (vehicle/hour)		**59.29**	**112.84**	**29.65**	**65.76**	**136.12**	**355.72**	**677.04**	**177.91**	**394.54**	**816.69**

4.4. Trip Distribution
There are four scenarios to obtain OD matrix:

- The 2024 scenario without industrial estate (2024-TKI),
- The 2024 scenario with industrial estate (2024-KI),
- The 2029 scenario without industrial estate (2029-TKI), and
- The 2029 scenario with industrial estate (2029-KI).

The trip distribution outside the industrial estate is sourced from the Semarang Raya traffic model from the FT UI Transportation Laboratory study in 2022. For the trip distribution from/to the industrial estate, it is only expected that TAZ 4, 5, 13, 15, 20, and 23 will generate trips to another TAZ. The trip distribution from/to industrial estate refers to the proportion of trip generation from each TAZ.

For the scenario with industrial park, the Tegineneng Industrial Estate was used as a reference for trip distribution from/to industrial estate by knowing the additional number of vehicle trips due to the presence of industrial estates. The number of trips after the presence of industrial estates in 2022 will change to 1,495 times the number of trips before development. As many as 33.13% of trips are the result of the existence of the industrial area [17].

4.5. Trip Assignment
The trip assignment stage is carried out using a generalized cost approach where this value is determined based on equation (2) where trips between zones of traffic analysis (TAZ) will look for trips with the cheapest value. The following table presents the weight values or operational cost coefficients for all types of vehicles (A).

Table 4. Operational Cost Coefficient, *A*

Vehicle Type	MCY	CAR	LGV	MGV	HGV
A (min./km)	0.20	0.30	0.30	0.35	0.40

Source : FT UI Transportation Laboratory (2022)

Furthermore, to obtain vehicle travel time, calculations are carried out using the capacity restrained (CR) approach as contained in equation (3). In this study, the coefficient values α and β are used by default from BPR, which are respectively 0.15 and 4 which are used for all types of road sections.

4.6. Validation

Model validation is carried out by comparing the vehicle travel time in the model to the actual vehicle travel time. By using mean average percentage error (MAPE), the accuracy of the models performed on the CUBE software is known as in Table 5.

Table 5. Model Validation of Travel Time

A	B	Travel Time (min.)		%Error
		Cube	Google Maps	
1002	1113	8.40	9	6.68%
1087	1102	7.94	7	13.45%
1087	1031	8.20	11	25.46%
1114	1117	4.44	6	26.01%
1108	1098	2.69	3	10.28%
		MAPE		16,38%

5. Analysis

5.1. Trip Distribution

To analyze the trip distribution, desire lines can be used which describe the relationship between the origin and destination of a movement. The thicker the desire lines, the greater the number of trips and vice versa.

5.1.1. Non-Goods Vehicle. As an example, we use the results of desire lines on motorcycle mode (MCY). It can be seen that in 2024 the presence of industrial estates will not have such a significant impact. However, in 2029 there will be a significant increase if there is an industrial area.

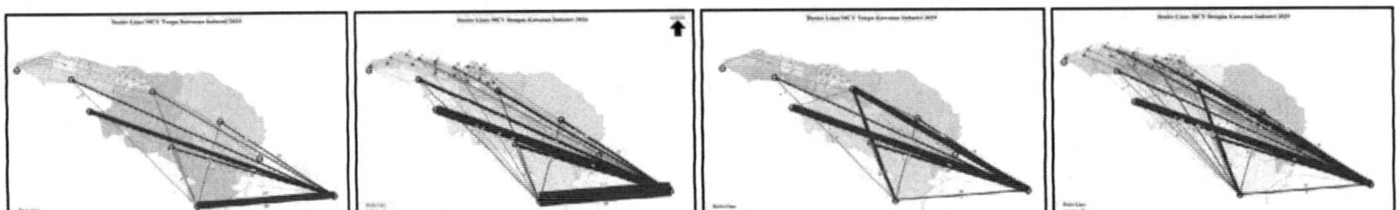

(a) 2024-TKI (b) 2024-KI (c) 2029-TKI (d) 2029-KI

Figure 3. Desire Lines on Motorcycle Mode (MCY).

5.1.2. Goods Vehicle. As an example, the results of desire lines are used in the heavy goods vehicle (HGV) mode. Similar results are also seen for non-goods vehicle modes, except that freight vehicles have a greater impact on trips from/to industrial areas than non-goods vehicles in both years.

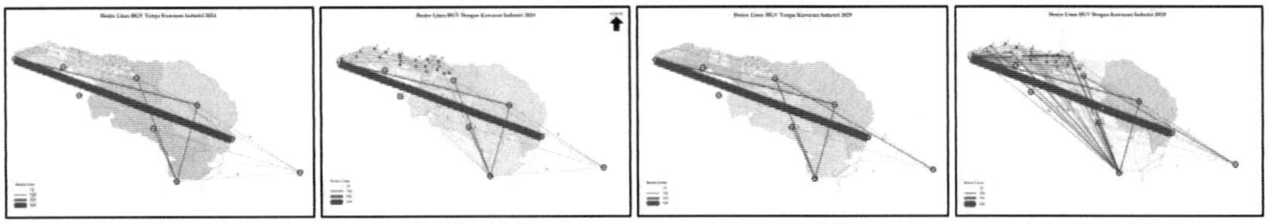

(a) 2024-TKI (b) 2024-KI (c) 2029-TKI (d) 2029-KI

Figure 4. Desire Lines on Heavy Goods Vehicle Mode (HGV).

5.2. Road Performance

In 2024, industrial estate has not had a significant impact on a road performance. But in 2029, it has had a significant impact on a road performance with a decrease in speed and an increase in degree of saturation or volume-capacity ratio (VCR).

5.2.1. Travel Speed. The following is an illustration of the speed of the road where the green color indicates a faster speed and the red color indicates the opposite.

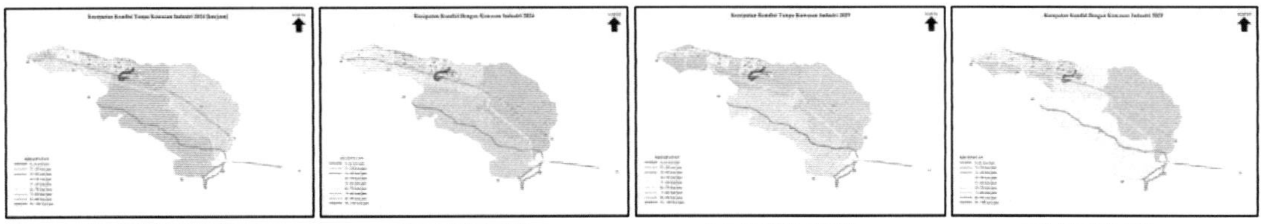

(a) 2024-TKI (b) 2024-KI (c) 2029-TKI (d) 2029-KI

Figure 5. Travel Speed on Road Segment.

5.2.2. Volume-Capacity Ratio (VCR). The following is an illustration of the VCR of the road where the green color indicates a faster speed and the red color indicates the opposite.

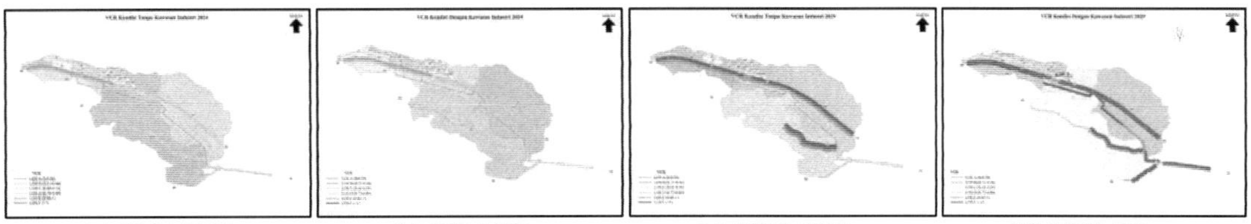

(a) 2024-TKI (b) 2024-KI (c) 2029-TKI (d) 2029-KI

Figure 6. VCR on Road Segment.

5.2.3. Recapitulation. The following is a recapitulation of road performance. It can be seen that the performance on toll road and primary arterial road 2/2 UD on average has entered poor and almost bad performance.

Table 6. Road Performance Recapitulation

Road Category (Type)	Measured Road Performance	Observation Year			
		2024-TKI	2024-KI	2029-TKI	2029-KI
Industrial Estate Road (2/2 UD)	Speed [km/hour]	36.31	36.31	35.86	35.40
	VCR	0.03	0.04	0.22	0.29
	Level Of Service	A	A	B	B
Industrial Estate Road (4/2 D)	Speed [km/hour]	49.11	49.10	49.08	48.01
	VCR	0.07	0.11	0.11	0.35
	Level Of Service	A	A	A	B
Toll Road-Main Road (2/1 UD)	Speed [km/hour]	83.94	83.46	77.79	72.88
	VCR	0.78	0.80	0.98	1.08
	Level Of Service	D	D	E	F

Road Category (Type)	Measured Road Performance	Observation Year			
		2024-TKI	2024-KI	2029-TKI	2029-KI
Toll Road-Ramp/Gate (1/1 UD)	Speed [km/hour]	92.00	91.98	91.97	72.88
	VCR	0.07	0.13	0.18	0.52
	Level Of Service	A	A	A	C
Primary Arterial Road (2/2 UD)	Speed [km/hour]	36.11	35.88	35.79	30.95
	VCR	0.32	0.38	0.38	0.72
	Level Of Service	B	B	B	C
Primary Arterial Road (4/2 UD)	Speed [km/hour]	45.04	45.02	45.03	43.87
	VCR	0.20	0.24	0.21	0.46
	Level Of Service	A	B	B	C
Local Road	Speed [km/hour]	27.86	27.86	26.64	26.44
	VCR	0.24	0.24	0.42	0.44
	Level Of Service	B	B	B	B

5.3. The Do-Something Scenario

The author simulates conditions in 2029 with an industrial estate with a do-something scenario by adding capacity or widening the toll road from 4/2 D to 6/2 D and on the primary arterial road from 2/2 UD to 4/2 UD. The results showed that after the scenario was carried out, there was an increase in performance on toll roads and primary arteries.

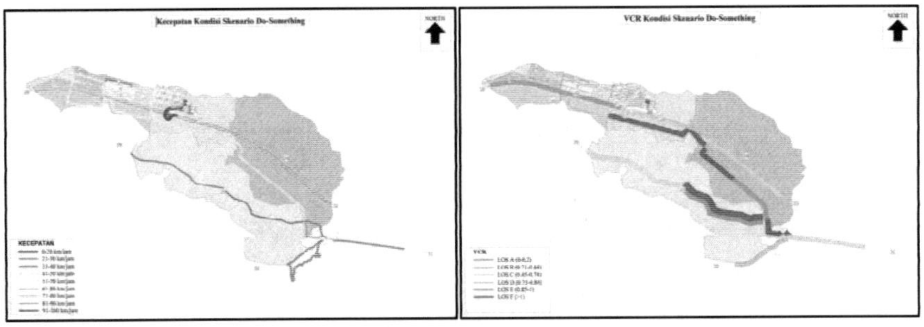

Figure 7. Speed and VCR in the Do-Something Scenario.

Table 7. Road Performance Recapitulation in the Do-Something Scenario

Road Category (Type)	Measured Road Performance	Kondisi Tahun Pengamatan	
		2029-KI	2029-KI-DS
Toll Road-Main Road (2/1 UD)	Speed [km/hour]	72.88	84.90
	VCR	1.08	0.72
	Level Of Service	F	C
Primary Arterial Road (4/2 UD)	Speed [km/hour]	30.95	35.85
	VCR	0.72	0.34
	Level Of Service	C	B

Another alternatives that could be considered in order to improve road performance in the future due to the KIT Batang development are utilization of containerized rail or sea transport to minimize the use of trucks and to provide mobility solutions for the workers (such as BRT, employee shuttle service, etc.). These alternatives are possible to be applied due to the development of Plabuan railway station and shipping port, and advanced transportation systems based on transit management solutions are also planned to be developed in the KIT Batang area. These solutions worth to be explored and analyzed with more data.

6. Conclusion

- The development of the Batang Integrated Industrial Estate phase 1 with an assumption of 3% occupancy has not had a significant impact on road performance around the industrial area in 2024.
- The development of phase 1 of the KIT Batang with the assumption of 18% occupancy will have a significant impact on the performance of roads around the industrial area in 2029.
- Overall, in 2024 and 2029, goods vehicles (LGV, MGV, and HGV) have a more impact on trip generation from within industrial areas than non-goods vehicles (MCY and CAR).
- To obtain improved performance on the Semarang – Batang Toll Road from F to C, it is necessary to widen the road from 4/2 D to 6/2 D.
- To obtain improved performance on Pantura Arterial Road from C to B, it is necessary to widen the road from 2/2 UD to 4/2 UD.

References

[1] Kementrian Perindustrian Republik Indonesia 2022 *Tangkap Relokasi Industri, Kemenperin Terus Pacu Pembangunan KIT Batang* (kemenperin.go.id)
[2] Tamin O Z 2000 *Perencanaan dan Pemodelan Transportasi, Edisi Kedua* (Bandung: Institut Teknologi Bandung)
[3] Direktorat Jenderal Bina Marga 1997 *Manual Kapasitas Jalan Indonesia 1997* (Jakarta: Bina Marga)
[4] Institute of Transportation Engineers 2017 *Trip Generation Manual 10th Edition*
[5] Putri A V and Kamaluddin M 2022 *Prarancangan Pabrik Cellulose Murni dari Bagasse-Ampas Tebu dengan Kapasitas 20.000 Ton/Tahun* (Yogyakarta: Universitas Gadjah Mada)
[6] Brahmasuta A and L I G B 2022 *Prarancangan Pabrik Selulosa Asetat dari Cotton dan Asetat Anhidrid Kapasitas 35000 ton/tahun* (Yogyakarta: Universitas Gadjah Mada)
[7] Setyawan R F 2009 *Efisiensi Teknis pada Bangunan Industri* (Depok: Universitas Indonesia)
[8] Arifani M F and Siswanto H M 2022 *Pra Rancangan Pabrik Garmen dengan Studi Kelayakan Produksi Kain Kerudung Kapasita 1.458.133 Pcs/ Tahuns* (Universitas Islam Indonesia)
[9] Kementrian Perindustrian Republik Indonesia 2016 *Industri Kompresor AC Mobil Senilai Rp 1,15 Triliun Beroperasi di Cikarang* (kemenperin.go.id)
[10] Mutharuddin M and Muharnmadun H 2019 *Analisis Dampak Lalu Lintas Kawasan Industri Studi Kasus PT Banten Energy International, Kecamatan Kramatwatu Kabupaten Serang* (Warta Penelitian Perhubungan) 25(3) https://doi.org/10.25104/warlit.v25i3.716
[11] Karim Y 2022 *Riset JLL: Okupansi Kawasan Industri Cikarang Capai 93 Persen* (merdeka.com)
[12] Triono D and Nabilah F 2022 *Analisis Pasar Tingkat Okupansi Kawasan Industri MM2100 Cikarang Barat Kabupaten Bekasi* (Syntax Literate: Jurnal Ilmiah Indonesia) 7(12) https://doi.org/https://doi.org/10.36418/syntax-literate.v7i12.11590
[13] Dinas Penanaman Modal dan Pelayanan Terpadu Satu Pintu Provinsi Jawa Tengah 2019 *Perencanaan Penyediaan Infrastruktur Pendukung Kawasan Industri di Jawa Tengah*
[14] Maria et al 1996 *Perilaku Masyarakat di Lingkungan Kawasan Industri Pulogadung*
[15] Komisi VII DPR RI 2022 *Pertumbuhan Kawasan Industri Beri Efek Ganda Bagi Ekonomi Nasional* (DPR.Go.Id) https://www.dpr.go.id/berita/detail/id/37451/t/Pertumbuhan+Kawasan+Industri+Beri+Efek+Ganda+Bagi+Ekonomi+Nasional.
[16] Yogatama B K 2015 *EJIP kaji kembangkan kawasan industri di luar Jawa* (industri.kontan.co.id)
[17] Octavia A 2022 *Dampak Rencana Pengembangan Kawasan Industri Strategis di Kecamatan Tegineneng terhadap Pola Pergerakan di Kabupaten Pesawaran* (Bekasi: Politeknik Transportasi Darat Indonesia – STTD)

IOP Conf. Series: Earth and Environmental Science 1294 (2024) 012005 doi:10.1088/1755-1315/1294/1/012005

Origin-destination matrix estimation in Solo agglomeration with the maximum likelihood method

T Agustin[1,*], **M Millah**[1], **F A Briliansyah**[1], and **N F Ulya**[1]

[1]Civil Engineering Department, Faculty of Engineering, Universitas Sebelas Maret, Surakarta, Indonesia
[*]Corresponding author, e-mail: tuti.agustin@staff.uns.ac.id

Abstract. The movement of vehicles is one form of fulfilling community needs. High vehicle movement can cause transportation problems, one of which is congestion. To overcome this problem, planning a transportation system is essential. Using the EMME/4 program, we analyzed the distribution of movements in the Origin Destination Matrix (MAT) using a gravity model. Using the User's Equilibrium method, the study aims to determine the parameter β, data validity value (R^2), and movement patterns. The research study area was carried out in Solo Agglomeration with 19 zones. The results indicate a β value of 0.0796 ($R^2 = 0.817$), with an estimated Solo Agglomeration movement of 34428 passenger-car-units per hour in 2023. The most extensive distribution of movements occurred within internal movements, making up 62.85%, while the smallest were external-external movements, accounting only for 4.01%.

1. Introduction

The growth of urban activity likely triggers land redevelopment. This kind of development is believed to be a significant component of city evolution [1]. Besides its positive effect, there is also a negative impact, such as the explosion of traffic movement. Traffic is inseparable from the needs of people's lives. According to [2], the demand for movement is a derivative need, where every movement occurs to fulfill community needs. If the movement that happens is not controlled correctly, it can cause transportation problems. These problems can be minimized or prevented through one of the transportation models, namely the estimation of the movement distribution represented by the Origin-Destination Matrix (OD-Matrix). There is a different method for estimating the O-D matrices. The maximum likelihood model is used by [3] with total trip attractions and trip productions of the study zones are known. [4] develops a new statistical method for estimating the matrix with random link-choosing proportions. Moreover, [5] and [6] use observed link flows and the prior to estimate the matrix with the Bayesian Network model. The goal of those methods in estimating the matrices is to predict the travel demand.

This study aims to determine how traffic moves throughout the Solo Agglomeration area. As the center of the national administration, commercial hub, trade core, and service center, Solo Agglomeration becomes a pull factor for people to go to the area [7]. In addition, the site is highly urbanized and has distinct characteristics in each region [8]. This research is expected to realize a long-term transportation system as a guideline in determining strategies for preventing or overcoming transportation problems related to vehicle movement.

To achieve the objectives of this study, a gravity parameter is needed, namely the β parameter. To estimate the trip distribution, this parameter will be used to generate the cost function. The β is obtained by parameter calibration utilizing the approach of maximum likelihood with the impedance function of negative exponential and the trip generation-attraction constraint.

2. Study Method

The location of the study is in the Solo Agglomeration area of 5677 km^2. This region comprises seven districts: Surakarta City, Klaten, Boyolali, Sukoharjo, Sragen, Karanganyar, and Wonogiri Regency. The road networks involved in this study are the toll road, arterial, and collector road networks with primary access to regional movements. The Solo Agglomeration area is divided into 13 internal and six external zones. The internal zones are divided based on the breakdown of districts or cities. In contrast, the external zones are divided based on towns or communities following the administrative area of Solo Agglomeration.

The primary source for the research is obtained from the survey results of traffic flow during observation, while secondary data is collected from related agencies or other sources. Both data types are processed into an estimation process assisted by the 'Equilibre Multimodal, Multimodal Equilibrium' (EMME)/4 program. EMME/4 is a program with high CPU performance in modeling road networks and multimodal requests [9]. Through the EMME/4 program, road network modeling will have a road network database, including capacity, free flow speed, and free flow journey time. The road network database was obtained by referring to 'Manual Kapasitas Jalan Indonesia' (MKJI) 1997 [10]. The EMME/4 program yields a matrix as its final output, which will be employed for MAT estimation through the Maximum Likelihood method and the negative exponential resistance function. Still, before that, it is necessary to calibrate the β parameter as a parameter that describes the average travel cost [2]. This β parameter will be used to compute the resistance function with a negative exponential approach within the gravity formula. After the MAT model results are obtained, the accuracy of the data is tested with the determination coefficient (R^2) based on the comparison of observed and modeled traffic flows. Figure 1 provides a more thorough explanation of the research flow.

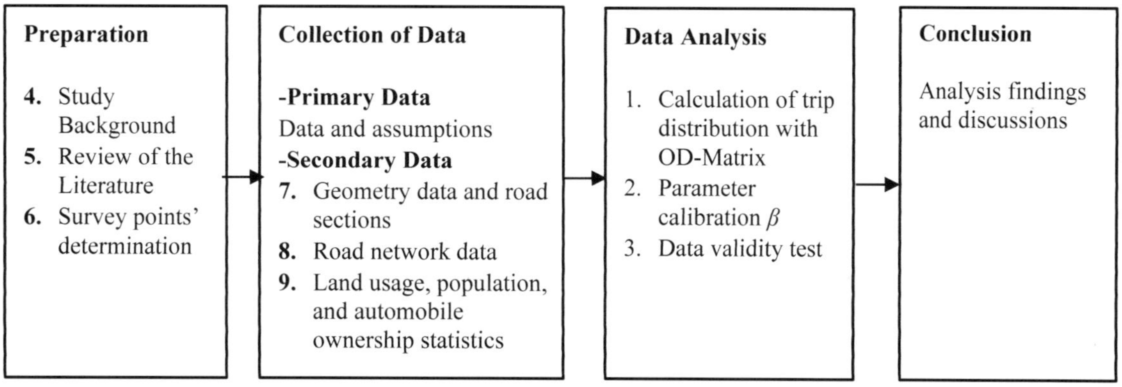

Figure 1. Research Phases

3. Maximum Likelihood (ML) Formulation

This study used the ML method in the calibration process for the β parameter. Based on combined probability density theory, for example, a random variable $x_1, x_2, x_3, \ldots x_n$, which is a stochastic variable characterized by its probability density function, which is expressed by $f(x, \theta)$ [2], [11]. The value of θ is an unknown variable. The probability of getting $x_1, x_2, x_3, \ldots X_n$ is stated in Equation 1 below.

$$L(x_1, x_2, \ldots x_n; \theta) = f(x_1; \theta), f(x_2; \theta), \ldots f(x_n; \theta) \quad (1)$$

The formula for the function L is multiplicative, so it will be easier if expressed in logarithmic form like this following Equation 2.

$$Log_e L(x_1, x_2, \ldots x_n; \theta) = \sum_{i=1}^{n} Log_e f(x_n; \theta) \quad (2)$$

Obtaining the value of θ can be done by reducing the function L to θ equal to zero as in Equation 3.

$$\frac{\partial L(x_1,x_2,..x_n;\theta)}{\partial\theta} = 0 \quad (3)$$

The derivation of the multinomial distribution through the ML method is stated in Equation 4 below.

$$\frac{N!}{a_1!a_2!a_3!\,...a_s!}\,p_1^{a_1}p_2^{a_2}p_3^{a_3}\ldots p_s^{a_s} \quad (4)$$

The p-value is expressed as a parameter signified by θ so that the ML function θ follows a multimodal distribution stated in Equation 5:

$$L(\theta) = k.[p_1(\theta)]^{a_1}.[p_2(\theta)]^{a_2} \ldots .[p_s(\theta)]^{a_s} \quad (5)$$

The value of $\frac{N!}{a_1!a_2!a_3!\,...a_s!}$ is expressed in k constant. If, for example, the p_{id} value is the probability of obtaining specific data from the distribution of trips between zones, the p_{id} value can be stated in Equation 6:

$$P_{id} = \frac{T_{id}}{T} \quad (6)$$

$$T = \sum_{i=1}^{n}\sum_{d=1}^{n}\hat{T}_{id} \quad (7)$$

The ML function to obtain values of $\hat{T}_{id}$ for every origin zone (i), and destination zone (d), is formulated in Equation 8.

$$L = c\prod_{i=1}^{n}\prod_{d=1}^{n}p_{id}^{\hat{T}_{id}} \quad (8)$$

The ML method's objective function is to maximize Equation 8. Then, by maximizing Equation 2 with the total movement limit, the following Equation 9 will be obtained.

$$\sum_{i=1}^{n}\sum_{d=1}^{n}T_{id} - T = 0 \quad (9)$$

Based on a pure logarithmic function, there will be a maximization of the L_1 *function, which* becomes:

$$L_1 = \sum_{i=1}^{n}\sum_{d=1}^{n}(\hat{T}_{id}Log_e p_{id}) - \theta(\sum_{i=1}^{n}\sum_{d=1}^{n}T_{id} - T) + Log_e c \quad (10)$$

By inserting Equation 6 into Equation10, we get:

$$L_1 = \sum_{i=1}^{n}\sum_{d=1}^{n}\left(\hat{T}_{id}Log_e\frac{T_{id}}{T}\right) - \theta(\sum_{i=1}^{n}\sum_{d=1}^{n}T_{id}) + \theta T + Log_e c \quad (11)$$

With the total movement limit, the value of θ is assumed to be 1. When the constant k is from Equation 5, the objective purpose of the ML method will be expressed in mathematical formula 12 as follows.

$$L_2 = \sum_{i=1}^{n}\sum_{d=1}^{n}(\hat{T}_{id}Log_e c\,T_{id} - T_{id}) \quad (12)$$

4. Analysis Findings and Discussion

The EMME/4 program and Microsoft Excel were utilized to support the T_{id} estimation results in this investigation. The parameter is calculated using the negative exponential resistance function and the KM method's 6th iteration, which results in a value of 0.0796. The parameter's value can indicate the typical trip cost on the resistance function or $f(C_{id})$. Transportation planners and policymakers can use the *β parameter* to estimate MAT in the following year with an overview of the Solo Agglomeration area to identify which road sections require handling in accessibility changes in Solo Agglomeration. The Solo Agglomeration research area had a total movement of 34428 pcu/hour as determined by the 2023 MAT estimation technique. Internal-internal, internal-external, external-internal, and external-external movements are among those obtained. The following Figure 2 shows the details of the number of movements.

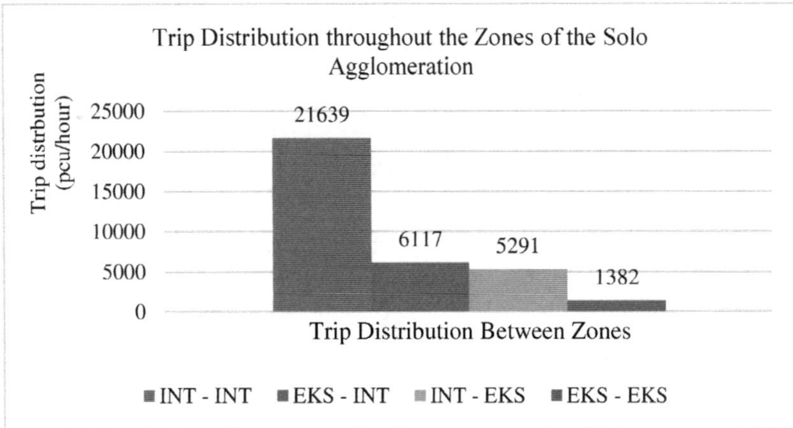

Figure 2. Movement Across Zones in Solo Agglomeration Region

Based on the movement across zones, the internal-internal movement shows the most significant movement between zones. This is caused by how widely movements are distributed among districts and cities. If the Solo Agglomeration study covers the movement, it is regarded as an internal movement of Solo Agglomeration. Zone D to zone C transition is one instance of movement. Zone C defines an external-internal movement as one coming from zone D to zone C. However, when observed from within that entity, the movement is seen as an internal movement within Solo Agglomeration. Given that Solo Agglomeration comprises seven districts/cities, each of the most significant district or city-level motions is an external-internal trip. Thus, the Solo Agglomeration's internal movement came to dominate. The internal-external zones experience less movement than the external-internal zones. This is because zone C contains large- and small-scale industrial and office spaces, attracting workers from inside and outside the Solo Agglomeration. Even though there were more moves due to this, they were still less frequent than those between internal zones.

The slightest movement is produced when motion occurs in the external-external zone. It can be observed in zones G, H, I, and J, all external zones with low values for generation and attraction. This results from people not wanting to engage in morning activities during prime time (06.30 – 07.30 WIB). To make the movement distribution study procedure simpler, given the size of the Solo Agglomeration, the initial 19 zones were reduced to 10 zones (zones A–J). As seen in Figure 3, a desire line represents movement outcomes across zones. The movement increases as the line becomes thicker, and vice versa. According to Figure 3, zones D (Sukoharjo Regency) and C (Surakarta City) are mainly where the movements with the highest flows occur. Zone C's sizable central business district is the reason for this. This fact is reinforced by an earlier study [12], which indicates that Surakarta City has a burgeoning health services industry and social activities that significantly impact the neighborhood.

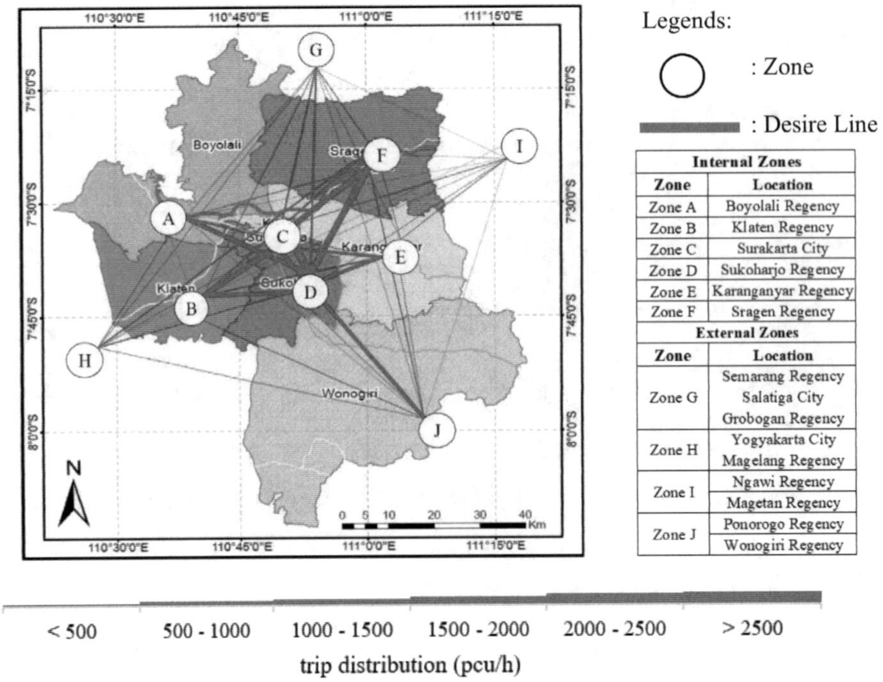

Figure 3. Desire Line, Estimated Results for Solo Agglomeration Movement in 2023

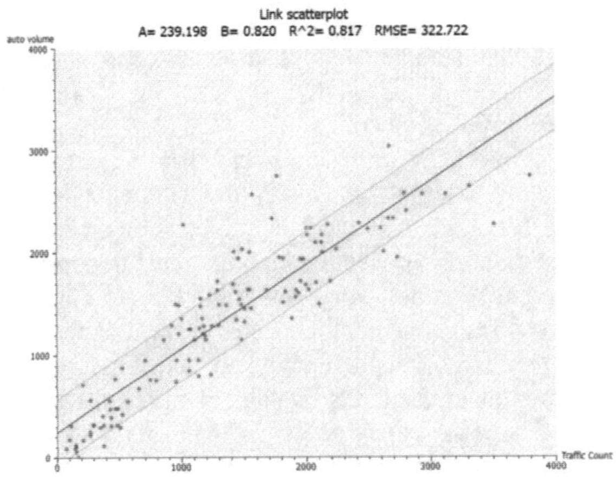

Figure 4. Scatter Plot: Testing the Traffic Flow's Validity

Moreover, zone D contains a substantial residential sector, especially for the construction of new housing, with as many as 1294 settlements in 2019 – 2021 [13]. A validity test is performed after receiving the MAT modeling result to compare traffic flow data from observations to traffic flow data from the modeling results. An R^2 value of 0.817 was obtained with an error of 18.3 %. Figure 4 displays the R2 value derived from the EMME/4 algorithm.

5. Conclusion

The parameter's value is calculated as 0.0796 based on the outcomes of the estimation of the distribution of movement in the Solo Agglomeration using the ML approach and the negative exponential resistance function. The data validity value achieved with a coefficient of determination (R^2) is 0.817 based on the

MAT gravity model. The overall movement per hour with the generation-pull restriction is 34428 pcu. The internal-internal zone movements, which account for 21639 pcu/hour (62.85%) of these total movements, have the most considerable estimated movement dispersion. Due to the sizeable central business district (CBD) in zone C and the sizable residential region in zone D, this movement has a high value. Further studies that include the collector and local road networks are required to strengthen the accuracy of this research and make the findings a greater reality.

References

[1] Y. L. Wang *et al.*, *The Mechanism of Urban Land Redevelopment Impact on Traffic Systems, ICCTP 2011 Towar. Sustain. Transp. Syst. - Proc. 11th Int. Conf. Chinese Transp. Prof.*, pp. 2571–2580, 2011, doi: 10.1061/41186(421)257.

[2] O. Z. Tamin, *Perencanaan, Pemodelan, dan Rekayasa Transportasi: Teori, Contoh Soal, dan Aplikasi*, 4th ed. Bandung: ITB, 2008.

[3] H. Spiess, *A maximum likelihood model for estimating origin-destination matrices, Transp. Res. Part B Methodol.*, vol. 21, no. 5, pp. 395–412, Oct. 1987, doi: 10.1016/0191-2615(87)90037-3.

[4] H. P. Lo *et al.*, *Estimation of an origin-destination matrix with random link choice proportions: A statistical approach, Transp. Res. Part B Methodol.*, vol. 30, no. 4, pp. 309–324, Aug. 1996, doi: 10.1016/0191-2615(95)00036-4.

[5] K. Perrakis *et al.*, *A Bayesian approach for modeling origin–destination matrices, Transp. Res. Part A Policy Pract.*, vol. 46, no. 1, pp. 200–212, Jan. 2012, doi: 10.1016/J.TRA.2011.06.005.

[6] L. Cheng *et al.*, *A Bayesian network model for origin-destination matrices estimation using prior and some observed link flows, Discret. Dyn. Nat. Soc.*, vol. 2014, 2014, doi: 10.1155/2014/192470.

[7] R. Noviani and L. Muta'ali, *Dinamika Karakteristik Tingkat Kekotaan Di Wilayah Metropolitan Solo Raya Tahun 1990-2015*, 2019.

[8] I. H. Pasaribu, *Kajian Proses Perkembangan Pusat Pelayanan dan Pusat Permukiman di Kawasan Metropolitan Solo Raya., UNDIP*, 2017.

[9] INRO Consultan Inc, *EMME/4*. Bentley, Canda, 2019.

[10] Direktorat Bina Marga, *Indonesian Highway Capacity Manual Project (IHCM)*, vol. 1, no. I. 1997.

[11] N. Trisnawati *et al.*, *Forecasting of origin-destination matrix of freight transportation of Surakarta with maximum likelihood method, AIP Conf. Proc.*, vol. 1977, no. June, 2018, doi: 10.1063/1.5042989.

[12] H. I. Samosir *et al.*, *Analisis Potensi Sektor Ekonomi di Kota Solo dan Semarang Provinsi Jawa Tengah dan Kota Yogyakarta Provinsi Daerah Istimewa Yogyakarta, Syntax Idea*, vol. 3, 2021.

[13] A. N. Aini *et al.*, *Kajian Pola Persebaran Permukiman Di Kecamatan Kartasura Kabupaten Sukoharjo, Desa-Kota*, vol. 4, no. 2, p. 241, 2022, doi: 10.20961/desa-kota.v4i2.63661.241-257.

Acknowledgments

The authors would like to express their gratitude to Universitas Sebelas Maret for providing financial support for this research. The funding was received from the RKAT PTNBH Universitas Sebelas Maret Fiscal Year 2023 under the RISET GRUP HIBAH RESEARCH scheme (HGR-UNS RESEARCH)-B, with the research assignment agreement letter number 228/UN27.22/PT.01.03/2023.

IOP Conf. Series: Earth and Environmental Science 1294 (2024) 012006 doi:10.1088/1755-1315/1294/1/012006

Accessibility and Land Use Effect of Residential Area with Different TOD Typology to Value Creation

*Vani Arliani[1,2], Ade Sjafruddin[3], Idwan Santoso[3], and Haryo Winarso[4]

[1]Doctoral Program of Civil Engineering, Faculty of Civil and Environmental Engineering, Institut Teknologi Bandung, Indonesia.
[2] Civil Engineering, Faculty of Infrastructure Planning, Universitas Pertamina, Jakarta, Indonesia.
[3] Civil Engineering, Faculty of Civil and Environmental Engineering, Institut Teknologi Bandung, Indonesia.
[4]School of Architecture, Planning and Policy Department, Institut Teknologi Bandung, Indonesia.
*vani.arliani@gmail.com

Abstract. TOD (Transit Oriented Development) is the planning design focusing on integrated transportation and land use in an area with certain radii. There are many transits node which is being designed as TOD, which varies into many typologies. Indication of TOD measurement from TOD ideal, can be measure from accessibility factor and land use factor. There is an indication that each TOD area will give different output (value creation) related to accessibility and land use function. Many studies states about the TOD in different city. where this paper will analyse many areas within the same city. The method is analysing the area into TOD index using AHP, then measuring the value creation with summation of real land and building priceThen analyse the correlation between TOD index and value creation. The study case locations is in Lebak Bulus, Dukuh Atas and Kelapa Gading. The identification of value creation will be useful as an input to the urban planners or policy makers for designing more targeted strategies. The discoveries made in comparative context differ from those made in prior studies. This paper is thus believed to make a new contribution to the existing TOD categorized with additional input related to value creation inside.

Keywords: TOD, accessibility, land use, value creation.

1. Introduction

Accessibility is a measure of comfort or convenience regarding the way land use locations interact with each other and the 'easy' or 'difficult' location being reached through the transportation network system (Black, 1981). Tamin (2000), add the definition, that accessibility is a concept that combines a geographic land use management system with a transportation network system that connects it. Then, there's a definition of accessibility defined by Geurs and van Wee (2004) that accessibility as the extent to which land use and transport systems enable a person able to reach activities or destinations using transport mode.

Accessibility and land use interaction has been taken as variables that will develop a characteristic in an area. Former researcher defined accessibility to transit is generally define as an environment variable that refer to locations, that describe in distance travelled or transit time of facilities and availability of transit services in the neighbourhood (Xu et al., 2015, 2018). Many academics and professionals research in many countries has been published many factors name accessibility as "by-transit accessibility" to quantify the ease of reaching important destinations by the transit system (Dimitriou, 1992; Armstrong and Rodriguez, 2006). Some studies correlate the land use and accessibility, in this case refer to external accessibility (Xie, 2021; Luti Consulting, 2016; Sharma and Newman, 2020).

Aligned with TOD concept which has characteristics as multifunctional land use (mixed-use), there is an integration with the public transport network around the transit location, and availability of non-motorized transportation facilities (bicycle or pedestrian lanes). Previous research place TOD as the solution better city planning and transformation the city with sustainability concept. The problem in many countries, both developed and developing countries is the restriction in TOD financing. An option to find a method that can recognize the potential value creation in an area, will be a solution for TOD financing. Hopefully, the method will help the government to design the policy to reach the TOD ideal in an area, and able to create how to finance the TOD.

Value creation and stakeholder responses will increase as the increment of land and building price in a TOD area. This paper will see the TOD criteria appropriation from accessibility factor and land use factor. We will be dividing the accessibility into two parts, internal accessibility, and external accessibility. Internal accessibility will observe how the resident in TOD area using non-motorized transportation to reach the transit node. External accessibility will see how the movement from one transit node to another. Accessibility factor also observe the activities as the effect of the new transit mode station and see the people mobility at the area. Later, observing how the transit node grown up the economic scale and enrich the diversity function in the area.

The methodology in this paper is to know the potential transit node for becoming the TOD area. In Jakarta the prospective TOD area is an area which has transit node in it. After knowing the transit node as the centre, then we can start to observe an area around with 800 meters radii. After that, we can analyse the area and give the TOD index, that measuring from the criteria of TOD (accessibility factor and land use factor). The next step is categorizing the area into eight function of land use and from the data of land and building price in the area. Value creation calculate from the cumulative calculation of the prices in certain TOD area. Latest step is to see the corelation between TOD index and value creation.

After deciding the location for case study and collecting the data, this paper attempts to fill the gap and conducts a comparative analysis of TOD among three residential areas in Jakarta (Indonesia) with different accessibility type. We arrange the paper as follows: Section 2 reviews the previous literature, Section 3 Value creation analysis, section 4 Main Results Summarized

2. Literature Review

The research criteria will be determined from previous literatures, which has been widely give many factors to analyse the TOD typology from the TOD ideal concept. There are still gaps from previous studies where this paper will be observing the TOD index from the interaction between accessibility and land use. There is a prediction that some element in determining the TOD characteristic that will form some specific value creation. For this research, the TOD criteria which has centre in a transit node on some residential area will analyse with external accessibility factor, internal accessibility factor, and land use factor.

2.1. TOD Typologies

TOD typologies has been analyse under different transport modes, such as "bus rapid transit (BRT), subway, railways and combination has been studied in prior study (Case et al., 2019; Chorus and Bertolini, 2011; Kamruzaman et al., 2014; Kumar et al., 2020; Li et al., 2019; Lyu et al., 2016; Monajem and Nosratian, 2015; Nasri and Zhang, 2014; Reusser et al., 2008; Li et al., 2019; Lyu et al., 2016; Zhang et al., 2019; Vale et al., 2018)". This paper will analyse a transit node which has different public transport. Lebak Bulus area is the centre of MRT and BRT station. Dukuh Atas is the centre of MRT, BRT, KRL station. Kelapa Gading is the centre of LRT station.

Models, such as "the node-place model by (Bertolini, 1999), and the Butterfly model by (Delta Metropolis Association, 2014), suggested that node should be in balance with the place for all station areas. Based on that suggestion, various stations were created depending on that balance and stations were slotted into the 'best-fitting' type". The model in this paper will analyse if there is a correlation between TOD index and value creation.

2.2. Value Creation

"The value impact range at different stations may vary owing to different accessibilities (Tang, et al., 2021). TOD planning will generate many huge profits with some typical physical characteristics are measured using the TOD Index (Singh, 2015). Many studies have learned that values of transit area happened by gaining the increase of land value around the transit station. Transit oriented rather than transit adjacent land uses are critical to maximize land value uplift (Bartholomew and Ewing, 2011)".

"Belzer and Autler (2002) defined the benchmark of TOD success are depends on location efficiency, value recapture, liveability, financial return, choice, efficient land use pattern at regional level. Renne & Wells (2005) lack of comprehensive tools for measurements of TOD. Evans & Pratt (2007) proposed TOD index as a potential device for considering the degree to which the project is intrinsically oriented towards transit. Also, Evans & Pratt (2007) stated that measuring TOD-ness will not only help in better TOD planning but also in TOD projects evaluation". If the identification of TOD index can be calculated, and value creation measurement can be calculated, then we can try to find the model between TOD index and value creation.

3. Methodology

This paper will see the TOD criteria from accessibility and land use. We will be dividing the accessibility into two parts, internal accessibility, and external accessibility. The flow chart of methodology for this paper, as follows:

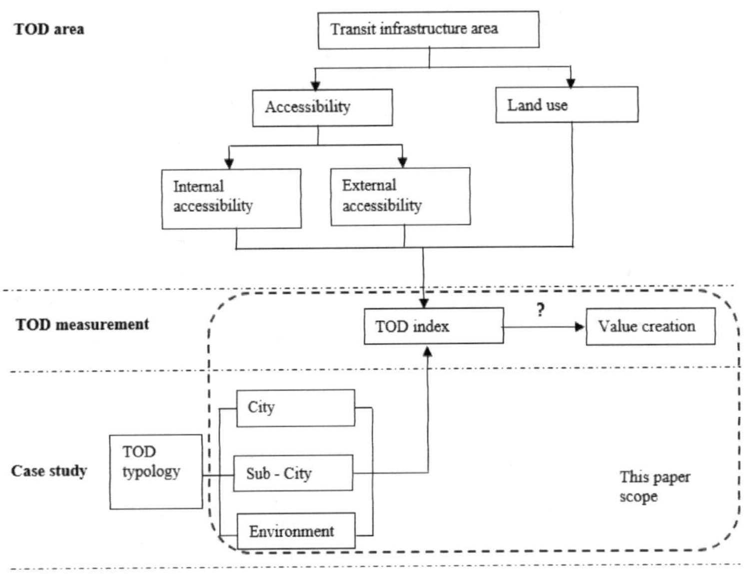

Figure 1. This paper framework

The method to formulate the TOD index in an area is by taking into account external accessibility, internal accessibility (first & last mile connectivity) and land use.
The TOD index formula is as follows:

$$TOD\ Index = \sum \{f_1(external\ accessibility), f_2(internal\ accessibility), f_3(landuse)\} \quad (1)$$

The calculation starts with analyse transit node as the centre and the area taken as an 800 meters radii circular area, with a transit node as the centre. We can analyse the level of TOD ideal from the criteria with the main indicators consist of external accessibility, internal accessibility (first & last mile travel), and land use. TOD ideal level can be measure by multidimensional aspects based on the characteristics or typical potential of an area using index concept. Range of TOD ideal will be between 0 to 1. The most ideal an area into TOD ideal categories, the result will be near to 1.

The first step, to analyse a transit node, can be seem through the figure 2 below, TOD describe as a circle in some area with many functions of building inside. TOD is a concept where a multifunctional building taken into one area with many diversities and density inside. So, there must be accessibility in one region between many deferent functions of building.

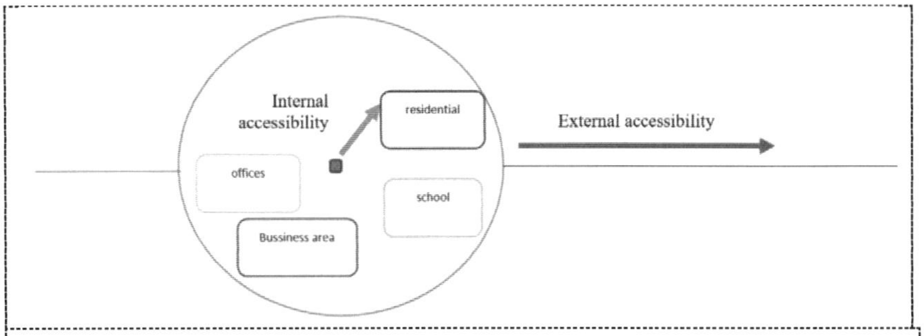

Figure 2. Accessibility in TOD consist of external and internal accessibility.

After measuring the TOD index, we will start to calculate the value creation. To analyse the value creation, there will be categorization for an area in 800-meter radii which divide into eight categories, such as residential, offices building, educational area, health facilities, business area, public facilities, open space zone and mixed-use areas. Each category in an area will be valuing based on the real transaction price for land and building above it. The result will be from the summation price that taken as the value which has been created in a transit node. We will see if there is any appropriation between TOD ideal index (level) to the value creation that has been created or there is no correlation at all.

3.1. TOD Typology

As already state by the Minister of ATR/BPN Republic of Indonesia, the TOD typologies classified as city centre, sub city and environment. From this reference, Dukuh Atas is classified as city TOD, Lebak Bulus classified as sub city TOD, while Kelapa Gading categorized as environment typology TOD. The paper will analyse some different typology.

The paper will take three places as the study case area. The locations within DKI Jakarta province is as follow:

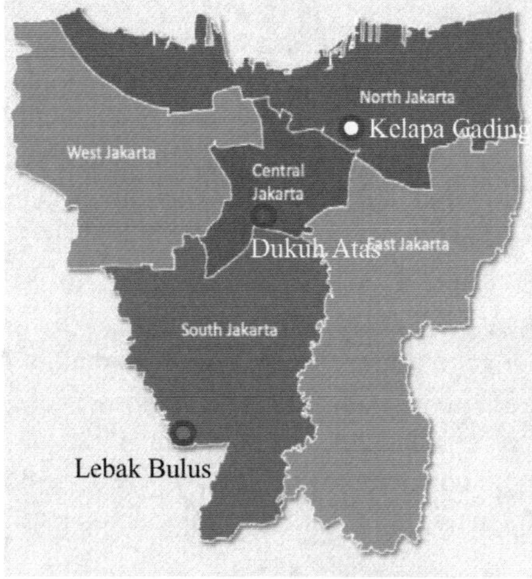

Figure 3. Study case location

Lebak Bulus location as a gate for Jakarta from South Tangerang city and Depok city. The situation in Lebak Bulus in between Sub City and City typology. While, Dukuh Atas is between border of South Jakarta city and Center Jakarta city. Kelapa Gading is the residential complex in North Jakarta city. Related to TOD typology, following are the categories of the three areas:

- Sub city : Lebak Bulus → Residential-type (landed house)
- City centre : Dukuh Atas → Residential-type (vertical building)
- City centre : Kelapa Gading → Residential-type (landed house)

The measurement of appropriation for Lebak Bulus and Kelapa Gading into TOD ideal categories based on external accessibility, internal accessibility, and land use:

Table 1. Identification of TOD Level with Accessibility and Land Use Categories

Criteria	Factor	Sub Factor	Unit	Location			Min	Maks	Standards
				Lebak Bulus	Dukuh Atas	Kelapa Gading			
External Accessibility	Public Transport Capacity (sit and stand)	Multimodes	passanger/hour	3550	4622	135	135	4.622	Maximum capacity in a transit node
Internal *Accessibility (first & last mile connectivity)*	Distance to transit	Average route distance from station to residential quarters	meter	842,31	820,47	835,63	0	800	ITDP (<500 - 1000)
		Average route distance from station to work places/offices	meter	785	595	1047	0	800	ITDP (<500 - 1000)
		Average route distance from station to hotels	meter	653	746	0	0	800	ITDP (<500 - 1000)
		Average route distance from station to health facilities	meter	710	0	0	0	800	ITDP (<500 - 1000)
		Average route distance from station to educational facilities	meter	723	1025	1536	0	800	ITDP (<500 - 1000)
		Average route distance from station to shopping	meter	596	783	500	0	800	ITDP (<500 - 1000)
	Pedestrian infrastructure	Proportion of pedestrian area to catchment area	%	0,041920	0,057760	0,034680	0	60	-Comparing TOD Sites by Walkability Indicators, Marc Schlossberg, 2004 -ITDP
		Physical obstacles in pedestrian lane	unit	7	5	7	0	20	-
		Wayfinding in pedestrian lane	meter	10	12	2	0	20	-
	First and last mile connectivity	bikesharing	unit/100 meter	0	10	0	0	115	ITDP.org (Jakarta matter)
		shuttle service (sag)	unit/200 meter	0	5	0	0	20	-
	Bicycle and micromobility infrastructure	Proporsi jalur sepeda terhadap *catchment area*	%	0,00764	0,008460	0	0	1	-
		Keselamatan jalur sepeda	-	0	8	0	0	10	Developing a bikeability index to enable the assessment of TOD nodes, Hartanto & Brussel et al. 2017
		Kenyamanan jalur sepeda		0	7	0	0	10	Developing a bikeability index to enable the assessment of TOD nodes, Hartanto & Brussel et al. 2017
		Proporsi area parkir sepeda terhadap *catchment area*	m^2	0	0.000846	0	0	1	-
	On demand microtransit	ojek	Unit/hour	20	20	5	0	20	-
		angkot	Unit/hour	6	3	0	0	10	-
Land Use	Density	Population density	$people/km^2$	16.865	12.138	12.652	0	29.320	-
		Job density	$employee/km^2$	31000,15	38356,44	25801,07	0	41667	-
	Diversity	Proportion of residential area	Rasio to catchment area	0,471	0,3598	0,6308	0	1	-
		Proportion of public offices area	Rasio to catchment area	0,104	0,1142	0,1493	0	1	-
		Proportion of industrial area	Rasio to catchment area	0,0362	0,0721	0,0058	0	1	-
		Proportion of educational facilities	Rasio to catchment area	0,0768	0,0044	0,0028	0	1	-
		Proportion of health facilities	Rasio to catchment area	0.0033	0	0	0	1	-
		Proportion of public facilities area	Rasio to catchment area	0,0525	0,0479	0.035	0	1	-
		Proportion of mixed-used area	Rasio to catchment area	0.0188	0,0753	0,0451	0	1	-
		Proportion of green open space area	Rasio to catchment area	0,0757	0.1516	0.0433	0	1	-
	Floor Area Ratio	Floor to land area reatio	-	3	5	2	0,5	3	TOD Overlay District, Fremont
							2	5	Permen ATR/BPN No 16/2017

The result for measurements in accessibility and land use function for the two residential areas are as follow:

Table 2. Measurement of TOD Level

Criteria	Factor	Sub Factor	Unit	Data		
				Lebak Bulus	Dukuh Atas	Kelapa Gading
External Accessibility	Public Transport Capacity (sit and stand)	Multimodes	passanger/hour	0.289	0.380	0
Internal *Accessibility (first & last mile connectivity)*	Distance to transit	Average route distance from station to residential quarters	meter	0,007	0,007	0,0067
		Average route distance from station to work places/offices	meter	0,042	0,056	0,03160
		Average route distance from station to hotels	meter	0,008	0,007	0
		Average route distance from station to health facilities	meter	0,009	0,000	0
		Average route distance from station to educational facilities	meter	0,012	0,008	0,0055
		Average route distance from station to shopping	meter	0,020	0,015	0,02394
	Pedestrian infrastructure	Proportion of pedestrian area to catchment area	%	0,000	0,000	1,29703E-05
		Physical obstacles in pedestrian lane	unit	0,070	0,098	0,070
		Wayfinding in pedestrian lane	meter	0,011	0,013	0,002108
	First and last mile connectivity	bikesharing	unit/100 meter	0,000	0,002	0
		shuttle service (sag)	unit/200 meter	0,000	0,011	0
	Bicycle and micromobility infrastructure	Proportion of bicycle lane to catchment area	%	0,000	0,000	0
		Bicycle lane safety	-	0,000	0,020	0
		Bicycle lane convenience		0,000	0,009	0
		Proportion of bicycle parking area to catchment area	m^2	0.000	0.000	0
	On demand microtransit	Motorbike as public transport *(ojek)*	Unit/hour	0,073	0,073	0,0182
		Public transport *(angkot)*	Unit/hour	0,019	0,009	0
Land Use	Density	Population density	people/km^2	0,017	0.012	0.012
		Job density	employee/km^2	0,055	0,069	0,046
	Diversity	Proportion of residential area	Rasio to catchment area	0,003	0,002	0,003419441
		Proportion of public offices area	Rasio to catchment area	0,001	0,001	0,000932169
		Proportion of industrial area	Rasio to catchment area	0,000	0,000	0,000028072
		Proportion of educational facilities	Rasio to catchment area	0,001	0,000	2,19542E-05
		Proportion of health facilities	Rasio to catchment area	0.000	0.000	0
		Proportion of public facilities area	Rasio to catchment area	0,000	0,000	0,000182952
		Proportion of mixed-used area	Rasio to catchment area	0.000	0.000	0.000296866
		Proportion of green open space area	Rasio to catchment area	0,000	0,001	0,000217955
	Floor Area Ratio	Floor to land area reatio	-	0,068	0,123	0,04092
			Index Score	**0,704**	**0,915**	**0,263**

From the result above, it stated that Dukuh Atas has the maximum score (0.915), following by Lebak Bulus has 0.704 means many criteria to TOD Ideal criteria has been fulfil in Lebak Bulus = 0.704. In the contrary Kelapa Gading has score 0.263 (score (maximum score 1). This is because the public transport facility, integrity area around many public transports has increase dramatically in Lebak Bulus. TOD principle is optimizing utilisation in public transportation.

3.2. Value creation in each area

The TOD principal indicates will give optimization to the value creation in an area. By knowing the potential value creation in area, it will give an opportunity to the stakeholder and also shareholders in responding and taken the policies. The biggest and optimal responses will give the optimal value that can be consider as optimal TOD index and optimal TOD application.

After knowing the index in specific area, it will show what is the potential value creation in it. To analyse the value creation, we separate an area into eight categories: residences, offices, educational, health centre, business, public facilities, green open space (Ruang Terbuka Hijau) and mixed use area.
Following are the types of the TOD Lebak Bulus and TOD Kelapa Gading:

Dukuh Atas TOD

Figure 4. Dukuh Atas Situation

Dukuh Atas location is in the center of DKI Jakarta province, which lies in the border of South Jakarta city and Center Jakarta city. In Dukuh Atas there are a lot of vertical residential building. Many people who work in Jakarta prefer to live there, because of accessibility and travel time to reach the offices.

Currently there are MRT (BNI city), KRL (Sudirman station) and BRT station (Dukuh Atas) have been operates there. Soon LRT station (Dukuh Atas) will operate to serve Dukuh Atas to Cibubur (LRT Jabodebek). Actually, people who live in apartments around Sudirman have been applied the concept of TOD, which most people using public transportation or walking to reach their offices. The development of convenient trottoir is the lead in Jakarta. Also, the facility such as sag way and parking area near station have been available. Traffic jam in Sudirman street is very routine and wasting time a lot. side that, there are many commuters who come to Dukuh Atas to do their work activities there, they come in the morning and go home in the afternoon. Most of the commuters use KRL, because KRL have the precise time arrival and no traffic jam at all.

Measuring TOD for knowing the value creation, first, there must be an area that separate into some categorized as follows:

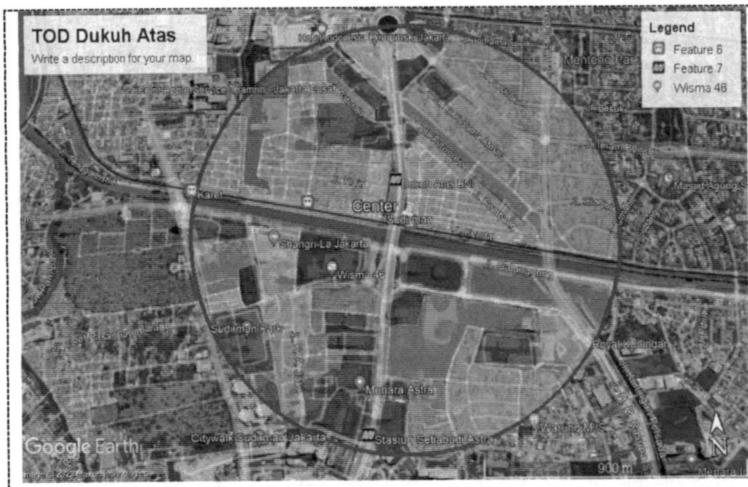

Figure 5. Dukuh Atas TOD

From the land and building real price, the calculation of value creation is as follow:

Table 3. Value creation in Dukuh Atas TOD

Dukuh Atas

Zone	Area Proportion	Value Creation		
		Land Price	Building Price	Total Asset
Residential	43,60%	51.047.241.449.605	71.252.990.260.746	122.300.231.710.351
Offices	13,83%	17.074.951.165.902	79.665.155.929.580	96.740.107.095.482
Educational	0,53%	156.162.187.500	284.340.000.000	440.502.187.500
Health	0,00%	0	0	0
Bussiness	8,73%	11.674.263.620.343	102.009.824.560.000	113.684.088.180.343
Public facilties	5,81%	2.161.107.042.460	1.217.940.500.000	3.379.047.542.460
Open Space Zone	18,37%	18.259.920.748.692	18.560.000.000	18.278.480.748.692
Mixed-used	9,12%	8.421.053.905.422	149.158.731.500.000	157.579.785.405.422
Total		**108.794.700.119.924**	**403.607.542.750.326**	**512.402.242.870.250**

Lebak Bulus TOD

Figure 6. Lebak Bulus Situation

Lebak Bulus position as a south gate to enter Jakarta from the surrounding city behind. The border of Lebak Bulus is Ciputat (South Tangerang, Banten province) and Cinere (Depok, West Java province). Lebak Bulus is residential type in sub-urban in city area with high population density with large family size. Even most of the residential in Lebak Bulus consist of landed house, there are some apartments as vertical house has been available. Landed house within 800-meter radii, including Lebak Lestari,Pondok Indah and Adhyaksa housing complex.

Recently there are MRT and BRT station in Lebak Bulus, that affected most people who lived inside the area taken public transportation for their daily activities. If the mobility of the people transforms into public transportation it will align with TOD principal.

To attract people who live in the area do walking, the government and stakeholder must increase the quantity and quality of pedestrian lane has successfully trigger the people to use public transport. Measuring TOD for knowing the value creation, first, there must be an area that separate into some categorized as follows:

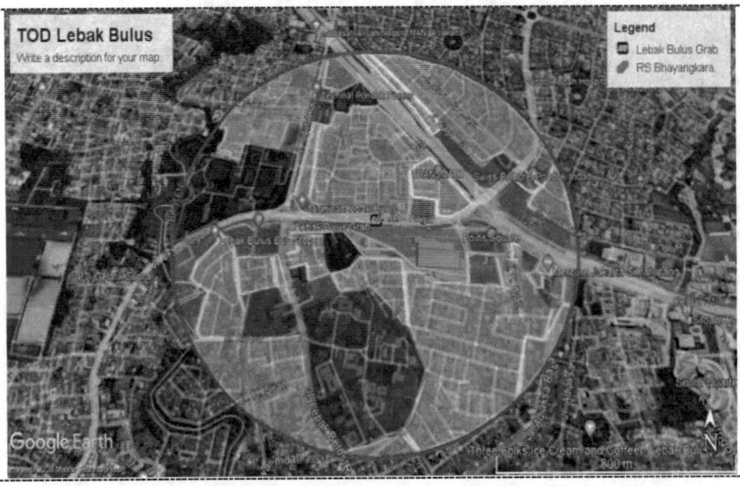

Figure 7. Lebak Bulus TOD

From the land and building real price, the calculation of value creation is as follow:

Table 4. Value creation in Lebak Bulus TOD

Lebak Bulus

Zone	Area Proportion	Value Creation		
		Land Price	**Building Price**	**Total Asset**
Residential	56,18%	21.407.823.370.400	27.347.448.030.526	48.755.271.400.926
Offices	12,41%	4.771.977.589.524	2.517.110.000.000	7.289.087.589.524
Educational	9,16%	4.162.537.204.495	1.358.875.000.000	5.521.412.204.495
Health	0,40%	198.000.001.980	1.389.000.000.000	1.587.000.001.980
Bussiness	4,32%	5.078.006.347.016	2.051.771.610.733	7.129.777.957.749
Public facilties	6,26%	4.853.516.390.879	667.107.500.000	5.520.623.890.879
Open Space Zone	9,03%	6.195.186.010.025	0	6.195.186.010.025
Mixed-used	2,24%	2.286.633.956.096	1.253.115.205.870	3.539.749.161.966
Total		**48.953.680.870.415**	**36.584.427.347.129**	**85.538.108.217.544**

<u>Kelapa Gading TOD</u>

Figure 8. Kelapa Gading Situation

Kelapa Gading position is the city centre of Jakarta. This area is a well-arranged housing complex. In 800-meter radii with LRT station as the centre, there is only health facilities function that is not available.

Kelapa Gading is residential type in city centre area with medium population density. This area is one of high-class area where the people have good economical level in Jakarta. The latest public transportation facility in Kelapa Gading is LRT. A little bit far from this transit node, there is also the BRT station.

The habit and lifestyle of the people is using their own vehicle. Need a policy to ask the people for willing to use public transportation for their activities.

Measuring TOD for knowing the value creation, first, there must be an area that separate into some categorized as follows:

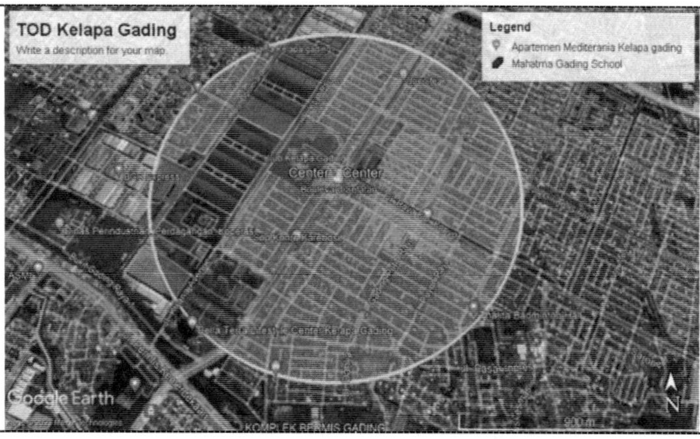

Figure 9. Kelapa Gading TOD

From the land and building real price, the calculation of value creation is as follow:

Table 5. Value creation in Kelapa Gading TOD

Kelapa Gading	Zone	Area Proportion	Value Creation		
			Land Price	Building Price	Total Asset
	Residential	69,17%	25.509.824.152.875	21.401.618.074.857	46.911.442.227.732
	Offices	16,37%	8.414.747.550.000	3.792.894.000.000	12.207.641.550.000
	Educational	0,30%	135.877.000.000	216.250.000.000	352.127.000.000
	Health	0,00%	0	0	0
	Bussiness	0,63%	416.232.000.000	531.274.050.000	947.506.050.000
	Public facilties	3,84%	2.066.572.430.074	827.625.000.000	2.894.197.430.074
	Open Space Zone	4,74%	2.438.638.950.000	0	2.438.638.950.000
	Mixed-used	4,95%	2.637.095.401.605	3.353.325.312.500	5.990.420.714.105
	Total		**41.618.987.484.554**	**30.122.986.437.357**	**71.741.973.921.911**

From the previous data we have known that the summary result as follows:

Table 6. Result

	TOD level (index)	Value Creation (IDR)
Dukuh Atas	0.915	512.402.242.870.250
Lebak Bulus	0.704	85.538.108.217.544
Kelapa Gading	0.263	71.741.973.921.911

The corelation chart between TOD indexes (level) and value creation shown below:

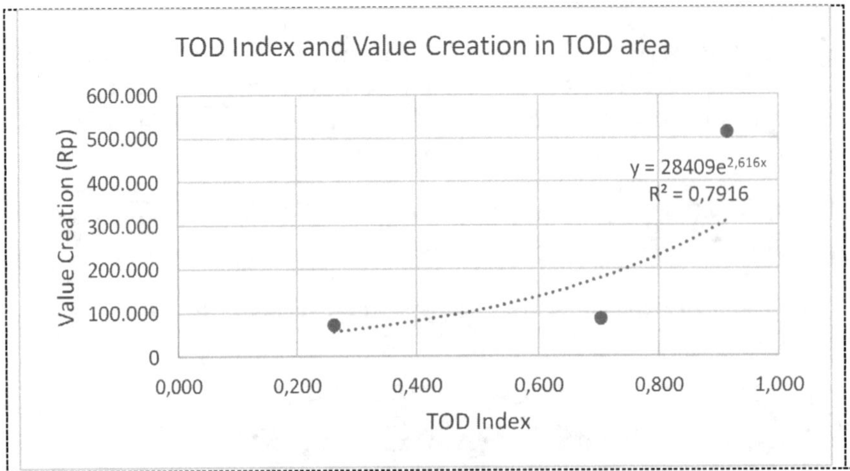

Figure 10. Correlation of TOD Index with Value Creation

From above data, it has been proven that the higher TOD level into TOD ideal criteria, has given the higher value creation. Dukuh Atas which has the most index to TOD ideal for 0.915, has given the biggest value creation Rp 512.402.242.870.250. Lebak Bulus with level TOD conformity 0.704 to TOD ideal has given value creation Rp 85.538.108.217.544. On the other hand, Kelapa Gading with TOD level 0.263 has given value creation Rp 71.741.973.921.911.
The type of correlation is not linear but logarithmic, with the equation is:

$$y = 28409e^{2.616x} \tag{2}$$

The R^2 value is 0.7916, means TOD valid has the correlation to value creation. The main result we want to prove if there is correlation between TOD level with value creation, and it has been proven.

4. Conclusion

The ranking applies where the area with a higher index means it has the highest level of conformity to the TOD criteria. The area with the dense population, has many activities and variable land uses with appropriate accessibility will have the highest level to TOD ideal. Even though the type of land use same as residential area, it will give different value creation because of better index of internal and external accessibility. From this research, it's clearly that Dukuh Atas is the higher, then Lebak Bulus is higher index than Kelapa Gading in TOD, gave the higher level of value creation.

In a simple way, we can find the value of land in TOD area, (both sell or lease price) will give a higher price for the place in TOD area with good accessibility (internal and external accessibility). In other words, if we increase the accessibility of an area (both internal and external accessibility), it will increase the land value and potential for being a good TOD. The measurement of value creation will be useful as an input to the urban planners or policy makers for designing more targeted strategies and as an opportunity to TOD financing option.

The model correlation between TOD index and value creation in this paper, need further studies which applied in the different areas, to confirm the result.

5. References

[1] Brovarone; Vitale, E; et al. 2018. *Promoting TOD Through Regional Planning. A Comparative Analysis of Two European Approaches*

[2] Carlton, I. 2007. *Histories of TOD-Perspective on the Development of the TOD Concept*

[3] Cervero, R; et al. 2014. *BRT TOD: Leveraging Transit Oriented Development with bus rapid transit investments.*

[4] Corte and Gaudio. 2014. A Literature Review on Value Creation and Value Capturing in Strategic Management Studies

[5] Clacherty, Alistair. 2011. Creating and Capturing Value Around Transport Nodes - Transit Oriented Development Success Stories

[6] Deboosere, R; et al. 2018. *Accessibility-Oriented Development.*

[7] Ewing, R; et al. 2019. *Comparative Case Studies: Trip and Parking Generation at Orenco Station TOD, Portland Region and Station Park TAD, Salt Lake City Region.*

[8] ITDP-British Embassy Mexico City. (unspecifically year) *TOD - Regenerate Mexican Cities to Improve Mobility* (Accessed October 2019).

[9] Kumar, P; et al. 2020. *Identification of Neighborhood Typology for Potential Transit Oriented Development.*

[10] Li, Z; et al. 2019. *Transit Oriented Development among Metro Station Areas in Shanghai, China: Variations, Typology, Optimization and Implications for Landuse Planning.*

[11] Lobo, A & partners. 2012. *TOD Guide for Urban Communities, EMBARQ, Mexico.*

[12] Luti Consulting. 2016. *Transit and Urban Renewal Value Creation*

[13] Martinez, L.M, et al. 2014. *Measuring and Evaluating the Impacts of TOD measures - Searching for Evidence of TOD characteristics in Azambuja train line.*

[14] Noland, Robert. B. & partners (October 2014). *Measuring Benefits of Transit Oriented Development. MNTRC-RUTGERS* [Accessed October 2019]

[15] Papa and Bertolini. 2015. *Accessibility and TOD in European Metropolitan Areas.*

[16] Renne, J. L; et al. 2019. *The Opportunities and Tensions of Historic Preservation and Transit Oriented Development (TOD.*

[17] Su, et al. 2020. *TOD Typologies Around Metro Station Area in Urban China : A Comparative Analysis of Five Typical Megacities for Planning Implications.*

[18] Suzuki, Hiroaki, et al. 2015. *Financing Transit Oriented Development with Land Values.*

[19] Viruly et al. 2017. *Value Creation Around Transport Infrastructure in South Africa : The Case of Gautrain*

[20] Yan-yan, Chen et al. 2016. *An Evaluating Method of Public Transit Accessibility for Urban Area Based on GIS.*

[21] Zhou, J; et al. 2019. *Can TODness Improve (Expected) Performances of TODs? An Exploration Facilitated by Non-traditional Data.*

[22] Yang, K & Pozani, D. 2018. *A Decade of TOD Policies in Brisbane, Australia: Development and Land Use Impacts.*

Acknowledgments

The author would like to express the deepest gratitude to Lembaga Pengelolaan Dana Pendidikan (LPDP/Indonesia Endowment Fund for Education) under the Ministry of Finance Republic of Indonesia for funding this study, and much gratitude to Institut Teknologi Bandung for opportunities to study. Any opinions, models and conclusions are those of authors and not reflecting the stand of the institution.

IOP Conf. Series: Earth and Environmental Science 1294 (2024) 012007 doi:10.1088/1755-1315/1294/1/012007

The model validation for trip assignment using all or nothing and user's equilibrium method

Amirotul M. H. Mahmudah[1*], W Hartono[1], R Lestari[2], S Rachmawati[2], and D V Septiasworo[2]

[1]Lecturer of Department of Civil Engineering, Universitas Sebelas Maret, Surakarta 57126, Indonesia,
[3]Student of Civil Engineering Undergraduate Program, Universitas Sebelas Maret, Surakarta 57126, Indonesia

*Corresponding author's e-mail: amirotulmhm@staff.uns.ac.id

Abstract. In transportation planning, an accurate transportation model is crucial. The trip-based approach is widely used, which relies on the four-step model (FSM) for transportation modeling. The model generates an origin-destination (O-D) Matrix as the starting point, which is then used to determine traffic flow on the road network through a trip assignment. This approach is superior because the (O-D) Matrix is a result of the interaction between land use and accessibility. In Boyolali Regency, research was conducted using traffic volume data that reflected the conditions at the beginning of 2023. The gravity model was used to determine trip distribution for this study, and the Least Squares method and negative exponential function calibrated the model. The study aim is to compare the performance of All or Nothing and User's Equilibrium assignment methods using the EMME/3 software. The results showed that All or Nothing had a higher R^2 value of 0.913 compared to User's Equilibrium's 0.911, suggesting that the All or Nothing assignment method is more suitable for Boyolali Regency. The research predicted that the total movement of Boyolali Regency in 2023 would be 12,354 pcu/hour, with most of the movement being external-external. As Boyolali Regency is a transit area, vehicles only pass through it.

1. Introduction

The balance between transportation supply and travel demand is the main objective of transportation planning. Transportation planning involves the process of making decisions about future regional road infrastructure developments. Estimation of demand is an essential process in transportation planning. Some tool analysis can be used to estimate demand, one of which is transportation modeling [1]. One of the most popular methods used in transportation modeling is the four-step model (FSM), which is used to estimate traffic flows in accordance with a network and travel demand representation. This model is commonly used in developing transportation policy [2]. A sequential FSM is a way developed to handle the complexity of road networks. In the first step, trips are represented as trip ends, trip productions, and trip attractions, this step is called trip generation. Next, in trip distribution, trip production is distributed to match the distribution of trip attraction and to represent travel impediments (time and/or cost), resulting in a trip table. Next, in mode choice, the trip tables are essentially calculated to represent the relative proportion of trips by alternative modes. Finally, in route selection, the trip tables are assigned to the transportation network [3].

In the second stage of the current FSM, there were various methods employed. Among these, two were the Wardrop user's equilibrium and the All or Nothing approach. The user's equilibrium concept was introduced in 1976 and was applied to road networks in cities such as Winnipeg and the Parkway East. This approach was discussed in numerous articles, focusing on its implications for usage [4]–[7]. The user's equilibrium is sensitive to traffic congestion. In addition, another method called The All or Nothing method has been applied in many pieces of research since All or Nothing phenomena still widely occurred [8]. By implementing the All or Nothing traffic assignment in logistic transport routes, urban planners can optimize their planning and design using this algorithm. This results in a more efficient and effective logistic transport route for transportation and logistics [9]. All or Nothing is not suitable for areas with heavily congested road networks [10].

In general, study locations are often identified where user's equilibrium and the All or Nothing method are analyzed for traffic assignment. However, comparing these methods can result in a more reliable model. Therefore, the aim of this research is to compare performance resulting from traffic assignment using the All or Nothing and User's Equilibrium method to get a reliable model. By integrating a valid and accurate basic transportation model with FSM, transportation planners can make more precise predictions and design more effective transportation plans. Implementing this test in Boyolali Regency is expected to guide transportation planners in having an excellent model to consider transportation policy in transportation planning.

1.1. Gravity Model

The gravity model is a commonly used and well-known model due to its simplicity. It states that the movement between the starting point (i) and the destination (d) is directly proportional to the size of the starting point (O_i) and the size of the destination (D_d), while being inversely proportional to the distance between the two. This can be expressed mathematically in Equation 1.

$$T_{id} = A_i . O_i . B_d . D_d . f(C_{id}) \tag{1}$$

dimana,

T_{id}	=	The number of trips from origin zone i to destination zone d,
A_i	=	The balancing factor for each origin zone i,
B_d	=	The balancing factor for each destination zone d,
O_i	=	Total moves from home zone i,
D_d	=	Total moves from home zone d,
$f(C_{id})$	=	Barrier function or measure of accessibility between zones of origin and destination.

1.2. Least Squares Estimation Method

The Least Squares estimation method is a technique for calibrating the β parameters in the gravity model. Its main approach is to minimize the square of the difference between the observed data and the modeling results for calibrating unknown parameters. Equation 2 presents the system of simultaneous equations with unknown parameter β.

$$\frac{\partial S}{\partial \beta} = f = \sum_{i=1}^{N} \sum_{d=1}^{N} \left[\frac{2}{\hat{T}} \left(T_{id} - \hat{T}_{id} \right) \cdot \frac{\partial T_{id}}{\partial \beta} \right] \tag{2}$$

1.3. Traffic Assignment

In the trip distribution stage, traffic assignment entails taking an origin-destination matrix and inputting it into the road network to determine the current flow on each road segment, and the total number of trips on the network being analyzed. Traffic assignment can be performed using various methods, some of which are listed below.

1.3.1. All or Nothing. This assignment method assumes that individuals will select a route based solely on the shortest distance without considering the impact of traffic congestion. Under the All or Nothing loading model, all travellers will have a uniform understanding of the optimal route (the shortest, fastest, or most cost-effective). However, this approach is not suitable for areas with heavily congested road networks [9]–[11].

1.3.2. User's Equilibrium. The first principle of the Wardrop Equilibrium Law is the self-regulation of traffic when congestion is present. This principle takes into account the impact of traffic jams and assumes that drivers are unable to alter their routes in order to reduce travel expenses during periods of congestion.

1.4. Statistical Test Indicator

To assess the performance of each estimation method, statistical test indicators were used. The coefficient of determination (R^2) was the indicator used, and it was calculated using Equation 3. This helped to compare the results of the different methods in terms of their statistical indicators.

$$R^2 = \frac{\Sigma\left(\hat{Y}_i - \bar{Y}_i\right)^2}{\sum(Y_i - \bar{Y}_i)^2} \quad (3)$$

The range of R^2 values is based on presented in Table 1.

Table 1. Indicator Scale (R^2) [12].

R^2	Scale
0,80 – 1,00	Very high
0,60 – 0,80	High
0,40 – 0,60	Neither high or low
0,20 – 0,40	Low
0,00 – 0,20	Very low (invalid)

1.5. EMME (Equilibre Multimodal Multimodal Equilibrium)

EMME, which stands for Equilibre Multimodal or Multimodal Equilibrium, is a professional software that is utilized to fulfil transportation modelling needs. It serves as a travel demand modelling system that can forecast transportation on an urban, regional, and national level. The research being conducted in this instance specifically employs the EMME/3 software.

2. Research Method

2.1. Research Location

This study was conducted in Boyolali Regency and used primary arterial, collectors, and toll roads. The region was divided into 15 zones, consisting of eight internal and seven external zones. Internal zones were created through zone aggregation, merging multiple zones based on traffic movement distribution or area size. This study combined several districts to form each internal zone since not all roads reach every sub-district. External zones were located along the border of the Boyolali Regency. A map of the region used in this study can be found in Figure 1.

Figure 1. Boyolali Regency Map.

2.2. *Research Data*

While, in theory, travel is linked to people's activities, one possible way to start collecting data is to focus on the starting and ending points of trips instead of specific activities. Additionally, traffic flow data for certain roads on the road network can be the first input for this model. This research incorporates both primary and secondary data. The primary data comprises traffic flow statistics gathered from 15 survey points located in the field during morning peak hours. The survey lasted 2 hours, from 06.00 to 08.00 WIB, with readings taken at 15-minute intervals. The secondary data was obtained from various agencies in Boyolali Regency and included administrative maps, road network data and maps, road section data, population data, and land function data.

3. Results and Discussion

This study obtains the parameter β through the Least Squares estimation method. The resulting value for β is 0.0702, which can be used to describe the average trip cost in the resistance function. This value indicates that the average travel cost is higher than the initial assumption, with a β parameter value of 0.08. The β parameter value obtained from this study can be used to estimate (OD) Matrix in the future [13].

Based on the study conducted, it has been estimated that the total movement estimation of the Boyolali Regency by 2023 would be 12,354 pcu/hour. The study reveals that the most significant movements in the region are related to external-external travel, as depicted in Figure 2. Additionally, the figure also displays movements between zones within the Boyolali Regency.

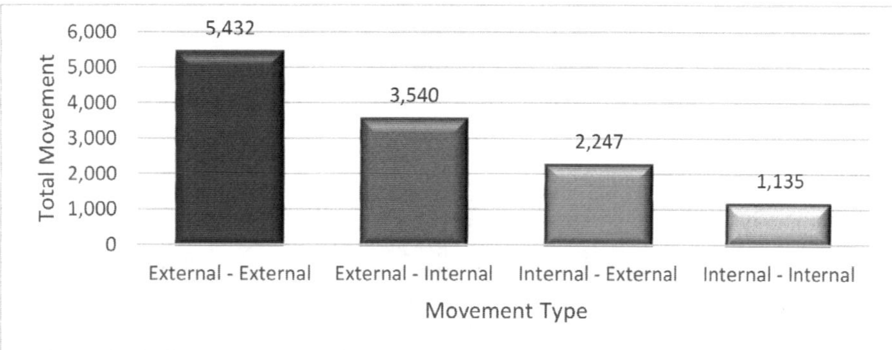

Figure 2. Total movement inter-zona (pcu/h).

External movements accounted for the largest share of movement in Boyolali Regency, with a total of 5,432 pcu/hour or 43.97% of the total movement. This is due to the fact that Boyolali Regency serves as a transit area where vehicles simply pass through.

In Boyolali Regency, the external-internal movement was the second most significant movement, accounting for 3,540 pcu/hour or 28.65% of the total movement. This movement was because many residents who live outside the Boyolali Regency work there. Boyolali Regency has various industrial areas, likely to attract labour from outside the region. The presence of industrial and trade areas also causes people from outside the region to enter, which affects the external-internal movements within the region.

The Internal-External movement is the third largest movement in Boyolali Regency, with 2,247 pcu/hour or 18.19% of the total movement. This movement is mainly caused by residents of Boyolali Regency who work outside of the area. The main attraction for this movement is towards Surakarta City and its surrounding areas, which has seen progress in the economic, cultural, and tourism sectors and increased population.

The smallest type of movement within Boyolali Regency is internal-internal movement. In 2023, the total number of internal movements in the regency is expected to be 1,135 per hour, which accounts for 9.19% of all movements in the area. These internal movements are primarily for school commutes in the morning, as many schools are close to students' homes within Boyolali Regency.

Figure 3 displays the desire line, which portrays the results of the movement.

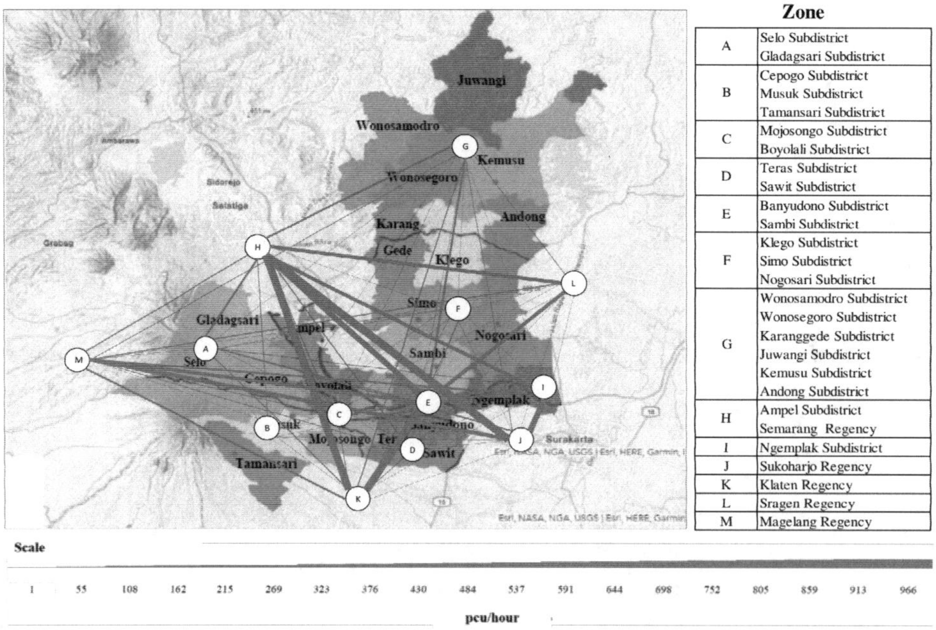

Zone

A	Selo Subdistrict Gladagsari Subdistrict
B	Cepogo Subdistrict Musuk Subdistrict Tamansari Subdistrict
C	Mojosongo Subdistrict Boyolali Subdistrict
D	Teras Subdistrict Sawit Subdistrict
E	Banyudono Subdistrict Sambi Subdistrict
F	Klego Subdistrict Simo Subdistrict Nogosari Subdistrict
G	Wonosamodro Subdistrict Wonosegoro Subdistrict Karanggede Subdistrict Juwangi Subdistrict Kemusu Subdistrict Andong Subdistrict
H	Ampel Subdistrict Semarang Regency
I	Ngemplak Subdistrict
J	Sukoharjo Regency
K	Klaten Regency
L	Sragen Regency
M	Magelang Regency

Figure 3. The Desire Line of total trip of Boyolali Regency in 2023.

After the trip distribution process, the (OD) Matrix is charged to the road network during the trip assignment process. In this study, EMME/3 software facilitates the All or Nothing and User's Equilibrium loading methods. The validity of traffic loading is determined by comparing the R^2 value of each loading method.

In this study, traffic flow was modeled using two loading methods: All or Nothing and User's Equilibrium. The traffic flow obtained from the All or Nothing method was assigned using the EMME/3 software in the first iteration, based on the shortest path. For the User's Equilibrium method, the traffic flow was modeled through trip assignment using EMME/3 software in auto-assignment mode.

Traffic assignment using the All or Nothing and User's Equilibrium loading methods produces an R^2 of 0.913 and 0.911, as shown in Figure 4 and Figure 5. The value of R2 is obtained by comparing observed and modeled traffic flows using EMME/3 software. The R^2 generated from these two methods is included in the category of very high validity. The modelled traffic flows using the All or Nothing and User's Equilibrium loading methods are highly similar to the observed traffic flows.

Road users in Boyolali Regency still take the shortest route from origin to destination without considering alternatives. One of the factors that cause Boyolali Regency to have a tendency towards the All or Nothing method of imposition is that most of the roads in Boyolali Regency have yet to experience congestion. Therefore, the All or Nothing loading method for areas that are not jammed is correct. Another factor that makes Boyolali Regency tend towards the All or Nothing approach of imposition is the road network, which is less complex than in big cities, so the All or Nothing loading method can still be used.

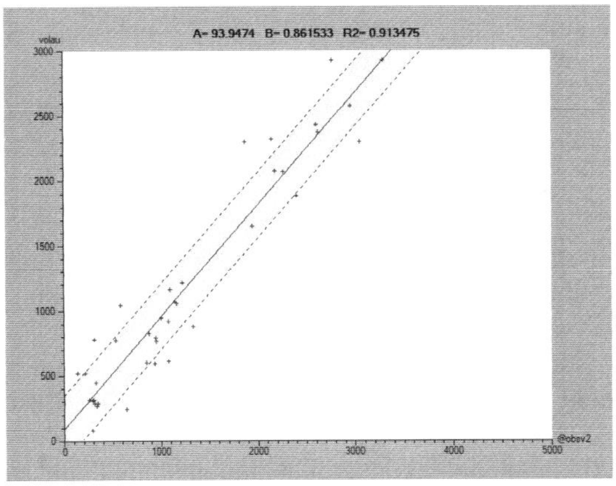

Figure 4. Comparison of Traffic Flows from Field Observation and Developed Model Using All-or-Nothing Assignment Method.

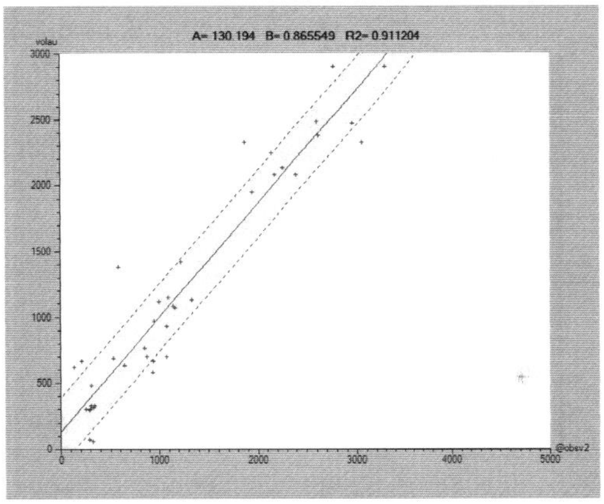

Figure 5. Comparison of Traffic Flows from Field Observation and Developed Model Using User's Equilibrium Assignment Method.

In comparison, when analysing the R^2 values of the All or Nothing method, it was discovered that the values for metropolitan areas were lower than those of big cities. In the Bandung metropolitan areas, the R^2 value ranged from 0.519 to 0.557 [14], [15], whereas for this research with the location study in Boyolali as a big city, it was 0.913. In the case of the user's equilibrium method, the R^2 value for the Bandung Metropolitan area varied from 0.76 to 0.791[14][15], while for the big cities (Surakarta, Sukoharjo, Boyolali), it was between 0.80 to 0.91 [16]–[19]. These findings align with prior research.

To confirm that Boyolali Regency has yet to experience congestion, this study calculates the degree of saturation (DS) for the collector, arterial, and toll roads. The result is shown in the Figure 6. The survey was conducted during the morning peak hour. Of the 126 roads surveyed, only 12 had congested traffic.

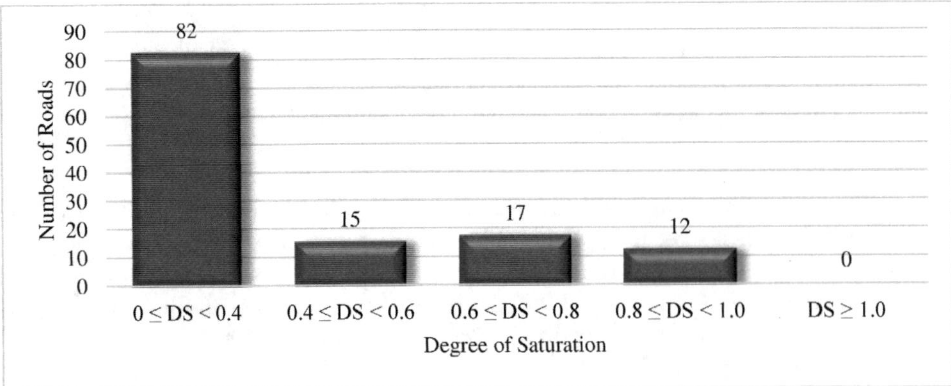

Figure 6. The Degree of Saturation of Roads in Boyolali Regency.

4. Conclusion

The value of the β parameter obtained in this study is 0.0702. According to the gravity model, the estimated traffic movement distribution in Boyolali Regency for 2023 is 12,354 pcu/hour. As Boyolali Regency serves as a transit area, the most significant traffic movement will be from one external zone to another. All or Nothing and User's Equilibrium methods have high validity, with R^2 values of 0.913 and 0.911, respectively. The R^2 value for loading using the All or Nothing method is higher compared to the User's Equilibrium method. This suggests that Boyolali Regency leans towards the All or Nothing method. Some people still prefer the shortest route, the trend towards the All or Nothing method could be due to the lack of congestion.

5. References

[1] G. Yadav and S. D. Ghodmare, "Transportation Planning Using Conventional Four Stage Modeling : An Attempt for Identification of Problems in a Transportation System," *Int. J. Sci. Res. Sci. Technol.*, vol. 8, no. 4, pp. 556–567, 2021, doi: 10.32628/ijsrst218482.

[2] S. Shahriari, E. N. Robson, J. Wang, V. V. Dixit, S. T. Waller, and T. H. Rashidi, *Integrating a computable general equilibrium model with the four-step framework*, vol. 50, no. 4. Springer US, 2023.

[3] M. G. McNally, "Handbook of Transport Modelling Article information : About Emerald www.emeraldinsight.com," 2007.

[4] M. Florian and S. Nguyen, "Application and Validation of Equilibrium Trip Assignment Methods.," *Transp. Sci.*, vol. 10, no. 4, pp. 374–390, 1976, doi: 10.1287/trsc.10.4.374.

[5] D. E. Boyce and B. N. Janson, "The Effect on Equilibrium Assignment of Different Congestion Functions," *Transp. Res. Part A Gen.*, vol. 15, no. A, pp. 223–232, 1981.

[6] B. N. Janson, S. P. T. Thint, and C. T. Hendrickson, "Validation and Use of Equilibrium Network Assignment for Urban Highway Reconstruction Planning," *Transp. Res. Part A Gen.*, vol. 20, no. 1, pp. 61–73, 1986, doi: 10.1016/0191-2607(86)90016-6.

[7] M. Florian and D. W. Hearn, "Traffic assignment: equilibrium models," vol. 215, pp. 305–315, 2001.

[8] H. Suprayitno, "Manual Validation and Calibration Method for All-or-Nothing Traffic Assignment," in *International Seminar on Science and Technology*, 2016, pp. 29–36.

[9] C. Hui, "Application Study of All-or-Nothing Assignment Method for Determination of Logistic Transport Route in Urban Planning," *Comput. Model. New Technol.*, vol. 18, no. 12C, pp. 932–937, 2014.

[10] O. Z. Tamin, *Perencanaan dan Pemodelan Transportasi*, Second edi. Bandung: Penerbit ITB, 2000.

[11] E. S. Lee and P. G. Oduor, "Using multi-attribute decision factors for a modified all-or-nothing

traffic assignment," *ISPRS Int. J. Geo-Information*, vol. 4, no. 2, pp. 883–899, 2015, doi: 10.3390/ijgi4020883.

[12] K. Mohamad, "THE FATE AND TRANSPORT OF NUTRIENTS USING WATERSHED MODEL IN TROPICAL AGRICULTURAL WATERSHED," Universiti Teknologi Malaysia, 2019.

[13] I. Ekowicaksono, F. Bukhari, and A. Aman, "Estimating Origin-Destination Matrix of Bogor City Using Gravity Model," *IOP Conf. Ser. Earth Environ. Sci.*, vol. 31, no. 1, pp. 1–5, 2016, doi: 10.1088/1755-1315/31/1/012021.

[14] O. Z. Tamin and R. Suyuti, "Kajian Peningkatan Akurasi Matriks Asal-Tujuan yang Dihasilkan dari Data Arus Lalulintas pada Kondisi Keseimbangan," *ITB J. Sci.*, vol. 39, no. 1&2, pp. 23–39, 2007.

[15] R. Suyuti and O. Z. Tamin, "Penggunaan Model Gravity-Opportunity (Go) Dalam Estimasi Matriks Asal-Tujuan (Mat) Menggunakan Data Arus Lalulintas," *J. Transp.*, vol. 7, no. 1, pp. 45–56, 2007.

[16] E. M. Zada, Syafi'i, and S. J. Legowo, "Estimasi Matrik Asal Tujuan (MAT) dari Data Arus Lalu Lintas dengan Metode Estimasi Kuadrat Terkecil Menggunakan Piranti Lunak Emme/3 (Studi Kasus Kota Surakarta)," *18th Int. Symp. Indones. Inter-University Transp. study Forum - 18*, no. August, 2015.

[17] Syafi'i, M. B. Santoso, and D. Handayani, "The origin-destination matrix for freight transportation with Bayesian inference method," *AIP Conf. Proc.*, vol. 1977, pp. 040018-1-040018–8, 2018, doi: 10.1063/1.5042988.

[18] E. M. P., Syafii, and S. J. Legowo, "Estimasi Parameter Model Gravity Dengan Metode Entropi-Maksimum Dan Fungsi Hambatan Eksponensial Negatif (Studi Kasus Kabupaten Sukoharjo)," *J. Matriks Tek. Sipil*, vol. 9, no. 1, pp. 53–60, 2021, doi: 10.20961/mateksi.v9i1.51907.

[19] J. O. Rum, Syafii, and J. S. Legowo, "Estimasi Parameter Model Gravity dengan Metode Penaksiran Kuadrat Terkecil dan Fungsi Hambatan Eksponensial Negatif (Studi Kasus Kabupaten Sukoharjo)," *J. Matriks Tek. Sipil*, vol. 9, no. 2, pp. 84–90, 2020, doi: 10.20961/mateksi.v9i2.52588.

Acknowledgement

We want to thank to the Chancellor of Sebelas Maret University, who has given the decision that this research is funded by RKAT PTNBH Universitas Sebelas Maret Fiscal Year 2023 through a Research scheme *PENELITIAN HIBAH GRUP RISET (PENELITIAN HGR-UNS) B* with Research Assignment Agreement Number: 228/UN27.22/PT.01.03/2023.

IOP Conf. Series: Earth and Environmental Science 1294 (2024) 012008 doi:10.1088/1755-1315/1294/1/012008

Assessing the impact of Medan-Stabat toll road on carbon monoxide pollutant in arterial roads

Ridwan Anas[1], Zaid Perdana Nasution[2] and Adam Hasnan[1]

[1]Department of Civil Engineering, Faculty of Engineering, Universitas Sumatera Utara, Medan 20155

[2] Department of Environmental Engineering, Faculty of Engineering, Universitas Sumatera Utara, Medan 20155

ridwan.anas@usu.ac.id

Abstract. Carbon monoxide is the result of the combustion of motorized vehicles which can pollute the air and will cause negative impacts to the health of human beings. The development of a toll road is initiated in the routes of an arterial road which have been causing congestion or terrible traffic jams, and is expected directly to reduce the amount of traffic jams and to increase the speed of travel, and indirectly to reduce air pollution in the arterial roads. The objective of this paper is to predict the effects of the Medan-Stabat Toll Road operation to the CO pollutant contents in the arterial roads. The results indicate that the level of pollution is reduced up to 10 % and 25% due to traffic shifting on the arterial roads to the toll road.

1. Introduction

Air contamination has changed the natural characteristics of the atmosphere which have been caused by chemical, physical or biological materials from both natural processes and human activities who have been fulfilling their needs both in the production and in the transportation sector [1]. The contribution of the transport sector to carbon monoxide (CO) pollutants has been proved by many previous research [2]. According to World Research Indonesia in 2018, land transportation contributes approximately 46% of the entire air pollution, then the industrial sector which also has contributed approximately 43% of the air pollution. Air contamination as a result of industrial and transportation activities will cause negative impacts to the ecosystem [3].

The objective of toll roads development in Indonesia is to increase regional connectivity in order to support the logistic sector. Until 2022, Indonesia has operated 69 toll highways as long as 2,578 km [4]. The development of a toll road in general is initiated in the routes of an arterial road which has caused a congestion or has a high traffic, with the objective to shift the traffic from the arterial road to the toll road. Therefore, the operation of a toll road is expected directly to reduce the load of traffic and increase the travel speed, and indirectly reduce air pollution in the arterial roads due to the active operation of the Medan-Stabat Toll Road. This paper aim is to predict the influences of the Medan-Stabat Toll Road operation to the CO pollutant contents in the arterial road. The limitation of this paper does not consider changes in average speed on arterial roads due to the Medan-Stabat toll road not operating.

The emission of CO which derives from transportation activities can be calculated by using the vehicle volume approach, and the type and speed of the vehicle [5]. Therefore, it is identified the

IOP Conf. Series: Earth and Environmental Science 1294 (2024) 012008 doi:10.1088/1755-1315/1294/1/012008

aggregation value of the pollutant formed in CO data is expected to provide information for the stakeholders, and will be the basic consideration of developing the toll road towards the quality of air in the future.

2. Literature Review

2.1. Environment impact of road transportation

Emission is a substance resulted from a certain activity which enters the ambient air and has a potential aspect as a pollutant. After having been evaluated from the effect to the perspective of public health, The main toxic pollutant is emitted to the environment from the fuel combustion which contains carbon monoxide (CO), nitrogen dioxide (NO_2), hydrocarbon (HC) and particulates (PM). The land transportation sector has contributed 18% of the total amount of CO pollutants [6], which contains high concentrated substances that contains toxic substances [7]. Along with the economic growth, transportation activities which have been developed rapidly and significantly may cause air pollution [8] [9]. The most amount of pollutant caused by motorized vehicles is CO, as a result of fuel combustion. The highest calculated concentration of CO pollutant at distance 20 m from the road [10] and will be the main source of contamination in the urban areas [11].

The Emission Factor is one value which has described the relationship between the quantity from the emission released by the air with a certain activity related with the source of pollutant [12]. These factors usually are usually acknowledged as the weight of the pollutant, such as illustrated in the table below.

Table 1. The emission factor based on the type of vehicle [12].

Types of Vehicles	Emission Factor (grkm^{-1})				
	CO	**HC**	NO_X	PM_{10}	SO_2
Motorcycle	14.00	5.90	0.29	0.24	0.008
Passenger Car (gasoline)	40.00	4.00	2.00	0.01	0.026
Passenger Car (solar)	2.80	0,20	3.50	0.53	0.440
Bus	11.00	1.30	11.90	1.40	0.930
Truck	8.40	1.80	17.70	1.40	0.820

From the table above it can be illustrated that the type of vehicle and the contents of pollutant emission released, while the next table can illustrate the aggregation of the vehicle emission based on the average speed of the vehicle.

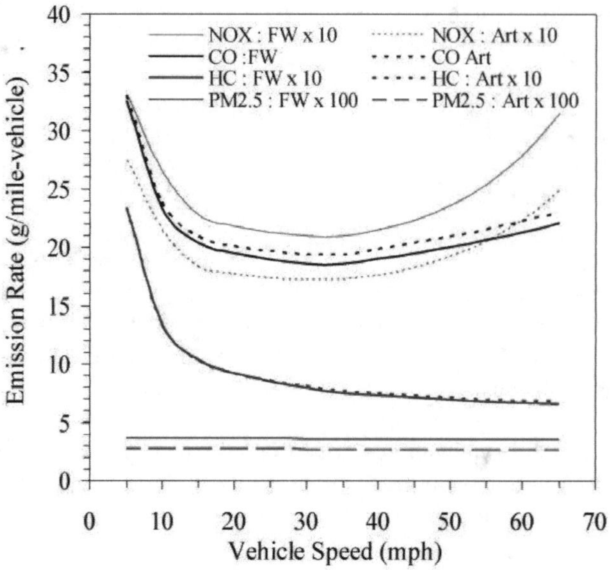

Figure 1. Emission based on the type and speed of the vehicle [13].

2.2 Predictive Models of Carbon Monoxide Pollutant

To predict the emission of the vehicle released by the vehicle considers many factors such as the surrounding air temperature, the length of the journey, the type of vehicle, the age and the fuel used and the speed of the vehicle. Many Prediction Models have been initiated based on the solid correlation between the use of fuel with the content of emission, because the fuel will be one main agent in the process of combustion [14]. An, et al. (1997) has created an equation to predict the CO pollutant which has been released by the vehicle with an equation as mentioned below.

$$EOE_{CO} = a_{CO}FC \tag{1}$$

Where:

EOE_{CO} : engine out CO-emission (gr km^{-1})
a_{CO} : constant ($g_{CO}\, g_{Fuel}^{-1}$)
FC : fuel consumption (gr km^{-1})

The relationship between fuel and emission is acquired from the empirical data from the developed nations, where in several cases of certain pollutant emission will not depend physically to the consumption level of the fuel [15]. ETSU (1997) which has also created an equation below to predict the emission related with the speed of the vehicle:

$$EOE = a1 + a2S + a3S^{a4} + a5e^{a6S} \tag{2}$$

Where:

EOE : Engine out emission in gkm^{-1}
i : Type of emission (which are HC, CO, NO_X, SO_2, Pb, PM or CO_2)
a1 until a6 S : Model parameter which varies based on the type of emission and
S : The speed of the vehicle in km $hour^{-1}$

Indonesia has applied a guideline in predicting the size of emission which is the Line Source Model and the Osaka Municipal Government [16]. The Line Source Model is based on the assumption that the emission of pollutants in the highway is unlimited until every dot has a similar distance from the side of the road and will receive pollution with the equal amount of concentration, with the equation as written below.

$$C_{(x,y,z)} = \frac{Q}{\mu\sigma_y\sigma_z} exp\left\{\frac{-H^2}{2\sigma^2}\right\} exp\left\{\frac{-y^2}{2\sigma_y^2}\right\} \tag{3}$$

Where:

C : The concentration of pollutant (gr m^{-3})
Q : Emission force (gr s^{-1})
H : The height in the source of emission (m)
(x,y,z) : Receptor coordinates
σ : Deviation standards
z : 1.5 m, in the disperse case ≈ 0 (ground level)
y : 0, $\sigma_y \approx 1$

The Osaka Municipal Government (OMG) Model has mentioned the emission released by the motorized vehicle in on road as an imaginary volume which flies above the surface of the road.

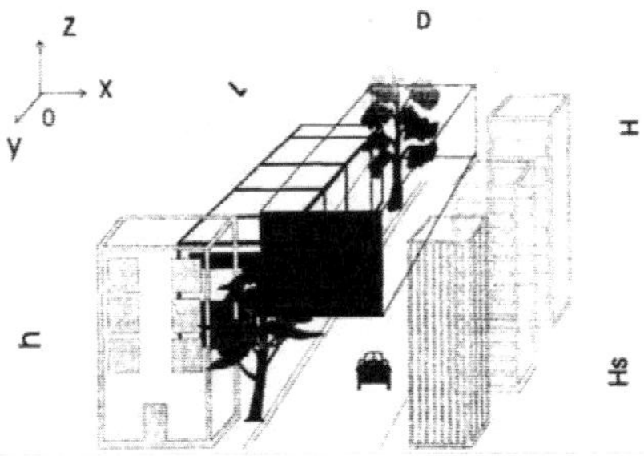

Figure 2. The Imaginary Pollutant Volume above the surface of the road [16].

The concentration of pollutant in the OMG model can be calculated with the equation as elaborated below:

$$C = \frac{Qp}{u}\left[\frac{(m+1)(n+1)}{K_yK_zX} exp\left\{-\frac{u_a(m+1)y^2}{4K_yX}\right\}\right]^{1/2}\left[exp\left\{-\frac{u_a(n+1)(z-z')^2}{4K_zX}\right\} + exp\left\{-\frac{u_a(n+1)(z+z')}{4K_zX}\right\}\right] \quad (4)$$

Where:

C : Concentration of pollutant (gr m^{-3})
Qp : The emission force (gr s^{-1})
(x,y,z) : Receptor coordinates (m)
u_a : The Speed of the sewer advection (ms^{-1})
Ky Kz : The constant diffuses in the direction of y and z (m^2s^{-1})
L : The parameter scale (10 m)
m,n : Empirical parameter (m=0.86; n=1)

The model which has been adopted considers the number of vehicles and the speed of the vehicles which has a close relationship with the concentration of pollutants. The model selection is based on the height and density of the construction.

3. Methodology

After Medan-Stabat toll road is operated, the traffic in arterial road will shifting into the toll road. The idea is to predict the impact of diverted traffic on vehicle emission in the arterial road, by using the scenario with or without toll road operated. The research location of the Soekarno-Hatta and Tengku Amir Hamzah arterial roads with the consideration is the arterial roads which exists on the side of the Medan-Stabat toll road.

On both the arterial roads, the average height of a building could reach 8.2 meters with a density rate of 0.831, based on that matter the prediction of a CO gas emission will use a Lined Source Model Equation.

In this research a scenario is conducted with and without the operation of the Medan-Stabat toll road, The size of reducing the emission of CO pollutants on the arterial road is assumed to reduce caused by the shifting of vehicles from arterial road to the toll road. In order to identify the concentration of CO pollutant on both the roads, an equation is required [16]:

1. The CO emission rate, which can be acquired by including the average speed of the vehicle to the roads with an equation as elaborated below:

$$qCO = 867.92V^{-0.8648} \tag{5}$$

Where:
q : The flow of emission (gr km^{-1})
V : The average speed of the vehicle (km hr^{-1})

2. The CO emission force (Q), is the mass of the pollutant that is dispersed from the source of emission per unit of time. The Volume of Vehicle evaluated is the volume during the peak hour with the type of large vehicle equivalent to the passenger loaded -vehicle. The size of the emission can be acquired from the equation mentioned below:

$$Q = nq \tag{6}$$

Where:
Q : The CO emission force (gr sec^{-1})
n : The number of vehicles per second
q : The Flow of emission (gr km^{-1})

4. Result

The basis of traffic analysis were data collected directly on T. Amir Hamzah and Soekarno-Hatta streets (arterial road) and Medan-Stabat toll road. The Analysis on the traffic volume during the on-peak hours with the composition of the vehicles (Light Vehicle with < 2800 CC and Heavy Vehicle > 3000 CC).

Table 2. Traffic volume and average speed

No	Street Name	Traffic Volume (Veh hr^{-1})			Avarage Speed (km hr^{-1})		
		Motorcycle	Light Vehicle	Heavy Vehicle	Motorcycle	Light Vehicle	Heavy Vehicle
1	Soekarno-Hatta	4,301	681	104	48	45	40
2	T. Amir Hamzah	2,198	283	87	47	42	40
3	Medan-Stabat Toll Road	-	990	81	-	95	52

The volume during the on-peak hours of each category of the vehicle is equivalent to the unit of passenger loaded-vehicles furthermore will be normalized with the emission multiplication factor.

Table 3. Normalizing traffic volume with emission multiplier.

No	Street Name	Traffic Volume (Vehicle sec^{-1})			Normalize Traffic Volume with Emission Factor			Total
		Motorcycle	Light Vehicle	Heavy Vehicle	Motorcycle	Light Vehicle	Heavy Vehicle	
					(0.6)	(0.76)	(1.93)	PCU sec^{-1}
1	Soekarno-Hatta	1.195	0.189	0.029	0.043	0.717	0.144	0.916
2	T. Amir Hamzah	0.611	0.079	0.024	0.023	0.366	0.060	0.473

3	Medan-Stabat Toll Road	-	0.275	0.022	-	-	0.209	0.252

The CO emission rate is the size of pollutant masses released by one vehicle per kilometre and acquired by including the variable of the average speed of the vehicle to the roads predicted by the equation (5):

Table 4. CO emission rate of each vehicle.

No	Street Name	Avarage Speed ($km\ hr^{-1}$)	Emission Rate ($gr\ km^{-1}$)
1	Soekarno-Hatta	44.33	32.69
2	T. Amir Hamzah	43	33.56
3	Medan-Stabat Toll Road	73.5	21.11

The CO emission force during on-peak hours is acquired by using the equation (6), can see in the table below:

Table 5. CO emission force.

No	Street Name	Emission Force ($gr\ s^{-1}$)
1	Soekarno-Hatta	29.954
2	T. Amir Hamzah	15.866
3	Medan-Stabat Toll Road	5.323

From the table above, it is identified that the total of CO emission force for the Soekarno-Hatta Street with the amount around 29.954 gr s^{-1} For the Tengku Amir Hamzah Street is with the total of CO emission force amounts up to 15.866 gr s^{-1}. For the Medan-Stabat toll road the amount of CO emission force amounts up to 5.323 gr s^{-1}.

The concentration of pollutants is the amount of contaminated substance released to the air by the traffic within the volume unit. In order to identify the amount of concentration of the CO pollutants on each road, an equation is applied (6), with the results of the calculation elaborated below:

Table 6. Total concentration of CO pollutants.

No	Street Name	Q ($gr\ s^{-1}$)	μ ($m\ s^{-1}$)	σz m	σy m	CO $g\ m^{-3}$	CO ($\mu g\ m^{-3}$)
1	Soekarno-Hatta	29.954	5.5	13	11.72	0.0114	11,373
2	Tengku Amir Hamzah	15,866	2.7	23	20.80	0.0039	3,909
3	Medan-Stabat Toll Road	5,323	2.7	23	20.80	0.0013	1,312

The concentration of CO pollutants in the arterial roads amounts up to 11,373 μg m^{-3} and 3,909 μg m^{-3}. The concentration of the CO pollutants without the operation of the toll road amounts up to 12,685 μg m^{-3} and **5.221** μg m^{-3}, is acquired with the assumption that the concentration of CO pollutants in

Soekarno-Hatta Street and T. Amir Hamzah Street by adding the total amount of the concentration of CO pollutant in the Medan-Stabat toll road.

5. Conclusion

Based on the result of the analysis and the discussion on the level of CO pollutant concentration caused by the traffic in the Soekarno-Hatta Street and the Tengku Amir Hamzah Street, it is concluded that:

The prediction of the concentration of CO pollutant caused by the traffic in the Soekarno-Hatta Street amounts up to 12,685 μg m^{-3} and the T. Amir Hamzah Street amounts up to 5.221 μg m^{-3}. The concentration of CO pollutant in the research location has not exceed the standard limit of the quality of national ambient air which amounts up to 30.000 μg m^{-3}.

The number and the average speed of vehicles have a strong correlation to the amount of total concentration of CO pollutant. After the operation of the Stabat-Binjai Toll Road, the level of pollution is reduced amounts up to ±10 % and ±25 % due to the traffic shifting on the arterial roads to the toll road. The risk of CO affecting human health is greater in the arterial roads compared to the toll roads, because the toll roads have wider lanes and are far from inhabited areas. In the toll roads, CO can spread immediately so that the concentration of carbon monoxide tends to be smaller.

References

[1] Corbitt R A 2004 *Air Pollution in Standard Handbook of Environmental Engineering Second Edition* (New York: McGraw-Hill)
[2] Duc K N and Duy V N 2018 *Energy Sustainable Dev* **43** p67
[3] Demir Abdullah 2015 *Social and Behavioral Sciences* **195** p2610
[4] Santosa Wimpy Parikesit Danang Wardhana Yuki M.A Dewanti Makmur Amelia Safrilah Kurniawan Dwi Ardianta 2023 *Perspektif Kebijakan dalam Pembangunan Jalan Tol di Indonesia* (Jakarta: Kementerian PU) p8
[5] Kementerian Lingkungan Hidup (KLH) 2010 *Pelaksanaan pengendalian pencemaran udara di daerah* (Jakarta: Kementerian Lingkungan Hidup) p22
[6] European Environment Agency (EEA) 2022 *Transport and environment report 2021* (Luxembourg: Publications Office of the European Union) pp17-26
[7] Medhi F Roohollah R Sadegh H and Ali R 2015 *Journal of Atmospheric Pollution Research* **6** p550
[8] Sengkey S L Jansen F and Wallah S 2011 *Jurnal Ilmiah Media Engineering* **1** p119
[9] Nagpure A S and Gurjar B R 2014 *Novel Combustion Concepts for Sustainable Energy Development* (New Dehli: Springer) p405
[10] Kijewskaa Kinga Wojciech Konickia Stanisław Iwan 2016 *Transportation Research Procedia* **14** p1543
[11] Olsson L 1994 *Science of The Total Environment* (Solna: Swedish Environmental Protection Agency) p 27
[12] Ismayanti 2011 *Pengantar Pariwisata* (Jakarta: Grasindo)
[13] Batterman Stuart Zhang A Kai Kononowech Robert 2010 *Environmental Health* **9** p29
[14] An F Barth M Norbeck F and Ross M 1997 *Transportation Research Record* **1587** p52
[15] European Environment Agency (EEA) 2022 *Transport and environment report 2021* (Luxembourg: Publications Office of the European Union)
[16] Kementerian Lingkungan Hidup (KLH) 1999 *Pedoman Teknik Tata Cara Prediksi Polusi Udara Skala Mikro Akibat Lalu Lintas* (Jakarta: Kementerian Lingkungan Hidup)

Traffic sign comprehension among motorcyclists: the effect of sign display and sign type

D Maulina[1*], D Y Irwanda[1], P Zahra[1], D Gittanty[1], T Faulina[1], F Putri[1]

[1]Faculty of Psychology, Universitas Indonesia

*Correspondence concerning this manuscript should be addressed to Dewi Maulina, Faculty of Psychology, The University of Indonesia, Depok Campus 16424, Indonesia. Tel.: +62 217863520; fax: +622184993910. E-mail address: dewi.maulina@ui.ac.id

Abstract. The previous studies showed an inconsistent result for the effect of sign type on sign comprehension in different sign display. Therefore, the current study aimed to examine the effect of sign display (symbol+text or symbol-only) and sign type (warning sign, regulatory sign, or command sign) on traffic sign comprehension. A mixed experimental design was conducted on 100 young adult motorcyclists in the Jabodetabek area with two years of riding experience. In each sign, participants were measured their reaction time to comprehend the sign (sign comprehension time) and the accuracy of the meaning of the sign (sign comprehension level). The result showed that the effect of sign display depends on the type of traffic signs both in sign comprehension time and comprehension level. Based on the results, it can be inferred that further training and socialization are needed to increase riders' comprehension of all types of traffic signs.

1. Introduction

Each year, approximately 1.3 million people die due to traffic accidents, as one of the main causes of death in the world, including in Indonesia [1]. The number of traffic accidents in Indonesia from 2015 to 2019 increased by 4,87 percent each year [2]. Among all road users, motorcyclists as the largest road users, contributed to the most traffic accidents in the last decades [2–4]. This high number of traffic accidents among motorcyclists could be caused by the lack of discipline in obeying traffic regulations, the lack of road safety knowledge, and the low level of traffic signs comprehension [5–7].

The limited knowledge of comprehending traffic signs could be one of the crucial factors that contributed to accidents among motorcyclists [8,9]. Previous research showed that Indonesian motorcyclists, especially those of young age, still need to improve their knowledge regarding traffic signs and regulations [4]. Traffic signs are an important factor to control and improve traffic safety [10]. Road users are encouraged to comprehend all the roads by using traffic signs. According to Sodikin et al. [11], drivers' comprehension of road signs is crucial because good comprehension indicates that the sign can effectively communicate the message to the driver. However, Dewar et al. [12] stated that symbolic signs encountered on the road often are misunderstood by the driver. For example, some of the traffic signs are often interpreted with the opposite meaning. In Ben-Bassat and Shinar's study [13], participants comprehended incorrectly some of the traffic signs, because they had difficulties in comprehending the sign that led them to misinterpret the sign. For example, signs of "no entry for motorcycles" and signs of "maximum speed limit" were often misinterpreted by more than 10% of participants. Misinterpreting traffic signs can be dangerous and increase the risk of traffic accidents, thus a comprehensive understanding of the traffic sign is needed.

The characteristics of traffic signs are the most important aspects that will determine the rider's understanding and comprehension of the traffic sign. This includes the type of traffic sign, shape and color of a traffic sign, familiarity, and sign display[13]. In many countries, including Indonesia, the traffic signs are usually displayed in symbol only, text only, or a combination of both [14,15].

The effect of sign display on sign comprehension has been found in several studies. For example, Shinar and Vogelzang [16] examined the effect of adding text to traffic signs on the speed and accuracy of the driver in understanding traffic signs, especially warning signs, regulatory signs, and information signs. The result of this study showed that the sign that is displayed with symbols and text would improve the correctness and reduce the time to comprehend the sign, especially in less familiar signs. Therefore, it can be concluded that comprehension could be improved by adding text to the symbol, thus decreasing the likelihood of misinterpreting the sign.

Another study by Maulina et al. [17] also found that the effect of sign display on sign comprehension level and time in warning sign depended on how familiar the sign was for the motorcyclists. If the participants were unfamiliar with the signs, the symbol and text condition had a faster sign comprehension time. Meanwhile, if the signs are familiar, a faster reaction was found in the symbol-only condition. Their study also showed that symbols and text conditions had a significantly higher level of sign comprehension level than symbol-only conditions, especially for unfamiliar signs. The result of this study is in line with Koyuncu and Amado [18], which found that the best reaction time was when the traffic sign was displayed with symbol only and symbol with text rather than text only. In their study, they also used warning signs. Based on the previous studies' results, it could be concluded that sign display has a significant effect on sign comprehension. In general, past studies showed that the highest sign comprehension was when the traffic signs were displayed in symbol+text condition, especially with less familiar traffic signs.

One of the theories that could explain the effect of sign display on sign comprehension is the Integrative Model of Text and Picture Comprehension (or ITPC model) proposed by Schnotz [19]. This model combines the concepts of several cognitive models: multiple memory systems [20], working memory [21,22], and dual coding [23]. According to this model, the verbal information (text) and pictorial information (symbol) are going through different channels before being transmitted and processed in the working memory. Comprehending text and pictorial information are active processes of coherence formation. In this process, coherent knowledge was built from that external information and from the individual's prior knowledge. However, the mental mode construction would be less optimal if the individual has fewer sources of information. This situation happens when there is only one source of information, i.e., text only or picture only, or when the individual does not have prior knowledge. As a result, comprehending the information would be more difficult. Therefore, having another source of information, for example, combining text and symbol, provides an additional route for mental model construction thus enhancing comprehension.

In sign comprehension, the symbol-only signs are not understood immediately by road users especially if the sign contains complex symbols which could add extra load to the comprehension process [24]. Thus, adding text to the symbol will improve accuracy and lessen the time for the rider to understand the sign. In other words, text and symbols can complement each other to create meaning when combined. Combining the text with the traffic signs will expand the meaning of the accompanying symbol in the traffic sign when the said symbol is not sufficient to be comprehended by the driver [25].

Moreover, adding text to the symbol would be more beneficial in helping the driver to learn the meaning of the traffic sign correctly when the sign is less familiar or new [16].

According to Ou and Liu [26], the effect of sign display on sign comprehension also depends on the sign type. An important thing to note from the previous studies on sign display is each study used different types of traffic signs, for example, warning signs [16–18], regulatory signs [16], and information/guidance signs [16]. Therefore, it could be assumed that sign type is another important factor that needs to be considered in effecting sign comprehension.

According to the Regulation of the Minister of Transportation of the Republic of Indonesia Number 13 Year 2014 [15], there are four different types of traffic signs, i.e. warning signs, command signs, regulatory signs, and guide/information signs. Warning signs are usually used to give awareness to the driver about the potential danger or dangerous places on the road and the nature of the danger. Prohibition Signs state the actions that are prohibited to be performed by road users. Regulatory signs state the orders of road rules and policies to road users. While guidance signs are intended to guide road users and give other necessary information.

Few studies in various countries have proved the effect of sign type on sign comprehension. However, those studies used different sign displays. For example, Al-Madani and Al-Janahi's [27] study among UEA drivers and Yuan and colleagues' [28] studies among drivers in China found that the warning signs were only answered correctly by half of the participants. Another study by Maulina et al. [29] also found that warning signs were less comprehended than regulatory and command signs among Indonesian motorcyclists. Furthermore, this study proved that sign comprehension training is effective for improving sign comprehension among motorcyclists, especially for warning signs. The similarity between these three studies [27–29] was using symbols only as the sign display. Thus, it could be concluded that the comprehension of the warning signs displayed with symbols only was considered very low among road users. This could be due to the warning signs tend to have difficult symbols and thus are hard to understand [30].

On the contrary, Ou and Liu [26] found that warning signs and regulatory signs were better comprehended than guide signs among Taiwanese and Vietnamese motorcyclists. In this study, all signs were displayed by symbol and text. Meanwhile, Sodikin et al. [11] showed that the understanding of warning, regulatory, and command+guide signs, which were displayed by symbol only, was evenly distributed among road users in Indonesian two-wheelers and four-wheelers.

The inconsistent results from the previous studies showed that the effect of sign type on sign comprehension is still inconclusive. Such discrepant results could be due to the different sign display that was used by those studies [11,26–29,31]. Therefore, the current study aimed to further examine the interaction of sign display and sign type in affecting sign comprehension.

2. Current study

In the present study, we investigated the effect of sign display and sign type on sign comprehension among motorcyclists. There are a few modifications from the previous studies. First, previous studies used different road users as participants, for example, Koyuncu and Amado [18] used general road users, while Shinar and Vogelzang [16] and Wontorczyk and Gaca [31] used car or professional drivers, and Maulina et al. [17,29,32] focused on motorcyclists. This current research would be focused on motorcyclists, as they have the highest accident rate in Indonesia.

Second, the country where the sign is placed could also contribute to sign comprehension [33,34]. Even though traffic signs have the same symbol, they could have different meanings in different countries. For example, the "no bicycles allowed" sign in Europe has a similar design to the "bicycle only" sign in Latin America [35,36]. Thus, it is possible to obtain a different result by conducting a study in a different country from the previous studies, especially in non-WEIRD (Western, Educated, Industrialized, Rich, and Democratic) countries [37,38].

Third, very few studies have examined the combination of sign display and sign type on sign comprehension. Previous studies have proven that sign comprehension was influenced by only sign display [16–18] or sign type [27–29], individually. By replicating and combining sign display and sign type in a single study, we can examine the certainty of past findings, find an interaction effect, and boost future studies' innovation, especially in the current wave of Open Science where false positive findings are highly prevalent [39].

Fourth, previous studies measured sign comprehension in different ways. For example, few studies used sign comprehension time [18], level [11,26–29,32], and both [16,17]. In the current study, sign comprehension would be measured by both sign comprehension time and level so that the speed and accuracy in comprehending the sign could be seen comprehensively.

Lastly, familiarity with the sign is one important factor in ergonomic guidelines for the sign to be easily comprehended [13,26,36]. Familiar and meaningful traffic signs will be better understood than unusual and unfamiliar ones [34,40,41]. Therefore, this study will use unfamiliar traffic sign to control prior experience with the traffic sign.

Based on previous studies, a combination of symbol+text would lead to better comprehension [16–18,32]. Therefore, it would be predicted that traffic sign comprehension in traffic signs with symbol+text will be higher than the traffic signs with symbol-only. Furthermore, every sign type has its own characteristics, i.e. warning signs are generally harder to understand as they have more complex and abstract symbols than other types of traffic signs [28,30]. Thus, it is expected that sign comprehension in warning signs will be lower than in regulatory signs and command signs. Finally, we hypothesized that there is an interaction between sign display and sign type on traffic sign comprehension among motorcyclists.

3. Method

3.1 Participants

Participants consisted of 104 licensed riders (76 males and 28 females) who live in the Jakarta, Bogor, Depok, Tangerang, and Bekasi (Jabodetabek) areas. The inclusion criteria for participants were riders who had a minimum of two years of riding experience and active riders with a minimum of riding five days a week. Participants aged in the range of 19 to 24 years old (M_{age}=19.96, *SD* =1.10) were randomly divided into two groups. The road sign with symbol and text group consisted of 52 participants (39 males and 13 females), with an age average of 19.92 (*SD* =0.92). The road sign symbol-only group consisted of 52 participants (37 males and 15 females), with an age average of 20.00 (*SD* = 1.25). An online questionnaire (Google Form platform) was used to recruit participants by using a convenience sampling technique.

The minimum detectable effect size of interest was f = 0.125 (or η^2p= 0.015), calculated based on a sensitivity power analysis using G*Power [42]. In the calculation, we used ANOVA: repeated measures, within-between interaction, with an alpha of 0.05, a power of 0.80, the current sample size (*N* = 104), 2 groups, and 3 measurements. The current study adhered to ethical standards in the 1964 Declaration of Helsinki. Informed consent was provided to the participants before the experiment and a reward would be given after the experiment.

3.2 Design

This current study used a mixed design: 2 (sign display: symbol+text vs symbol-only) x 3 (sign type: warning signs vs regulatory signs vs command signs), with sign type being a within-subject factor. The dependent variable is sign comprehension, which was measured by the reaction time to comprehend the sign (sign comprehension time) and the accuracy of the meaning of the sign (sign comprehension level). The sign familiarity was held constant by using only the unfamiliar signs.

3.3 Measurement and Apparatus

3.3.1 Computer and Website. The research was conducted in a computer laboratory, by using a PC, with a 17 inches monitor. A website called Road Sign Avenger (http://roadsignapp.herokuapp.com) was used to conduct the experiment. An example of a website display is shown in figure 1.

IOP Conf. Series: Earth and Environmental Science 1294 (2024) 012009 doi:10.1088/1755-1315/1294/1/012009

3.3.2 Stimulus. Fifteen traffic signs, consisting of five regulatory signs, five command signs, and five warning signs, were used in the current experiment. There were two variations of sign display, which were symbol only and symbol+text. Examples of sign display variation can be seen in Figure 1.

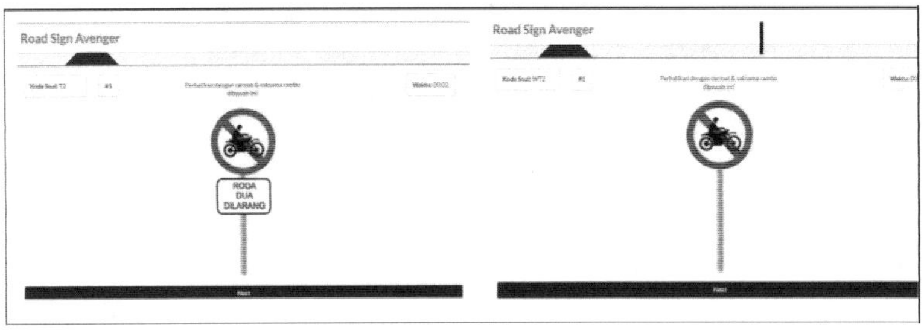

Figure 1. Website view (symbol + text vs symbol only).

3.3.3 Sign Comprehension Test. Sign comprehension time was measured by the time interval between the sign was presented to the participant and their response in pressing the 'next' button on the computer keyboard. The response to the 'next' button indicated that the participant had comprehended the meaning of the road sign. The sign comprehension time was measured in seconds. Meanwhile, the sign comprehension level was measured by the accuracy of the participant's answer in the fields provided on the computer screen when they were asked about the meaning of each sign. Ben-Bassat and Shinar's [10] scoring system was used for the sign comprehension level, with a score of +2 for well-written and complete answers, a score of +1 for correct but incomplete answers, a score of 0 for the wrong answer, and a score of -1 for the opposite answer.

3.4 Procedure

In the preparation phase, a survey of 60 licensed riders with the same characteristics as the experiment phase was conducted. The survey was intended to determine the types of traffic signs that will be used in the experiment. To control prior knowledge and familiarity with the traffic signs, we select 30 traffic signs, consisting of 10 regulatory signs, 10 command signs, and 10 warning signs. Participants were asked to rate the familiarity of each sign with a 10-point Likert scale, from 0 (never seen at all) to 10 (always see). Based on survey results, we selected 15 signs that consist of 5 signs for each type of traffic sign. All the signs were chosen based on the lowest mean in each type. We also conducted a pilot study with 20 participants to ensure that the procedures and manipulations were sufficient enough for participants. From the pilot study, we revised the instructions and website appearance.

In the experimental phase, participants had to fill in the informed consent before the study began. The participants did the experiment using a PC individually. There were two groups of participants, which were the sign with symbol+text group or symbol-only group. After the experimenter explained the instructions, participants began the experiment which consisted of 15 traffic signs as the stimulus. In each trial, participants were given a stimulus of traffic signs from the computer. Then, once the participants comprehended the meaning of the sign, they were instructed to press the 'next' button as soon as possible. After they pressed the button, there was a field on the next slide and participants had to type the meaning of the sign. This sequence lasted until the next 15 signs. All the types of traffic signs (warning signs, regulatory signs, and command signs) were displayed to each participant with a counterbalanced sequence. Finally, participants were debriefed and received rewards after completing the experiment.

4. Results

4.1 Sign comprehension time

Table 1. Mean and standard deviation of sign comprehension time (N=104).

Sign Display	Sign Type							
	Regulatory sign		Command sign		Warning sign		Total	
	M	*SD*	*M*	*SD*	*M*	*SD*	*M*	*SD*
Symbol + Text (n = 52)	23.98	14.42	30.87	18.34	27.40	12.22	27.42	12.80
Symbol-only (n = 52)	20.85	11.99	26.46	15.05	31.75	25.49	26.35	14.54
Total (N = 104)	22.41	13.29	28.66	16.84	29.58	20.01		

A two-way mixed ANOVA analysis was conducted to analyze the sign comprehension time scores from all the experimental groups. The result showed no statistically significant main effect for sign display ($F(1, 102) = 0.16$, $p = 0.693$, $\eta 2_{partial} = 0.002$), a statistically significant main effect of sign type ($F(2, 204) = 10.78$, $p < 0.001$, $\eta 2_{partial} = 0.096$) and a statistically significant interaction between sign display and sign type ($F(2, 204) = 3.96$, $p = 0.020$, $\eta 2_{partial} = 0.037$). This means the effect of sign display on sign comprehension depended on the sign type. Further simple effect analysis showed that in the symbol+text condition, a statistically significant difference was only found between the regulatory sign and command sign ($p = 0.001$), in which the regulatory sign ($M = 23.98$, $SD = 14.42$) was comprehended significantly faster than the command sign ($M = 30.87$, $SD = 18.34$). Meanwhile, in the symbol-only condition, a statistically significant difference was found between all sign types (all $ps < 0.05$). The regulatory sign ($M = 20.85$, $SD = 11.99$) was comprehended significantly faster than the command sign ($M = 26.46$, $SD = 15.05$), and the warning sign ($M = 31.75$, $SD = 25.49$), whereas the command sign was significantly faster than warning signs (see figure 2). Thus, the symbol-only sign would be comprehended fastest in regulatory sign, while the longest comprehension time was in warning sign.

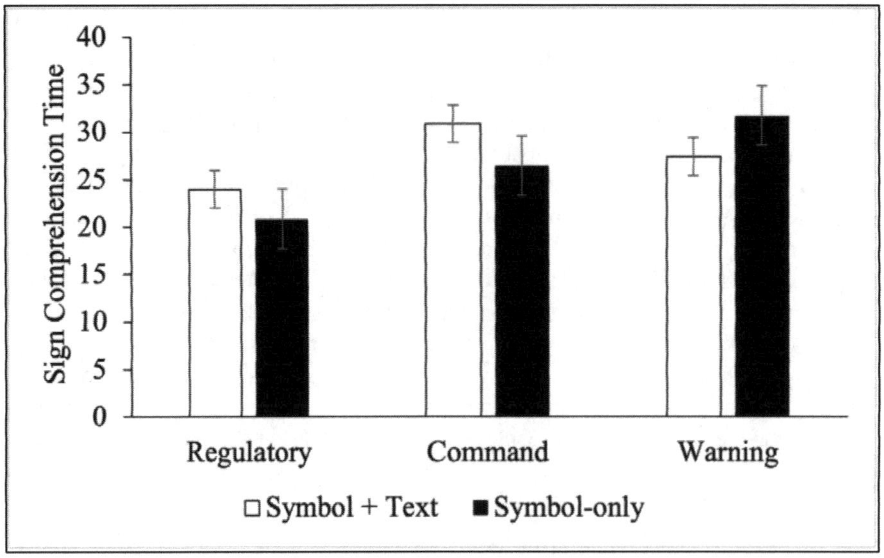

Figure 2. The average sign comprehension time on each sign type for symbol+text and symbol-only condition

4.2 Sign comprehension level

Table 2. Mean and standard deviation of sign comprehension level (N=104).

Sign Display	**Sign Type**							
	Regulatory sign		Command sign		Warning sign		Total	
	M	*SD*	*M*	*SD*	*M*	*SD*	*M*	*SD*
Symbol + Text (n = 52)	8.08	2.53	7.31	1.55	6.06	1.64	7.15	1.24
Symbol-only (n = 52)	8.46	1.94	6.87	2.11	4.48	1.93	6.60	1.27
Total (N = 104)	8.27	2.25	7.09	1.85	5.27	1.95		

A two-way mixed ANOVA analysis was also conducted to analyze the sign comprehension level scores from all the experimental groups. The results showed a statistically significant main effect of display ($F(1, 102) = 4.90$, $p = 0.029$, $\eta 2_{\text{partial}} = 0.046$), a main effect of sign type ($F(2, 204) = 68.31$, $p < 0.001$, $\eta 2_{\text{partial}} = 0.401$), and interaction between sign display and sign type ($F(2, 204) = 7.25$, $p = 0.001$, $\eta 2_{\text{partial}} = 0.066$). Following this interaction, additional simple effects analyses were run, and the result showed that there was a statistically significant difference in comprehension level between regulatory sign, command sign, and warning sign in both symbol+text condition and symbol-only condition (all $ps < 0.05$). The highest comprehension level was found in regulatory sign ($M_{symbol+text} = 8.08$, $SD_{symbol+text} = 2.53$; $M_{symbol\text{-}only} = 8.46$, $SD_{symbol\text{-}only} = 1.94$), while the warning sign ($M_{symbol+text} = 6.06$, $SD_{symbol+text} = 1.64$; $M_{symbol\text{-}only} = 4.48$, $SD_{symbol\text{-}only} = 1.93$) received the lowest comprehension level (see figure 3). Furthermore, simple effect analysis also revealed that there was no statistically significant difference in sign comprehension level between symbol+text and symbol-only (all $ps > 0.05$) in both regulatory and command sign. A distinct finding was seen for the warning sign, in which symbol+text was statistically significantly comprehended correctly than the symbol-only condition ($p < 0.05$; see table 2.).

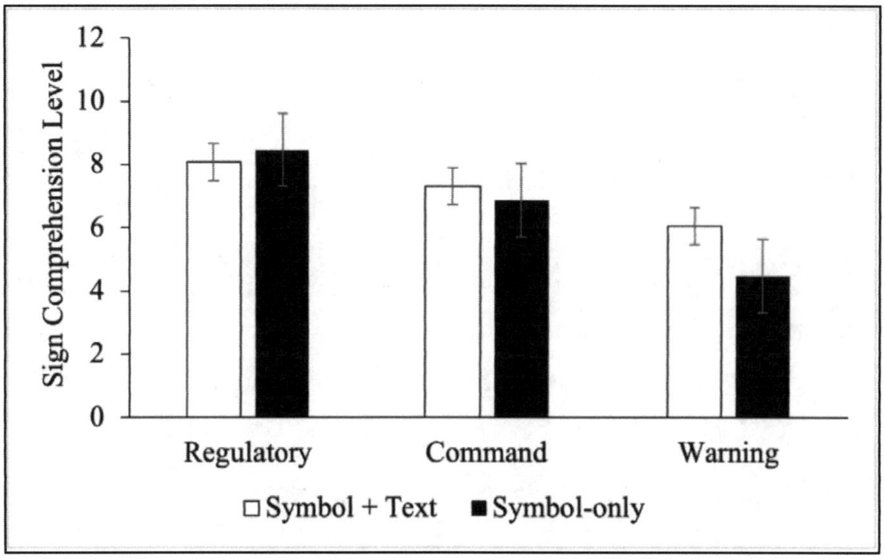

Figure 3. The average sign comprehension level on each sign type for symbol+text and symbol-only condition

Based on all the significant results above, it could be concluded that the current study (N= 104) had good power because the sensitivity power for each result was higher than the minimum (η^2_{partial} = 0.015).

5. Discussion

The study aimed to investigate the effect of two sign displays (symbol+text vs symbol only) and three types of traffic signs (regulatory sign vs command sign vs warning sign) on sign comprehension among motorcyclists. Our study found that sign display had a main effect only on sign comprehension level, but not on sign comprehension time. Moreover, a significant main effect of sign type on sign comprehension time and level was also found. Finally, an interaction effect between sign display and sign type was found to be significant in both sign comprehension time and level.

In sign comprehension time, the current study found that the regulatory sign was comprehended the fastest while the slowest was the command sign in the symbol+text condition. A similar result was found in the symbol-only sign condition in which the regulatory sign was comprehended the fastest while the warning sign needed the longest time to comprehend. Based on these results, the regulatory sign was the sign type that is comprehended the fastest in symbol+text and symbol-only conditions. This result is in line with a study by Ou and Liu [26] that sign comprehension was better for the common sign in Taiwan which is the regulatory signs. The possible explanation for why the regulatory sign could be comprehended faster than other signs is related to the characteristic of the regulatory sign which is it contains more concrete symbols. Concrete symbols consist of objects that are often seen in people's daily lives, so the information is already stored in their memory. Thus, it would take a little time for the person to recognize and comprehend the information. Past studies indicated that recognition of concrete stimuli (e.g. faces and everyday objects) was better than abstract stimuli, for example, pictures of snow crystals and inkbolt patterns [43–46]. Therefore, in this study, the sign comprehension time in unfamiliar regulatory sign would still be faster than other signs because it contains concrete symbols that could help the riders to quickly comprehend the meaning.

Another interaction pattern between sign display and sign type was found in the sign comprehension level. In both symbol+text and symbol-only conditions, the highest sign comprehension level was found in the regulatory sign, while the lowest was found in the warning sign. The symbol in the warning sign was considered more abstract and less familiar than other signs, so the riders need more effort to comprehend the warning signs [28,29]. Complex and unfamiliar symbol could cause confusion to the riders because there is no prior knowledge in their memory that could help the riders to comprehend the meaning of the symbol on the sign. This resulted in the rider not knowing and/or misunderstanding the meaning of the sign. This explains why in this study, the comprehension level of warning sign, which consist of abstract symbol, is lower than other signs. One way for the riders to accurately comprehend an abstract symbol is by making them more familiar with the symbol. According to Shen et al. [44], a complex and abstract symbol will become more familiar if it is exposed for a long period of time. Therefore, increasing the familiarity of the symbol can make the sign easier to be identified and decrease the error rate.

Furthermore, in the current study, the result showed that in the warning sign, symbol+text condition produced a significantly higher sign comprehension level than symbol-only condition. This result was in line with a previous study by Maulina et al. [32] that found sign comprehension level for the warning sign in symbol+text and text-only conditions was higher than in symbol-only condition. As mentioned before, the warning sign consists of an abstract symbol. Adding text to the abstract and unfamiliar symbol can help the riders to comprehend the meaning of the sign because signs with no text elicit confusion for participants to interpret the symbols. The information could be remembered more easily when it contains symbol and text because that combination enhances the meaning of each symbol and text [47]. In unfamiliar signs, riders need to associate the symbol and text to comprehend the meaning of the sign, thus the sign comprehension level of the symbol+text sign will be higher than the symbol-only sign.

It is important to note several limitations to the study. First, participants in this study only involve young riders (19-24 years old). Age is an important factor in riders' sign comprehension and younger drivers (age less than 25 years old) had the lower traffic signs' familiarity and comprehensibility than older riders [48]. Therefore, future study needs to include age in examining the effect of the sign characteristic on sign comprehension. Second, the current study used unfamiliar traffic signs. Since familiarity is one of the important factors that contribute to riders' ability to comprehend the meaning of the sign, future studies should compare the comprehension of familiar and unfamiliar signs to test the consistency of the findings. Lastly, our study was conducted in laboratory experimental design, thus a field experiment study is needed to enhance the external validity of the study. Adding environmental context, which is done in field experiment, had a significant positive effect on sign comprehension because the information given by a road sign is context-dependent [49,50]. Thus, it can increase the driver's understanding of traffic signs.

In conclusion, the low comprehension of unfamiliar signs can be overcome by providing additional text on the unfamiliar signs or through training. Sign comprehension training for unfamiliar signs, especially warning signs, was proven to be effective in increasing the sign comprehension level among motorcyclists [29]. Training could be given periodically as a series of processes for the rider's license.

6. References

[1] World Health Organization (WHO) 2022 Road traffic injuries. https://www.who.int/news-room/fact-sheets/detail/road-traffic-injuries.

[2] Central Bureau of Statistics (Badan Pusat Statistik) 2022 *Land transportation Statistics 2021 (Statistik Transportasi Darat 2021)*

[3] Jusuf A, Nurprasetio I P and Prihutama A 2017 Macro Data Analysis of Traffic Accidents in Indonesia *Journal of Engineering and Technological Sciences* **49** 132–43

[4] Soimun A, Leliana A, Ulmi E I, Ziantono D H and Widyastuti H 2020 Analysis of Student Comprehension of Traffic Signs (Analisis Pemahaman Pelajar pada Rambu Lalu Lintas) *Jurnal Teknologi Transportasi Dan Logistik* **1** 91–100

[5] Mauludi A A, Djunaidi Z and Arif L S 2021 Perilaku Berisiko Sebagai Faktor Penyebab Kecelakaan Pada Pengemudi Sepeda Motor Komersial: Systematic Review *Jurnal Keselamatan Transportasi Jalan (Indonesian Journal of Road Safety)* **8** 12–25
[6] Kitamura Y, Hayashi M and Yagi E 2018 Traffic problems in Southeast Asia featuring the case of Cambodia's traffic accidents involving motorcycles *IATSS Research* **42** 163–70
[7] Sugiyanto G and Santi M Y 2016 Karakteristik Kecelakaan Lalu Lintas dan Pendidikan Keselamatan Berlalulintas Sejak Usia Dini: Studi Kasus di Kabupaten Purbalingga *Semesta Teknika* **18** 65–75
[8] Al-Rousan T M and Umar A A 2021 Assessment of Traffic Sign Comprehension Levels among Drivers in the Emirate of Abu Dhabi, UAE *Infrastructures (Basel)* **6** 122
[9] Ishartomo F, Suhardi B and Rohani J M 2020 Ergonomic principles in traffic signs comprehension: A literature review p 030055
[10] Ben-Bassat T and Shinar D 2015 The effect of context and drivers' age on highway traffic signs comprehension *Transp Res Part F Traffic Psychol Behav* **33** 117–27
[11] Sodikin, Munawar A and Setiadji B H 2016 Drivers' Comprehension of the Traffic Signs *International Journal of Science and Research (IJSR)* **5** 534–8
[12] Dewar R E, Kline D W and Swanson H A 1994 Age differences in comprehension of traffic sign symbols *Transp Res Rec*
[13] Ben-Bassat T and Shinar D 2006 Ergonomic Guidelines for Traffic Sign Design Increase Sign Comprehension *Human Factors: The Journal of the Human Factors and Ergonomics Society* **48** 182–95
[14] The United Nations Educational S and C O (UNESCO) 1968 *Convention on Road Traffic* (Vienna: The United Nations Educational, Scientific and Cultural Organization (UNESCO))
[15] Ministry of Transportation of the Republic of Indonesia (Kementerian Perhubungan Republik Indonesia). 2014 *Regulation of the Minister of Transportation of Republic of Indonesia Number 13 Year 2014 concerning Traffic Signs (Peraturan Menteri Perhubungan Republik Indonesia Nomor PM 13 Tahun 2014 tentang Rambu Lalu Lintas)* (Jakarta: Ministry of Transportation of the Republic of Indonesia (Kementerian Perhubungan Republik Indonesia).)
[16] Shinar D and Vogelzang M 2013 Comprehension of traffic signs with symbolic versus text displays *Transp Res Part F Traffic Psychol Behav* **18** 72–82
[17] Maulina D, Andini E, Tanri D, Kamsanih N, Mazaya S and Zhafirah T 2022 Traffic Signs Comprehension among Motorcyclists: The Effect of Sign Type and Sign Familiarity *Humaniora* **13** 81–90
[18] Koyuncu M and Amado S 2008 Effects of stimulus type, duration and location on priming of road signs: Implications for driving *Transp Res Part F Traffic Psychol Behav* **11** 108–25
[19] Schnotz W 2014 Integrated Model of Text and Picture Comprehension *The Cambridge Handbook of Multimedia Learning* (New York: Cambridge University Press) pp 72–103
[20] Atkinson R C and Shiffrin R M 1971 The Control of Short-Term Memory *Sci Am* **225** 82–91
[21] Baddeley A D 1986 *Working memory* (Oxford: Clarendon Press)
[22] Baddeley A 2000 The episodic buffer: a new component of working memory? *Trends Cogn Sci* **4** 417–23
[23] Paivio A 1986 *Mental representations: A dual coding approach* (Oxford: Oxford University Press)
[24] Guo D, Zhang S, Wright K L and McTigue E M 2020 Do You Get the Picture? A Meta-Analysis of the Effect of Graphics on Reading Comprehension *AERA Open* **6** 233285842090169
[25] Bartłomiejczyk M 2013 Text and image in traffic signs *Linguistica Silesiana* **34** 111–31
[26] Ou Y-K and Liu Y-C 2012 Effects of sign design features and training on comprehension of traffic signs in Taiwanese and Vietnamese user groups *Int J Ind Ergon* **42** 1–7

[27] Al-Madani H and Al-Janahi A-R 2002 Assessment of drivers' comprehension of traffic signs based on their traffic, personal and social characteristics *Transp Res Part F Traffic Psychol Behav* **5** 63–76
[28] Yuan L, Ma Y -f., Lei Z -y. and Xu P 2015 Driver's Comprehension and Improvement of Warning Signs *Advances in Mechanical Engineering* **6** 582606–582606
[29] Maulina D, Siregar E S, Rachma T A, Nashria S A and Irwanda D Y 2022 How effective is training for improving traffic sign comprehension? Examining the interaction between training and sign type among motorcyclists *IATSS Research* **46** 614–22
[30] Lesch M F, Horrey W J, Wogalter M S and Powell W R 2011 Age-related differences in warning symbol comprehension and training effectiveness: effects of familiarity, complexity, and comprehensibility *Ergonomics* **54** 879–90
[31] Wontorczyk A and Gaca S 2021 Study on the Relationship between Drivers' Personal Characters and Non-Standard Traffic Signs Comprehensibility *Int J Environ Res Public Health* **18** 2678
[32] Maulina D, Shafira M D A and Santoso G P 2023 Effect of Signs Types on Level of Traffic Signs Understanding of Motorcyclists *Psympathic : Jurnal Ilmiah Psikologi* **10** 19–28
[33] Linkov V and Zámečník P 2020 Cultural Differences-Induced Mistakes in Driving Behaviour: An Opportunity to Improve Traffic Policy and Infrastructure *Mistakes, Errors and Failures across Cultures* (Cham: Springer International Publishing) pp 605–19
[34] Zhang T and Chan A H S 2013 Traffic sign comprehension: a review of influential factors and future directions for research *Proceedings of the International MultiConference of Engineers and Computer Scientists* (Hong Kong: International MultiConference of Engineers and Computer Scientists.)
[35] Ben-Bassat T, Shinar D, Caird J K, Dewar R E, Lehtonen E, Sinclair M, Zakowska L, Simmons S, Liberman G and Pronin M 2021 Ergonomic Design Improves Cross-Cultural Road Sign Comprehension *Transp Res Part F Traffic Psychol Behav* **78** 267–79
[36] Shinar D, Dewar R E, Summala H and Zakowska L 2003 Traffic sign symbol comprehension: a cross-cultural study *Ergonomics* **46** 1549–65
[37] Brady L M, Fryberg S A and Shoda Y 2018 Expanding the interpretive power of psychological science by attending to culture *Proceedings of the National Academy of Sciences* **115** 11406–13
[38] Hurtado S and Chiasson S 2016 An Eye-tracking Evaluation of Driver Distraction and Unfamiliar Road Signs *Proceedings of the 8th International Conference on Automotive User Interfaces and Interactive Vehicular Applications* (New York, NY, USA: ACM) pp 153–60
[39] Open Science Collaboration 2015 Estimating the reproducibility of psychological science *Science (1979)* **349**
[40] Bañares J R, Caballes S A, Serdan M J, Liggayu A T and Bongo M F 2018 A comprehension-based ergonomic redesign of Philippine road warning signs *Int J Ind Ergon* **65** 17–25
[41] Chan A H S and Ng A W Y 2010 Effects of sign characteristics and training methods on safety sign training effectiveness *Ergonomics* **53** 1325–46
[42] Faul F, Erdfelder E, Lang A-G and Buchner A 2007 G*Power 3: A flexible statistical power analysis program for the social, behavioral, and biomedical sciences *Behav Res Methods* **39** 175–91
[43] Kroll J F and Merves J S 1986 Lexical access for concrete and abstract words. *J Exp Psychol Learn Mem Cogn* **12** 92–107
[44] Shen Z, Xue C and Wang H 2018 Effects of Users' Familiarity With the Objects Depicted in Icons on the Cognitive Performance of Icon Identification *Iperception* **9** 204166951878080
[45] Bellhouse-King M W and Standing L G 2007 Recognition Memory for Concrete, Regular Abstract, and Diverse Abstract Pictures *Percept Mot Skills* **104** 758–62
[46] Goldstein A G and Chance J E 1971 Visual recognition memory for complex configurations *Percept Psychophys* **9** 237–41

[47] Hou Y, Yang W and Sun Y 2017 Do pictures help? The effects of pictures and food names on menu evaluations *Int J Hosp Manag* **60** 94–103
[48] Taamneh M and Alkheder S 2018 Traffic sign perception among Jordanian drivers: An evaluation study *Transp Policy (Oxf)* **66** 17–29
[49] Ben-Bassat T 2013 The effect of context and ergonomic design of traffic signs on driver comprehension – a preliminary evaluation *Proceedings of the Human Factors and Ergonomics Society Annual Meeting* **57** 1943–7
[50] Bazire M and Tijus C 2009 Understanding road signs *Saf Sci* **47** 1232–40

IOP Conf. Series: Earth and Environmental Science 1294 (2024) 012010 doi:10.1088/1755-1315/1294/1/012010

Counterflow riding behavior of motorcyclists: The role of moral obligation and variables in The Theory of Planned Behavior

J N Sirait[1] and G A Santoso[1]

[1] University of Indonesia, Faculty of Psychology, Depok, Indonesia

E-mail: jemima.netanya@ui.ac.id (J N Sirait), guritnaningsih.ny@ui.ac.id (G A Santoso)

Abstract. Motorcycles are one of the most popular modes of transportation in Indonesia because of their various benefits. However, these benefits are followed by increased traffic violations, one of which is counterflow riding behavior. This study aimed to examine the more consistent relationship among variables relating to The Theory of Planned Behavior, namely attitude, subjective norm, and perceived behavioral control, as well as the variable of moral obligation with counterflow riding behavior. A correlational study was conducted on 151 motorcyclists aged 18 – 64, domiciled in Jabodetabek (Jakarta, Bogor, Depok, Tangerang, and Bekasi), with active driver's licenses, and have done counterflow riding behavior. The results showed that attitude and moral obligation significantly correlated to counterflow riding behavior. Among the two variables, moral obligation is more consistently correlated with counterflow riding behavior than attitude. These findings imply that serious and consistent efforts are needed from law enforcement officials to educate and promote sanctions in curbing the counterflow riding behavior of motorcyclists.

1. Introduction

There has been a significant increase in motorcycles in Indonesia in the last five years. In 2018, there were 106 million units of motorcycles [1]. While in 2021, the number reached 120 million units of motorcycles [1]. A higher number of violations followed a large number of motorcycles. Pusat Informasi Kriminal Nasional [2] reported that in 2021, 1.3 million out of 1.7 million traffic violations were motorcyclists. Of all those violations, about 230,000 were counterflow riding behavior [3]. Then, in 2022, traffic violation by motorcyclist increased into 4.2 million out of 5.5 million [2]. These data showed a serious violation phenomenon, especially counterflow riding behavior have been done by motorcyclist. According to Maulina [4], counterflow riding behavior is one of Indonesia's most typical risky riding behaviors.

Engaging in risky riding behavior can elevate the chances of an accident occurring or intensify the extent of harm sustained during an accident [5]. Risky riding behavior is divided into errors and violations [6]. Errors are failures to display behavior planned to achieve a goal [6 & 7], whereas violations are actions that are usually deliberate, considered unsafe, and often illegal [7]. Counterflow riding behavior is considered as violation because it is against traffic rules [8]. This behavior can lead to traffic irregularities, endangering others and oneself, congestion, and accidents [9]. There are several reasons why motorcyclists do this behavior, such as their intention to reach their destination quickly, follow other motorcyclists, the absence of law enforcement officials, save time and fuel, or avoid law enforcement [9 & 10].

In general, there are several factors that influence traffic violation behavior. First is the internal factors of the motorcyclist, which include demographics. One of demographic factors that influences traffic violation behavior is gender. Male drivers were found to be more aggressive and more likely to break traffic rules than female drivers [11 & 12]. On the other hand, research in Bandung, Yogyakarta, and Surabaya found that the behavior of repeated violations of male drivers is less than female drivers [13]. Although, this is not the case in metropolitan cities such as Jakarta [13]. Another internal factor that influences traffic behavior is age. Previous studies have found that young adult motorcyclists were more likely to engage in risky riding behaviors [13 & 14]. Young riders tend not to consider risk perception when making decisions [12]. Young age is a sign of a less experienced rider [7]. Novice riders are willing to violate traffic because of cognitive impulsivity [15]. In accordance with that, research by Leong et al. [12] found that motorcyclists with 1 to 5 years of riding experience are more likely to engage in risky riding behaviors.

The next factor that influences traffic violation behavior is external factors. Various elements, including road geometry, asphalt quality, engine capacity, and time, influence the perceived behavioral control, thereby impacting a motorcyclist's inclination to partake in counterflow riding behavior [13]. In the context of counterflow riding behavior, road design where the place of the U-turn is farther than the destination encourages motorcyclists to engage in the behavior [4]. Research also found that congested roads and vacant opposite lanes also influence the decision of riders to engage in counterflow riding behavior [4]. A literature review found that weather and temperature are related to motorcyclists' behavior violating traffic rules [16], such as running red lights when the weather is hot [17]. Maulina's study [4] found that motorcyclists engage in more counterflow riding behavior when the weather is sunny and decreases when the weather is rainy.

The next external factor that influences traffic violation behavior is time pressure. Motorcyclists engage in counterflow riding behavior with the excuse of wanting to arrive at the destination quickly, for example, to go to work [4 & 18]. Study by Xu et al. [15] found that time pressure affects the intention of motorcyclists to violate traffic rules. Another external factor is descriptive norms [15]. Descriptive norms or social norms are informal rules that are largely unwritten and determined acceptable, appropriate, and mandatory actions in a particular group or society [19]. This means motorcyclists engage in counterflow riding behavior simply because they follow other motorcyclists [9 & 20]. Moreover, if the motorcyclists know that counterflow riding behaviors usually occur on the road they are on, then the motorcyclist will tend to engage in the counterflow riding behavior [4 & 20]. It also means that there is a factor of familiarity with the road that the counterflow riding behavior occurs on [4]. When motorcyclists see other violating riders not getting penalized, this can be social pressure for those motorcyclists to do the same [13]. Other external factors can include shortening mileage, driving on one-way roads, saving fuel, and the absence of law enforcement officials [9 & 18].

Violations or risky riding behavior are generally studied using the Theory of Planned Behavior [8] [13 & 21]. According to Ajzen [22], the central element of the Theory of Planned Behavior revolves around an individual's intention to execute a particular behavior. Intention can also be referred to as a motivational factor or the amount of effort a person is willing to exert to perform a behavior. The three aspects that determine a person's intentions are attitude, subjective norm, and perceived behavioral control. Attitude, which refers to the positive or negative evaluation of behavior [22], was found to increase intentions to engage in counterflow riding behavior [13, 21, 23]. However, Prahastomo's [24] study on motorcyclists in Jakarta and Depok found that attitude did not affect the intention to do counterflow behavior. Subjective norm, which refers to the perceived social pressure to perform or not perform a behavior [22], were also found to increase the intention to engage in counterflow riding behavior [13, 21, 23]. Previous research on motorcyclists in Jabodetabek also found that subjective norms, especially descriptive norms, are predictors of rider's intent to engage in counterflow riding behavior [9]. On the contrary, Prahastomo [24] found that subjective norm does not affect the intention to engage in counterflow riding behavior. Next, perceived behavior control, which refers to the perception of ease or difficulty in performing a behavior [22], was also found to increase the intention to engage in counterflow riding behavior [13 & 24]. However, Kumphong and Surinaud's study [21] found that perceived behavior control does not affect intentions to engage in counterflow riding behavior.

In addition to the three factors above, Ajzen [22] stated that moral obligation can be considered as factor that influences behavioral intentions in the context of unethical or unregulated behavior. Moral obligation refers to the perceived responsibility to perform or not perform a particular behavior [22]. Elliott and Thomson [25] found that moral obligation is related and can predict risky driving behavior, particularly speeding behavior. A study by Bazargan-Hejazi et al. [26] also found that moral obligation has a negative relationship with risky driving behavior, namely driving while texting.

Counterflow riding behavior in Indonesia is generally studied using the Theory of Planned Behavior [8, 13, 24] but has never considered the role of moral obligation. Therefore, this research aimed to see if moral obligation correlates with the counterflow riding behavior of motorcyclists in Jabodetabek. Moreover, the various results of previous studies also encourage to examine further the correlation between The Theory of Planned Behavior factors, namely attitude, subjective norm, and perceived behavioral control, with the counterflow riding behavior of motorcyclists in Jabodetabek. This research focuses on examining the relationships of the variables, specifically with behavior and not intentions, because of the possibility that intentions are not always channeled into behavior [27]. The results of this study can be useful for traffic policy makers in developing efforts to reduce violations of counterflow riding behavior in motorcyclists.

2. Literature Review

2.1. Counterflow Riding Behavior

Counterflow riding behavior occurs when a person intentionally or unintentionally drives against the predetermined direction [28]. Similarly, Kadeha et al. [29] stated that counterflow riding behavior is the movement of vehicles in the direction opposite to the set direction. Counterflow riding behavior is classified as risky riding behavior, more precisely, violation. Violations refer to acts of violating traffic practices or rules that are considered unsafe and often illegal [7 & 8]. Violations are divided into intentional and unintentional violations [6 & 7]. In this case, counterflow riding behavior is an example of intentional violation.

2.2. The Theory of Planned Behavior

The Theory of Planned Behavior is a commonly utilized theory in social psychology for analyzing violations in riding behavior. This study analyses four factors i.e., attitude, subjective norm, perceived behavioral control, and moral obligation.

Attitude denotes the extent to which one evaluates a behavior positively or negatively [22]. This constitutes that if a person evaluates their behavior more positively, the intention to perform said behavior becomes greater. Consequently, negative evaluation of one's behavior causes lesser intention to perform said behavior. Ajzen and Fishbein [27] stated that attitudes toward behavior include instrumental and experiential aspects. The instrumental aspect is an evaluation of how useful a behavior is, while the experiential aspect is an evaluation of how pleasant or boring a behavior is [27]. Sometimes, a behavior has both positive instrumental and experiential aspects. However, it is not uncommon to find a behavior with one aspect of the attitude that is positive, but the other aspect of the attitude is negative [27]. This is found in counterflow riding behavior, where motorcyclists find the behavior beneficial for the instrumental aspect but frightening or harmful for the experiential aspect [4].

Individual attitudes towards counterflow riding behavior can be used as a reference in making decisions to engage in counterflow riding behavior. When the motorcyclist believes the behavior's possible consequences are positive, the individual will be motivated to perform the behavior [23]. This aligns with previous study, which found that attitude increases the intention to engage in counterflow riding behavior [21]. Previous studies in the broader context have also found that attitude is one of the factors and predictors of risky riding behavior [13, 23, 30]. This is a reference for this research to hypothesize that attitude is positively and significantly related to counterflow riding behavior. An individual's positive evaluation of counterflow riding behavior can increase the tendency of motorcyclists to engage in the behavior.

Subjective norm relates to the perceived societal pressure, either endorsing or discouraging the engagement in a behavior [22]. Correspondingly, this aspect refers to the extent to which a person is driven to conform to certain important people or groups [31]. The higher the perceived social pressure, the intention to conform such approved behavior will also be higher. Conversely, the lower the perceived social pressure, the person will have less intention to conform to the approved behavior. Ajzen and Fishbein [27] stated that there are two types of norms, i.e., injunctive norms and descriptive norms. Injunctive norm refers to perceptions of behavior that should be performed in each situation [8 & 27]. A person will behave according to injunctive norms out of importance or seek social approval from others [32]. Whereas descriptive norm refers to the perception that others perform or do not perform certain behaviors [8] [27]. This means that if many people perform a behavior in a particular context, one can assume that the behavior makes sense to perform in that context [27]. Individual subjective norms can be used as a reference in deciding to engage in counterflow riding behavior. When the social pressure felt by individuals is strong, individuals will be driven to engage in the behavior. This was found in previous research where individual subjective norms increased the intention to engage in counterflow riding behavior [21]. Irta et al. [8] found that descriptive norms, specifically, increase intention and can predict counterflow riding behavior. Even in a broader context, it was found that subjective norm was positively and significantly associated with risky riding behavior, although the effect was not as great as other aspects [13] [23]. These findings provide a reference for this study to hypothesize that subjective norm positively and significantly correlates to counterflow riding behavior. The social pressure that individuals feel from those around them, or the behavior of other motorcyclists influences their decisions to engage in counterflow riding behavior. Individual subjective norms can increase the tendency of motorcyclists to behave according to those norms.

Perceived behavioral control refers to an individual's assessment of how effortless or burdensome it is to perform a behavior [22]. This constitutes that if a person perceives the behavior as effortless, the intention to perform said behavior increases. Consequently, the intention decreases as the behavior is perceived as burdensome. If a person perceives that behavior is easy, the greater the intention to do the behavior. Conversely, the intention to perform a specific behavior is weaker if the individual perceives that the behavior is difficult to perform. It is assumed that perceived behavioral control results from reflection on past experiences and expected obstacles [22]. Perceived behavioral control consists of perceived capacity and perceived autonomy [27]. Capacity indicates both the competence or conviction to carry out a behavior and the level of effort or challenge involved in executing that behavior [27 & 32]. Autonomy refers to an individual's belief that their ability to perform a behavior relies on their own actions and choices [27 & 32].

In the context of driving, perceived behavioral control can explain and predict risky driving behaviors involving intentional violations [7]. This is in line with research by Yanuvianti et al. [23], which found that perceived behavioral control increases the intention of motorcyclists to engage in risky riding behaviors. Past studies have additionally shown that this factor contributes to an increased intent to partake in risky riding behaviors [13, 24, 30, 33]. These findings are a reference to hypothesize that perceived behavioral control is positively and significantly related to counterflow riding behavior. Consideration of how easy or challenging and the belief in one's ability to perform counterflow riding behavior plays a role in one's decision to partake in it. The greater the perception of individual behavioral control, the more likely the rider is to engage in the behavior.

In addition to the three aspects, Ajzen [22] stated that in certain contexts, not only subjective norms need to be considered but also moral norms. Moral norms relate to the perception of a moral duty or obligation, either to engage in or refrain from a specific behavior [22]. Moral obligation serves as a moral guideline to determine whether a behavior is acceptable or not [25 & 34]. Moral obligation can also influence behavioral intention, just as attitude, subjective norm, and perceived behavioral control [22 & 34]. Moral obligation can be of influence when behavior is considered unethical or not in accordance with the rules. Moral obligation is often associated with experiencing guilt, making ethically incorrect choices, and behaving in a manner inconsistent with your principles [26]. Yunianti et al. [35] interjected in regards to moral obligation, that it is the result of a person's conscience and not coercion from outsiders. In other words, a person with a high moral obligation will always be obedient and will not be affected by other factors [31]. Therefore, moral obligation is considered in this study as well.

Moral obligation was found to have an influence on rule-abiding behavior. A study by Yunianti et al. [35] found that moral obligation influence rule compliance, specifically paying taxes. The same study found that the better the perception of the moral obligation to pay taxes, the more taxpayer complies in paying taxes. This means that the better the perception of moral obligation to a rule, the more compliance will be. Bazargan-Hejazi et al. [26] also found that moral obligation has a negative relationship with risky driving behavior, namely driving while texting. The results of these studies become a reference to hypothesize that moral obligations are negatively and significantly related to violation behavior, in this case, counterflow riding behavior. An individual's moral obligation becomes a factor in deciding whether or not to partake counterflow riding behavior. The higher the moral obligation to not engage in the behavior, the more it influences the motorcyclist's behavior. Motorcyclists' moral obligation can increase the tendency to behave in accordance with the existing rule, in this case, not engaging in counterflow riding behavior.

3. Methodology

3.1. Research Participants

Participants for this research were 151 motorcyclists domiciled in the Jabodetabek area, aged over 18, having an active driver's license, and having engaged in counterflow riding behavior in the past. The sampling technique used was a non-random sampling method, specifically accidental sampling. Data collection was conducted online using Google Form.

3.2. Research Design

The design of this study was correlational quantitative research. Five variables were used such as attitude, subjective norm, perceived behavioral control, moral obligation, and counterflow riding behavior. Counterflow riding behavior was measured through the participant's weekly average engagement in the behavior.

3.3. Research Instruments

This research used an instrument developed by Prahastomo [16] to measure attitude, subjective norm, and perceived behavioral control. Attitude refers to how positive is participant's evaluation towards counterflow riding behavior. Attitude items obtained an alpha coefficient of $\alpha = 0.563$ and cr_{IT} in the range of 0.234 – 0.449. Subjective norm refers to the social pressure participant's experience to engage in counterflow riding behavior. Subjective norm items obtained an alpha coefficient of $\alpha = 0.721$ and cr_{IT} in the range of 0.479 – 0.611. Perceived behavioral control refers to participant's perception on how easy to engage in counterflow riding behavior. Perceived behavioral control items obtained an alpha coefficient of $\alpha = 0.626$ and cr_{IT} of 0.457 on both items. To measure the moral obligation, this research used the moral obligation dimension on the instrument developed by Yunianti et al. [35] to measure compliance with paying taxes. Moral obligation refers to participant's willingness to not engage in counterflow riding behavior regardless of other motorcyclist behavior. Moral obligation items obtained an alpha coefficient of $\alpha = 0.670$ and cr_{IT} in the range of 0.295 – 0.514. Overall, items were revised to fit the context of the counterflow riding behavior. Several new items were also added to make the number of items adequate. The instrument is a self-report questionnaire with items quantities as follows: four for attitude, three subjective norm, two for perceived behavioral control, and six for moral obligation, answered using a Likert scale of 1 (strongly disagree) to 4 (strongly agree). The participant's variable score is obtained by summing the answers from all items for each variable.

4. Result

Among 151 participants, majority of them were domiciled in Jakarta (47.7%) and Depok (22.5%), in the age range of 21 – 45 years (80.8%), and males (62.9%). Out of all participants 19.9% had been ticketed, and 5.3% had experienced an accident when they engaged in counterflow riding behavior. The majority of participants, on a weekly average, engaged in 1 – 3 counterflow riding behavior (89.4%) and mostly at night (45.7%). More than half of the participants (54.3%) engaged in the behavior to avoid further U-turns.

Table 1. Overview of Attitude, Subjective Norm, Perceived Behavioral Control, and Moral Obligation Scores.

Variables	Scale Median	*M*	*SD*	Minimum	Maximum
Attitude	2.5	2.11	2.433	4	16
Subjective Norm	2.5	2.09	2.343	3	12
Perceived Behavioral Control	2.5	2.27	1.648	2	8
Moral Obligation	2.5	3.93	3.014	8	24

As seen in Table 1, participants' attitude score (M = 2.11; SD = 2.43) for counterflow riding behavior tends to be low. Participants' subjective norm score (M = 2.09; SD = 2.34) and perceived behavioral score (M = 2.27; SD = 1.65) for counterflow riding behavior tends to be low as well. On the other hand, participants' moral obligation score (M = 3.93; SD = 3.01) for counterflow riding behavior tends to be high.

Pearson's correlation analysis found a statistically significant and positive correlation between attitude and counterflow riding behavior (r = 0.146, p < 0.005, r^2 = 0.021, one-tail). The result also showed that the correlation between the two variables is categorized as a small effect size according to Cohen [39], who stated that the small effect size category is r^2 = 0.01. The calculation r^2 means that 2.1% of the counterflow riding behavior variance can be explained by attitude. Subsequently, results displayed a statistically significant and negative correlation between moral obligation and counterflow riding behavior (r = -0.214, p < 0.001, r^2 = 0.046, one-tail). This correlation between the two variables is categorized as small effect size according to Cohen [39], who states that the small effect size category is r^2 = 0.01. The result of the r^2 calculation means that 4.6% of the counterflow riding behavior variance can be explained by moral obligation. In contrast, the correlation between subjective norm and counterflow riding behavior showed a positive correlation but not statistically significant (r = 0.123, p > 0.005, one-tail). The same result also found between perceived behavioral control and counterflow riding behavior, which were positively correlated, but the correlation was not significant (r = 0.107, p > 0.005, one-tail).

Table 2. Correlation Between Variables.

Variables		1	2	3	4	5
1 Attitude	***r***	1				
2 Subjective Norm	***r***	0,457^b	1			

3 Perceived Behavioral Control	*r*	0,412[b]	0,476[b]	1		
4 Moral Obligation	*r*	-0,509[b]	-0,358[b]	-0,547[b]	1	
5 Weekly Average of Counterflow Riding Behavior	*r*	0,146[a]	0,123	0,107	-0,214[b]	1
	r^2	0,021	0,015	0,012	0,046	

[a] Significant at p < 0,05 (1-*tailed*).
[b] Significant at p < 0,01 (1-*tailed*).

5. Discussion

The current study showed that most motorcyclists who engage in counterflow riding behavior are domiciled in Jakarta and Depok. This is related to the density of motorcycles in the two major cities so that counterflow riding behavior occurs more [4]. In addition, the majority of participants who engaged in counterflow riding behavior are 21 – 45 years old, which falls into the young adult category [40]. This finding aligns with study by Susilo et al. [12] and Goh et al. [23], where young adult motorcyclists were found to be more inclined to partake in risky riding behaviors. Furthermore, the majority of motorcyclists who engage in counterflow riding behavior were male rather than female. This is in line with the findings of Leong et al. [22] and Chang and Yeh [11], that male drivers showed more aggressive tendencies and more likely to violate traffic rules in contrast to female drivers.

The results showed that out of the four variables, only attitude and moral obligation were correlated with counterflow riding behavior. Attitude was positively correlated with counterflow riding behavior. This means that the more positive the motorcyclist's evaluation of the counterflow riding behavior, the higher the likelihood of performing the behavior. Conversely, the more negative the motorcyclist's evaluation of the counterflow riding behavior, the likelihood to perform this behavior will be lower. Previous studies displayed similar results [12,13,15], which found that attitude was positively correlated with risky riding behavior, with one of them being counterflow riding behavior. Although the majority of participants had negative evaluations towards the counterflow riding behavior, the result showed participants engaged in the behavior at least 1 – 3 times per week. This is in accordance with the effect size analysis, which shows attitude only has a small effect on counterflow riding behavior. Attitude does not play enough role in a motorcyclist's decision to partake or not in counterflow riding behavior.

Based on the data, motorcyclists engaged in this behavior to avoid further U-turns, save time, avoid traffic jams, and avoid law enforcement. In addition, the majority of participants were never ticketed or had an accident for engaging in the behavior. A similar result was found in the study of Leong et al. [22], where motorcyclists perceived accidents as the most critical negative consequence of counterflow riding behavior. However, they also felt positive consequences of the behavior, such as feeling satisfied to show off and saving time [22]. A study by Yanuvianti et al. [15] also found that motorcyclists engaged in risky riding behavior because of the perceived benefits and did not have the intention to harm themselves or others. There is always the possibility of experiencing negative consequences of counterflow riding behavior, but the possibility is smaller than the positive consequences experienced when engaging in the behavior. These advantages and the lack of negative consequences further motivate motorcyclists to engage in the behavior.

Of all the variables measured, moral obligation had a negative and significant relationship with counterflow riding behavior. This means that the higher the motorcyclists' desire to conform their behavior to existing norms and well-regarded rules, the more likely they are not engaging in the counterflow riding behavior. Conversely, the lower the motorcyclists' desire to conform their behavior to well-regarded norms and rules, they are more likely to engage in the behavior. The majority of participants had positive moral obligation scores, meaning they desired to conform to well-regarded norms. A similar result was found in the study of Bazargan-Hejazi et al. [18], where moral obligation was negatively and significantly correlated with willingness to engage in risky driving behaviors, specifically texting while driving. Experiencing guilt due to not adhering to established norms and existing laws was a significant factor in the inclination to send and read text messages while driving. [18]. Yunianti et al. [35] similarly observed that increased moral obligation correlates with greater adherence to regulations, for example paying taxes.

In reality, the engagement in counterflow riding behavior should be low. However, the participants engage in the behavior at least 1 – 3 times weekly. This is due to the fact that there are little to no negative consequences for motorcyclists that engages in the behavior. The result also showed that the majority of participants had never been ticketed or had an accident because they engaged in the counterflow riding behavior. This is also supported by the effect size of the relationship between moral obligation and counterflow riding behavior that falls into the small category. It can be interpreted that moral obligation plays little role in the decision to engage or not in the counterflow riding behavior, even though its role is greater than attitude.

The results also showed that the majority of participants scored positively on attitude and moral obligation. However, all participants had engaged in the behavior at least once per week. The participants' attitude and moral obligation indicated their desire not to engage in the behavior. However, the advantages and the absence of negative consequences played a greater role in the decisions made.

Out of the four variables, the two variables that were not significantly related were subjective norm and perceived behavioral control. This means that motorcyclists did not feel social pressure from those around them to engage in the counterflow riding behavior because they did not condone it. It can be concluded that the decision to engage in the behavior is purely based on the motorcyclist alone, not the influence of their referents, even if the referents are considered an important person or group to them, such as best friends/close friends, relatives, or co-workers. Most participants were young adults (20 – 45 years), where in this phase of life, the main psychosocial issue is forming intimate relationships with others [40]. Even if important people or groups disagree with maintaining intimate relationships and gaining benefits from counterflow riding behavior, motorcyclists may say they will conform to expectations, though this is not the case.

The findings also indicated that perceived behavioral control, did not factor into the decision to participate in counterflow riding behavior. This could be related to data showing that the majority of participants engaged in the behavior at night when law enforcement officers rarely patrol. Based on the data (experience of accidents and being ticketed because of counterflow riding behavior), capacity aspect of perceived behavior control may be more influential than autonomy in the context of counterflow riding behavior.

This study had some limitations. First, counterflow riding behavior is a traffic violation behavior, so there is a possibility that participants tried to show certain impressions, such as faking good. It requires high transparency from participants. Second, the data was collected online, so researchers can only rely on the honesty of participants when filling out questionnaires. To avoid this, data collection can be done in the field to obtain participants that have engaged in counterflow riding behavior. This can be due to the participants' seriousness when completing the questionnaire. Filling out an online questionnaire also means that the condition of participants when filling out the questionnaire is unknown. It is possible that participants were tired, rushed, or did not read items carefully. It is recommended for future research to include self-checker questions to ensure participants read and filled out the questionnaire seriously. This also affects the psychometric properties of instruments, where some variables' instrument has not reached the standard limit of alpha coefficient 0.7. Hence, further revisions must be done to these instruments. In addition, the measurement of the weekly average of counterflow riding behavior was categorized into two groups (1 – 3 times and more than three times in one week). Such categories caused the description of participants' counterflow riding behavior to be less detailed. It would be better to use continuous scale in future research to measure weekly average of counterflow riding behavior. Finally, in this study the participants were only from Jabodetabek area. This means the participants were less representative, thus affecting the external validity of the current study. Future research can be conducted with motorcyclists from different areas, even smaller cities. Regarding participants characteristics, future research should also be conducted with specific group age, especially late adolescents to instill moral obligations.

6. Conclusion

This study showed that out of the four variables, only attitude and moral obligation correlated significantly with counterflow riding behavior. Therefore, the positive consequences felt by motorcyclists when engaging in counterflow riding behavior must be minimized. This means negative consequences, namely speeding tickets, must be done consistently. Thus, motorcyclists must see and believe in the negative consequences of counterflow riding behavior, rather than just knowing it. It is also important to accompany this effort by providing education about the importance to obey traffic regulation.

References

[1] Badan Pusat Statistik Perkembangan Jumlah Kendaraan Bermotor Menurut Jenis (Unit) *2019-2021 Badan Pusat Statistik*

[2] Pusat Informasi Kriminal Nasional Langgar Lantas *Kepolisian Republik Indonesia*

[3] *Pusiknas Bareskrim Polri 2021 Jurnal tahunan Pusiknas Bareskrim Polri Jurnal Data Pusat Informasi Kriminal Nasional Bareskrim Polri*

[4] Maulina D 2014 *Pengaruh script mengendara berisiko persepsi jarak dan persepsi risiko terhadap perilaku mengendara berisiko pada pengendara sepeda motor dalam situasi pro risk dan anti risk* Universitas Indonesia

[5] Eby D W 2004 Driving, Risky *Encyclopedia of Applied Psychology, Three-Volume Set* **1 627–632** 10.1016/B0-12-657410-3/00697-8

[6] Reason J, Manstead A, Stephen S, Baxter J and Campbell K 1990 Errors and violations on the roads: A real distinction? *Ergonomics* **33 10–11 1315–1332** 10.1080/00140139008925335

[7] Shinar D 2017 *Traffic safety and human behavior* (2nd ed.) (Bingley: Emerald Publishing Limited)

[8] Irta A Z, Maulina D and Santoso A G 2021 Peran norma deskriptif dan kepribadian terhadap perilaku melawan arah pada pengendara sepeda motor *Psympathic: Jurnal Ilmiah Psikologi* **8 1 01–20** 10.15575/psy.v8i1.10748

[9] Yusnan O D Y 2021 Pelanggaran melawan arah lalu lintas oleh pengendara sepeda motor (Studi kasus Jl. Kaharuddin Nasution Kota Pekanbaru Provinsi Riau) Universitas Indonesia

[10] Sofyanida T and Yanuvianti M 2016 Studi Deskriptif Mengenai Perilaku Mengemudi Agresif pada Mahasiswa Pengendara Sepeda Motor di Kota Bandung **943–949**

[11] Chang H L and Yeh T H 2007 Motorcyclist accident involvement by age, gender, and risky behaviors in Taipei, Taiwan *Transportation Research Part F: Traffic Psychology and Behaviour* **10 2 109–122** 10.1016/j.trf.2006.08.001

[12] Leong L V, Mohd Shafie S A, Gooi P K and Goh W C 2021 Assessing self-reported risky riding behavior of motorcyclists at unsignalized intersections for sustainable transportation *Sustainability* **13 16 9144** 10.3390/su13169144

[13] Susilo Y O, Joewono T B and Vandebona U 2015 Reasons underlying behaviour of motorcyclists disregarding traffic regulations in urban areas of Indonesia *Accident Analysis & Prevention* **75 272–284** 10.1016/J.AAP.2014.12.016

[14] Goh W C, Leong L V and Cheah R J X 2020 Assessing significant factors affecting risky riding behaviors of motorcyclists *Applied Sciences* **10 18 6608** 10.3390/app10186608

[15] Xu Y, Li Y and Jiang L 2014 The effects of situational factors and impulsiveness on drivers' intentions to violate traffic rules: Difference of driving experience *Accident Analysis and Prevention* **62 54–62** 10.1016/j.aap.2013.09.014

[16] Ma C, Yang D, Zhou J, Feng Z and Yuan Q 2019 Risk riding behaviors of urban e-bikes: A literature review *International Journal of Environmental Research and Public Health 2019*

16 13 2308 10.3390/IJERPH16132308
[17] Zhang Y and Wu C 2013 The effects of sunshields on red light running behavior of cyclists and electric bike riders *Accident Analysis and Prevention* **52 210–218** 10.1016/j.aap.2012.12.032
[18] Dandona R, Kumar G A and Dandona L 2006 Risky behavior of drivers of motorized two wheeled vehicles in India *Journal of Safety Research* **37 2 149–158** 10.1016/j.jsr.2005.11.002
[19] Cislaghi B and Heise L 2018 Theory and practice of social norms interventions: Eight common pitfalls *Globalization and Health* **14 1** 10.1186/s12992-018-0398-x
[20] Irta A Z 2021 *Peran norma deskriptif dan trait kepribadian terhadap perilaku melawan arah pada pengendara sepeda motor di Jakarta* Universitas Indonesia
[21] Kumphong J and Surinaud E 2021 A study of factors affecting wrong-way riding behaviors among motorcyclists: The case of urban arterial road in Khon Kaen City, Thailand *The 26th National Convention on Civil Engineering* **1–8**
[22] Ajzen I 1991 The Theory of Planned Behavior *Organizational Behavior and Human Decision Processes* **50 2** 10.1016/0749-5978(91)90020-T
[23] Yanuvianti M, Coralia F and Qodariah S 2020 *Proceedings of the 2nd Social and Humaniora Research Symposium (SoRes 2019)*
[24] Prahastomo A 2019 *Kajian intensi pelanggaran lalu lintas oleh pengendara sebagai dampak penerapan manajemen lalu lintas* Universitas Indonesia
[25] Elliott M A and Thomson J A 2010 The social cognitive determinants of offending drivers' speeding behaviour *Accident Analysis and Prevention* **42 1595–1605** 10.1016/j.aap.2010.03.018
[26] Bazargan-Hejazi S, Teruya S, Pan D, Lin J, Gordon D, Krochalk P C, and Bazargan M 2016 The theory of planned behavior (TPB) and texting while driving behavior in college students *Traffic Injury Prevention* **18 1 56–62** 10.1080/15389588.2016.1172703
[27] Ajzen I and Fishbein M 2010 Predicting Changing Behavior *Taylor & Francis Group*
[28] Kemel E 2015 Wrong-way driving crashes on French divided roads *Accident Analysis and Prevention* **75 69–76** 10.1016/j.aap.2014.11.002
[29] Kadeha C, Haule H, Ali M S, Alluri P and Ponnaluri R 2021 Modeling Wrong-way Driving (WWD) crash severity on arterials in Florida *Accident Analysis and Prevention* **151 105963** 10.1016/j.aap.2020.105963
[30] Cristea M and Gheorghiu A 2016 Attitude, perceived behavioral control, and intention to adopt risky behaviors *Transportation Research Part F: Traffic Psychology and Behaviour* **43 157–165** 10.1016/j.trf.2016.10.004
[31] Bobek D D and Hatfield R C 2003 An investigation of the Theory of Planned Behavior and the role of moral obligation in tax compliance *Behavioral Research in Accounting* **15 1 13–38**. 10.2308/bria.2003.15.1.13
[32] Conner M 2020 Theory of planned behavior *Handbook of Sport Psychology* **1–18** 10.1002/9781119568124.ch1
[33] Tunnicliff D J, Watson B C, White K M, Hyde M K, Schonfeld C C and Wishart D E 2012 Understanding the factors influencing safe and unsafe motorcycle rider intentions. *Accident Analysis & Prevention* **49 133–141** 10.1016/j.aap.2011.03.012
[34] Conner M and Armitage C J 1998 Extending the theory of planned behavior: A review and avenues for further research *Journal of Applied Social Psychology* **28 15 1429–1464** 10.1111/j.1559-1816.1998.tb01685.x
[35] Yunianti L N, Putri N K, Sudibyo Y A and Rafinda A 2019 The influence of awareness, moral obligations, tax access, service quality and tax sanctions on taxpayer compliance in paying motor vehicle tax *Journal of Accounting and Strategic Finance* **2 1 1–13** 10.33005/jasf.v2i1.20
[36] Kaplan R M and Saccuzzo D P 2013 *Psychological testing: Principles, applications, and issues* (8th ed.) (Australia: Wadsworth and Cengage Learning)

[37] Taber K S 2018 The use of Cronbach's Alpha when developing and reporting research instruments in science education *Research in Science Education* **48 6 1273–1296** 10.1007/s11165-016-9602-2
[38] Nunnally J C and Bernstein I H 1994 *Psychometry theory* (3rd ed.) (New York: McGraw-Hill)
[39] Cohen J 1988 *Statistical power analysis for the behavioral sciences* (2nd ed.) (New York: Routledge)
[40] Papalia D E, Olds S W and Feldman R D 2009 *Human Development* (11th ed.) (New York: McGraw-Hill Education)

IOP Conf. Series: Earth and Environmental Science 1294 (2024) 012011 doi:10.1088/1755-1315/1294/1/012011

Identification of crash-prone segment on Cipularang toll road

S Nabila[1], **A Kusumawati**[1], **and W Weningtyas**[1]
[1]Department of Civil Engineering, Bandung Institute of Technology, Bandung, Indonesia

E-mail: nabilasyahla1211@gmail.com **and** aine@itb.ac.id

Abstract. Traffic safety is an important issue in the field of transportation. Toll roads are one of the infrastructures that are prone to crashes, because on toll roads drivers can drive their vehicles at high speeds so there is a high risk of having an crash, as happens every year on Cipularang Toll Road. This research will determine the crash-prone segments on Cipularang Toll Road section using the network screening method. Crash and traffic data for the last five years (2018-2022) were obtained from PT Jasa Marga Purbaleunyi. Identification of crash-prone segments carried out using ten performance indicators contained in HSM (2010), namely average crash frequency, crash rate, critical rate, Equivalent Property Damage Only average crash frequency, relative severity index, excess predicted average crash frequency using method of moment, Level of Service of Safety, excess predicted average crash frequency using SPF, probability of specific crash type exceeding threshold proportion, and expected average crash frequency with Empirical Bayes adjustment. The analysis results show that the five segments with the highest priority for treatment are KM 91-92, KM 92-93, and KM 86-87 (Jakarta-bound direction), KM 120-121 and KM 101-102 (Bandung-bound direction).

1. Introduction

Land transportation is one of the solutions to fulfill human needs by moving goods or services with the infrastructure used, namely roads. Road infrastructure is usually prone to crashes, one of which is the freeway or toll road. With the "freeway" status, it does not guarantee that traffic crash problems can be overcome. Crash mostly happens because freeways spur drivers to use high speeds to exceed the permitted limits. One of the factors that cause crashes is road factors (geometric) and the lack of road safety furnitures. It is necessary to do road safety management [1].

One of the toll roads that are prone to crashes is the Cipularang Toll Road (Cikampek – Purwakarta – Padalarang). Crashes on this toll road occur every year, causing the formation of several crash-prone points on this toll road. On this toll road, front and rear crash crashes generally occur. This resulted in a high number of victims who indicated serious or minor injuries to death.

Several studies on the ranking of crash-prone segments on the Cipularang Toll Road section have been carried out using various methods. In this study, crash-prone segments will be identified with the data from 2018 to 2022 using the network screening method, which refers to the Highway Safety Manuals (2010).

The network screening method will obtain comprehensive results because the lowest number of crashes in a segment can be adequately identified due to RTM bias considerations. So that the ranking results will show whether there has been a change in the crash-prone segment from previous studies, which indicates effectiveness in handling crashes on the Cipularang Toll Road section before.

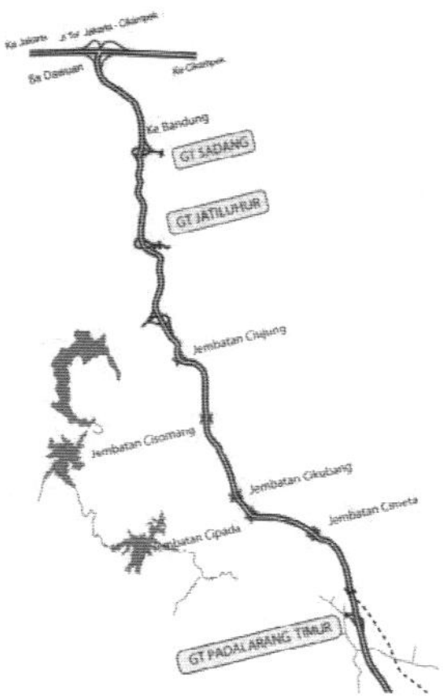

Figure 1. Cipularang Toll Road Section.

2. Literature Review

Several studies about the analysis of crashes on the Cipularang Toll Road have already been carried out. The studies also analyzing crash prediction modeling and crash cost.

2.1. Crash Prone Segment Identification

There are several studies regarding the identification of crash-prone segments on the Cipularang Toll Road using a variety of different methods. In 2017, Damanhuri [3] research started by identifying vehicle travel behavior, blackspot location geometry, and the influence of travel behavior on blackspot location geometry. Then in 2019 Hanafi et al. [4] analyzing the crash-prone segment by using Korlantas weighting method and Deni with Mayani [5] analyzing the crash-prone segment using a descriptive method. Meanwhile, in the research of Andre Jonathan et al. [6] in 2020, a one-way Annova comparative test was carried out by including the crash rate calculation in finding crash-prone locations. In general, the results of previous research indicated that the crash-prone segment on the Cipularang Toll Road section is located at KM 91 to KM 93 Jakarta-bound direction.

2.2. Crash Prediction Model

Research conducted by Kusumawati and Rakhmat (2011) [7] used the Generalized Poisson (GP) regression model which is generalized to the standard Poisson regression model. The performance of the model will be compared with the model developed using the Negative Binomial (NB) regression model. Parameters used in GP crash prediction modeling and NB regression include average daily traffic, the number of horizontal curves, and the existence of a median with certain specifications. Here are the basic equations used for modeling.

$$\mu = k \times (LHRT)^{\alpha} \exp(\gamma_1 X_1 + \gamma_2 X_2 + \ldots) \tag{1}$$

Where,

μ	= Expected crashes (incidents/3 years)
LHRT	= Annual Average Daily Traffic (3 years average in vehicles/day)
k, α, β, γ	= Estimated parameters
$X_1, X_2, \ldots$	= Various geometric/road environment factors

In the research of Rakhmat et al. [8] a model was developed to identify and rank hazardous road locations using a nonlinear regression model. The most appropriate model for modeling the number of traffic crashes on the Purbaleunyi Toll Road section, with crash data for 2010 period, is the binomial negative regression model. This model relates the number of traffic crashes with exposure and risk factors developed using statistical methods. Here is the resulting model equation.

$$\mu = LHRT^{0,484} \exp(0,242D + 0,243MED - 3,238) \tag{2}$$

Where,

μ = Expected crashes (incidents/3 years)
LHRT = Annual Average Daily Traffic (3 years average in vehicles/day)
D = Degrees of curvature (radians)
MED = Existence of a median with a width < 2.5 m and a height < 1.75 m (1 if the median meets these specifications, 0 otherwise)

Then in the research of Kusumawati et al. (2013) [9] an crash prediction model was developed for case studies of the Palimanan - Kanci, Tangerang - Merak, and Purbaleunyi toll roads. This study aims to develop a general crash prediction model for toll roads in Indonesia by considering the relationship between crash frequency and traffic flow, as well as various road and environmental geometric characteristics. This model can also be used to identify blackspot locations on toll roads with crash data from 2007 to 2010. The following is the general equation resulting from this study.

$$\mu = AADT^{-0.521} \exp(0.103NTIK - 0.911W) \tag{3}$$

Where,

μ = Expected crashes (incidents/3 years)
AADT = Annual Average Daily Traffic (3 years average in vehicles/day)
NTIK = Number of horizontal curves
W = Track width (m)

2.3. Crash Cost

In Meydita and Kusumawati [10] research on 2013, an analysis of the crashes cost that occurred on the toll road was carried out using the human capital method approach. This approach will calculate the crash costs based on direct and indirect costs. The classification of crash costs is based on the number of vehicles involved and the level of crash fatality, namely fatal, serious injury, light injury, or material loss.

The crash costs in this study are higher because what is calculated is the unit crash cost for toll roads while in Pd T-02-2005-B the unit crash costs for urban and inter-urban roads which can involve motorcycles, the losses are materially cheaper. Apart from that, this research also takes into account the costs of delays or congestion so that it is more comprehensive compared to Pd T-02-2005-B. The following is a recapitulation of toll road crash costs resulting from this study.

Table 1. Indonesian Toll Road Crash Costs.

Crash Classification	Crash Unit Cost (2013)
Fatal	Rp 1,063,849,963.00
Heavy Injury	Rp 290,880,669.00
Light Injury	Rp 170,642,494.00
Material Loss	Rp 214,735,187.00

Sources : Meydita and Kusumawati (2013)

3. Methodology

The data processing uses the network screening method based on the American Association of State Highways and Transportation (2010). The data to be processed includes crash and traffic data obtained from PT Jasa Marga Purbaleunyi in the period of 2018 to 2022. The network screening method will review a transportation network that is intended to identify and rank the most likely sites to realize a reduction in the frequency of crashes by implementing preventive measures.

The network screening method consists of five processes which can be explained as follows.

a. Determining the focus of this analysis, which is to identify and rank sites that have the potential for improvement in reducing the number of crashes.
b. Identifying Networks and Establishing Reference Populations. In this study, the reference population was determined from the start which is the Cipularang Toll Road section from KM 67 to KM 121.6, which passes through the Cikampek, Purwakarta, and Padalarang areas.
c. Determining performance measures that are selected based on the availability of existing data. Based on HSM (2010), all performance measures require crash data and road information for categorization. Things that make it different include the need for traffic data on several performance measures, including crash rates, critical rates, level of service of safety, and others. In addition, there are several performance measures that require calibrated safety performance measures and overdispersion parameter data, such as excess predicted average crash frequency using Safety Performance Functions (SPFs) and Equivalent Property Damage Only (EPDO) average crash frequency with EB adjustment. Other considerations include the EPDO weighting factor and relative severity indices contained in several performance measures.
d. Determining a Screening Method. In this analysis, the simple ranking method is used to rank 1-km crash-prone segments on Cipularang Toll Road Bandung-bound direction (A-direction) and Jakarta-bound direction (B-direction).
e. Screen and Evaluation of Results. The segment most likely to reduce the frequency of crashes is the most suitable for the selected performance measures. In this study, the selection of crash-prone segment will be limited to five segments throughout the Cipularang Toll Road.

4. Results

The steps taken before determining crash-prone segments are determining performance measures. In determining performance measures there are some criteria for selecting them. One of them is the availability of data and input. The required data is available, including crash and traffic volume data. In addition, a safety performance function (SPF) and crash cost data are needed, which is one of the criteria for data availability. The SPF equation used is contained in equation (2) while the crash cost data is shown in Table 1.

Based on these considerations, there are ten performance measures that can be used in this study. The following are the performance measures used to rank crash-prone segments on the Cipularang Toll Road section. For a more detailed calculation method, see the Highway Safety Manual (2010) [2].

4.1. Identification and Ranking of Crash Prone Segments

4.1.1. Average Crash Frequency

In a ranking based on the average crash frequency, the segment that has the largest total number of crashes in the 5-year range, from 2018 to 2022, will be given the highest ranking. The following is the ranking of ten most hazardous segments based on the average crash frequency.

Table 2. Average Crash Frequency Data Ranking Results.

Rank	Crash Frequency	
	Segment	Total
1	B91-92	22
2	B92-93	12
3	A101-102	9
4	A120-121	9
5	A112-113	8
6	A116-117	8
7	B86-87	8
8	A113-114	7
9	B84-85	7
10	A84-85	6

4.1.2. Crash Rate

For performance measure crash rate, crash frequency will be measured by traffic volume. The exposure used for sections or segments is road vehicles or VMT (Vehicle-Miles Travelled) over a five-year period. The following is the calculation and ranking results based on the crash rate in each segment, which is limited to ten crash-prone segments.

Table 3. Crash Rate Data Ranking Results.

Rank	Segment	C	LHRT	L (km)	R
1	B91-92	22	13158.4	1	91.61
2	B92-93	12	13158.4	1	49.97
3	A101-102	9	13425.7	1	36.73
4	A120-121	9	13425.7	1	36.73
5	B86-87	8	13158.4	1	33.31
6	A112-113	8	13425.7	1	32.65
7	A116-117	8	13425.7	1	32.65
8	B84-85	7	13158.4	1	29.15
9	A113-114	7	13425.7	1	28.57
10	B76-77	6	11917.3	1	27.59

4.1.3. Critical Rate

The ranking is based on the crash rate at each location compared to the critical crash rate using the methods from the Highway Safety Manual. The following is the result of the calculation and ranking based on the difference in the crash rate and critical crash rate in each segment.

Table 4. Critical Rate Data Ranking Results.

Rank	Segment	C	LHRT	L (km)	R	Rc	Difference	Over/Under Critical
1	B91-92	22	13158.4	1	0.92	0.28	-0.63	Over Critical
2	B92-93	12	13158.4	1	0.50	0.28	-0.22	Over Critical
3	A101-102	9	13425.7	1	0.37	0.28	-0.08	Over Critical
4	A120-121	9	13425.7	1	0.37	0.28	-0.08	Over Critical
5	B86-87	8	13158.4	1	0.33	0.28	-0.05	Over Critical
6	A112-113	8	13425.7	1	0.33	0.28	-0.04	Over Critical
7	A116-117	8	13425.7	1	0.33	0.28	-0.04	Over Critical
8	B84-85	7	13158.4	1	0.29	0.28	-0.01	Over Critical
9	A113-114	7	13425.7	1	0.29	0.28	0.00	Over Critical
10	B76-77	6	11917.3	1	0.28	0.29	0.02	Under Critical

4.1.4. Equivalent Property Damage Only (EPDO) Average Crash Frequency

Weighting factor is assigned to crashes based on severity, namely fatal, serious injury, minor injury, or property damage only (PDO). The weighting factor is calculated relative to the cost of the crash resulting in loss of property or PDO. Crash costs based on severity will produce an EPDO value. Later the segments will be ranked from high to low EPDO scores. Sites at the highest rank can be selected for further investigation. So that the following is the result of calculating the EPDO value and its ranking for all segments on both lines.

Table 5. EPDO Average Crash Frequency Data Ranking Results.

Rank	Segment	Fatal	Heavy Injury	Light Injury	PDO	EPDO
1	B91-92	7	12	2	1	53.5
2	A120-121	4	5	0	0	26.6
3	A112-113	4	4	0	0	25.2
4	B86-87	4	3	1	0	24.7
5	B92-93	2	8	2	0	22.3
6	B87-88	3	3	0	0	18.9
7	A113-114	2	5	0	0	16.7
8	A84-85	2	4	0	0	15.3
9	B88-89	2	4	0	0	15.3
10	B90-91	2	4	0	0	15.3

4.1.5. Relative Severity Index

In determining the Relative Severity Index or RSI it is necessary to determine the cost of an crash for each level of fatality and the total cost of all crashes for each location. Then the average crash cost per segment is compared with the overall average crash cost. The resulting RSI value will show segments that have more than the average or less than the average. The following is the result of processing the RSI value in each segment which is ranked based on the average RSI value per segment.

Table 6. RSI Data Ranking Results.

Rank	Segment	Fatal	Heavy Injury	Light Injury	PDO	RSI	RSI Average	Check
1	A85-86	2	0	0	0	Rp6,042,435,821.94	Rp3,020,717,910.97	High
2	A79-80	2	1	0	0	Rp6,867,368,497.94	Rp2,289,122,832.65	High
3	B115 – 116	2	1	0	0	Rp6,867,368,497.94	Rp2,289,122,832.65	High
4	A111 – 112	2	0	1	0	Rp6,525,961,700.52	Rp2,175,320,566.84	High
5	A87 – 88	2	2	0	0	Rp7,693,301,173.94	Rp1,923,325,293.49	High
6	A90 – 91	1	1	0	0	Rp3,846,650,586.97	Rp1,923,325,293.49	High
7	A112 – 113	4	4	0	0	Rp15,386,602,347.89	Rp1,923,325,293.49	High
8	B70 – 71	1	1	0	0	Rp3,846,650,586.97	Rp1,923,325,293.49	High
9	B87 – 88	3	3	0	0	Rp11,539,951,760.91	Rp1,923,325,293.49	High
10	B102 – 103	1	1	0	0	Rp3,846,650,586.97	Rp1,923,325,293.49	High

Overall, the average RSI value is IDR 1,270,569,363.00. So that there are 40 segments that have an average RSI value per segment higher than the average RSI. The average RSI value per segment will be affected by the total number of crashes in that segment and the number of crashes with the highest costs.

4.1.6. Excess Predicted Average Crash Frequency Using Method of Moment

The average crash frequency in each segment adjusted for the variance in crash data and the average crash frequency for the reference population. The observed and adjusted average crash frequency for that location is compared to the average crash frequency for the entire segment. The following is a recapitulation of the calculation of the PI value for each segment per kilometer.

Table 7. PI Data Ranking Results.

Rank	Segment	C	$(N_{obs,i}-N_{obs,rp})^2$	$N_{obs(adj)}$	PI
1	B91-92	22	349.69	15.06	11.76
2	B92-93	12	75.69	8.77	5.47
3	A101-102	9	32.49	6.88	3.58
4	A120-121	9	32.49	6.88	3.58
5	A112-113	8	22.09	6.26	2.96
6	A116-117	8	22.09	6.26	2.96
7	B86-87	8	22.09	6.26	2.96
8	A113-114	7	13.69	5.63	2.33
9	B84-85	7	13.69	5.63	2.33
10	A84-85	6	7.29	5.00	1.70

4.1.7. Excess Predicted Average Crash Frequency Using Safety Performance Functions (SPFs)

In the data processing, a comparison will be made between the average crash frequency observed with the average crash frequency from the SPF. The difference between the observed and predicted crash frequencies will be the excess of the crash frequencies estimated using the SPF. If the excess frequency is more than zero, the segment will have the highest ranking based on the excess value. The following is the result of calculating and ranking the segments based on the excess value.

Table 8. Excess Predicted Average Crash Frequency Using SPFs Data Ranking Results.

Rank	Segment	C	LHRT	D (Rad)	MED	μ	Excess
1	B91-92	22	13158.4	3.54	1	6.97	15.03
2	A120-121	9	13425.7	0.00	1	2.99	6.01
3	A101-102	9	13425.7	0.95	1	3.76	5.24
4	B92-93	12	13158.4	3.54	1	6.97	5.03
5	A116-117	8	13425.7	0.41	1	3.30	4.70
6	B86-87	8	13158.4	0.52	1	3.36	4.64
7	A113-114	7	13425.7	0.41	1	3.30	3.70
8	A112-113	8	13425.7	1.91	1	4.74	3.26
9	B103-104	6	13158.4	0.00	1	2.96	3.04
10	A106-107	6	13425.7	0.00	1	2.99	3.01

4.1.8. Level of Service of Safety (LOSS)

The predicted average crash frequency for segments with similar characteristics is predicted from the SPF calibrated to local conditions. The observed crash frequency was compared with prediction of average crash frequency. Each location is placed into one of the four LOSS classifications on HSM (2010) [2]. The following is the result of calculating the LOSS value for all segments per kilometer.

Table 9. LOSS Data Ranking Results.

Rank	Segment	C	μ	σ	LOSS
1	B91-92	22	6.97	3.839	IV
2	A101-102	9	3.76	2.456	IV
3	B86-87	8	3.36	2.271	IV
4	A113-114	7	3.30	2.245	IV
5	A116-117	8	3.30	2.245	IV
6	A120-121	9	2.99	2.101	IV
7	B92-93	12	6.97	3.839	III
8	A96-97	6	5.18	3.077	III
9	A112-113	8	4.74	2.888	III
10	B84-85	7	4.49	2.777	III

4.1.9. Probability of Specific Crash Types Exceeding Threshold Proportion

A threshold proportion (pi*) will be selected for each population based on the proportion of types of crash or severity. Then the variance value will be calculated for the reference to calculate the alpha and beta parameters used to find the target crash probability. This probability value will be the basis for the ranking of the segment. On the probability calculation, data of fatal and injury crashes will be used. The ranking based on the probability shown in the following table.

Table 10. Target Crash Probability Data Ranking Results.

Rank	Segment	P
1	B91-92	98.6%
2	A120-121	98.4%
3	A101-102	98.4%
4	B92-93	98.3%
5	A116-117	98.3%
6	B86-87	98.3%
7	A113-114	98.2%
8	A112-113	98.2%
9	B103-104	98%
10	A106-107	98%

4.1.10. Expected Average Crash Frequency with Empirical Bayes Adjustment (EB)

It will be weighted from the observed average crash frequency and the expected crash frequency from SPF with the EB method. The following is the result of the calculation and ranking process based on the estimated average crash frequency adjusted for EB in the last year, which is 2022.

Table 11. Expected Average Crash Frequency with EB Adjustment Data Ranking Results.

Rank	Segment	C	w_{total}	$N_{exp,1(total)}$	$N_{exp,n(total)}$
1	B91-92	22	0.037	6.790	3.33
2	A101-102	9	0.064	1.940	3.03
3	A120-121	9	0.079	1.909	2.98
4	A112-113	8	0.051	1.837	2.87
5	A116-117	8	0.072	1.797	2.81
6	A113-114	7	0.072	1.671	2.61
7	A96-97	6	0.047	1.624	2.54
8	A84-85	6	0.055	1.611	2.52
9	A88-89	6	0.058	1.604	2.51
10	A108-109	6	0.061	1.563	2.44

IOP Conf. Series: Earth and Environmental Science 1294 (2024) 012011 doi:10.1088/1755-1315/1294/1/012011

4.2. Selection of Top 5 (Five) Segments Ranking Results

The following is a recapitulation of the ranking results from all performance measures, which are limited to 10 segments. PM on the table refers to Performance Measures and the number refers to the order in the previous section.

Table 12. Crash Prone Segment Ranking Recapitulation (Part 1).

Rank	PM-1	PM-2	PM-3	PM-4	PM-5
1	B91-92	B91-92	B91-92	B91-92	A85-86
2	B92-93	B92-93	B92-93	A120-121	A79-80
3	A101-102	A101-102	A101-102	A112-113	B115 – 116
4	A120-121	A120-121	A120-121	B86-87	A111 – 112
5	A112-113	B86-87	B86-87	B92-93	A87 – 88
6	A116-117	A112-113	A112-113	B87-88	A90 – 91
7	B86-87	A116-117	A116-117	A113-114	A112 – 113
8	A113-114	B84-85	B84-85	A84-85	B70 – 71
9	B84-85	A113-114	A113-114	B88-89	B87 – 88
10	A84-85	B76-77	B76-77	B90-91	B102 – 103

Table 13. Crash Prone Segment Ranking Recapitulation (Part 2).

Rank	PM-6	PM-7	PM-8	PM-9	PM-10
1	B91-92	B91-92	B91-92	B91-92	B91-92
2	B92-93	A120-121	A101-102	A120-121	A101-102
3	A101-102	A101-102	B86-87	A101-102	A120-121
4	A120-121	B92-93	A113-114	B92-93	A112-113
5	A112-113	A116-117	A116-117	A116-117	A116-117
6	A116-117	B86-87	A120-121	B86-87	A113-114
7	B86-87	A113-114	B92-93	A113-114	A96-97
8	A113-114	A112-113	A96-97	A112-113	A84-85
9	B84-85	B103-104	A112-113	B103-104	A88-89
10	A84-85	A106-107	B84-85	A106-107	A108-109

Considering all existing performance measures and the frequency of ranking of each performance measure, here are the five most crash-prone segments on the Cipularang Toll Road section, using crash data and traffic data from 2018 to 2022.

1) B91 – 92 (KM 91 – 92 Jakarta-bound direction)

Figure 2. Crash Prone Segment KM 91 – 92 Jakarta-Bound Direction.

2) B92 – 93 (KM 92 – 93 Jakarta-bound direction)

Figure 3. Crash Prone Segment KM 92 – 93 Jakarta-Bound Direction.

3) A120 – 121 (KM 120 – 121 Bandung-bound direction)

Figure 4. Crash Prone Segment KM 120 – 121 Bandung-Bound Direction.

4) A101 – 102 (KM 101 – 102 Bandung-bound direction)

Figure 5. Crash Prone Segment KM 101 – 102 Bandung-Bound Direction.

5) B86 – 87 (KM 86 – 87 Jakarta-bound direction)

Figure 6. Crash Prone Segment KM 86 – 87 Jakarta-Bound Direction.

5. Discussions

Segment ranking is based on the frequency of fulfilling each performance measure and being in the top five positions. As previously explained, the segment higher up on the list will be considered as the segment which will most likely to benefit from preventive measures intended to reduce the frequency of crashes or severities.

Ranking must comply with existing performance measures by prioritizing the top five rankings in each ranking result. B91-92 segment occupies the first position because out of ten performance measures, this segment occupies the first position in eight performance measures. In the second rank is the B92-93 segment, which has the second highest rank for all performance measures.

In third and fourth place, segments A120-121 and A101-102 are in the same position. However, the A120-121 segment is ranked higher because the level of compliance with performance measures and its position in the top five rankings is higher than the A101-102 segment, as in the EPDO Average Crash Frequency and Excess Predicted Average Crash Frequency Using SPFs. Based on these considerations, the A120-121 segment received a higher ranking because it could benefit from the preventive measures provided.

The results of the ranking that have been carried out are in accordance with previous research. This shows that the location of crash-prone segments on the Cipularang Toll Road has stayed the same. It is also indicated that the treatment provided before has not been successful in dealing with crashes on the Cipularang Toll Road section.

6. Conclusions

Based on the results of ranking crash-prone segments using the network screening method, the most crash-prone segments on the Cipularang Toll Road section based on data from 2018 to 2022 are KM 91 – 92 Jakarta-bound direction, KM 92 – 93 Jakarta-bound direction, KM 120 – 121 Bandung-bound direction, KM 101 – 102 Bandung-bound direction, and KM 86 – 87 Jakarta-bound direction. In terms of previous studies, there has been an increase in the number of crashes and the number of crash victims. But the crash-prone locations are still between KM 91 to KM 93. This shows previous treatments applied to the locations were not able to improve the road safety condition that the locations remain to be the most crash-prone locations.

The use of the network screening method helps in determining crash-prone segments in a more comprehensive manner. Because with RTM bias considerations, all segments with lowest frequency can be identified properly. In addition, the existence of various performance measures used in the rating process helps to review the selection of crash-prone segments from several exposures used.

7. Recommendations

Suggestions that can be given for further research regarding the ranking of crash-prone segments is to carry out an analysis using the relationship between the number of crashes and the types of vehicles involved. Research can be continued by determining alternative recommendations for crash-prone segment.

8. Acknowledgments

This paper and research would not have gone smoothly without the help and direction of supervisors, Aine Kusumawati, S.T., M.T., Ph.d. and Dr. Eng. Widyarini Weningtyas, S.T., M.T. I would also like to thank PT Jasa Marga Purbaleunyi for their assistance in completing the data needed for this research.

9. References

[1] Herawati. (2014). Karakteristik dan Penyebab Kecelakaan Lalu Lintas di Indonesia Tahun 2012. *Warta Penelitian Perhubungan*, *26*(3), 133. https://doi.org/10.25104/warlit.v26i3.875
[2] American Association of State Highway and Transportation Officials. (2010). Highway Safety Manuals. AASHTO.
[3] Damanhuri, M. (2017). *Studi Identifikasi Lokasi Daerah Rawan Kecelakaan Lalu Lintas (Blackspot) pada Ruas Jalan Tol Cipularang Provinsi Jawa Barat.* Universitas Pendidikan Indonesia.
[4] Hanafi, H., Rusgiyarto, F., dan Pratama, R. (2020). Analisis Tingkat Keselamatan Jalan Tol Berdasarkan Metode Pembobotan Korlantas (Studi Kasus: Jalan Tol Cipularang). *Jurnal Teknik: Media Pengembangan Ilmu Dan Aplikasi Teknik*, *18*(2), 49. https://doi.org/10.26874/jt.vol18no2.106
[5] Setiawan, D., dan Asima, M. (2019). Pemetaan Risiko Kecelakaan Lalu Lintas di Ruas Jalan Tol Cipularang. *Jurnal Teknik Sipil*, *15*(2), 100–113. https://doi.org/10.28932/jts.v15i2.1923
[6] Sihombing, A. J., dan Widyastuti, H. (2021). Analisa Kecelakaan Lalu Lintas di Ruas Jalan Tol Cipularang, Purwakarta. *Jurnal Teknik ITS*, *9*(2). https://doi.org/10.12962/j23373539.v9i2.57996
[7] Kusumawati, A., dan Rakhmat, L. A. (2011). Modelling Traffic Crash Occurrence at Purbaleunyi Toll Road - Indonesia using Generalised Poisson Regression Model. *Proceedings of the 4th ASEAN Civil Engineering Conference2.*
[8] Rakhmat, L. A., Kusumawati, A., Frazila, R. B., dan Hendarto, S. (2013). Pengembangan Model Prediksi Kecelakaan Lalu Lintas pada Jalan Tol Purbaleunyi. *Jurnal Teknik Sipil*, *19*(3), 277. https://doi.org/10.5614/jts.2012.19.3.8
[9] Kusumawati, A., Upahita, D. P., dan Hendarto, S. (2013). Development of Crash Prediction Model for Rural Toll Road Sections. *Proceedings of EASTS, Vol 9, 9*, 1–15.
[10] Meydita, N., dan Kusumawati, A. (2013). Toll Road Traffic Crash Cost Analysis. *SIBE, 232–255.*

IOP Conf. Series: Earth and Environmental Science 1294 (2024) 012012 doi:10.1088/1755-1315/1294/1/012012

A review of accident data for traffic safety studies in Indonesia

N S Kusumastutie[1,2], B. Patria[1*], S. Kusrohmaniah[1], and T D Hastjarjo[1]

[1] Faculty of Psychology, Universitas Gajah Mada, Yogyakarta, Indonesia
[2] Road Transportation System Engineering Program, Politeknik Keselamatan Transportasi Jalan, Tegal, Indonesia
*patria@mail.ugm.ac.id

Abstract. Accident data is a crucial indicator of traffic safety. This study investigates the use of accident data in traffic safety studies in Indonesia. The analysis was conducted on studies that met the eligibility criteria, which included using traffic accident data in Indonesia and articles published in Indonesian and English language journals or proceedings. The search was conducted on four databases: Garuda, Neliti, Google Scholar, and Scopus. The final selection resulted in 50 articles. The data analysis indicates the need for improvement in the number of studies, the utilization of data elements in investigations, supplement data, and data analysis techniques. In addition to providing recommendations to optimize the use of traffic accident data in future studies, this study also emphasizes the importance of improving the accuracy of traffic accident data.

1. Introduction

Worldwide traffic safety is still very concerning. Traffic accidents have caused in 1.35 million fatalities yearly, making them the leading cause of death among children and young adults [1]. The same concern is also present in Indonesia. After experiencing a decrease from 116,411 in 2019 to 100,028 in 2020, traffic accidents rose again to 103,645 in 2021 [2]. The number of traffic accidents indicates traffic safety [3]. Although it is prone to errors [4] and has some other limitations [3], data on traffic accidents form the basis for most decisions to improve traffic safety in various countries [4]. Traffic accident data determines accident-prone locations and identifies factors contributing to accidents. This information is then utilized to implement interventions for enhancing traffic safety.

Kweon [5] mentioned two types of traffic accident data: traffic safety data and supplement data. Generally, police accident data is used as the primary source of traffic safety data. Additionally, hospital accident data and safety survey data are also included in it. In addition to the traffic safety data, supplement data can be used to enhance the understanding of factors contributing to the occurrence and outcomes of accidents. Supplement traffic data include roadway and traffic data, license and registration data, travel survey data, and sociodemographic and economic data.

Referring to the classification provided by Kweon [5] above, this review intends to investigate how this accident data is employed in traffic safety studies in Indonesia. These studies were obtained through a literature search conducted in several databases. The findings of this study are expected to offer valuable insights for upcoming research aimed at using traffic accident data to enhance traffic safety.

This study employed the traditional review approach. A traditional review aims to determine what has been achieved, enabling the consolidation and summarization of past efforts and identifying gaps [6]. Several traffic safety studies have used this approach [7]–[9]. However, it is worth noting that traditional reviews have distinct characteristics from systematic literature reviews [10]. These characteristics can be considered weaknesses of traditional reviews, making them prone to publication bias and the file-drawer problem [6], [11]. Nevertheless, when executed meticulously, this type of review can offer an extensive overview of the current state of research within a short time frame and at a low cost [11]. This review is based on Weed's guidelines [12] to address these weaknesses.

2. Methods

2.1. Literature search

The present study aimed to answer how previous studies have utilized traffic accident data in Indonesia. This review was conducted following the guidelines from Weed [12], which include determining the statement of purpose, search methods, criteria for evaluating the quality of studies, methods for summarizing evidence, and criteria for conclusion and recommendations. As mentioned above, using guidelines in writing a traditional review is necessary to avoid publication bias and the file-drawer problem.

In the literature search, several eligibility criteria were established. Studies had to involve utilizing traffic accident data within the Indonesian region. The traffic accident data classification referred to Kweon [5]. Studies needed to be published in journals or proceedings and written in Indonesian or English. The requirement for studies to be published in journals or proceedings served as a means to assess the quality of the research. No specific timeframe criteria were applied in the search.

The literature search was conducted in two stages. The first stage was conducted in October 2022 and was carried out in Garuda, Neliti, and Google Scholar databases. The keyword used was "data kecelakaan lalu lintas." For the search on Google Scholar, only results from the first five pages were considered. The second search stage was conducted in the Scopus database in July 2023. The keyword used was "traffic accident" AND "Indonesia." The purpose of the second stage was to complement the results of the first stage.

2.2. Data analysis

The articles obtained from the search were then selected based on the eligibility criteria. The relevant articles were abstracted and analyzed descriptively. The results were then depicted in tables and graphics. Furthermore, a narrative summary was provided to explain and relate the selected studies to the review objective. Conclusions and recommendations were then provided based on the findings.

3. Results

3.1. Overview of the studies

The search found 50 articles that met the eligibility criteria as the final result. These articles comprised 35 articles from the first-stage search (national publications) and 15 from the second-stage search (international publications). Table 1 shows the distribution of selected article sources. These studies range from 2008 to 2022 (Figure 1), indicating that the utilization of traffic accident data in road safety studies has been ongoing from 2008 to the present. Figure 2 depicts the geographical distribution of these studies. Most studies were conducted in regency areas (66%), with the fewest in sub-district areas (2%). Nationally-scaled studies were also limited (8%). The supplementary material, which includes summarized abstracts of the selected articles, is accessible at https://osf.io/r3ujp/.

Table 1. The distribution of selected article sources.

Sources	*N*	%	Selected studies
National proceedings	3	6	[13], [14], [15]
National journals	32	64	[16], [17], [18], [19], [20], [21], [22], [23], [24], [25], [26], [27], [28], [29], [30], [31], [32], [33], [34], [35], [36], [37], [38], [39], [40], [41], [42], [43], [44], [45], [46], [47]
International proceedings	7	14	[48], [49], [50], [51], [52], [53], [54]
International journals	8	16	[55], [56], [57], [58], [59], [60], [61], [62]

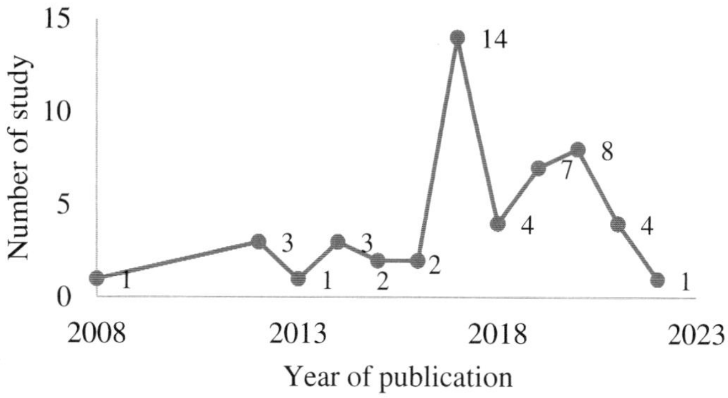

Figure 1. The number of studies based on the year of publication.

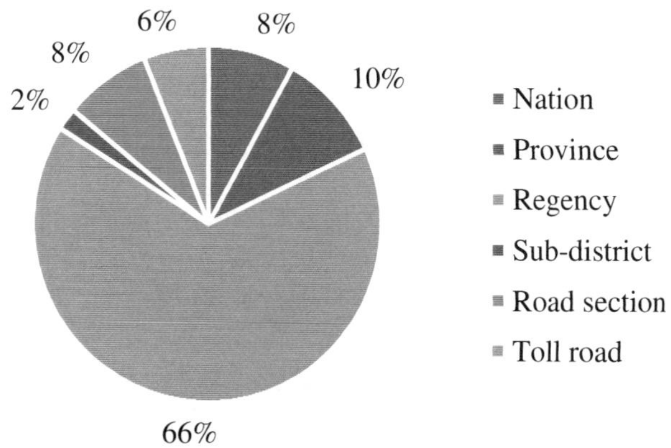

Figure 2. Area of the studies.

3.2. Traffic accident data

Figure 3 illustrates the sources of traffic accident data used in the selected studies. Traffic accident data was obtained from six institutions: the police, the Ministry of Transportation, the Komite Nasional Keselamatan Transportasi (abbreviated as KNKT, which stands for National Transportation Safety Committee), the Badan Pusat Statistik (abbreviated as BPS, which stands for Central Statistics Agency), toll road companies, and hospitals. Most of the data was sourced from recorded accidents by the police (n=38). Two studies referred to BPS data, but it should be noted that the traffic accident data in BPS publications originated from police data. Furthermore, four studies used data from multiple institutions, for instance, police and hospitals (*n*=2) and police and BPS (*n*=2).

Interestingly, some studies utilized specific accident data, such as accidents on toll roads (*n*=3) and accidents involving public transportation with specific criteria handled by the KNKT (*n*=1). The KNKT is a non-structural institution tasked with investigating traffic accidents involving public transportation, with specific criteria including a minimum of eight fatalities, attracting widespread public attention, causing significant damage to infrastructure, exhibiting repetition in vehicle types or specific locations, and resulting in environmental pollution [63].

One study [20] utilized the Health and Demographic Surveillance System (HDSS) survey data. The HDSS is a surveillance system that collects periodic data on population transitions, health status, and social transitions starting from 2015 [64]. HDSS is conducted by the Faculty of Medicine, Public Health, and Nursing at Universitas Gadjah Mada in collaboration with the Sleman District Government. This study [20] utilizes HDSS data to investigate the relationship between demographic characteristics and accident injuries. The result revealed that age is associated with accident injury status.

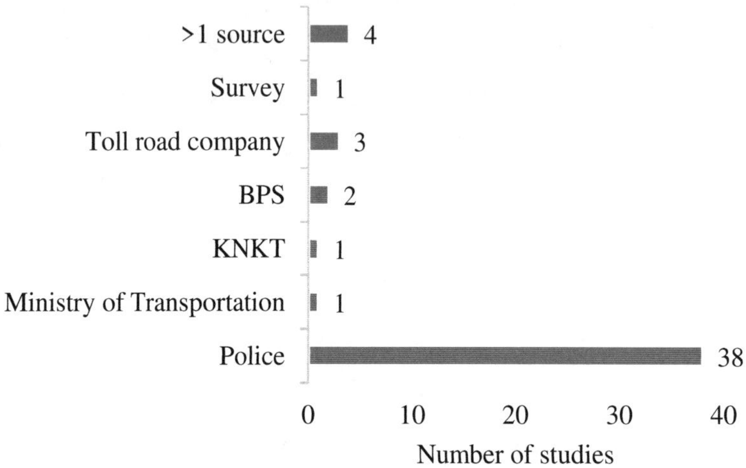

Figure 3. Data sources.

Figure 4 shows the data duration in the selected studies. Most studies utilized traffic accident data within a relatively long timeframe, ranging from three to five years. Some studies used data with a timeframe of more than five years (*n*=11), indicating consistency in data recording.

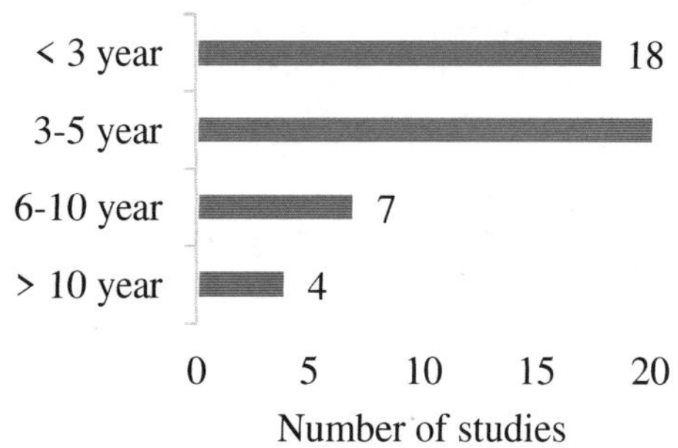

Figure 4. Data duration.

Referring to Kweon [5], the data elements are divided into information regarding location, environment, vehicle, driver, injury severity, and others. Location information includes province/regency names, road names, and location coordinates. Environment information includes time and date, weather conditions, lighting, crash types, road conditions and classifications, land use, and traffic conditions. Vehicle information includes vehicle types, brands, vehicle movements and maneuvers before the crash, and an estimate of property damage. Driver information depicts individuals involved in the accident, including drivers, motorcyclists, pedestrians, and cyclists. The data enclose gender, age, profession, address, education, marital status, economic status, role, and types of violations. Injury severity includes minor, moderate, severe, and fatalities. Lastly, the "others" category includes the number of accidents, chronology, and causes of the accidents. Table 2 displays the data elements used in the selected studies. Injury severity is the most common data element in the selected studies, while driver information is the least common.

Table 2. Data elements.

Data elements	Number of studies	Common variables
Location information	29	Road name
Environment information	30	Time and date
Vehicle information	24	Vehicle types
Driver information	20	Age and gender
Injury severity	39	Injury severity
Others	25	Number of accidents

Supplement data can be used to enhance understanding of the factors contributing to the occurrence and outcomes of accidents [5]. Seventeen (34%) studies utilized supplement data. Supplemental secondary data included annual traffic volume and geometric data obtained from toll road company databases (e.g., [48]). Supplemental observational data included traffic volume, road geometry, and speed (e.g., [24]).

3.3. Studies' outcomes

As shown in Figure 5, most studies utilized traffic accident data to determine the characteristics of traffic accidents (n=13) and identify black spots (n=15). Previous studies also frequently used traffic accident data to identify patterns of accidents (n=11). Studies that generate accident models to investigate factors related to accidents and calculate accident costs tend to receive less attention.

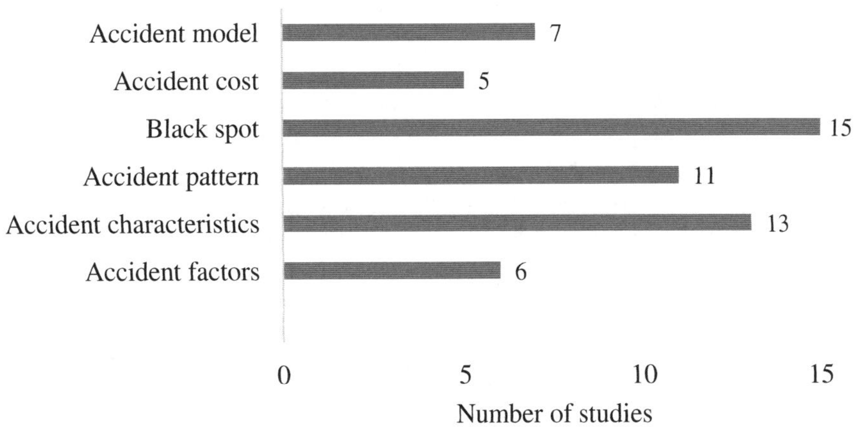

Figure 5. Previous studies' outcomes.

Previous studies described accident characteristics to provide an overview of the frequency of accidents based on specific criteria, such as the day or month of occurrence, time of occurrence, collision type, severity level, vehicle type, and the age and gender of road users involved (e.g., [16]). Generally, studies aiming to identify black spots or black sites also investigated these accident characteristics. Previous studies use several commonly used methods in determining black spots or black sites, including Z-score and Cussum (e.g., [31]) or Equivalent accident number and UCL (e.g., [61]). Previous studies attempted to identify accident patterns using data mining analysis, primarily employing K-means clustering. These analyses were conducted not only to cluster locations based on vulnerability (e.g., [14]) but also to cluster the occurrence of accidents based on time (e.g., [15]).

4. Discussion

This study investigates the utilization of accident data in traffic safety studies in Indonesia. The results showed that researchers' interest in using traffic accident data still needs to be improved. The first publication was in 2008, reached its highest number of publications in 2017, and experienced a decrease in the number until 2022. Previous studies' coverage areas were mainly local, while studies using national data were minimal.

Like other countries, the police are the primary source of traffic accident data [5]. The KNKT and toll road companies provide accident data according to their respective authorities. The use of survey data needs to be improved, as only one study utilizing HDSS data has been found so far [20]. Moreover, this survey is regional in scale, not national, and not explicitly focused on traffic accidents [64]. The presence of studies using data with a timeframe of more than ten years indicates that the documentation has been sustained. It is worth noting that traffic accident data recorded in BPS publications date back to 1992 [2].

Furthermore, only a few studies utilize supplement data and optimize the various available data elements in accident data. Supplement data is typically used in studies to identify black spots, involving traffic observations and road conditions at specific locations to determine appropriate interventions. The utilization of driver data is relatively low compared to other data elements. Although supplementing data and using various data elements is driven by the research questions, considering both in data analysis may provide a more comprehensive understanding.

Moreover, only a limited number of studies have specifically analyzed accidents involving certain vehicle types, with four studies focusing on motorcycles [20], [41], [59], [62] and one study on buses [43]. This result indicates a need for more attention to specific traffic accident cases to enable in-depth analysis. Specific analysis has yet to be found for other vulnerable road users, such as pedestrians.

Based on the review results, several recommendations are suggested to optimize the use of traffic accident data in future studies. Firstly, there is a need for an accessible database for various traffic safety studies, an example, similar to the Highway Safety Information System (HSIS) in the United States [5]. This database should meet various studies' data needs, from policy studies to engineering and behavioral studies. It should be nationally integrated and incorporate multiple sources.

Traffic accident data documented by the police is already accessible to the public through publications by BPS [2]. The KNKT regularly provides investigation reports on its website [65]. These publications should be further enhanced to be more "researcher-friendly," such as by providing comprehensive data elements for researchers to request access. The provision of disaggregated datasets should, of course, comply with research ethics regulations. The relevant agency's website can also display basic aggregate data analysis, such as frequency, rate, and trend analysis [5]. These analyses can be added to the traffic accident statistics displayed on several agency websites [66]–[68]. This way, researchers and the public can quickly obtain a general overview of traffic safety.

Moreover, the utilization of traffic survey data also needs to be optimized. In addition to HDSS, there is The Indonesian Family Life Survey (IFLS), a national-scale longitudinal survey, with this sample representing approximately 83% of the Indonesian population [69]. Questions related to traffic accident experiences can be added to the survey questionnaire. Furthermore, accidents recorded in self-reports can be linked to official databases to enhance data validity [70]. However, researchers need to be mindful of the limitations of self-reporting, such as reporting biases. They must take appropriate measures to address social desirability and incorrect memory recall.

Second, considering that traffic accident data is rich in the necessary information to be used as a basis for improving traffic safety, its utilization needs to be optimized. Future studies analyze specific collision types involving specific road users (such as vulnerable road users like pedestrians or cyclists) or specific locations (such as urban areas or intersections). Generally, vulnerable road users have received less attention from researchers [70]. Some previous studies were found to be related to motorcycle accidents, as this type of accident is dominant in Indonesia[1].

The use of supplement data should be encouraged. For example, in studies on interventions in black spots, the observation of traffic conflicts can be used as supplement data along with observations of road and traffic conditions. The traffic conflict technique shows promise as a complementary indicator to accident data, and it has also been applied in several studies in Indonesia [71].

Third, there is a need for appropriate data analysis techniques. Various studies have employed data mining techniques for examining accident data, encompassing classification techniques like decision trees, clustering techniques, and association rule mining [72]. For example, to tackle heterogeneity in accident data, which may conceal specific patterns, Depaire et al. [73] applied latent class clustering to detect more homogeneous traffic accident categories. After forming the clusters, injury analysis is performed for each cluster. De Oña et al. [74] also recommend using two or more data analysis techniques. The data heterogeneity can also be addressed by improving data quality through noise reduction, such as by using an algorithm called NoiseCleaner, suggested by Deb and Lew [75]. The review conducted by Gutierrez-Osorio and Pedraza [7] on using analytic algorithms and machine learning methods in traffic safety studies can be further referenced.

Lastly, it is important to underline caution when utilizing traffic accident data. Several studies have raised issues regarding the accuracy of accident data [4], [70], [76], [77]. There are two types of errors in accident data: reporting and recording errors [4], [77]. Several studies have found that the most inaccurate data pertains to road characteristics, severity levels of accidents, and accident locations. Since traffic accident data forms the backbone of safety systems, errors can render safety improvement efforts ineffective [4]. With these errors, accident patterns become more challenging to identify, the identification of accident-prone locations becomes inaccurate, and the determination of parameters responsible for accidents becomes imprecise.

Several efforts can be made to address the potential errors in accident data. Improving the reporting and recording system of accident data can be achieved by enhancing the skills and motivation of personnel, automating the recording process, utilizing modern tools, and increasing public awareness to report accidents to the police [4]. Additionally, the black box in vehicles can be utilized to assess the accuracy of accident data [77].

The study has provided an overview of how accident data is used in traffic safety studies in Indonesia and offers insights for future research. However, the study has limitations, primarily due to its non-systematic nature. It is an initial exploration of utilizing accident data and should be followed up with a systematic literature review. Additionally, insights from using accident data in other countries should be sought.

The limitations of the accessed database, chosen keywords, and the exclusion of gray literature are also limitations of this study. These limitations have resulted in a limited number of analyzed studies. Nevertheless, the findings of this study offer valuable insights and can be followed up with a systematic literature review, as suggested above.

5. Conclusion

This study has provided an overview of the use of accident data in traffic safety studies in Indonesia. The findings show the need for improvement in the number of studies, using data elements in investigations, supplement data, and data analysis techniques. This study has also provided recommendations, particularly regarding the need for a comprehensive database that is easily accessible to the public and researchers and optimizing the utilization of traffic accident data. Finally, this study emphasizes the importance of improving the accuracy of traffic accident data.

Acknowledgements
NSK would like to acknowledge the financial support provided by the Indonesia Endowment Fund for Education (LPDP) of the Ministry of Finance of the Republic of Indonesia through doctoral scholarships and research funding.

Avability of data and material
All data generated and/or analyzed during this study are included in this publication. Any unpublished material is available from the corresponding author upon request.

Disclosure statement
No potential conflict of interest was reported by the authors.

References

[1] World Health Organization, *Global Status Report On Road Safety 2018*. Geneva: World Health Organization, 2018.

[2] Badan Pusat Statistik, "*Jumlah Kecelakaan, Korban Mati, Luka Berat, Luka Ringan, dan Kerugian Materi 1992-1994*," 2023. https://bps.go.id/indicator/17/513/10/jumlah-kecelakaan-korban-mati-luka-berat-luka-ringan-dan-kerugian-materi.html (accessed Jul. 10, 2023).

[3] A. Laureshyn and A. Varhelyi, "*The Swedish Traffic Conflict Technique Observer's Manual.*" Lund University, Lund, 2018.

[4] A. Ahmed, A. F. M. Sadullah, and A. S. Yahya, "Errors in accident data, its types, causes and methods of rectification-analysis of the literature," *Accid. Anal. Prev.*, vol. 130, no. February 2017, pp. 3–21, 2019, doi: 10.1016/j.aap.2017.07.018.

[5] Y.-J. Kweon, "Crash data sets and analysis," in *Handbook of Traffic Psychology*, B. E. Porter, Ed. London: Academic Press, 2011, pp. 97–105.

[6] M. J. Grant and A. Booth, "A typology of reviews: An analysis of 14 review types and associated methodologies," *Health Info. Libr. J.*, vol. 26, no. 2, pp. 91–108, 2009, doi: 10.1111/j.1471-1842.2009.00848.x.

[7] C. Gutierrez-Osorio and C. Pedraza, "Modern data sources and techniques for analysis and forecast of road accidents: A review," *J. Traffic Transp. Eng. (English Ed.*, vol. 7, no. 4, pp. 432–446, 2020, doi: 10.1016/j.jtte.2020.05.002.

[8] W. Young, A. Sobhani, M. G. Lenné, and M. Sarvi, "Simulation of safety: A review of the state of the art in road safety simulation modelling," *Accid. Anal. Prev.*, vol. 66, pp. 89–103, 2014, doi: 10.1016/j.aap.2014.01.008.

[9] S. M. S. Mahmud, L. Ferreira, M. S. Hoque, and A. Tavassoli, "Application of proximal surrogate indicators for safety evaluation: A review of recent developments and research needs," *IATSS Res.*, vol. 41, no. 4, pp. 153–163, 2017, doi: 10.1016/j.iatssr.2017.02.001.

[10] M. D. J. Peters, C. M. Godfrey, P. McInerney, Z. Munn, A. C. Tricco, and H. Khalil, "Chapter 11: Scoping Reviews," in *JBI Manual for Evidence Synthesis*, E. Aromataris and Z. Munn, Eds. JBI, 2020.

[11] M. Blettner, W. Sauerbrei, B. Schlehofer, T. Scheuchenpflug, and C. Friedenreich, "Traditional reviews, meta-analyses and pooled analyses in epidemiology," *Int. J. Epidemiol.*, vol. 28, no. 1,

pp. 1–9, 1999, doi: 10.1093/ije/28.1.1.

[12] D. L. Weed, "Methodologic guidelines for review papers," *JNCI J. Natl. Cancer Inst.*, vol. 89, no. 1, pp. 6–7, 1997, doi: 10.1093/jnci/89.1.6.

[13] N. N. Hayati and S. Sulistyono, "Kecelakaan lalu lintas pada Jalur Pantura Surabaya-Tuban," in *Seminar Nasional Teknik Sipil*, 2012, no. March, pp. 251–261.

[14] M. A. Ismail, "Aplikasi data mining dalam pengelompokan korban kecelakaan lalu lintas dengan algoritma k-medoids berbasis desktop," in *Prosiding SNASTIKOM: Seminar Nasional Teknologi Informasi & Komunikasi*, 2021, pp. 28–34.

[15] E. Wicaksono, Kusrini, and E. T. Lutfi, "Analisis data kerawanan kecelakaan lalu lintas menggunakan metode k-means (Studi kasus Polres Bantul)," in *Seminar Nasional Teknologi Informasi dan Multimedia 2017*, 2017, vol. 5, no. 1, pp. 110–114.

[16] K. B. Ajie, H. Sulistio, and A. Suharyanto, "Integrasi sistem data kecelakaan lalu lintas di Kabupaten Jember," *Rekayasa Sipil*, vol. 11, no. 1, pp. 57–64, 2017, doi: 10.21776/ub.rekayasasipil.2017.011.01.8.

[17] E. Haryono, A. Widodo, and S. Abusini, "Kajian model automatic clustering-fuzzy time series-markov chain dalam memprediksi data historis jumlah kecelakaan lalu lintas di Kota Malang," *J. Sains Dasar*, vol. 2, no. 1, pp. 63–71, 2013.

[18] L. Iswari and E. G. Ayu, "Pemanfaatan algoritma k-means untuk pemetaan hasil klasterisasi data kecelakaan lalu lintas," *Teknoin*, vol. 21, no. 1, pp. 1–13, 2015, doi: 10.20885/teknoin.vol21.iss1.art7.

[19] I. R. Maharsi, M. A. Mukid, and Y. Wilandari, "Peramalan jumlah kecelakaan di Kota Semarang tahun 2017 menggunakan metode runtun waktu," *J. Gaussian*, vol. 6, no. 3, pp. 301–316, 2017.

[20] A. T. Mariana and F. S. T. Dewi, "Cedera akibat kecelakaan lalu lintas di Sleman, DIY: Data HDSS 2015 dan 2016," *Ber. Kedokt. Masy.*, vol. 34, no. 6, pp. 230–235, 2018.

[21] A. D. Saputra, "Studi tingkat kecelakaan lalu lintas jalan di Indonesia berdasarkan data KNKT (Komite Nasional Keselamatan Transportasi) dari tahun 2007-2016," *War. Penelit. Perhub.*, vol. 29, no. 2, pp. 179–190, 2017, doi: 10.1016/j.injury.2011.11.002.

[22] R. Saragih, "Implementasi apriori pada data kecelakaan lalu lintas dalam pencarian relasi antar variabel pelaku LAKA," *Algoritm. J. Ilmu Komput. dan Inform.*, vol. 6341, no. November, pp. 12–17, 2017.

[23] R. Saragih and J. N. Sitompul, "Analisis perbandingan data mining mengidentifikasi pola keterkaitan variabel kecelakaan lalu lintas di Polresta Kota Medan," *J. Inf. Syst. Dev.*, vol. 4, no. 1, pp. 39–45, 2019, [Online]. Available: https://ejournal.medan.uph.edu/index.php/isd/article/view/281.

[24] Y. Y. N. Adlina and S. Nurlaela, "Analisis faktor kecelakaan lalu lintas Surabaya berdasarkan perspektif tata ruang melalui pemodelan spasial," *J. Tek. ITS*, vol. 10, no. 1, pp. 13–20, 2021.

[25] I. F. Anshori and Y. Nuraini, "Pengelompokan data kecelakaan lalu lintas di Kota Tasikmalaya menggunakan algoritma k-means," *J. Responsif Ris. Sains dan Inform.*, vol. 2, no. 1, pp. 118–127, 2020, doi: 10.51977/jti.v2i1.198.

[26] W. Aprianti and J. Permadi, "K-means clustering untuk data kecelakaan lalu lintas jalan raya di Kecamatan Pelaihari," *J. Teknol. Inf. dan Ilmu Komput.*, vol. 5, no. 5, p. 613, 2018, doi: 10.25126/jtiik.2018551113.

[27] A. M. Suswanto and A. A. G. Kartika, "Perencanaan jalan berkeselamatan di Kabupaten Tuban," *J. Tek. ITS*, vol. 10, no. 2, pp. 265–272, 2021.

[28] N. N. Pinata, I. M. Sukarsa, and N. K. D. Rusjayanti, "Prediksi kecelakaan lalu lintas di Bali dengan XGBoost pada Python," *J. Ilm. Merpati (Menara Penelit. Akad. Teknol. Informasi)*, vol. 8, no. 3, p. 188, 2020, doi: 10.24843/jim.2020.v08.i03.p04.

[29] R. Fitria, W. Nengsih, and D. H. Qudsi, "Implementasi algoritma FP-Growth dalam penentuan pola hubungan kecelakaan lalu lintas," *J. Sist. Informasu*, vol. 2, no. 13, pp. 118–124, 2017.

[30] D. Wicaksono, R. A. Fathurochman, B. Riyanto, and Y. I. Wicaksono, "Analisis kecelakaan lalu lintas (Studi kasus - Jalan Raya Ungaran - Bawen)," *J. KARYA Tek. SIPIL*, vol. 3, no. 2, pp. 345–355, 2014.

[31] I. G. . P. A. Swari, P. A. Suthanaya, and I. N. W. Negara, "Analisis biaya dan penanganan lokasi rawan kecelakaan akibat kecelakaan lalu lintas di Kota Denpasar," *J. Spektran*, vol. 2, no. 2, pp. 24–30, 2014, doi: 10.24843/spektran.2014.v02.i02.p04.

[32] G. Sugiyanto, B. Mulyono, and M. Y. Santi, "Karakteristik kecelakaan lalu lintas dan lokasi black spot di Kabupaten Cilacap," *J. Tek. Sipil*, vol. 12, no. 4, pp. 259–266, 2014, doi: 10.24002/jts.v12i4.634.

[33] A. S. Aribowo, "Analisa asosiatif data mining untuk mengetahui pola kecelakaan lalu lintas," *Telematika*, vol. 8, no. 2, pp. 2–7, 2015, doi: 10.31315/telematika.v8i2.458.

[34] T. Yandi, F. Lubis*, and Winayati, "Analisis karakteristik kecelakaan lalu lintas pada Jalan Yos Sudarso Kota Pekanbaru," *J. Tek.*, vol. 14, no. 1, pp. 17–21, 2020, doi: 10.31849/teknik.v14i1.3141.

[35] S. U. Utama, R. Magetsari, and V. Pribadi, "Estimasi prevalensi kecelakaan lalu lintas dengan metode capture-recapture," *Ber. Kedokt. Masy.*, vol. 24, no. 1, pp. 16–26, 2008, [Online]. Available: http://journal.ugm.ac.id/index.php/bkm/article/view/3603.

[36] T. K. Titus and M. Jajuli, "Clustering data kecelakaan lalu lintas di Kecamatan Cileungsi menggunakan metode k-means," *Gener. J.*, vol. 6, no. 1, pp. 1–12, 2022, doi: 10.29407/gj.v6i1.16103.

[37] G. Sugiyanto and A. Fadli, "Identifikasi lokasi rawan kecelakaan lalu lintas (blackspot) di Kabupaten Purbalingga, Jawa Tengah," *J. Tek. Sipil dan Perenc.*, vol. 19, no. 2, pp. 128–135, 2017.

[38] N. Utomo, "Analisa faktor penyebab kecelakaan lalu lintas pada segmen Jalan By-Pass Krian-Balongbendo (Km. 26+000-Km. 44+520)," *J. Tek. Sipil KERN*, vol. 2, no. 2, pp. 73–84, 2012, [Online]. Available: http://ejournal.upnjatim.ac.id/index.php/kern/article/download/1370/1135.

[39] F. Carina, "Analisis karakteristik kecelakaan dan penanganan lokasi rawan kecelakaan lalu lintas di Kota Palembang," *J. Penelit. Dan Kaji. Tek. Sipil*, vol. 5, no. 1, pp. 24–31, 2017.

[40] M. E. Bolla, J. T. R. N. Blegur, and R. Raman, "Analisis karakteristik dan biaya kecelakaan lalu

lintas di Kota Kupang," *J. Tek. Sipil*, vol. IV, no. 1, pp. 53–64, 2015.

[41] A. A. Anisarida and W. Santosa, "Korban kecelakaan lalu lintas sepeda motor di Kota Bandung," *J. HPJI*, vol. 5, no. 2, pp. 129–136, 2019, doi: 10.26593/jh.v5i2.3373.129-136.

[42] A. D. Ermawati, G. Sugiyanto, and E. W. Indriyati, "Penentuan lokasi rawan kecelakaan lalu lintas dengan pendekatan fasilitas perlengkapan jalan di Kabupaten Purbalingga," *Din. Rekayasa*, vol. 15, no. 1, p. 65, 2019, doi: 10.20884/1.dr.2019.15.1.258.

[43] I. Farida, W. Santosa, and A. C. Sutandi, "Karakteristik dan biaya kecelakaan lalu lintas di Kabupaten Garut," *J. Transp.*, vol. 19, no. 2, pp. 143–150, 2019, doi: 10.26593/jt.v19i2.3471.143-150.

[44] K. Imtihan and H. Fahmi, "Analisis dan perancangan sistem informasi daerah rawan kecelakaan dengan menggunakan Geographic Information Systems (GIS)," *J. Manaj. Inform. Sist. Informasi)*, vol. 3, no. 1, pp. 2614–1701, 2020, [Online]. Available: http://e-journal.stmiklombok.ac.id/index.php/misi.

[45] A. Rompis, J. Mallo, and D. Tomuka, "Kematian akibat kecelakaan lalu lintas Kota Tomohon tahun 2012-2014," *e-CliniC*, vol. 4, no. 1, 2016, doi: 10.35790/ecl.4.1.2016.10837.

[46] M. Mahmudi, M. Fahmi, N. Elfurqany, S. Sarah, and A. Rusyana, "Analisis korespondensi pada korban kecelakaan lalu lintas berdasarkan provinsi di Indonesia," *JMPM J. Mat. dan Pendidik. Mat.*, vol. 4, no. 1, pp. 49–60, 2019, [Online]. Available: http://journal.unipdu.ac.id:8080/index.php/jmpm/article/view/1617.

[47] L. Susiana, I. T. Utami, and J. Junaidi, "Penerapan metode boosting pada cart untuk mengklasifikasikan korban kecelakaan lalu lintas di Kota Palu," *Nat. Sci. J. Sci. Technol.*, vol. 8, no. 2, pp. 106–109, 2019, doi: 10.22487/25411969.2019.v8.i2.13536.

[48] A. Irfan, R. Al Rasyid, and S. Handayani, "Data mining applied for accident prediction model in Indonesia toll road," *AIP Conf. Proc.*, vol. 1977, 2018, doi: 10.1063/1.5043013.

[49] G. Sugiyanto, Y. A. Prasetyo, E. W. Indriyati, Y. Yanto, and M. Y. Santi, "Determining of black spot location in Purbalingga Regency using road geometric approach," *IOP Conf. Ser. Earth Environ. Sci.*, vol. 847, no. 1, 2021, doi: 10.1088/1755-1315/847/1/012019.

[50] A. Sandhyavitri, Zamri, S. Wiyono, and Subiantoro, "Three strategies reducing accident rates at black spots and black sites road in Riau Province, Indonesia," in *Transportation Research Procedia*, 2017, vol. 25, pp. 2153–2166, doi: 10.1016/j.trpro.2017.05.415.

[51] R. Watrianthos, M. Bobbi Kurniawan, Kusmanto, S. Budiman, and B. Ulya, "Mapping of traffic accidents in Labuhanbatu Regency using GIS support," *J. Phys. Conf. Ser.*, vol. 1566, no. 1, 2020, doi: 10.1088/1742-6596/1566/1/012104.

[52] A. Sandhyavitri, J. Wira, and A. Martin, "Green technology as a strategy in managing the black spots in Siak Highway, Indonesia," *IOP Conf. Ser. Mater. Sci. Eng.*, vol. 345, no. 1, 2018, doi: 10.1088/1757-899X/345/1/012037.

[53] K. Syahputri, R. M. Sari, I. Rizkya, U. Tarigan, I. Siregar, and T. A. Farhan, "Clustering the vulnerability of traffic accidents in Medan city with fuzzy c-means algorithm," *IOP Conf. Ser. Mater. Sci. Eng.*, vol. 801, no. 1, 2020, doi: 10.1088/1757-899X/801/1/012030.

[54] W. Budiawan, S. Saptadi, Sriyanto, C. Tjioe, and T. Phommachak, "Traffic accident severity prediction using naive bayes algorithm - A case study of Semarang Toll Road," *IOP Conf. Ser. Mater. Sci. Eng.*, vol. 598, no. 1, 2019, doi: 10.1088/1757-899X/598/1/012089.

[55] D. Supriyatno, S. W. Mudjanarko, and H. Dwiatmoko, "Traffic accident model reviewed from the factors of driving behaviour of Surabaya-Gempol Toll Road," *Int. J. Integr. Eng.*, vol. 12, no. 8, pp. 278–287, 2020.

[56] S. P. Santosa, A. I. Mahyuddin, and F. G. Sunoto, "Anatomy of injury severity and fatality in Indonesian rraffic accidents," *J. Eng. Technol. Sci.*, vol. 49, no. 3, pp. 412–422, 2017.

[57] A. Jusuf, I. P. Nurprasetio, and A. Prihutama, "Macro data analysis of traffic accidents in Indonesia," *J. Eng. Technol. Sci.*, vol. 49, no. 1, pp. 132–143, 2017.

[58] G. Sugiyanto, "The cost of traffic accident and equivalent accident number in developing countries (Case study in Indonesia)," *ARPN J. Eng. Appl. Sci.*, vol. 12, no. 2, pp. 389–397, 2017.

[59] P. A. Suthanaya, "Analysis of fatal accidents involving motorcycles in low income region (Case Study of Karangasem Region, Bali-Indonesia)," *Int. J. Eng. Res. Africa*, vol. 19, no. June, pp. 112–122, 2016, doi: 10.4028/www.scientific.net/JERA.19.112.

[60] G. Sugiyanto, R. Suciningtyas, Jajang, and M. Y. Santi, "Strategies for handling black spot area to increased road safety," *ARPN J. Eng. Appl. Sci.*, vol. 15, no. 9, pp. 1101–1110, 2020.

[61] G. Sugiyanto, A. Fadli, and M. Y. Santi, "Identification of black spot and equivalent accident number using Upper Control Limit method," *ARPN J. Eng. Appl. Sci.*, vol. 12, no. 2, pp. 528–535, 2017.

[62] S. Abusini and L. Ambarwati, "Motorcycle accident model on the road section of highlands region by using Genelarized Linear Model," *Int. J. Civ. Eng. Technol.*, vol. 8, no. 10, pp. 1249–1258, 2017.

[63] Pemerintah Republik Indonesia, "Peraturan Pemerintah Republik Indonesia Nomor 62 Tahun 2013." 2013.

[64] Faculty of Medicine Publik Health and Nursing, "Tentang HDSS Sleman," 2023. https://hdss.fk.ugm.ac.id/tentang-hdss-sleman/#toggle-id-8 (accessed Jul. 10, 2023).

[65] Komite Nasional Keselamatan Transportasi, "Laporan Investigasi," 2023. https://knkt.go.id/investigasi (accessed Jul. 11, 2023).

[66] Badan Pusat Statistik, "Tabel Dinamis Subjek Transportasi," 2023. https://bps.go.id/subject/17/transportasi.html#subjekViewTab5, (accessed Jul. 11, 2023).

[67] Kepolisian Republik Indonesia, "Statistika Laka Lantas," 2023. https://pusiknas.polri.go.id/laka_lantas (accessed Jul. 11, 2023).

[68] Komite Nasional Keselamatan Transportasi, "Laporan dan Informasi Statistik," 2023. https://knkt.go.id/statistik (accessed Jul. 11, 2023).

[69] The Indonesia Family Life Survey, "The Indonesia Family Life Survey (IFLS)," 2023. https://www.rand.org/well-being/social-and-behavioral-policy/data/FLS/IFLS.html (accessed Jul. 10, 2023).

[70] N. A. Kamaluddin, C. S. Andersen, M. K. Larsen, K. R. Meltofte, and A. Várhelyi, "Self-reporting traffic crashes-a systematic literature review," *Eur. Transp. Res. Rev.*, vol. 10, no. 26, pp. 1–18, 2018, [Online]. Available: https://doi.org/10.1186/s12544-018-0301-0.

[71] N. S. Kusumastutie and P. Rusmandani, "A Brief Review: Traffic Conflict Techniques and The Challenges of The Studies in Indonesia," *MATEC Web Conf. conCERN-2 2018*, vol. 270, pp. 1–6, 2019, doi: 10.1051/matecconf/201927003004.

[72] M. John and H. Shaiba, "Apriori-based algorithm for Dubai road accident analysis," *Procedia Comput. Sci.*, vol. 163, pp. 218–227, 2019, doi: 10.1016/j.procs.2019.12.103.

[73] B. Depaire, G. Wets, and K. Vanhoof, "Traffic accident segmentation by means of latent class clustering," *Accid. Anal. Prev.*, vol. 40, no. 4, pp. 1257–1266, 2008, doi: 10.1016/j.aap.2008.01.007.

[74] J. De Oña, G. López, R. Mujalli, and F. J. Calvo, "Analysis of traffic accidents on rural highways using Latent Class Clustering and Bayesian Networks," *Accid. Anal. Prev.*, vol. 51, pp. 1–10, 2013, doi: 10.1016/j.aap.2012.10.016.

[75] R. Deb and A. W. C. Liew, "Noisy values detection and correction of traffic accident data," *Inf. Sci. (Ny).*, vol. 476, pp. 132–146, 2019, doi: 10.1016/j.ins.2018.10.002.

[76] R. Elvik and A. B. Mysen, "Incomplete accident reporting: Meta-analysis of studies made in 13 countries," *Transp. Res. Rec.*, vol. 1665, no. 99, pp. 133–140, 1999.

[77] Y. Chung and I. Chang, "How accurate is accident data in road safety research? An application of vehicle black box data regarding pedestrian-to-taxi accidents in Korea," *Accid. Anal. Prev.*, vol. 84, pp. 1–8, 2015, doi: 10.1016/j.aap.2015.08.001.

IOP Conf. Series: Earth and Environmental Science 1294 (2024) 012013 doi:10.1088/1755-1315/1294/1/012013

Addressing Indonesia's biggest road safety challenge: Reducing motorcycle deaths

Kardina N.S Ayuningtyas[1,2], Raphael Grzebieta[1,3], Jake Olivier[1,4], Carlo Caponecchia[1]

[1] Transport and Road Safety Research Centre, School of Aviation, UNSW Sydney, High Street, Kensington, NSW 2052, Australia
[2] Civil Engineering Department, Institut Teknologi Bandung, Jalan Ganesha 10, 40132, Bandung, Indonesia,
[3] Department of Forensic Medicine, Monash University at the Victorian Institute of Forensic Medicine, 65 Kavanagh Street, Southbank Victoria 3006, Australia,
[4] Department of Statistics, School of Mathematics and Statistics, UNSW Sydney, High Street, Kensington, NSW 2052, Australia

k.ayuningtyas@unsw.edu.au

Abstract. Motorcyclists are the largest contributor to road fatalities in Indonesia, with the main causal factors found to be speeding and not wearing a helmet. One of the most efficient ways to reduce road deaths in Indonesia would be to reduce speeding behaviour and increase helmet wearing by strengthening speed and helmet wearing enforcement combined with education. Stronger enforcement combined with education have been shown to be highly effective in reducing fatalities in other countries. This study assesses the benefits of stronger enforcement on speed limit and helmet wearing. It also explores what are the most effective strategies to strengthen enforcement using fixed and mobile speed cameras, point to point speed cameras and AI technology that can be used to enforce helmet wearing rates. The paper concludes with an estimation of the potential benefits if such stronger enforcement were implemented in Indonesia.

1. Introduction

Road crashes are the primary cause of fatalities in Indonesia for children and young adults (1). Among all road users in Indonesia, motorcycles account for the largest contributor to fatalities at 81%, followed by cars at 8% (1). Moreover, national crash data (1) shows that 68.74% of motorcyclists killed in fatal road crashes did not have a motorcycle license.

National data regarding contributing factors in road crashes for Indonesia as a whole is not available whereas there is data available for Bandung City (2). The primary causal factor for road crashes involving injury or death in Bandung City was found to be speeding, accounting for around 25% of all crashes (2). Neglecting to give way to pedestrians was also a major contributor accounting for 20% of fatalities (2). A 2019 study by Kusumawati et al. showed that 40% of commuting motorcycles exceeded the speed limit on national six-lane two-way divided roads in urban areas. The proportion of speeding motorcycles between the hours of midnight and 6 am reached 63.21% (3). Additionally, Ayuningtyas et al. found that 56-71% motorcyclists had been speeding on urban roads, whereas only 8-18% had been speeding on rural roads (4). Considering that speeding is a major causal factor of crashes and motorcyclist fatalities, any reductions in speeding behaviour would likely reduce fatalities not only in Bandung City but in Indonesia generally.

In terms of reducing speeding behaviour and ensuring riders travel within speed limits, it is also important to consider appropriate speed limits as a fundamental aspect of enhancing road safety. Speed limits need to be carefully chosen to balance the need for mobility function and the human body's tolerance to kinetic energy in the event of a crash, taking into account road user interactions, potential crash types, road infrastructure, and other controlling aspects, e.g., weather. However, there has been some community debate regarding speed limits and how setting a higher speed limit can potentially yield increased societal financial benefits through reduced travel time. On the other hand, it is important to note that setting higher speed limits increases the risk of road crashes, leading to increased fatalities and serious injuries (5, 6). Therefore, setting acceptable 'safe' speed limit values and ensuring they are complied with, play a vital role in reducing fatalities among motorcyclists and improving overall road safety in Indonesia.

Personal protection for motorcyclists also appears to be another aspect of road safety contributing to Indonesia's road casualties. Interestingly, the issuing of speeding fines for motorcyclists who exceed the speed limit are closely linked to the use of helmets, which are designed to enhance the safety of riders and pillions. The effectiveness of helmets as a safety measure has been proven and well-establish (7). Correct helmet usage has been shown to reduce the risk of deaths by 42% and head injuries by 69% (8). In Indonesia, there is a law mandating the use of helmets for motorcyclists (9). However, the data indicates that the wearing rates of helmets among riders in Indonesia are around 71% (10). Thus, encouraging and enforcing the correct use of helmets can also play a significant role in reducing injuries and fatalities among motorcyclists in Indonesia.

Considering that motorcyclist crashes are a primary contributor to traffic fatalities in Indonesia, where speeding, not wearing a helmet and unlicensed riding appear to be major causal factors, addressing these issues may have significant injury/fatality reduction potential. Additionally, Indonesia's road safety plan has targets to ensure that half of all drivers and riders comply with speed limits, and all riders wear helmets that meet the Indonesian helmet standard by 2030 (1). The strategy includes specific campaigns targeting seven main risk factors, including helmet usage and speeding among them. The implementation of speed limit enforcement measures, along with continuous monitoring and evaluation, is also part of the strategy to enhance road safety (1).

Regarding travel speed, Indonesia's road safety strategy includes the development of guidelines for safe speed limits, enforcement measures to ensure compliance with speed limits, continuous monitoring of speed limits, and a speed control program for national roads and toll roads. Nevertheless, specific details regarding how these guidelines will be developed or what policies will be implemented, has not yet been provided.

One of the most effective ways to reduce speeding behaviour and increase helmet wearing rates is through stronger enforcement (11, 12). Studies have shown that riders indicated that the fear of being caught was the most significant factor influencing decision-making regarding driving speed and travelling within the speed limit (13). Moreover, evidence from several countries worldwide supports the notion that speeding behaviour significantly decreases and helmet usage increases when strong enforcement measures are in place. The effectiveness of enforcement is further enhanced when combined with education awareness programs (14). An additional likely benefit of stronger enforcement would be the reduction of unlicenced riders.

The primary objective of this paper is to explore previous research from the general literature, assessing the benefits and effectiveness of stronger enforcement on speed limits and helmet wearing compliance, particularly for fatality and serious injury reductions. Moreover, this study will assess the benefits of reducing speed limits and explore possible enforcement strategies for promoting compliance with speed limits and helmet usage, based on best practice from other countries.

While it should be noted that road fatalities are influenced by numerous factors, this study will specifically concentrate on the most common causal factor for road fatalities, which is speed, particularly concerning those crashes involving motorcycles, as they contribute significantly to Indonesia's road deaths. The study will also consider helmet wearing as a critical safety measure to mitigate the impact of fatal road crashes for motorcyclists.

2. Methodology

The methodology used for this study was to explore the current research literature, specifically those focussing on Low- and Middle-Income Countries (LMICs), regarding motorcycle usage and speeding and helmet usage, to determine what evidence-based interventions exist to reduce fatalities and serious injuries for this user group. Those benefits were then applied to Indonesian casualty data and information concerning speeding and helmet usage to estimate the potential gains regarding casualty reductions.

Regarding the causal factor of crashes, national data is unavailable. However, Bandung City causal data is available and thus is used here as a proxy for the national statistics to determine potential casualty reduction benefits.

2.1 Benefits of Reducing Speed Limits and Stronger Enforcement

The available evidence from systematic reviews and meta-analyses indicates that legislation and enforcement interventions have demonstrated significant impact, particularly in Low- and Middle-Income Countries (LMICs) (11). These findings suggest that implementing and enforcing road safety laws, including enforcing speed limits and a helmet wearing law in LMICs can lead to significant improvements in road safety outcomes.

2.1.1 Safe System Approach and Safe Speed Limits

Maintaining safe speeds underpins the Safe System Approach and is crucial in preventing crashes that can lead to fatal and serious injuries (FSIs). Setting safe speed limits and enforcing road rules are one of the key targets to achieving FSI reductions. Numerous studies have shown that lowering speed limits reduces the risk of FSIs, and conversely higher speed limits result in higher risk of FSIs (5, 6). However, determining the optimal speed limit to balance safe speed and mobility remains a challenging question.

It is often impractical to set the speed limit as low as possible while considering the need for efficient mobility. Nevertheless, several studies have been conducted to find the optimal safe speeds for speed limits where the risk of a fatality occurring is set at 10%. One of the most widely cited publication is the Wramborg proposal for safe speeds presented in Warsaw at a road safety conference held in 2005 (Figure 1). He proposed that the risk of a fatality involving a car collision should not exceed 10%. For head on collisions the impact speed should not exceed 70 km/h, whereas for intersections where a side impact may occur, the impact speed should not exceed 50 km/h. Regarding vulnerable road users such as pedestrians and cyclists, Wramborg suggested that motor vehicles should not be allowed to exceed 30 km/h on streets where they might be at risk (15). No supportive evidence or validating data was provided by Wramborg for these speed limits.

Interestingly, a 2019 study by Hussain et al. using a systematic review and meta-analysis approach to synthesise data from all prior relevant studies, also supports setting speed limits of 30-40 km/h for high pedestrian active areas (16) for a 10% risk of a fatality occurring at that speed (Figure 2). These recommended speed limits are commonly adopted by best practice countries that have achieved the lowest road fatality rates internationally.

The study by Hussain et al. (16) is so far the only extensive systematic review and meta-analysis carried out to determine the impact speed versus fatality risk curves for this road user group. It appears that no systematic reviews and meta-analyses have been carried out to determine whether the impact speed versus fatality risk curves for head-on collisions, frontal collisions and/or side impact crashes have been carried out.

2.1.2 Effect of Reducing Travel Speed on Fatality and Serious Injury Risk

Setting a safe speed limit for maximum travel speed is a critical factor in determining optimal speed limits that takes into account the human body's tolerance to impact speed and the need for mobility. When speeding occurs and exceeds a certain safe speed limit, the impact of a crash on the human body becomes significantly greater as a result of the increased kinetic energy the vehicle possess prior to the crash. This relationship between increasing or decreasing travel speed and the resulting change in casualties is illustrated in the power model shown in Figure 3, as demonstrated in Nilsson's 2004 study (17).

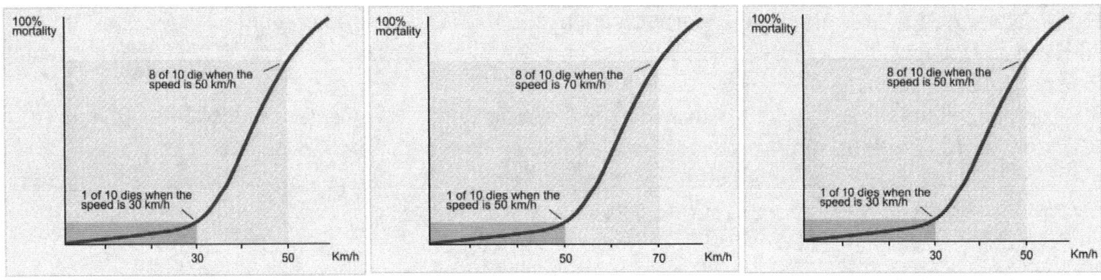

(a) Probability if car driver/passenger fatality by Head-on collision

(b) Probability of car driver/passenger fatality by Side impact collision

(c) Probability of Pedestrian/cyclist fatality by car collision

Figure 1. Probability of death in car collisions according to Wramborg (2005) (15).

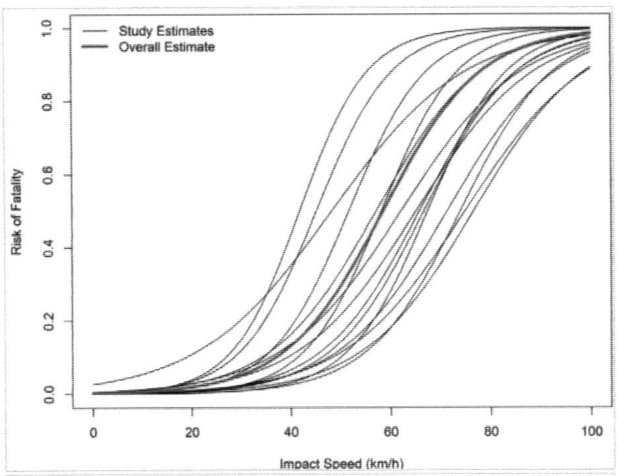

Figure 2. Probability of pedestrian fatality risk involving a car collision by Hussain (2019) (16).

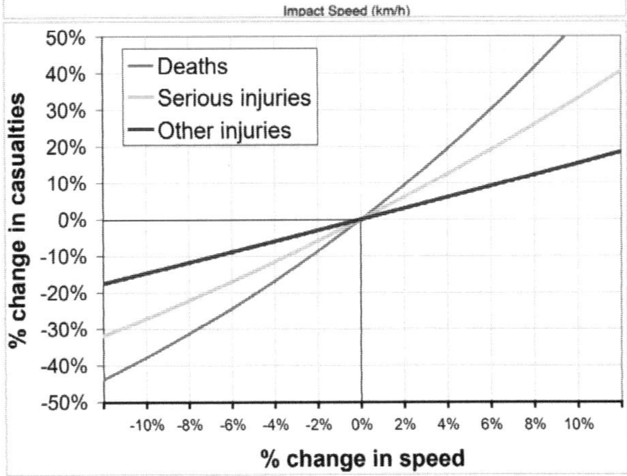

Figure 3. Relationship between change in travel speed and changes in casualty rates (17, 18).

Evidence supporting Nilsson's power model was confirmed by Elvik et al. (18) using a systematic review and meta-analysis method combining all relevant previous evidence-based studies. It illustrates how even a slight change in speed can lead to a significant increase in deaths. In fact, speed limit reduction is one of the single most economically efficient countermeasures available to regulators for reducing FSIs occurring in the road system as opposed to having to implement costly infrastructure treatments (13).

There are a number of anecdotal real-world examples that also demonstrate the relationship between travel speed change and subsequent percentage change in casualties even when the speed limit is set at high values. For instance, in 2011 Kansas, a state within the United States, the speed limit was increased on the selected set of freeway segments from 70 mph (≈112 km/h) to 75 mph (≈120 km/h) (a 7% increase). Through before and after studies, Shirazinejad et al. found total crashes increased 27% and

FSIs increased 35% after the speed limit was changed (19). This and other studies have found that for small speed changes, the percentage change in deaths is typically about four times the percentage change in speed (17, 18, 20).

Another example is a study in South Australia that examined the impact of a speed limit change from 110 km/h to 100 km/h on rural roads. It resulted in a 25.56% reduction of casualties and 27.40% fall in casualty crashes (21). Yet another study by Hoareau et al. in 2022 in South-east Queensland found an 88% reduction in fatal crashes and 23% reduction in serious injury crashes when the speed limit in urban environments where reduced from 60 km/h to 50 km/h (22).

2.2 Indonesia's Speed Limits Compared to Other Countries

Figure 4 shows a graph of the road fatalities per 100,000 inhabitants in 2020 for the thirty-four International Transport Forum (ITF) member countries. Among the ITF member countries, Norway, Sweden, Denmark, and Switzerland are countries with the lowest fatality rates, while Colombia, United States, Chile, and Argentina are countries with the highest fatality rates (23). In comparison, Indonesia's fatality rate of 8.78 per 100,000 would be considered very high and similar to Argentina's rate, i.e., it would be ranked in the bottom 10% worst performing ITF countries.

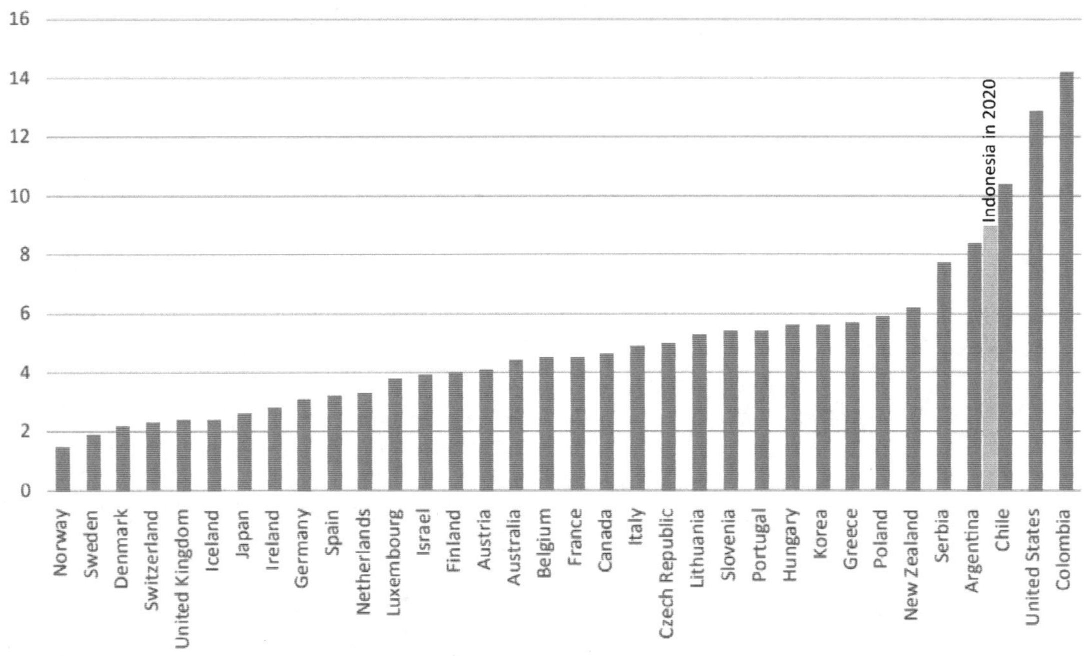

Figure 4. Road fatalities per 100,000 inhabitants in ITF member countries in 2021, and Indonesia in 2020 (1, 23).

Table 1 shows what the speed limits have been set by in those countries with the lowest fatality rate per 100,000 population and for those with the highest fatality rates. The speed limits between these countries do not vary significantly. Indonesia's speed limits appear to be similar to those countries with the lowest fatality rates. However, the fatality rates in Indonesia are more comparable to countries with the highest fatality rates. What this indicates is that the speed limits set in Indonesia would not appear to have much influence regarding fatality rates. Instead, the primary contributing factor would appear to be speeding behaviour of itself along with other contributing factors such as helmet wearing, and unlicenced riding.

Table 1. National speed limits in ITF member countries and Indonesia (23, 24).

Country	Speed Limit (km/h)		
	Urban	Rural	Motorways
Norway	50	80	90,100,110
Sweden	30,40,50	60,70,80,90,100	110,120
Denmark	50	70, 80, (90 for specific section)	110,130
Switzerland	50	80	120, (100 on expressway)
Colombia	50	90	120
[a]US, Nevada	72-105	113 or more	113 or more
Chile	50	90,100	100, 120
Argentina	40-60, 20-70	110	120-130
Indonesia	50 (30 on residential area)	80	min 60, max 100

[a]US speed limit is set by each state; Nevada speed limit was based on Official Driver Handbook Nevada, 2023 (25)

2.3 Benefits of Stronger Enforcement Regarding Speed Limits in Indonesia and Other Countries

Enforcement is crucial in ensuring the effectiveness of road safety laws, including compliance with speed limits. It is the key element in successfully implementing the Safe System Approach (12). Stronger speed enforcement has been shown to improve compliance with speed limits and subsequently reduce crashes and fatalities (26).

The fear of being caught and fined when speeding is one of the main reasons why drivers and riders avoid speeding (13). A study by Ayuningtyas et al. (2021) in West Java Province, Indonesia found that the main reasons that motorcyclists give when questioned why they speed is that they want to reach their destination faster and they know that the police are more concerned with licencing and registration documents rather than speeding (4).

There are numerous studies regarding the effectiveness of speed enforcement. DaCota in 2012 showed that there had been significant improvements resulting from speed enforcement, particularly on reducing both speeding behaviour and the number of crashes (12). However, the size of the reported effects of speed enforcement differ. For example, a meta-analysis study by Erke in 2009 using 45 studies found crashes at all levels of severity reduced by 18% crashes when speed was enforced (27). Pilkington and Kinra found that casualty and crash reductions in the immediate vicinity of speed camera locations varied between 17% and 71% for fatalities, 12% and 65% for injuries, and 5% and 69% for crashes, (28). A study by Soole et al found that strengthened speed enforcement had demonstrated positive effects on vehicle speeds, decreasing both the average and 85th percentile speeds by up to one-third (29).

It should be noted that the primary goal of speed enforcement is to discourage drivers/riders from speeding, not only in relation to future behaviour of those who have been apprehended but more importantly, for those who are not. Enforcement needs to evoke the perception in a driver's mind that anywhere at any time they can be caught speeding and will either have to pay a fine and/or possibly lose their licence. Furthermore, enforcement operations need to focus on roads, conditions and time where speeding has the most negative impact on road safety levels. Preferably, enforcement operations should be based on a comprehensive analysis of road crashes, their outcomes, and the role of speeding in the crash (13).

2.4 Benefits of Introducing a Strongly Enforced Helmet Law

2.4.1 Benefits of Wearing Helmet for Motorcyclists

Helmets help protect a motorcyclist against moderate impact forces to the head during a crash. The evidence supporting helmet use as an injury prevention measure is robust, with estimates indicating significant reductions in deaths and various types of head injuries. The World Health Organization notes that helmet use has been related to reductions in head injuries as follows: 28%-64% for fatal injuries, 58%-60% for head injuries, 47%-74% for brain injuries, 14%-63% for face injuries, and 14%-48% for neck injuries (14). Furthermore, wearing a full-face helmet offers additional benefits by protecting motorcyclists from various traffic noise-related health problems that can result in headaches, depression, tiredness, and fatigue problems (30).

However, the effectiveness of wearing a helmet also depends on helmet type and how it is fixed to the head. A helmet that is not properly fastened can be ejected from the head during a high-impact crash, therefore offering no protection. In comparison, motorcyclists who wear loosely fastened helmets increase their risk of head injury almost two times than those with firmly fastened helmet. Additionally, a full face helmet offers greater resistance to ejection compared to an open face helmet. Full face helmets also offer protection to the lower part of the head (jaw and neck) and therefore provides greater safety (14, 31).

2.4.2 Indonesia Helmet Wearing Rates Compared to Other Countries
Table 2 compares helmet wearing rates for ITF member countries with the highest and lowest road fatality rates (see Figure 4). However, it should be noted that national data related to helmet-wearing rates is not available in Indonesia. As a result, data from Bandung City was used as a proxy for national data.

Table 2. Helmet law and wearing rates in ITF member countries and Indonesia (23, 25, 32).

Country	Law (Powered Two-Wheelers)	Helmet Wearing Rates for Powered Two-Wheelers (R: Rider, P: Pillion)
Norway	Yes	100
Sweden	Yes	98
Denmark	Yes	100
Switzerland	Yes	100
Colombia	Yes	79R, 53P
[a]US, Nevada	Yes	65
Chile	Yes	95R, 87P
Argentina	Yes	69R, 42P
[b]Indonesia	Yes	77R, 50P

[a]US does not have national law for Two-wheelers **(23)**. Nevada USA has a wearing helmet law **(25)**
[b]Indonesia's helmet wearing rates for Bandung City data **(32, 33)**

Table 2 shows that countries with the lowest fatality rates had higher helmet wearing rates compared to countries with the highest fatality rates. In Indonesia's case, the helmet wearing rates are similar to those countries with highest fatality rates. Considering that motorcyclists are the largest contributors to road fatalities in Indonesia, and helmets have been proven to significantly enhance rider and pillion safety, it is important to increase the rate of helmet usage among motorcyclists. This is not only in line with Indonesia's road safety target of ensuring all riders wear helmets by 2030 but also helps mitigate the impact of road crashes on motorcyclists. To increase compliance with helmet laws, strong enforcement of the helmet law, rider licencing and speeding combine with education would be the most effective way to achieve this goal. Such a policy strategy would work synergistically to increase not only rider compliance with helmet laws but also compliance with other traffic laws.

2.4.3 Benefits of Education and Stronger Enforcement of Helmet Law
Data from both high and low-income countries demonstrates that achieving a compliance rate of over 95% with a mandatory helmet law is feasible through active enforcement (14). Furthermore, as enforcement efforts increase, there is a corresponding increase in compliance with helmet laws. However, the effectiveness of enforcement can be significantly enhanced when combined with education campaigns. According to the World Health Organization (WHO, 2023), enforcement combined with education is eight times more effective than enforcement alone, and approximately 12 times more effective than education alone (14).

A systematic review and meta-analysis was carried out by Lepard et al. (2021) examining what effect enforcement and education campaigns have on helmet use among motorcycle riders and pillions. Evidence strongly supports the notion that a combined approach, integrating both enforcement and education, is more effective in promoting helmet usage and improving road safety for motorcyclists.

Furthermore, Lepard et al. (2021) found mass media campaigns aimed at promoting helmet use are effective, but their impact becomes more pronounced when they are run for an extended duration, preferably lasting more than a year. Shorter campaigns may not be sufficient to bring about significant behavioural changes in helmet usage among motorcycle users (34).

To ensure the success of any helmet wearing campaign for motorcyclists, visible and sustained enforcement measures on the road are essential. The fear of penalties is often cited by motorcycle users as one of the motivating factors for wearing helmets. Actively enforcing helmet laws can serve as a strong deterrent and encourage compliance with the legislation, ultimately leading to increased helmet usage and a large reduction in rider/pillion casualties (35). Evidence from various countries also shows that the successful enforcement of mandatory helmet laws is feasible across different income levels. To ensure the effectiveness of such legislation, key elements include effective enforcement measures, ensuring the availability of affordable helmets, and conducting extensive education campaigns for both police and the community (14).

3. Analysis of Potential Benefits

3.1 Estimating Fatality Reduction with Stronger Speed Enforcement

Bandung City data indicates that around 24.63% of motorcycles exceeded the speed limit by more than 10%, whereas riders exceeding the speed limit by more than 20% was 18.6%, on national six-lane two-way divided road in urban areas (3). In other words, more than 40% of motorcyclists exceed the speed limit on these national roads (3).

Reducing the speed limit and enforcing against speeding behaviour, the reduction in deaths and serious injuries would be significant. Figure 3 shows that a reduction of 4% in travel speed would result in almost 20% reduction in deaths whereas a 10% change in speed would result in an even higher reduction of around 40%. It is important to note that the effect is not linear but follows a power line relationship, meaning that as speed decreases, the likelihood of a driver or a rider surviving a crash increase significantly.

Stronger speed enforcement has proven to result in a significant reduction in road FSIs. A study conducted by ICF Consulting before the EU enlargement of 2004 revealed that effective enforcement practices could prevent 5800 road deaths annually resulting from speeding (36). In Chicago, the installation of speed cameras to enhance enforcement between 2013 and 2014 led to an estimated 12% reduction in FSI crashes across treated segments, with a 15% decline in fatalities and severe injury crashes from 2015 to 2017 (37).

A 2007 report by Kfv found there was around 75% reduction in fatal crashes in France between 2003 and 2005 after the strict application of "zero tolerance" for speed enforcement with the introduction of up to 1000 fixed and 500 mobile speed cameras focussing on high crash risk locations. The number of road deaths dropped by more than 35% within the last five years from 8160 (2001) to 5318 (2005), which represents the best progress of any EU country over this period (36). Additionally, a 2000 study by Tay carrying out a before and after study (with case controls) focussing on stronger speed enforcement using speed cameras in Christchurch, New Zealand, found serious injury crashes reduced by 32.3% and 9.17% in total crashes (38). In Western Australia, mobile speed camera operations in regional areas led to an estimated reduction of 18% and 13% in fatal crashes within 500 and 1000 meters of the camera site for every 100 speed camera sessions undertaken per month (13).

There were approximately 25,270 road fatalities in 2021 in Indonesia of which around 81% are made up of riders and pillions, i.e., around 20,500 motorcyclists. Using Bandung City data as a proxy for the whole of Indonesia, speeding accounted for 25% of all crashes, where 24.63% of motorcyclists exceeded the speed limit by more than 10% and 18.6% of riders exceeded the speed limit by more than 20%. That means around 5,125 riders killed were involved in speeding crashes of which around 40% exceed the speed limit. Using the relationship shown in Figure 3, estimating about a 70%-80% reduction in fatalities if strong speed enforcement were introduced, approximately 3,500 lives would be potentially saved.

3.2 Estimating Fatality Reduction with Stronger Helmet Enforcement

Strong enforcing of helmet laws would lead to a substantial decrease in both fatal and serious injuries, where the nonfatal injury rate could decline by as much as approximately 20% (14). The evidence from various countries, particularly for LMICs (34), indicates the feasibility of successfully implementing mandatory helmet law across all nations, regardless of their income level (14), would result in significant FSI reductions. Peng et al. (2017) in their systematic review of 60 studies indicated that implementing universal helmet laws which increased helmet use by 47% resulted in a reduction of total deaths by 32% and injuries by 32% (39). Moreover, in Thailand, after the enforcement of a helmet law act, helmet wearing among motorcyclists increased five-fold resulting in a reduction in deaths by 41.4% (40). Similarly, a study in Malaysia estimated a 30% reduction in fatalities after the introduction and enforcement of their helmet law (41).

Again, considering the total number of rider/pillion deaths is around 20,500, where around only 77% of riders and 50% of pillions are wearing a helmet, the deaths resulting from head injuries could be estimated to be around 25%, i.e., around 4,000 deaths. If a helmet law was strongly enforced, the potential benefit of 30% to 40% in fatality reductions could possibly result in around 1,500 lives saved.

3.3 Best Practice for Increasing Speed and Helmet Law Enforcement using AI Technology

Considering the significant benefits of stronger enforcement of speed limits and helmet law compliance to reduce fatality rates, a strategy on how to strengthen enforcement is needed. Introducing speed cameras along with the promotion of enforcement activities represents one of the most impactful, evidence based, and cost-effective measures aligned with global best practices for reducing speeds, preventing fatalities, and minimising injuries (42-44). This is consistent with Indonesia's seven point risk factor road safety plan which includes introducing speed control and increasing helmet wearing (1).

The use of AI technology with speed cameras increases public perception that enforcement can happen anywhere and at any time. Indonesia has started to introduce AI technology for monitoring speeding behaviour on several toll and urban roads. Patriot et al.'s case study looking at Indonesian Toll roads found that after the installation of 75 speed camera units, drivers who obeyed the speed limit rose significantly to around 76.6% (45). Furthermore, numerous studies around the world have found the effectiveness of speed camera increases the level of compliance with speed limits resulting in reducing fatality rates. For example, intensifying enforcement and penalties using speed cameras in France during a one year observation for 400 speed cameras showed that fatal and injury crashes decreased by 40% to 65% (44). Additionally, the assessment of 20 speed cameras initially implemented in the state of New South Wales, Australia demonstrated a substantial 71% reduction in speeding and resulted in a reduction of 89% in fatalities at the treated locations (46).

Utilising speed camera enforcement becomes particularly suitable when crashes are concentrated on particular road segments and areas where excess speeding occurs, and when the traffic volume makes physical policing less efficient and less effective (12). Point-to-point speed camera enforcement is one of the most effective methods for speed enforcement, where the average speed of vehicles is measured over longer distances, ranging from 1km to over 100km in length (46). Considering the financial constraints associated with installing speed cameras, it is essential to prioritise their placement in strategic locations to maximise their effectiveness and efficiency. One approach to achieve this would be by conducting a thorough analysis to identify blackspot locations for motorcyclists in Indonesia, particularly in high-speed areas with high casualty rates.

AI technology used in speed cameras can also be utilised to enforce helmet-wearing laws, making it a more effective and cost-efficient use of traffic cameras. By integrating helmet-wearing enforcement with speed cameras, authorities can simultaneously target multiple traffic violations, such as speeding and failure to wear helmets (or seatbelts) (14, 34, 47). This approach can lead to increased compliance rates with various traffic laws and contribute to overall road safety improvement. Furthermore, speed cameras can also be integrated to penalise violators, in accordance with the Traffic Law in Indonesia for speeding and helmet wearing (9). Moreover, the funds gained from penalising violators can also be allocated to installing additional speed cameras, thereby expanding coverage to a wider area to enhance enforcement efforts.

It is also important to recognise that the effectiveness of enforcement can be further improved when complemented by education. These educational initiatives can take the form of campaigns and traffic training programs aimed at promoting responsible motorcycling and obtaining proper motorcycling licenses.

4. Conclusion and Recommendation

Motorcyclists are the largest contributor to road fatalities in Indonesia, with the main causal factors being speeding and not wearing helmets. There is an urgent need to increase compliance with speed limits and helmet usage to reduce fatality rates among motorcyclists in Indonesia, which would be consistent with the Indonesian government's road safety plan. Based on evidence and best practices from other countries and research studies, stronger enforcement has been found to be one of the most cost-effective ways to increase the level of compliance with speed limits and helmet wearing, resulting in significant fatality reductions of around 3,500 lives, i.e., about an 18% reduction in Indonesian fatalities overall.

Furthermore, it also has been proven that enforcement alone is around eight times less effective than enforcement combined with education and providing education alone without enforcement is twelve times less effective than when combined with enforcement. Reinforcing enforcement using AI technology in speed cameras, in combination with educational campaigns, is now becoming world's best practice in road trauma reduction. This enhances the public's perception that enforcement can occur anytime and anywhere, creating a deterrent effect that encourages compliance with speed limits, helmet laws, and other traffic regulations, leading to the consequent avoidance of fines or penalties.

One of the recommended types of speed cameras to install are point-to-point speed cameras. To maximise their effectiveness while considering the cost limitations, it is important to identify blackspot locations for motorcyclists, particularly areas with high instances of speeding behaviour. Conducting a study before and after installing speed cameras can help measure their impact on FSI reductions accurately. Media campaigns, license tests, and other educational initiatives can significantly enhance the effectiveness of enforcement measures when carried out together.

5. Acknowledgements

This paper is supported by the Indonesian Endowment Fund for Education (LPDP) scholarship.

6. References

[1] RUNK. Indonesian National Road Safety Plan 2022.
[2] BRSAR. Bandung Road Safety Annual Report 2017. Bandung, Indonesia; 2017.
[3] Kusumawati A, Ayuningtyas KNS, Ellizar E. The effect of speed limit violation on motorcycle crash rate: case study Bandung National Road. Journal of Indonesia Road Safety 2019;2(3):171-82.
[4] Ayuningtyas KNS, Kusumawati A, Ellizar E, editors. Motorcyclists Behavior Toward Speeding in West Java Province, Indonesia. Eastern Asia Society for Transportation Studies; 2021; Japan.
[5] Farmer CM. The effects of higher speed limits on traffic fatalities in the United States, 1993-20172019.
[6] Lori Mooren, Raphael Grzebieta, Job S. Speed-the Biggest and Most Contented Road Killer. Journal of the Australasian College of Road Safety. 2014;25(1):13-8.
[7] Austroads. Guide to Road Safety Part 4 : Safe People. Sydney: Austroads Ltd, Sydney NSW Australia; 2021.
[8] WHO. Global Status Report on Road Safety 20182018.
[9] UU. Traffic and Road Transportation. 2009.
[10] WHO. Global Status Report on Road Safety 2018. Geneva: World Health Organization; 2018.
[11] Tavakkoli M, Torkashvand-Khah Z, Fink G, Takian A, Kuenzli N, de Savigny D, et al. Evidence From the Decade of Action for Road Safety: A Systematic Review of the Effectiveness of Interventions in Low and Middle-Income Countries. Public Health Rev. 2022;43:1604499.
[12] DaCoTa. Speed Enforcement. 2012.
[13] Austroads. Guide to Road Safety Part 3: Safe Speed. Sydney2021. 48 p.
[14] WHO. Helmets. A road safety manual for decision-makers and practitioners Geneva: World Health Organization; 2023.
[15] Wramborg P, editor A New Approach to a Safe and Sustainable Road Structure and Street Design for Urban Areas. Road Safety on Four Continents 2005 5-7 October 2005; Warsaw, Poland.

[16] Hussain Q, Feng H, Grzebieta R, Brijs T, Olivier J. The relationship between impact speed and the probability of pedestrian fatality during a vehicle-pedestrian crash: A systematic review and meta-analysis. Accid Anal Prev. 2019;129:241-9.
[17] Nilsson G. Traffic safety dimensions and the power model to describe the effect of speed on safety. Buletin 221. 2004.
[18] Elvik R, Christensen P, Amundsen A. Speed and Road Accidents: an Evaluation of Power Model Oslo, Norway: Institute of Transport Economics; 2004.
[19] Shirazinejad R, Dissanayake S, Al-Bayati A, York D. Evaluating the Safety Impacts of Increased Speed Limits on Freeways in Kansas Using Before-And-After Study Approach. Sustainability. 2018;11(1).
[20] Austroads. Guide to Road Safety Part 3: Safe Speed2021.
[21] Long AD, Kloeden CN, Hutchinson TP, McLean AJ, editors. Reduction of Speed Limit from 110 km/h to 100 km/h on Certain Roads in South Australia: A Preliminary Evaluation. Australasian Road Safety Conference; 2015; Gold Coast, Queensland.
[22] Hoareau E, Newstead S, Oxley P, Cameron. An evaluation of the 50 km/h speed limits in South East Queensland. Victoria: Monash University Accident Research Centre, Clayton, VIC; 2002.
[23] ITF. Road Safety Annual Report 2022. Paris: OECD Publishing; 2022.
[24] PM. Procedures for Determining Speed Limits. In: Transportation Mo, editor. Jakarta2015.
[25] DMV. Nevada Official Driver Handbook. In: Vehicle DoM, editor. Nevada2023.
[26] Elvik R. Speed Limits, Enforcement, and Health Consequences. Annual Review of Public Health. 2012;33(1):225-38.
[27] Erke A, Goldenbeld C, Vaa T. Good practice in the selected key areas: Speeding, drink driving and seat belt wearing: Results from meta-analysis. Police Enforcement Policy and Programmes on European Roads; 2009.
[28] Pilkington P, Kinra S. Effectiveness of speed cameras in preventing road traffic collisions and related casualties: systematic review. BMJ. 2005;330(7487):331-4.
[29] Soole DW, Watson BC, Fleiter JJ. Effects of average speed enforcement on speed compliance and crashes: a review of the literature. Accid Anal Prev. 2013;54:46-56.
[30] Das CP, Goswami S, Swain BK, Das M. Effect of wearing helmet on traffic noise-induced health issues of motorcycle riders. Journal of Transport & Health. 2022;27.
[31] Yu WY, Chen CY, Chiu WT, Lin MR. Effectiveness of different types of motorcycle helmets and effects of their improper use on head injuries. Int J Epidemiol. 2011;40(3):794-803.
[32] JH-IIRU, Unpad. Data Technical Report - Round 8, Bloomberg Initiative for Global Road Sfaety. In: Padjajaran JHIIRUaU, editor. Bandung2019.
[33] BRSAR. Bandung Road Safety Annual Report 2018. Bandung2018.
[34] Lepard JR, Spagiari R, Corley J, Barthelemy EJ, Kim E, Patterson R, et al. Differences in outcomes of mandatory motorcycle helmet legislation by country income level: A systematic review and meta-analysis. PLoS Med. 2021;18(9):e1003795.
[35] Bao J, Bachani AM, Viet CP, Quang N, Nguyen N, Hyder AA. Trends in motorcycle helmet use in Vietnam: results from a four-year study. Public Health. 2017;144S:S39-S44.
[36] Kfv. Summary and Publication of Best Practices in Road Safety in the Member States. European Commisions; 2007.
[37] Tilahun N. Safety Impact of Automated Speed Camera Enforcement: Empirical Findings Based on Chicago's Speed Cameras. Transportation Research Record: Journal of the Transportation Research Board. 2022;2677(1):1490-8.
[38] Tay R, editor Do speed cameras improve road safety? Traffic and Transportation Studies; 2000.
[39] Peng Y, Vaidya N, Finnie R, Reynolds J, Dumitru C, Njie G, et al. Universal Motorcycle Helmet Laws to Reduce Injuries: A Community Guide Systematic Review. Am J Prev Med. 2017;52(6):820-32.
[40] Ichikawa M, Chadbunchachai W, Marui E. Effect of the helmet act for motorcyclists in Thailand. Accident Analysis & Prevention. 2003;35(2):183-9.

[41] Supramaniam V, van Belle G, Sung JFC. Fatal motorcycle accidents and helmet laws in Peninsular Malaysia. Accident Analysis & Prevention. 1984;16(3):157-62.
[42] Wilson C, Willis C, Hendrikz JK, Le Brocque R, Bellamy N. Speed cameras for the prevention of road traffic injuries and deaths. Cochrane Database Syst Rev. 2010(11):CD004607.
[43] Shaaban K, Mohammad A, Eleimat A. Effectiveness of a fixed speed camera traffic enforcement system in a developing country. Ain Shams Engineering Journal. 2023;14(10).
[44] GRSP. Speed Management. A Road Safety Manual for Decision-Makers and Practitioners. Geneva2008.
[45] Patriot T, Kusuma A, Nahry N. Analysis of Speed Limiting Using Speed Cameras: Case Study in Indonesian Toll Road. CSID Journal of Infrastructure Development. 2020;3(1).
[46] Job S, Cliff D, Fleiter, J.J., , Flieger M, Harman B. Guide for Determining Readiness for Speed Cameras and Other Automated Enforcement. Geneva, Switzerland: Global Road Safety Facility and the Global Road Safety Partnership; 2020.
[47] UNSRF. United Nations Road Safety Fund. 2019 Annual Report 2020.

Risk Factors associated with Stress-related Riding Behavior among Indonesian Motorcyclists

Renni Anggraini[1*], Muhammad Isya[2], Noer Fadhly[3], Cut Mutiawati[4]

[1,2,3,4]Department of Civil Engineering, Faculty of Engineering, Universitas Syiah Kuala.
Jln. Tgk Syech Abdurrauf No.7, Darussalam, Banda Aceh, INDONESIA – 23111
**Corresponding author*: renni.anggraini@usk.ac.id

Abstract. The risk of road accidents is influenced by various elements, particularly human factor. In Indonesia, motorcycles contribute significantly to accidents due to insufficient knowledge of traffic laws and a lack of awareness about safe riding practices. Stress-related riding behavior, which may lead to this incident, is closely linked to driving discipline. Therefore, this study aimed to examine stress-related behavior as the risk factor in road accidents among motorcyclists. To achieve this, a self-reported questionnaire was distributed to 300 motorcyclists in Banda Aceh City and surrounding areas in Aceh Province. Factor analysis was used to simplify the number of indicators included, after which they were prioritized in order of importance. The results showed the most influential indicators out of the 26 considered in this study. A total of 3 underlying factor were identified, comprising *anxious riding*, *aggressive riding*, and *anger*. By addressing the risk factor associated with stress-related riding behavior, road safety in Indonesia was enhanced.

1. Introduction

According to the World Health Organization, 2018 [1], the number of deaths associated with road traffic is rising globally, reaching 1.35 million in 2016. This makes road accident the 8th leading cause of mortality across all age groups, particularly children and young adults between 5 and 29 years. Over the last 15 years, the rate of road traffic deaths has remained relatively constant at around 18 per 100,000 people. Some progress has been made, as evidenced by a decrease from 135 per 100,000 vehicles in 2000 to approximately 64 in 2016. According to ITF, 2021 [2], data from 34 member countries of the IRTAD (International Traffic Safety Data and Analysis) Group, showed that the Covid-19 pandemic affected mobility patterns and the number of road fatalities in 2020. The number of road deaths decreased by 8.6% across all modes of transportation. However, the significant decrease in 2020 fell short of the target of 50% worldwide reduction in fatalities by 2020. This was set by the United Nations Decade of Action for Road Safety. Most of the data from member countries were obtained from Europe, Australia, and Japan. Meanwhile, information from motorcycle-dependent countries such as Indonesia remains scarce.

As reported by WHO (2018), the updated data from Indonesia showed that motorcycle users accounted for up to 74% of all road fatalities in the country. The excessive use of this transportation means, is most likely attributed to the high cost of a car, as well as the inadequacy of public transportation services and facilities [3]. Understanding factor contributing to violations and traffic accidents is critical in formulating policies to implement road safety on urban roads. Despite the influence of factor such as vehicles and roads [4], the human factor remains the most dominant [5].

According to [6], it accounts for approximately 80 and 95% of accidents [7] and accident-related factor, respectively.

Vivo et al. (2006) emphasized that human factors, including the psychological condition of motorcyclists, contribute to 60-80 percent of traffic accidents [8]. This point out the importance of a comprehensive understanding of human aspects for enhancing riding safety. Studies have shown that young motorcyclists often overestimate their riding abilities, showing higher confidence levels while perceiving lower accident risks [9] [10]. According to the report by Ulleberg and Rundmo (2003), a motorcyclist's attitude toward traffic safety can mediate the relationship between personality characteristics and riding behavior [11]. Furthermore, studies from [12] and [13] motorcyclists negligence and lack of discipline are the leading causes of traffic accidents.

The Traffic Regulations Discipline, as stipulated in Law Number 22 of 2009 regarding Road Traffic and Transportation [14], concerning the conduct of motorcyclists on the road. It emphasizes adherence to established laws and traffic regulations for both motorized and non-motorized users. This discipline is directly related to aggressive riding tendencies that often result in violation of traffic laws [15].

Tasca (2002) defined aggressive riding behavior as an intentional act that increases the risk of traffic accidents. This behavior is motivated by impatience, annoyance, hostility, and/or an attempt to save time [16]. Furthermore, it represents a dysfunctional pattern of social behavior that endangers public safety. The identification of specific aggressive driving behaviors includes tailgating, weaving in and out of traffic, improper passing (such as cutting in too close in front of the overtaken vehicle), passing on the road shoulder, making improper lane changes (failure to signal), failing to yield the right of way to other road users, preventing other drivers from passing, displaying unwillingness to extend cooperation to fellow motorists, struggling to merge or change lanes due to traffic conditions, and driving at speeds significantly exceeding the norm. This behavior often leads to frequent tailgating, abrupt lane changes, running stop signs, and running red lights.

Abojaradeh et al. (2014) reported that riding behavior is a significant factor in many traffic accidents, injuries, and fatalities [17]. Stress is a significant factor influencing riding behavior and is closely linked to individual characteristics. Given that riding is a complex task demanding concentration, individuals under this condition encounter difficulties in making sound decisions and are more prone to succumb to road rage. Furthermore, the condition is associated with an increased likelihood of engagement in serious accidents [18].

Thwe et al., (2017) investigated the relationship between riding stress, riding behavior, and accident in Myanmar [19]. The survey comprised 46 professional motorcyclists from certain companies in the country. In addition, it included both experimental and questionnaire surveys. The devices used were a heart rate sensor, a watch, and a video camera. Each participant attached a heart rate sensor around his/her chest, and a watch provided respiratory rate recording via its connection to the sensor. Valid data were subjected to heart rate variability (HRV) analysis. To represent stress levels, the difference between the average baseline condition and the most stressful segments was analyzed. The questionnaire survey included demographic information, unsafe riding behavior, the 26-item Driver Behavior Questionnaire (DBQ), and the 43-item riding behavior related to stress. The results showed that stress, as well as demographic factor, influence riding behavior and contribute to traffic accidents.

According to WHO (2018), citing data from the Indonesian National Police in 2016, a substantial 74% of the fatalities among road users in the country were comprised of two-wheeled and three-wheeled motorcyclists. Similar to other countries with a high population of motorcyclist, there are issues related to behavior on the road. Some studies have used the DBQ instrument to measure motorcycle-motorcyclists behavior. Suwarto et al., (2019) conducted a report on the fear of novice motorcyclists in Semarang City using DBQ. Based to the results, fear does not necessarily reduce lapses and errors while riding, but it decreases the intention of young motorcyclists to commit ordinary and aggressive violations [20]. Lady et al. (2020) used the DBQ instrument to investigate the effect of age, riding experience, and accident rate on motorcyclists riding behavior. The study showed a significant disparity in violations committed by motorcyclists under the age of 35 in Cilegon City when compared to their older counterparts [21]. Putranto and Alyandi (2019) explored deeper into this issue by utilizing the Indonesian Motorcycle Rider Behavior Questionnaire (IMRBQ), a tool derived from the DBQ. The results presented significant correlations between certain constructs in Indonesian Family Values, the

Indonesian Driver Behavior Questionnaire (IDBQ), and IMBRQ. Consequently, it was concluded that the family plays a pivotal role in instilling essential values that contribute to better riding and motorcycle conduct on the road [22]. In the context of Indonesia, report focusing on stress-related riding behavior has been limited. This study aims to investigate the risk factors influencing road accidents, with a specific emphasis on stress-related riding behavior of motorcyclists in Banda Aceh City, Indonesia. To assess this behavior, some indicators will be drawn from the study of Thwe et.al (2017).

2. Methods

2.1. Study Area

The investigation was conducted in Aceh Province, located in the northwestern region of Indonesia. Banda Aceh City, the provincial capital, was selected as the primary study area, with an inclusion of the surrounding areas, comprising Aceh Besar Regency. Based on Central Bureau Statistics data in 2022, the total population of Banda Aceh City and Aceh Besar Regency, were 257,635 and 414,490, respectively. The regions composed a total area of 55.85 km² and 2,882.83 km², with average population densities of 4,612.98 people/km² and 143.78 people/km², respectively.

2.2. Respondent, Samples, and Demographic

Multivariate analysis necessitates a sample size ranging from 100 to 200 [23], which should ideally be 5-10 times the number of indicators [24], or at least comprise10-15 participants per indicator (Field, 2009). In this particular study, the latter criterion was met, ensuring a robust foundation. With 26 indicators multiplied by 10 participants, the total amounted to 260, which was reasonably approximated to 300 samples. This quantity of samples proved to be suitable for conducting factor analysis [25]. Comrey and Lee (1992) further categorized factor analysis samples into three groups, namely 100 (indicating a poor sample size), 300 (indicating a good sample size), and 1000 (indicating an excellent sample size) [26].

According to demographic data, male motorcyclists constituted the majority of respondents (58.7%). A significant proportion fell within the age range of 20-29 (54.3%), followed by those aged 30-39 (21%), 17-19 (14%), 40-49 (7.7%), and 50-59 (2.7%) respectively. The statistics showed that 90.7% of participating motorcyclists had valid riding licenses. Additionally, about 31.3% of respondents who rode motorcycles had been engaged in severe crashes while 68.7% had never experienced a serious accident. More details are presented in Table 1.

Table 1. Descriptive statistics of respondents.

Characteristics	Classification	Percentage (%)
Gender	Male	58.7
	Female	41.3
Age	17-19 years	14
	20-29 years	54.3
	30-39 years	21
	40-49 years	7.7
	50-59 years	2.7
	≥ 60 years	0.3
Licence ownership	Yes	90.7
	No	9.3
Severe accident engagement	Yes	31.3
	No	68.7

2.3. Questionnaire Design

A self-reported questionnaire survey was conducted in the year 2021 among motorcyclists residing in Banda Aceh City and Aceh Besar Regency, Aceh Province, Indonesia. The questionnaire was designed to identify the risk factor associated with stress-related riding behavior, which may have contributed to a road accident. In Bahasa Indonesia, it was distributed to motorcyclists using a 4-point Likert scale, ranging from Never to Always [27]. To prevent respondents from becoming confused about the direction of their responses [28], this scale was deliberately chosen without a central tendency measure.

To collect data on riding behavior, the DBQ was adopted based on previous studies conducted in Indonesia and other countries ([17]; [29]; [30]; [31]; and [19]). For this report, 26 indicators were developed based on scenarios and situations commonly encountered by motorcyclists in the country, with the majority being adapted from Thwe et al., 2017 [19]. A five-point Likert scale (1=never, 2=rare, 3=sometimes, 4=often, and 5=always) was used by Thwe et al. (2017), while this study used a four-point Likert scale (1=never, 2=rare, 3=seldom, and 4=nearly all the time). Riding behavior of 42 professional motorcyclists in Myanmar was examined, focusing on 43 items of stress-related riding behavior. Given that this investigation centered on motorcyclists riding behavior, only a few indicators were applied, and others were selected based on their correlation with motorcycles and motorcyclists. Table 2 presents full details of the 26 indicators, along with the mean and standard deviation. It is important to note that these indicators exhibit some similarities, necessitating a new grouping based on exploratory factor analysis (EFA).

Table 2. Mean and standard deviation of each indicator.

No	Indicator	Mean	Standard Deviation
1	Worrying when riding in bad weather	3.530	0.580
2	Worrying when riding on a road that I've never been to	3.353	0.619
3	Worrying when riding on roads with poor road geometry and a poor asphalt surface	3.513	0.552
4	Worrying when riding with fragile, breakable, or expensive goods	3.490	0.520
5	Worrying when riding on hilly or winding roads	3.460	0.597
6	Worrying when riding on a road with several junctions	3.403	0.607
7	Worrying when riding at night	3.363	0.637
8	Worrying when riding on a road with vehicles recklessly parked	3.377	0.650
9	Worrying when riding in the wrong lane	3.217	0.604
10	Worrying when riding in a traffic jam	3.307	0.622
11	Annoyed when riding a motorcycle that is worn out or unpleasant	3.277	0.561
12	Annoyed when other motorcyclists do silly things	3.310	0.763
13	Forcing to ride without getting adequate rest or sleep	3.230	0.570
14	Drive aggressively when there is a psychologically demanding task	3.290	0.633
15	Annoyed when failed to overtake other vehicles	3.330	0.655
16	Annoyed when the traffic light turns red	2.997	0.715
17	Engage in dispute with another road user	3.410	0.591
18	When in a bad mood, ride aggressively	3.247	0.560
19	Ride fast until there is a sudden lane change	3.320	0.553
20	Annoyed with other motorists during traffic jam	3.303	0.565
21	Get nervous riding with people of high status (employer, etc.)	3.700	0.473
22	Angry when riding behind a slow-moving vehicle	2.483	0.563
23	Angry when riding in traffic jams with no U-turn	2.523	0.778
24	Flashing lights or horns in anger	3.060	0.514
25	Impatient during rush hour	3.210	0.758
26	Riding through when the traffic signal changes from yellow to red	1.977	0.557

2.4. *Exploratory Factor Analysis (EFA)*

According to [23], EFA is a multivariate analytical method used to examine the underlying patterns or relationships among a large number of variables. It aims to determine whether the information can be condensed or summarized in a smaller set of factor or components. Hair (2019) stated that the main goal of factor analytic methods is to discover a means of condensing the amount of information included in a large number of original variables into a smaller group of new, composite dimensions (factor) with minimal information loss. This can be achieved through either an exploratory or a confirmatory factor analysis. Based to [24], those underlying variables or dimensions were known as factor (or latent variables).

In EFA process, each variable or indicator will be examined for feasibility, to determine whether it is possible to be incorporated [24]. EFA method can be performed when the KMO MSA value is greater than 0.5 [32]. Meanwhile, according to Field (2009), the variable of a value less than 0.5 should be disregarded. After the screening test process, the core process of EFA includes the extraction of a set of existing variables (indicators) to form one or more *factor* (factor also termed as dimensions, components, or latent variables). The most widely used extraction method was Principal component analysis (PCA).

PCA was used when the objective was to summarize most of the original information (variance) in a minimum number of factor for prediction purposes. In contrast, common factor analysis was applied primarily to identify underlying factor or dimensions that reflect what the variables share in common. The most direct comparison between the 2 methods is by their use of the explained versus unexplained variance.

3. Results and Discussion

3.1. *Kaiser-Meyer-Olkin Measure of Sampling Adequacy (KMO-MSA), Bartlett's Test of Sphericity, and Extraction Method of Principal Component Analysis (PCA)*

The first stage of EFA is the screening of several variables to obtain those that meet the criteria for analysis. Table 3 shows the results of the KMO-MSA and Bartlett's test of sphericity as a suitability measure for EFA. KMO-MSA was 0.963 (> 0.5), and Bartlett's test of sphericity χ^2 (325) = 5759.859 (p < 0.05). This indicated that the correlations between items were sufficiently large, adequate, and suitable for EFA.

Table 3. KMO-MSA value, Bartlett's test of sphericity, and eigenvalues.

Component	*Initial Eigenvalues*		
	Total	*% of Variance*	*Cumulative %*
1	13.255	50.982	50.982
2	2.391	9.196	60.178
3	1.092	4.199	64.378
Kaiser-Meyer-Olkin (KMO) measure of sampling adequacy (MSA)			.963
	Bartlett's Test of Sphericity Approx. Chi-Square		5759.859
		df	325
		Sig.	0.000

3.2. *Factor Loadings*

A Principal Component Analysis (PCA) was performed on the 26 items using orthogonal rotation (varimax). Initially, an analysis was conducted to acquire eigenvalues for each component in the dataset. Out of these, 3 components exhibited values greater than 1, collectively accounting for 50.982% of the variance, as presented in Table 3. According to the suggestions of Field (2009), eigenvalues serve as an approach to determining the optimal number of factors for inclusion in the analysis. Another method was the scree plot, which was also considered. Field (2009) contends that eigenvalues associated with a variate signify its substantive significance. Since not all factors are retained in the analysis, and there exists a debate over the criterion for determining the statistical importance of a factor, it is prudent to retain only those with substantial eigenvalues.

Table 4 presents the factor loadings post-rotation alongside communalities, and the values were >0.5. Factor 1 comprises 10 indicators, making a substantial contribution to stress-related riding behavior factors. It accounts for 50.982% of the total explained variance, with an eigenvalue of 13.255. This highlights its significant association with stress-inducing riding situations, as evidenced by 80% of the indicators displaying factor loadings above 0.7. Specifically, these significant factor loadings were *worrying when riding on a road with many junctions* (0.785), *worrying when riding at night* (0.780), *worrying when riding on a road with recklessly parked vehicle* (0.778), *worrying when riding in traffic jam* (0.758), *worrying when riding in bad weather* (0.755), *worrying when riding on hilly roads* (0.753), *worrying when riding in a road that I've never been to* (0.749), and *worrying when riding on roads with poor geometry and asphalt surface* (0.728). Other lower significant indicators are *worrying when riding in the wrong lane* (0.697), and *worrying when riding with fragile goods* (0.588). These collectively showed an elevated level of apprehension among motorcyclists when dealing with specific environmental and traffic conditions, signifying a prevailing sense of anxious riding within this group.

Factor 2 comprised 11 items, explaining 9.196% of the variance, with an eigenvalue of 2.391. All items in this factor are strongly associated with aggressive riding, showing factor loadings ranging from 0.474 to 0.793. The most significant indicators, with factor loadings exceeding 0.7, include *riding fast until a sudden lane change is required (0.793), riding aggressively when in a bad mood* (0.757), *and becoming agitated with other motorists during a traffic jam and subsequently riding aggressively* (0.749). Other indicators exhibit lower significance for factor two, such as *reacting to other imprudent actions of motorcyclists* (0.680), *engaging in disputes with fellow road users* (0.667), *and failing to overtake other vehicles* (0.639), *forcing to ride without enough rest* (0.634), *when riding a worn-out motorcycle* (0.629), *when traffic light turns red* (0.580), *when there is a demanding task* (0.550), *and when riding with high status people* (0.474).

Factor 3, which we named "anger," comprised 5 items closely associated with the angry condition. This factor had the smallest impact on stress-related riding behavior factor, accounting for only 4.199% of the variance with an eigenvalue of 1.092. In comparison to the indicators listed in the previous two factors, the factor loadings of each, showed a lower contribution which is less than 0.70. These indicators included feeling angry when riding behind a slow-moving vehicle (0.683), expressing anger through flashing lights or horns (0.666), experiencing frustration in traffic jams without a U-turn option (0.663), impatience during rush hour (0.554), and proceeding through a yellow-to-red traffic signal (0.547). In a parallel study, Liew et al. (2022)[33] conducted a comprehensive examination of anger among Malaysian motorcyclists. Similar to the method used in this study, EFA was adopted to generate 8 factors consisting of 59 indicators. The most significant factor related to anger was labeled 'unsafe or inappropriate actions,' comprising 14 associated items such as challenges in changing lanes and an inclination to accelerate while overtaking. Overall, all 26 indicators used in our study exhibited factor loadings exceeding 0.4, affirming their significance and inclusion, with none necessitating removal.

Table 4. Factor loadings and communalities.

No	Code	Indicator	Factor Loadings	Communalities	Cronbach Alpha
		Factor 1: ANXIOUS RIDING			0.939
1	ANX7	Worrying when riding on a road with many junctions	0.785	0.721	
2	ANX8	Worrying when riding at night	0.780	0.698	
3	ANX9	Worrying when riding a road with vehicles recklessly parked	0.778	0.774	
4	ANX11	Worrying when riding in a traffic jam	0.758	0.671	
5	ANX2	Worrying when riding in bad weather	0.755	0.667	
6	ANX6	Worrying when riding on hilly or winding roads	0.753	0.711	

7	ANX3	Worrying when riding on a road that I've never been to	0.749	0.701	
8	ANX4	Worrying when riding on roads with poor road geometry and a poor asphalt surface	0.728	0.648	
9	ANX10	Worrying when riding in the wrong lane	0.697	0.561	
10	ANX5	Worrying when riding with fragile, breakable, or expensive goods	0.588	0.515	
		Factor 2: AGGRESSIVE RIDING			0.935
1	AGG13	Ride fast until there is a sudden lane change	0.793	0.742	
2	AGG12	When in a bad mood, ride aggressively	0.757	0.703	
3	AGG14	Annoyed with other motorists during traffic jam	0.749	0.687	
4	AGG6	Annoyed when other motorcyclists do silly things	0.680	0.689	
5	AGG11	Engage in dispute with another road user	0.667	0.733	
6	AGG9	Annoyed when failed to overtake other vehicles	0.639	0.659	
7	AGG7	Forcing to ride without getting adequate rest or sleep	0.634	0.642	
8	AGG4	Annoyed when riding a motorcycle that is worn out, unpleasant, or vibrates	0.629	0.499	
9	AGG10	Annoyed when the traffic light turns red	0.580	0.661	
10	AGG8	Ride aggressively when there is a psychologically demanding task	0.550	0.591	
11	ANX1	Get nervous riding with people of high status, such as employers, or elder people, including parents	0.474	0.454	
		Factor 3: ANGER			0.834
1	AGG1	Angry when riding behind a slow-moving vehicle	0.683	0.635	
2	AGG3	Flashing lights or horns in anger	0.666	0.470	
3	AGG2	Angry when riding in traffic jams with no U-turn	0.663	0.673	
4	AGG5	Impatient during rush hour	0.554	0.698	
5	AGG15	Riding through when the traffic signal changes from yellow to red	0.547	0.536	

4. Conclusions

In conclusion, this study collected data from 300 motorcyclists in Banda Aceh City and Aceh Besar Regency, Aceh Province, Indonesia, in order to identify the risk factor linked to stress-related riding behavior. The questionnaire included 26 indicators to describe the conditions of stress-related riding behavior of motorcyclists. By using the Exploratory Factor Analysis (EFA), the data was classified into three factor: *anxious riding*, *aggressive riding*, and *anger*, with eigenvalues surpassing 1.0 and a total variance of 64.378%. The communalities value for the 26 indicators exceeded the defined threshold of 0.30, ranging from 0.454 to 0.774. Factor loadings for all indicators ranged from 0.474 to 0.793. The Cronbach's alpha values for all three factors exceeded 0.60, affirming robust reliability and internal consistency. Significantly, anxious riding stood out as the primary determinant of stress-related riding behavior among motorcyclists in Aceh Province. This study presents initial findings regarding factors and indicators that potentially contribute to stress-related riding behavior within the Indonesian context. Future study may explore the intricate relationships between these indicators and accident experiences. This will be a central focus of upcoming investigations.

Acknowledgments

The authors are grateful to Universitas Syiah Kuala and the Ministry of Education, Culture, Research, and Technology of the Republic of Indonesia for funding this study, following a Letter of Agreement for Assignment on Penelitian H-Indeks (PHI), Fiscal Year of 2020, under Contract No. 269/UN11/SPK/PNBP/2020. The authors are also thankful to those who contributed to this study.

References

[1] World Health Organization 2018 Violence and Injury Prevention: Road Traffic Injury, online: http://www.who.int/violence_injury_prevention/road_traffic/en/, accessed on: 10.08.2019
[2] International Transport Forum, ITF, 2021 The impact of Covid-19, *OECD Publishing, Paris*
[3] Anggraini R, Sugiarto S, and Pramanda H 2017 Factors affecting a trip generation of motorcyclists for non-mandatory activities, *AIP Conf. Proc.* 1903 060011
[4] Directorate General of Highway, Ministry of Public Works 2012 Director General of Highways Instruction No: 02/IN/Db/2012 concerning Road Safety Engineering Technical Guidelines, Technical Guide 1, Road Safety Engineering, Ministry of Public Works, Directorate General of Highways, Republic of Indonesia
[5] Grayson, G.B., Maycock, G., 1988. *From proneness to liability*. In: Rothengatter, J.A, De Bruin, R (eds), Road User Behavior. Theory and Research. Van Gorcum, Assen, pp. 234–242
[6] Tangkudung E S W, Sampouw M, and Tjahjono T 2010 Persepsi Pengendara Sepeda Motor terhadap Keselamatan Berlalu lintas berdasarkan Theory Planned Behavior. *Jurnal Transportasi* **10,** 1, 1-12
[7] Sabey B E and Taylor H 1980 *The known risks we run: the highway* (Transport and Road Research Laboratory Supplementary Report 567). Crowthorne, TRRL, UK
[8] Vivo R, Bergomi M, Rovesti S, Bussetti P, and Guaitoli G M 2006 Biological and behavioral factors affecting driving safety. *J. of Preventive Medicine and Hygiene*, **47**, 69-73
[9] Dejoy D M 1989 The Optimism Bias and Traffic Accident Risk Perception. *Accid. Anal. And Prev*. **21**, 4, 333-340
[10] White J M, Cunningham C, and Titchener K 2011 Young drivers' optimism bias for accident risk and driving skill: accountability and insight experience manipulations. *Accid. Anal. And Prev*. **43**, 4, 1309-1315
[11] Ulleberg P and Rundmo T 2003 Personality, attitudes, and risk perception as predictors of risky driving behavior among young drivers. *Safety Science*, **41**, 427-443
[12] Charty T, Packham D, Rhodes-Defty N, Salter D, and Silock D 1993 Risk and Safety on the Roads: Perceptions and Attitudes. University of Newcastle Upon Tyne: Foundation for Road Safety Research
[13] Hoekstra T, and Wegman, F 2011 Improving the Effectiveness of Road Safety Campaigns: Current and New Practices. *IATSS Research Journal*, **34**, 80-86
[14] Gov of Indonesia 2009 Law of the Republic of Indonesia No. 22 of 2009 on Road Traffic and Transportation. Jakarta, Indonesia
[15] Soffania M 2018 The Relationship Between Aggressive Driving Behavior of Motorcycle Drivers and Traffic Accidents (Case Study on High School Students in Sidoarjo Regency). *In Bahasa Indonesia. The Indonesian J. of Public Health*, **13**(2), 222–233
[16] Tasca L 2002 A Review of the Literature on Aggressive Driving Behavior Research. Ontario Ministry of Transportation, Canada
[17] Abojaradeh M, Jrew B, Al-Ababsah H, Al-Talafeeh A 2014 The Effect of Driver Behavior Mistakes on Traffic Safety. *Civil and Env. Research*, 6 (1), 39-54
[18] Norris F H, Matthews B A, and Riad J K 2000 Characteriological, situational, and behavioral risk factors for motor vehicle accidents: a prospective examination. *Accident Analysis and Prevention*, **32** 505–515
[19] Thwe P.P, Yamamoto T, Sato H, and Morikawa T 2017 The Impact of Demographics, Riding Stress, Riding Behaviors Related to Stress, and Unsafe Riding Behaviors on Accident Involvement among Professional Drivers in Myanmar. *J. of the Eastern Asia Society for Transport. Studies*, **12** 1950-1964.

[20] Suwarno F, Hartono H, and Lukman L 2019 Pengaruh rasa takut terhadap profil perilaku pengendara usia remaja: Studi dengan Driver Behavior Questionnaire (DBQ). *Jurnal Rekayasa Sipil (JRS-UNAND)*, **15** (2), 129-139.

[21] Lady L, Rizqandini L.A, and Trenggonowati D.L 2020 Efek usia, pengalaman berkendara, dan tingkat kecelakaan terhadap driver behavior pengendara sepeda motor. *Jurnal Teknologi Universitas Muhammadiyah Jakarta*, **12** (1), 57-63.

[22] Putranto L.S, Alyandi Y.F 2019 The relationship between children education in the family and car driving and motorcycle riding behaviour in Indonesia. *The International Journal of Integrated Engineering*, **11** (6), 254-267.

[23] Hair J.F 2019 *Multivariate Data Analysis, Seventh Edition*. New Jersey: Pearson Prentice Hall.

[24] Field, A 2009 *Discovering Statistics Using SPSS. 3rd Edition*. Sage Publications Ltd., London

[25] Tabachnick B.G, and Fidell L.S 2013 *Using Multivariate Statistics. Sixth Editions*. Pearson Education, Courier Companies, Inc., United States of America. Value Pack ISBN-13: 9780205885664.

[26] Comrey A.L, and Lee H.B 1992 *Interpretation and Application of Factor Analytic Results*. In A.L. Comrey, & H. B. Lee (Eds.), A First Course in Factor Analysis (p. 2). Hillsdale, NJ: Lawrence Eribaum Associates.

[27] Hadi S 1991 *Analisis Butir untuk Instrumen Angket, Tes, dan Skala Nilai,* Yogyakarta: FP UGM

[28] Sugiarto S, Miwa T, Sato H, and Morikawa T 2017 Explaining differences in acceptance determinants towards congestion charging policies in Indonesia and Japan. *J. of Urban Planning and Dev.*, **143**(2), pp.1-12

[29] Bener A, Yildirim E, Bolat E, Özkan, T, and Lajunen T 2016 The Driver Behavior Questionnaire as an Accident Predictor in Cross-cultural Countries in Qatar and Turkey: Global Public Health Problem. *British Journal of Medicine & Medical Research.* **15**(7), 1-9.

[30] Ge Y, Qu W, Jiang C, Du F, Sun X, and Zhang K 2014 The effect of stress and personality on dangerous riding behavior among Chinese drivers. *Accident Analysis and Prevention,* **73**, 34-40

[31] Poo F, Taubman-Ben-Ari O, Ledesma R, Díaz-Lázaro C 2013 Reliability and validity of a Spanish-language version of the multidimensional riding style inventory." *Transportation Research Part F*, **17**, 75-87.

[32] Malhotra N.K, and Birks D 2006 *Marketing Research: An Applied Approach*. 3rd Edition, Prentice Hall, Upper Saddle River.

[33] Liew S, Hamidun R, and Hamzah A 2022 Factors that Cause Anger among Motorcyclists: Exploratory Factor Analysis. *International J. of Public Health Science*, **11**(3), 893-902.

IOP Conf. Series: Earth and Environmental Science 1294 (2024) 012015 doi:10.1088/1755-1315/1294/1/012015

Road safety improving in blackspot of Mount Bromo tourist road access, East Java

S Sulistyono[1], L Fikria[1], T M Maliq[2], A Hasanuddin[1], S Z Ishak[3]

[1] Civil Engineering Study Program, Universitas Jember, Jember-68121, East Java, Indonesia

[2] Building Technology Engineering Study Program, Universitas Jember, Jember-68121, East Java, Indonesia

[3] Malaysia Institute of Transport, Universiti Teknologi Mara, Selangor-40450, Malaysia

Corresponding author: sonya.sulistyono@unej.ac.id

Abstract. Mount Bromo in East Java is one of Indonesia's priority tourist attractions. Safe, secure, orderly, and smooth traffic access is needed to support tourism. Safe roads can be an essential part of realizing this. This study aims to prioritize handling blackspots to discover safe roads. The research was conducted on three access roads to Mount Bromo from Probolinggo: Jalan Raya Sukapura, Jalan Raya Bromo, and Jalan Raya Lumbang. Analysis of the probability value of accidents, the impact value of the severity of accident victims, and the risk value of accidents are used to determine handling priorities. The results of the analysis obtained priority handling for Blackspot 1 by increasing the width of the road shoulder, Blackspot 2 by providing a crossing integrated with APILL, Blackspot 3 by completing street lighting and increasing lane width, Blackspot 4 by completing street lighting, and Blackspot 5 by increasing lane width. Prioritizing the handling of accident-prone locations in the road and environmental aspects can be done by reviewing the risk value of accidents for each road sub-component. A safe road will be realized at the Blackspot location.

1. Introduction

Mount Bromo is one of Indonesia's ten priority tourism destinations [1]. Located on the border of four districts, namely Malang, Lumajang, Probolinggo, and Pasuruan, there are many routes to Mount Bromo. The best route is through the entrance from Probolinggo City and Regency because there are many public transportation options, close to toll road access and Probolinggo terminal and station. As one of Indonesia's priority tourism destinations, it requires smooth, safe, and comfortable traffic access [2]. Millions die in accidents yearly, resulting in substantial economic losses [3]. Based on accident data recorded in Integrated Road Safety Management (IRSMS), from 2020 to 2022, the number of traffic accidents on the route to Mount Bromo continued to increase. In 2021, there were 51 accidents, an increase of 31.37% from the previous year, and continued to increase in 2022 to 53 accidents. This figure may be smaller than the facts on the ground, and it is because not all accidents are reported to the Police [4].

Accidents are influenced by three main factors, namely human factors, vehicle factors, and road and environmental factors [2]. In general, accidents occur because transportation facilities and infrastructure, both in terms of roads, vehicles, and other supporting infrastructure, are inadequate for societal development [5]. As much as 27% of accidents are caused by the interaction between human and environmental factors [6]. Good roads should be designed to enable safe, comfortable, efficient, and effective movement of people and goods [7]. Roads should ensure mobility and, most importantly, be safe [8]. It is important to conduct road safety inspections on existing roads to ensure safer roads [9]. Road safety inspections are a very efficient and cost-effective way to improve road safety [10]. Planning for road safety improvement can be done by prioritizing the handling of accident-prone areas, then handling can be done on road segments based on the average risk value of each segment [11]. This study aims to describe handling prioritization to improve road safety at the tourist access to Mount Bromo, East Java.

2. Method

2.1 Study Sites
The location of this research is in a blackspot on access to Mount Bromo tourism in East Java through the Probolinggo entrance, namely:
1. Blackspot 1 di Jalan Raya Sukapura KM SBY 94+500 – 95+000;
2. Blackspot 2 di Jalan Raya Sukapura KM SBY 99+000 – 99+500;
3. Blackspot 3 di Jalan Raya Sukapura KM SBY 102+000 – 102+500;
4. Blackspot 4 di Jalan Raya Lumbang KM TGS 0+000 – 0+500;
5. Blackspot 5 di Jalan Raya Lumbang KM TGS 15+500 – 16+000.

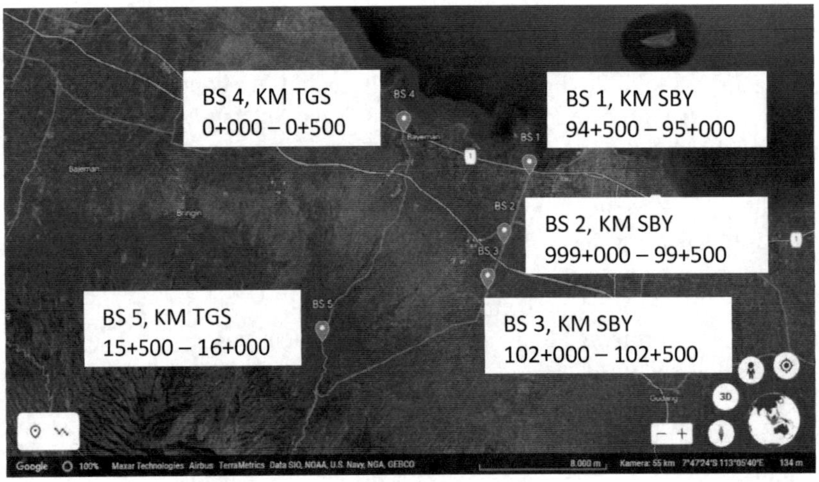

Figure 1. Map of the research location

2.2 Data Collecting
This research uses two types of data, namely primary and secondary data. Primary data in road safety inspection data includes road geometric data, pavement structure data, road section utilization data, complementary building data, road equipment, and traffic management and engineering. At the same time, secondary data is in the form of traffic accident data for 2020 - 2022.

2.3 Data Analysis
Improving road safety in Indonesia is a long-term program to improve living standards, enhance government effectiveness, and strengthen the management capabilities of government institutions [12]. Recent studies have not only focused on human error as the main cause of traffic accidents. This way of thinking is based on the fact that human error may be caused by traffic conditions and road systems that sometimes cannot be predicted by road users [13]. In realizing safe roads, it is important to conduct safety inspections, and then the results of road safety inspections will be analyzed using risk scores.

A road safety inspection examines an existing road to identify deficiencies, faults, and hazards that could cause an accident. A road safety inspection aims to ensure that the road and its components comply with applicable technical requirements and identify any features that may compromise safety [14]. Road safety inspection results are then analyzed using: (i) analysis of accident probability values, (ii) analysis of victim severity impact values, and (iii) analysis of risk values.

The probability of an accident value is determined based on the magnitude of the infrastructure deficiency value against the applicable technical requirements standard. The amount of road infrastructure deficiency against technical standards can be obtained with Equation 1.

$$X = \frac{a-b}{a} . 100\% \quad (1)$$

Where: X is the infrastructure deficiency against the technical standard, a is the road safety technical standard, and b is the measurement and observation results.

The probability values of the accident causes are shown in Table 1.

Table 1. Accident probability value

Measurement and observation results	Qualitative assessment	Quantitative assessment
Deficiency to technical standards < 10%	Never had an accident	1
Deficiency to technical standards 10% - 40%	Accidents occur up to 5 times a year	2
Deficiency to technical standards 40% - 70%	Accidents occur 5-10 times a year	3
Deficiency to technical standards 70% - 100%	There were 10-15 accidents in one year	4
Deficiency to technical standards >100%	Accidents occurred more than 15 times in one year	5

The severity impact value of accident victims is determined from accident data recorded in the Integrated Road Safety Management System (IRSMS). The accident data used is from at least the last three years [15]. The data collected is categorized based on the fatality level of the incident, namely fatal accidents (victims died), victims suffered serious injuries, victims suffered minor injuries, and no victims (only material losses) [15] [16]. The victim severity impact values are shown in Table 2.

Table 2. Accident victim severity impact value

Evacuation results of accident victims	Qualitative assessment	Quantitative assessment
The victim only suffered material losses.	Very light	1
Victims suffered minor injuries and material losses.	Light	10
Victims suffered serious injuries and material losses.	Medium	40
Victims were seriously injured and potentially died in the treatment process, as well as material losses.	Heavy	70
The victim died at the scene and material losses.	Very heavy	100

The risk value is obtained from Equation 2.

$$R = P . D \quad (2)$$

Where R is the risk value of the accident, P is the probability value of the accident cause, and D is the impact value of the victim's severity.

In accident risk assessment, the priority handling order is based on the risk category and the importance of handling [17]. The risk value and importance of handling are shown in Table 3.

Table 3. Value and risk category

Risk analysis		Level of importance of handling
Value	Category	
<125	Non-hazardous (NH)	Regular monitoring with scheduled road safety inspections at points of potential accidents.
125 - 250	Moderately hazardous (MH)	Unscheduled technical countermeasures are required based on the results of road safety inspections at the accident site and its surroundings.
250 - 375	Hazardous (H)	Scheduled technical countermeasures are required within two months from the approval of the road safety inspection results.
> 375	Very hazardous (VH)	Need total technical response with relevant stakeholders within two weeks of approval of the road safety inspection results.

3. Result and Discussion

3.1 Priority Order of Handling in Blackspot 1

Based on the analysis, there are three categories, namely Non-Hazardous (NH), Moderately Hazardous (MH), and Hazardous (H) in the KM SBY 94+500 - 95+000 segment as shown in Table 4.

Table 4. Risk value at Blackspot 1

No.	Sub-components	Unit	a	b	X	P	D	R	Risk category
1	Utilization of road right-of-way (Rumaja)	2 standard	2	1	50	3	100	300	H
2	Shoulder width	m	2	1,5	25	2	100	200	MH
3	Utilization of road courtesy space (Rumija)	2 standard	2	1,5	25	2	100	200	MH
4	Number of access persils	node	1	3	200	5	10	50	NH
5	Access way to the main road	2 way	2	0	100	4	10	40	NH
6	Warning signs	unit	1	0	100	4	10	40	NH
7	Utilization of road surveillance space (Ruwasja)	2 standard	2	1	50	3	10	30	NH
8	Sidewalk (100% = baik)	%	100	80	20	2	10	20	NH

The priority order of the main handling at Blackspot 1 based on Table 4 is the utilization of road right-of-way (Rumaja) sub-components with a risk value of 300 (category of Hazardous, H), shoulder width sub-components with a risk value of 200 (category of Moderately Hazardous, MH), and utilization of road courtesy space (Rumija) sub-components with a risk value of 200 (category of Moderately Hazardous, MH). Handling is done by controlling street vendors around KM SBY 94+800, widening the road shoulder to 2 m, and controlling illegal buildings for free space needs.

3.2 Priority Order of Handling in Blackspot 2

Based on the analysis, there are two categories, namely Non-Hazardous (NH) and Very Hazardous (VH) in the KM SBY 99+000 – 99+500 segment as shown in Table 5.

Table 5. Risk value at Blackspot 2

No	Sub-components	Unit	a	b	X	P	D	R	Risk category
1	Crosswalk	unit	1	0	100	4	100	400	VH
2	Number of access persils	node	1	8	700	5	10	50	NH
3	Access way to the main road	2 way	2	0	100	4	10	40	NH
4	Roadside markings	unit	1	0	100	4	10	40	NH

No	Sub-components	Unit	a	b	X	P	D	R	Risk category
5	Additional board signs	unit	2	0	100	4	10	40	NH
6	Traffic light (100% = working)	%	100	0	100	4	10	40	NH
7	Hectometer stakes	unit	5	0	100	4	10	40	NH
8	Stakes of road courtesy space (Rumija)	unit	10	1	90	4	10	40	NH
9	Side ditch (100% = working)	%	100	50	50	3	10	30	NH
10	Culverts (100% = working)	%	100	50	50	3	10	30	NH
11	Roadside drain	m	1	0,3	70	3	10	30	NH
12	Utilization of road right-of-way (Rumaja)	2 standard	2	1	50	3	10	30	NH
13	Utilization of road surveillance space (Ruwasja)	2 standard	2	1	50	3	10	30	NH
14	Utilization of road courtesy space (Rumija)	2 standard	2	1,5	25	2	10	20	NH

The main handling priority order at Blackspot 2, based on Table 5 is the crosswalk sub-components with a risk value of 400 (category of Very Hazardous, VH). Handling is done by providing zebra crossing facilities around Muneng Market.

3.3 Priority Order of Handling in Blackspot 3

Based on the analysis, there are two categories, namely Non-Hazardous (NH) and Moderately Hazardous (MH) in the KM SBY 102+000 – 102+500 segment as shown in Table 6.

Table 6. Risk value at Blackspot 3

No	Sub-components	Unit	a	b	X	P	D	R	Risk category
1	Lane width	m	3,5	2,4	31,43	2	100	200	MH
2	Access ways from the Persil to the main road	2 way	2	0	100	4	10	40	NH
3	Lane edge markings	unit	1	0	100	4	10	40	NH
4	Road lighting lamps	unit	10	0	100	4	10	40	NH
5	Hectometer stakes	unit	5	0	100	4	10	40	NH
6	Stakes of road courtesy space (Rumija)	unit	10	0	100	4	10	40	NH
7	Side ditch (100% = working)	%	100	50	50	3	10	30	NH
8	Roadside drain	m	1	0,5	50	3	10	30	NH
9	Utilization of road surveillance space (Ruwasja)	2 standard	2	1	50	3	10	30	NH
10	Shoulders	m	2	1,2	40	2	10	20	NH
11	Utilization of road courtesy space (Rumija)	2 standard	2	1,5	25	2	10	20	NH

The main handling priority order at Blackspot 3, based on Table 6 is the lane width sub-components with a risk value of 200 (category of Moderately Hazardous, MH). Handling is done by widening the lane to 3.5 m.

3.4 Priority Order of Handling in Blackspot 4

Based on the analysis, there are two categories, namely Non-Hazardous (NH) and Very Hazardous (VH) in the KM SBY 102+000 – 102+500 segment as shown in Table 7.

Table 7. Risk value at Blackspot 4

No	Sub-components	Unit	a	b	X	P	D	R	Risk category
1	Road lighting lamps	unit	10	0	100	4	100	400	VH
2	Number of access persils	node	1	4	300	5	10	50	NH
3	Access way to the main road	2 way	2	0	100	4	10	40	NH
4	Lane edge markings	unit	1	0	100	4	10	40	NH
5	Hectometer stakes	unit	5	0	100	4	10	40	NH
6	Stakes of road courtesy space (Rumija)	unit	10	0	100	4	10	40	NH
7	Shoulders	m	2	1,1	45	3	10	30	NH
8	Side ditch (100% = working)	%	100	50	50	3	10	30	NH
9	Culverts (100% = working)	%	100	50	50	3	10	30	NH
10	Utilization of road surveillance space (Ruwasja)	2 standard	2	1	50	3	10	30	NH
11	Utilization of road courtesy space (Rumija)	2 standard	2	1,5	25	2	10	20	NH
12	Roadside drain	m	1	0,95	5	1	10	10	NH

The main handling priority order in Blackspot 4, based on Table 7 is the street lighting sub-components with a risk value of 400 (category of Very Hazardous, VH). Handling is done by providing road lighting lamp facilities.

3.5 Priority Order of Handling in Blackspot 5

Based on the analysis, there are two categories, namely Non-Hazardous (NH) and Moderately Hazardous (MH) in the KM TGS 15+500 – 16+000 segment as shown in Table 8.

Table 8. Risk value at Blackspot 5

No	Sub-components	Unit	a	b	X	P	D	R	Risk category
1	Lane width	m	3,5	2,4	31,43	2	100	200	MH
2	Access ways from the Persil to the main road	2 way	2	0	100	4	10	40	NH
3	Lane edge markings	unit	1	0	100	4	10	40	NH
4	Road lighting lamps	unit	10	0	100	4	10	40	NH
5	Hectometer stakes	unit	5	0	100	4	10	40	NH
6	Stakes of road courtesy space (Rumija)	unit	10	0	100	4	10	40	NH
7	Side ditch (100% = Berfungsi)	%	100	50	50	3	10	30	NH
8	Roadside drain	m	1	0,5	50	3	10	30	NH
9	Utilization of road surveillance space (Ruwasja)	2 standard	2	1	50	3	10	30	NH
10	Shoulders	m	2	1,2	40	2	10	20	NH
11	Utilization of road courtesy space (Rumija)	2 standard	2	1,5	25	2	10	20	NH

The main handling priority order at Blackspot 5, based on Table 8 is the lane width sub-components with a risk value of 200 (category of Moderately Hazardous, MH). Handling is done by widening the lane to 3.5 m.

3.6 Recommendations for Handling Blackspot

Road safety improvements on blackspots as an effort to realize safe roads are shown in Table 9.

Table 9. Recommendations for handling at blackspot locations

No.	Components	Recommendation
Road Geometric		
1	Lane width	Increase lane width to 3.5m following technical requirements.
1	Shoulder condition	Protect large tree trunks, such as placing old tires or sandbags or painting tree trunks with reflective paint.
2	Side ditch	Increase the width of side ditches to 1 m following technical requirements.
3	Number of access persils	Controlling the number of access Persils to 1 every 500 m.
4	Access way to the main road	Control traffic through openings in side lanes to facilitate entry from local or neighborhood roads to main roads.
Complementary Buildings		
1	Culverts	Clearing trash/sediment that impedes water flow.
2	Roadside drain condition	Clearing trash/sediment that impedes water flow.
Utilization of Road Sections		
1	Width of Rumaja	Increase the width of Rumaja to 24 m following technical requirements.
2	Utilization of Rumaja	Curbing unauthorized activities on the road shoulder.
3	Width of Rumija	Increase Rumija width to 25 m following technical requirements.
4	Utilization of Rumija	Curbing unauthorized buildings in Rumija to provide free space.
5	Utilization of Ruwasja	Curb unauthorized buildings in the Ruwasja to provide free space.
Traffic Engineering and Management		
1	Presence of markings	Added lane edge markings.
2	Existence of signs	Installed warning signs for four junctions at KM 94+700, additional boards for caution signs at KM 99+000 and KM 99+200, additional boards for caution signs at KM 118+300, and warning signs (bends) at KM 118+200.
3	Existence of traffic signals	Performed maintenance on traffic signals.
4	Crosswalk	Adding zebra crossing at Muneng Market (KM 99+000).
Road Equipment Directly and Indirectly Related to Road Users		
1	Size and color of markings	Repainting the markings.
2	Size and color of signs	They are monitoring the condition of signs and replacing damaged or unclear signs.
3	Traffic signals	Perform maintenance on traffic signal lights.
3	Sidewalk	Adding guide tiles for the visually impaired.
3	Road lighting	Install street lighting every 50-60 m, especially at Blackspot 3, Blackspot 4, and Blackspot 5.
4	Directional stakes	Maintain tilting stakes, repaint stakes, and add reflectors.
5	Hectometer stakes	Install hectometer stakes every 100m on one side of the road.
6	Rumija stakes	Maintain and repaint Rumija stakes.

Based on Table 9, road safety improvements need to be made on almost all road components. Only the pavement structure aspects are functionally feasible and follow road safety technical standards. Seriousness is needed to improve road safety on Mount Bromo tourist access to realize a safe road.

5. Conclusions

The priority of handling blackspots on Mount Bromo tourist access is based on each blackspot's most significant value and risk category. In Blackspot 1, the shoulder, Rumija, and Rumaja sub-components are categorized as Moderately Hazardous (MH) with a risk value of 200. At Blackspot 2, the crosswalk sub-components are classified as Moderately Hazardous (MH) with a risk value of 200. At Blackspot 3, the street lighting sub-components are categorized as Very Hazardous (VH) with a risk value of 400. At Blackspot 5, the lane width sub-components are classified as Moderately Hazardous (MH) with a risk value of 200.

6. Acknowledgments

Thanks to the Institute for Research and Community Service (LP2M) of the Universitas Jember for internal research funding support through the KeRis Dimas 2023 grant scheme. Thanks to the Indonesian National Police Traffic Corps (Korlantas Polri) for supporting traffic accident data (IRSMS Data) and organizing the Focus Group Discussion. Thanks to the Traffic Unit of the Probolinggo Resort Police, who assisted in collecting field data.

7. References

[1] O. A. Sander and M. Christian, “Mengidentifikasi Determinan Keinginanan Berkunjung Lagi Pada 10 Bali Baru,” *J. Bus. Appl. Manag.*, vol. 14, no. 2, pp. 163–176, 2021.

[2] Y. Oktopianto and R. D. Anggara, “Penilaian Tingkat Risiko Keselamatan Jalan Pada Jalur Pariwisata,” *Borneo Eng. J. Tek. Sipil*, vol. 6, no. 1, pp. 55–62, 2022.

[3] A. Iqbal, Z. U. Rehman, S. Ali, K. Ullah, and U. Ghani, "Road Traffic Accident Analysis and Identification of Black Spot Locations on Highway," *Civ. Eng. J.*, vol. 6, no. 12, pp. 2448–2456, 2020.

[4] T. B. Kumar and C. J. Chowdary, "Road Safety Audit: a Case Study on NH-65," *Int. J. Eng. Technol.*, vol. 7, no. 2.1, p. 69, 2018.

[5] U. Enggarsasi and N. K. Sa’diyah, “Kajian Terhadap Faktor-Faktor Penyebab Kecelakaan Lalu Lintas Dalam Upaya Perbaikan Pencegahan Kecelakaan Lalu Lintas,” *Perspektif*, vol. 22, no. 3, pp. 228–237, 2017.

[6] Y. Huvarinen, E. Svatkova, E. Oleshchenko, and S. Pushchina, “Road Safety Audit,” *Transp. Res. Procedia*, vol. 20, pp. 236–241, 2017.

[7] Kementerian Pekerjaan Umum Direktorat Jenderal Bina Marga, *Panduan Teknis 2 Manajemen Hazard Sisi Jalan*. Jakarta, 2012.

[8] P. Colonna, P. Intini, N. Berloco, and V. Ranieri, "Integrated American-European Protocol for Safety Interventions on Existing Two-Lane Rural Roads," *Eur. Transp. Res. Rev.*, vol. 10, no. 5, 2018.

[9] A. K. Jameel and H. Evdorides, "Assessment of Road Safety Performance for Southeast Asian Countries," *J. Soc. Automot. Eng. Malaysia*, vol. 3, no. 3, pp. 246–259, 2019.

[10] S. Sanmithra, N. Namala, and M. S. Renuka, "Road Safety Audit of the Noida – Greater Noida Expressway," *Int. J. Res. Anal. Rev.*, vol. 6, no. 1, pp. 1079–1087, 2019.

[11] A. I. Hidayad, “Perencanaan Peningkatan Keselamatan Jalan Rute Wisata Watu Ulo,” Universitas Jember, 2023.

[12] Direktorat Jenderal Bina Marga, “Panduan Teknis 1 Rekayasa Keselamatan Jalan.” Jakarta, 2012.

[13] A. T. Mulyono, B. Kushari, and H. E. Gunawan, “Audit Keselamatan Infrastruktur Jalan (Studi Kasus Jalan Nasional KM 78-KM 79 Jalur Pantura Jawa, Kabupaten Batang),” *J. Tek. Sipil J. Teor. dan Terap. Bid. Rekayasa Sipil*, vol. 16, no. 3, p. 163, 2009.

[14] Kementerian Pekerjaan Umum Dan Perumahan Rakyat, *Desain Jalan Berkeselamatan*. Jakarta, 2016.

[15] S. Nuli, K. Rohan, and C. Joshnavi, "A Study on Road Safety Audit and Black Spot Identification," *IOP Conf. Ser. Earth Environ. Sci.*, vol. 1086, no. 1, pp. 1–8, 2022.

[16] K. H. Jadav, H. J. Padhya, and P. S. Shah, "Identification and Improvement of Black Spot for Collision : A Case Study of Road Stretch Surat," *Int. J. Res. Anal. Rev.*, vol. 7, no. 1, pp. 943–946, 2020.

[17] D. M. Effendi and O. Firdaus, "Analisis Keselamatan Jalan Pada Ruas Jalan Ahmad Yani Dalam Kota Pangkalpinang," *J. Fropil*, vol. 4, no. 2, pp. 87–100, 2016.

Spatial Analysis to Determine Black Spot Area in Kulon Progo Regency, Yogyakarta, Indonesia

W A Pratama[1] and N Mahmudah[1]
[1]Universitas Muhammadiyah Yogyakarta, Master Program in Civil Engineering, Bantul, Indonesia

Corresponding author: noor.mahmudah@umy.ac.id

Abstract. Nowadays the number of traffic accidents increase due to the surging volume of road traffic. The efforts to reduce number of accidents is important, especially to identify the location and causes the accidents. Therefore, this study aims to identify black spot areas, which are areas with a high concentration of traffic accidents, in Kulon Progo Regency in Yogyakarta, Indonesia. The method used is spatial analysis by combining analysis of traffic accident data and geographic information system (GIS) data using ArcGIS software. The study focused on provincial road consist of primary collector 2 and 3, as well as national road segments. The analysis of accident data allowed the identification of hotspots where accidents occur most frequently. Hotspot analysis methods such as Local Statistic Analysis and Kernel Density Estimation were used to assess the spatial patterns and clustering of accidents. The study found that the black spot area in Kulon Progo Regency was mainly concentrated on the National Road Segment, with fewer occurrences on the Provincial Road with function of Primary Collector 2 and minimal presence on the Primary Collector 3 Road Segment. This result can be used to develop strategies to mitigate accidents and improve road safety in the area.

1. Introduction

The escalating frequency of traffic accidents corresponds to the surging volume of human journeys undertaken. Conversely, human travel experiences a marked upswing due to burgeoning economic expansion and the proliferation of vehicle ownership [1]. According to data sourced from the World Health Organization (WHO), Indonesia witnessed 17 fatalities per 100,000 inhabitants during the course of 2019 [2]. Additionally, the Yogyakarta Police Traffic Unit has reported a surge of 2,480 traffic accident cases between 2021 and 2022 [3]. The effort to reduce the number of accidents is necessary due to the increasing number of mobility using road, hence it is important to know where and why the accidents occur [4]. Based on the previous research, it has already known that traffic accidents happen randomly and affected by several factors such as traffic volume, weather condition, and road geometric design [5].

Black spot area is described as the areas where traffic accidents concentrated [6]. Identification of black spot area could be conducted using Geographical Information System (GIS) as already done in Penang, Kuala Lumpur, Pakistan, and Beijing [7-11]. Effective spatial analysis methods are employed to identify black spot areas in Kulon Progo Regency. By combining geographic information systems (GIS) and accident data, researchers can pinpoint locations with a high frequency of accidents. This allows authorities to prioritize resources and implement targeted interventions to reduce the occurrence of accidents in these areas.

Based on [12], the previous studies, the lengths of crash hotspots are varied in different countries. Here are crash hotspots span approximately in some countries such as Belgium 100 meters [13]; Austria 250 meters [14]; Germany 100 meters [15]; Hong Kong100 meters [16, 17]; Flanders 100 meters for segment locations and 50 meters for intersections [18]; Norway 100 meters [14]; Hungary 100 meters [14]; Croatia 300 to 1000 meters [19].

Apart from length, other parameters in hotspot identification include the crash number threshold and period. The followings are treshold parameters used in different countries such as Belgium has ±3 crashes over 3 years [13]; Denmark has 4 crashes over 5 years [20]; Germany has 5 crashes over 3 years or 3 fatal crashes over 5 years [15]; Hungary has ±3 crashes over 3 years [14]; and Norway has ±4 crashes over 5 years [14]. These parameters are important and useful in identifying and analyzing crash hotspots in different countries [12].

This study aims to identify black spot areas on provincial and national road segments in Kulon Progo Regency by combining traffic accidents and the geographic information system (GIS) data in spatial analysis. A comprehensive exploration of spatial analysis methods employed in identifying black spot areas in study area. The process consists of identification, examine the factors that influence the occurrence of these areas, and discuss the significant impact that research findings have on enhancing road safety measures. The results of this study are expected to develop effective strategies and interventions to mitigate accidents and enhance road safety.

2. Methods

This study was carried out through several stages as shown on Figure 1. First stage is data collection on accident and road network, followed by data categorization, data analysis, and the creation of a hotspot map. It should be noted that this study primarily deals with secondary data. The accident data was acquired from the Police Traffic Unit of Yogyakarta, spanning the period from 2016 to 2018, encompassing attributes such as coordinates, severity classification, and the count of casualties. On the other hand, the road data was sourced from the Indonesia Geospatial Portal in the form of raster data. [21].

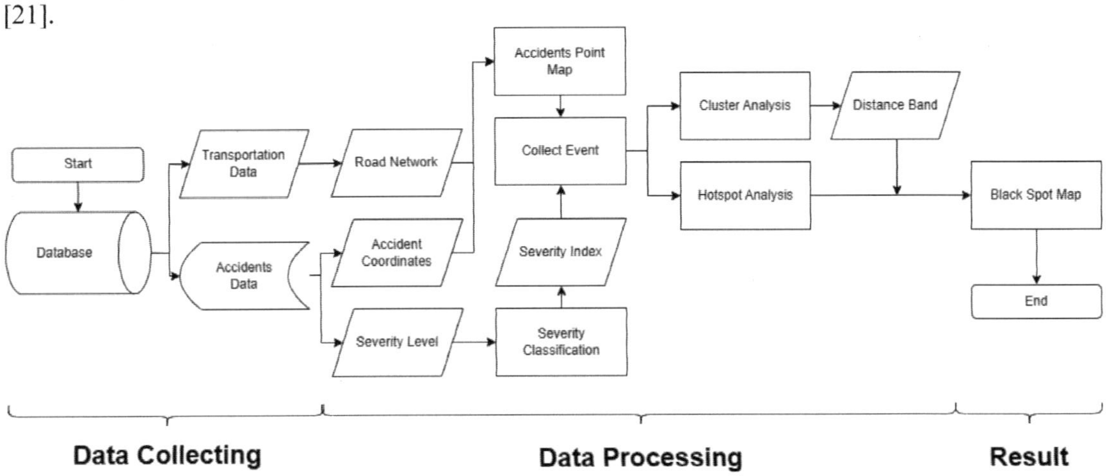

Figure 1. Flowchart of study stages.

2.1. Location

The main location of the study was on Kulon Progo Regency, which is located in Yogyakarta, Indonesia. Kulon Progo Regency is situated in the westernmost part of the Special Region of Yogyakarta Province. It shares borders with Central Java Province to the north and west, while the Indonesian Ocean is located to the south. Geographically, Kulon Progo Regency is mostly flat, although it is surrounded by

IOP Conf. Series: Earth and Environmental Science 1294 (2024) 012016 doi:10.1088/1755-1315/1294/1/012016

mountains, particularly in the northern region. The coordinates of the regency range between 7°38'42" and 7°59'3" South latitude, and 110°1'37" and 110°16'26" East longitude [22]. The map of Kulon Progo Regency could be seen on Figure 2.

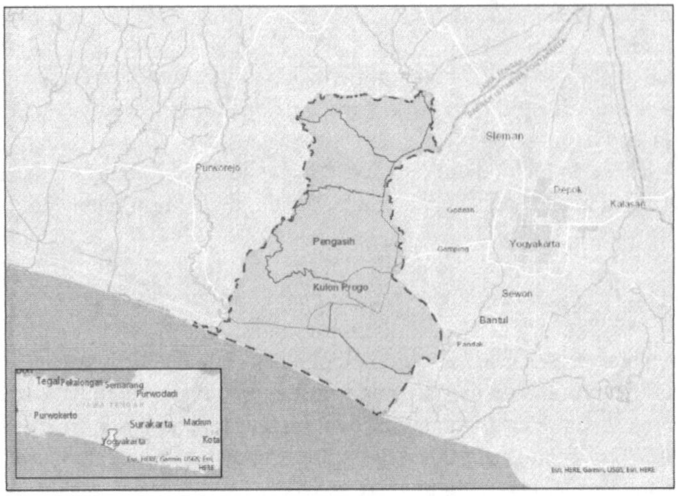

Figure 2. Study location: Kulon Progo Regency, Yogyakarta, Indonesia [23].

2.2. Accidents data

According to the data obtained from the Police Traffic Unit of Yogyakarta, between 2016 and 2018, there were a total of 1,478 recorded accident cases on the road segments in the Kulon Progo Regency as shown on Figure 3. Out of these cases, 112 resulted in fatalities [3]. The accident data includes various attributes such as the accident coordinates, severity level of the accidents, and the number of victims. In this study, the severity index of the victims is categorized into three levels using a risk-weighted score approach: death, heavy injury, and light injury. [24].

Table 1. exhibits the attribute table that was employed for projecting the accident points onto the map, in conjunction with the road data. Conversely, Figure 4. presents a comprehensive summary of the total number of avictims, categorized according to the severity level.

Table 1. Accidents data in Kulon Progo Regency, Yogyakarta, Indonesia.

Coordinates		Severity			Total Accidents
Longitude	Latitude	Fatal	Heavy	Slight	
414459	9146877	0	0	3	3
411333	9129912	0	0	2	2
396173	9128124	0	0	4	4
...	...	...	...	...	...

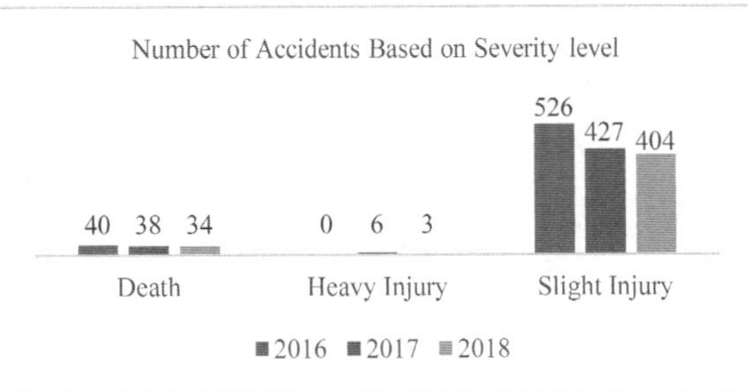

Figure 3. Number of accidents based on severity level.

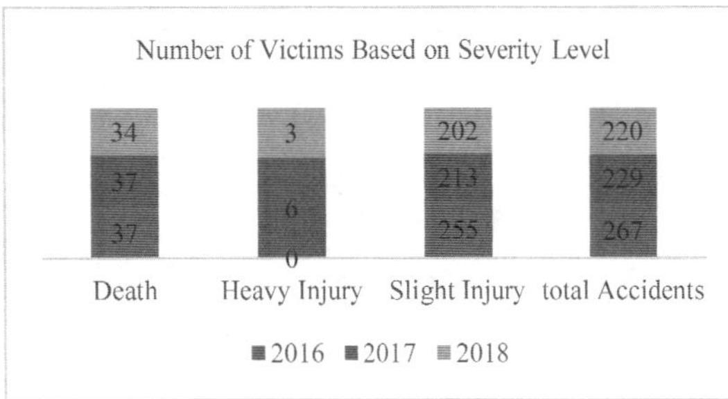

Figure 4. Number of victims based on severity level.

2.3. Road data

This study has specific focus on segments of provincial road consist of primary collector 2 and 3, as well as national road segments. The classification of road types was established according to the Governor decree of the Special Region of Yogyakarta No. 117/KEP/2016, which outlines the designation of collector roads 2 and 3 within the primary road network. [25]. Based on Law No. 2/2022 on 2nd revision of Law no. 38/2022, provincial road is devided into three functions, those are primary collector 2 and 3, and also strategical road [26]. Figure 5. illustrates the intricate road network of Kulon Progo Regency, wherein the segments highlighted in a light blue hue represent the collector 2 roads, the segments in a dark blue shade denote the collector 3 roads, and the segments in an orange hue signify the national roads.

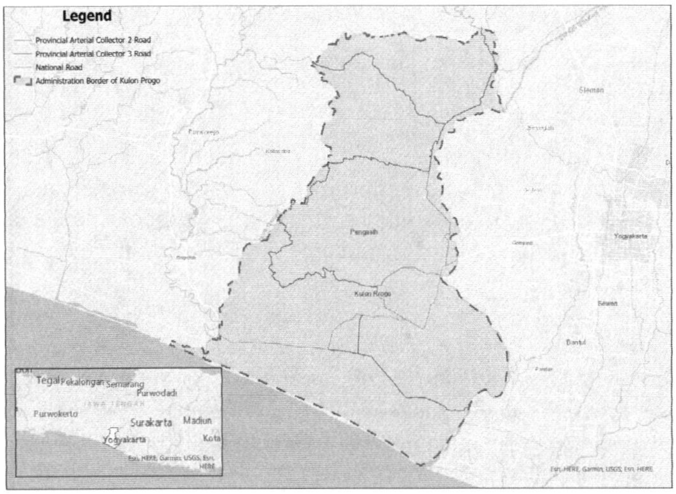

Figure 5. Provincial and national road segments of Kulon Progo Regency.

3. Result and discussion

3.1. Accidents poin map

The data pertaining to accidents and the road network were merged to create a comprehensive point map of accidents. This merging process was carried out using ArcGIS Pro 3.1.2, utilizing the WGS 1984 UTM Zone 49S coordinate system for projection. Figure 6a. to 6c. presents the accidents data spanning from 2016 to 2018, projected onto the primary focal road segment examined in this study which is in segments of provincial road consist of primary collector 2 and 3, as well as national road segments, Kulonprogo Regency, Yogyakarta. Subsequently, all the accidents points from the three-year dataset were consolidated into a single layer as displayed in Figure 6d.

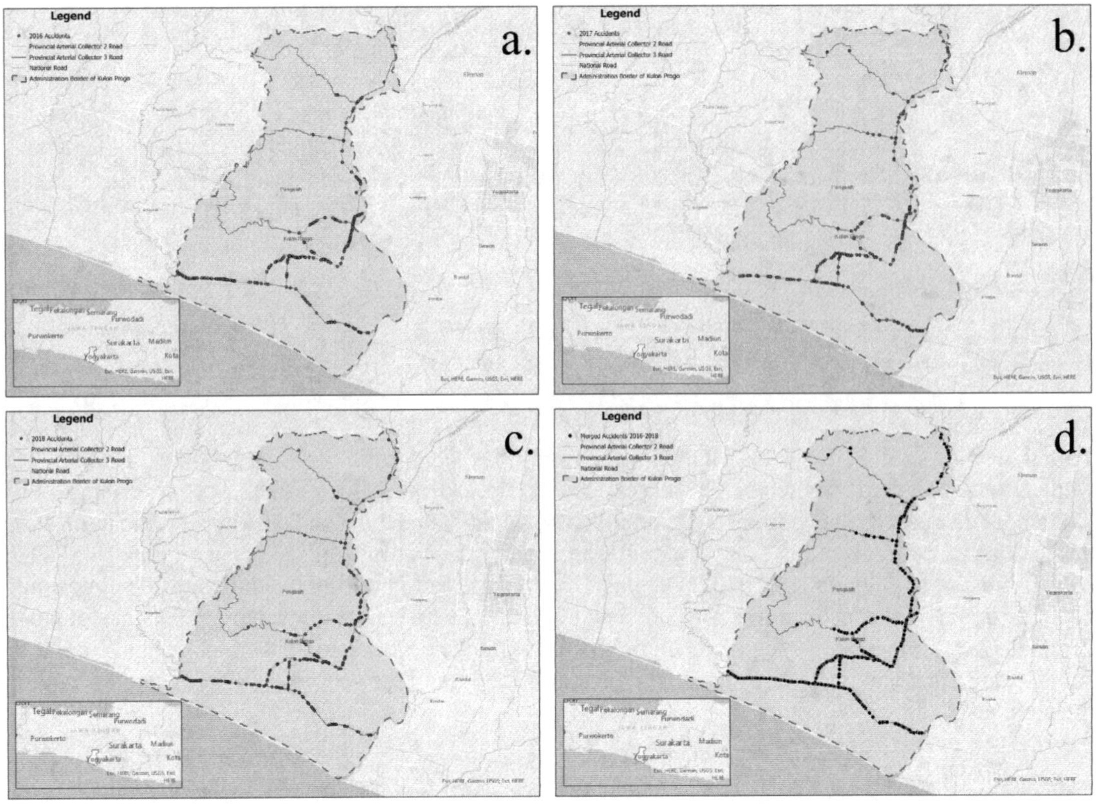

Figure 6. Accidents poin map (a. 2016 accidents data; b. 2017 accidents data; c. 2018 accidents data; merged accidents data).

Based on the projection of accidents data in Figure 6d., it is evident that a significant number of accidents have taken place in the focal road segment. This projection provides valuable insights into the frequency of accidents in this particular area, enabling us to identify hotspots where accidents occur most frequently. This information is crucial in formulating effective strategies to reduce the number of accidents in the identified areas.

3.2. Accidents cluster analysis

Clustering is the methodology employed to categorize vast quantities of data into distinct subsets, thereby rendering the data more comprehensible, utilizing a distance function. [27, 28]. Within ArcGIS Pro 3.1.2, there exist two geoprocessing tools that can be utilized to conduct cluster analysis. The manifestation of local spatial patterns may or may not coincide with the global manifestation. While it is conceivable for the local pattern to be congruent with the global indicator, there exist situations where the local pattern may be perceived as an anomaly that the global indicator would fail to detect. Furthermore, there may be instances where a handful of local patterns diverge from the overarching spatial trend [29]. In the domain of spatial analysis, it becomes crucial to juxtapose the results of localized spatial analysis with those of global spatial analysis.

3.2.1. Cluster and outlier analysis: Anselin Local Moran's I

Cluster and outlier analysis is geoprocessing tools that could be used to identify statistically hot spots, cold spot, and spatial outlier from the given set of weighted features [30]. Hot spot aids in locating areas with statistically significant high values relative to the predicted average by examining the spatial distribution of data points. In contrast, cold spot focuses on finding regions where a phenomenon does not concentrate or cluster. The Getis-Ord Gi* statistic, which determines z-scores for each site depending on its proximity to adjacent locations, is one often used measurement. Hot spots are those

IOP Conf. Series: Earth and Environmental Science 1294 (2024) 012016 doi:10.1088/1755-1315/1294/1/012016

with positive Z-scores, and cold spots are those with negative Z-scores. The technique of Local Moran I proves invaluable in examining developmental assessments, particularly in investigating the dynamics of core-periphery relationships and identifying the interconnectivity among distinct clusters of settlements encompassing individuals from neighboring micro-regions or counties. This analytical approach illuminates the intricate network of provider-receiver connections within these cohesive groups [31].

As evident from Figure 7a. to 7d., the findings of the localized spatial analysis reveal a clustering pattern among the accident points. The cluster analysis outcomes are represented by a spectrum of colors ranging from blue to red, indicating the level of clustering for each accident point. Figure 7a. to 7d. comprises four images that depict variations in severity-weighted analysis for conducting the localized spatial analysis. Directing our attention to the red circle, which signifies areas with a high frequency of accidents, we can preliminarily conclude that the red points may represent black spots in terms of severity weighting. Notably, the results of the localized spatial analysis indicate a higher concentration of clustering accident points in the death-weighted and slight-weighted analyses compared to the heavy-weighted analysis.

The findings of this comparative analysis endeavors to ascertain whether the localized hotspots align with the broader spatial trends observed in the global analysis. Through this comparison, we can evaluate whether the identified hotspots are integral components of larger clusters or if they represent local anomalies. Consistency in clustering patterns across different spatial scales furnishes compelling evidence of the presence and importance of these hotspots. It aids in substantiating the reliability and robustness of the localized spatial analysis results [29, 32].

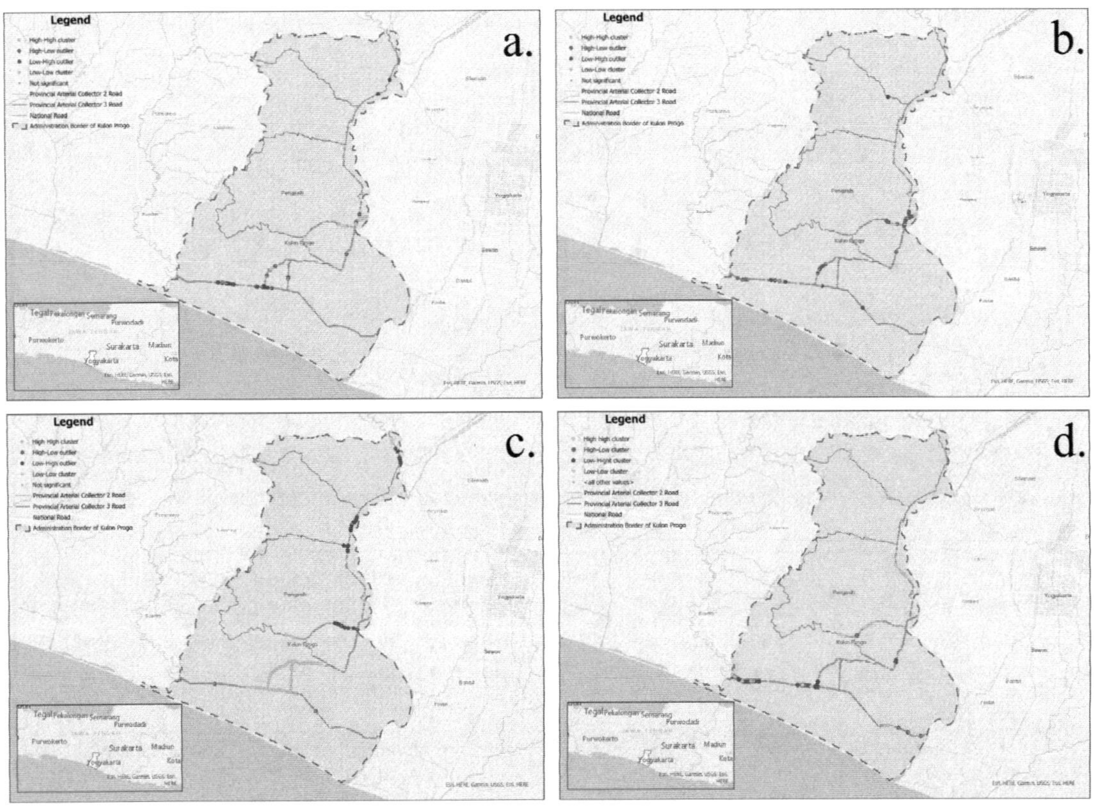

Figure 7. Cluster and Outlier Analysis: Anselin Local Moran's I (a. Total; b. slight injury; c. heavy injury; d. death).

If the hotspots identified through the localized analysis coincide with the clusters observed in the global analysis, it reinforces the understanding that these areas encounter a heightened frequency of accidents in comparison to their surroundings. This consistency affords a more comprehensive understanding of the spatial distribution of accidents and bolsters the development of targeted interventions and strategies to enhance road safety. Conversely, if disparities arise between the localized and global results, further inquiry becomes necessary to comprehend the underlying factors. It could indicate the existence of unique localized influences shaping accident patterns or potential limitations in the analysis methods employed. By comparing localized and global spatial results, we acquire a more profound comprehension of the spatial coherence and patterns of accidents.

Based on the localized spatial analysis conducted using severity-weighted analysis, the highest z-score for clustering accident points was obtained. Table 2. provides information that the most clustered data is derived from the death-weighted local spatial analysis, while the most randomized data is associated with the heavy-weighted analysis. The local analysis yielded a bandwidth (radius) value of approximately 1.675, which can be rounded to 1.7. For a more comprehensive depiction of the results, please refer to Figure 8.

Table 2. Z-score value of clustered accidents data from local spatial analysis using severity-weighted.

Severity-weighted	Clustering level	Z-score
slight-weighted	HH	1.78
heavy-weighted	HH	2.53
death-weighted	HH	1.68
death-weighted with merged accidents point	HH	2

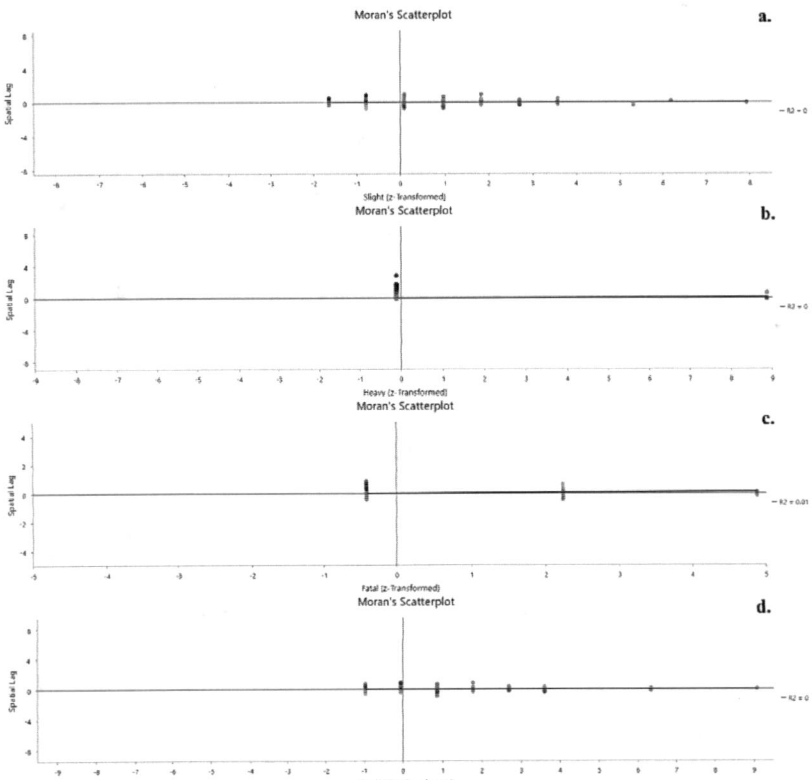

Figure 8. Moran's scatter plot (a. slight-weighted; b. heavy-weighted; c. death-weighted; d. death-weighted with merged accidents point).

3.2.2. Spatial autocorrelation: Global Moran's I

The Spatial Autocorrelation (Global Moran's I) tool is designed to evaluate the existence of spatial autocorrelation by considering both the spatial distribution of features and their associated attribute values. This tool proves particularly beneficial in ascertaining whether the observed pattern of features exhibits clustering, dispersion, or randomness present in the data [29]. Moran's I serves as a metric that spans from -1 to 1, wherein positive values indicate spatial clustering (the similarity of feature values among neighbouring features), negative values indicate spatial dispersion (the dissimilarity of feature values among neighbouring features), and values approximating zero imply a random pattern. By considering both the spatial locations and attribute values of features, it bestows valuable insights into the nature of spatial autocorrelation [33].

The outcomes of Global Moran's I analysis conducted on the accident data are presented in Figure 9. The results are shown separately for different severity levels of accidents. Figure 9a. represents the analysis results for slight injury accidents, Figure 9b. for heavy injury accidents, and the final figure depicts the analysis for fatal accidents. From the analysis, it can be concluded that fatal injury accidents exhibit a clustering pattern. This is indicated by the z-score value of 1.856747, which is the highest among all severity levels. The presence of a high z-score suggests that the occurrence of fatal accidents is not a random event, but rather forms a distinct spatial cluster.

In contrast, the analysis of accidents with severity levels other than fatal reveals that their occurrences are random. This conclusion is supported by the sequence of z-score values presented in Table 3., where the z-score for fatal accidents stands out as the only significant clustering result. These findings highlight the importance of distinguishing between different severity levels of accidents when analyzing their spatial patterns. The clustering of fatal accidents suggests the presence of specific factors or conditions that contribute to their higher occurrence in certain areas. Understanding these patterns can aid in the development of targeted interventions and measures to reduce the number of fatal accidents and improve overall road safety.

When performing a global Moran's analysis on a dataset, one statistical metric used to assess spatial autocorrelation is called the Z score. It determines whether similar values are significantly distributed or clustered in the geographic area under investigation. The Z score is determined by comparing the observed and expected values of the Moran's I statistic under the assumption of a random geographical distribution. Positive Z scores show positive spatial autocorrelation, which means that similar values are more likely to be close together. A negative Z score, on the other hand, denotes a dispersed pattern and shows negative spatial autocorrelation. The presence of spatial autocorrelation in the data is often indicated by a Z score larger than 1.96 (at a significance level of 0.05) [29,32,33].

Table 3. Z-score value of Global Moran's I analysis

Severity-weighted	Clustering level	Z-score
slight-weighted	Random	0.03
heavy-weighted	Random	0.56
death-weighted	Clustered	2

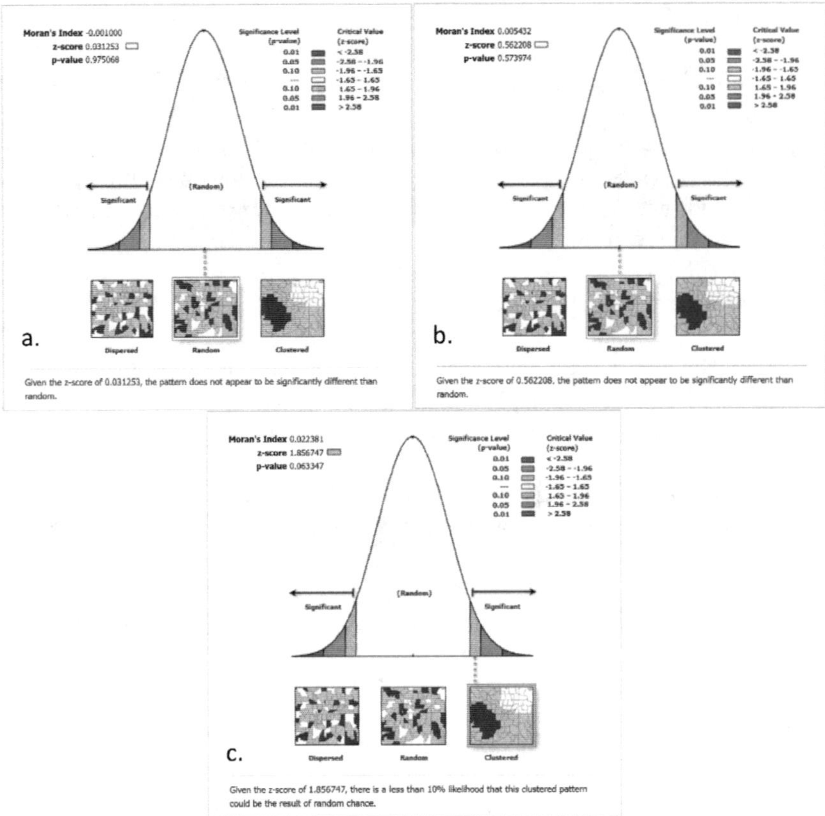

Figure 9. Spatial autocorrelation: Global Moran's I (a. slight injury; b. heavy injury; c. death).

3.2.3. Distance band (radius)

Both cluster analyses yield a z-score value that can be utilized as an indicator for determining the severity level of accidents, which can then be weighted for blackspot analysis in the subsequent stages of analysis. Figure 10. provides a comparison of the z-score values, highlighting the consistency of death injuries forming a cluster. Consequently, we can determine that the severity level of fatal accidents should be employed as the weighted-severity level in the blackspot analysis.

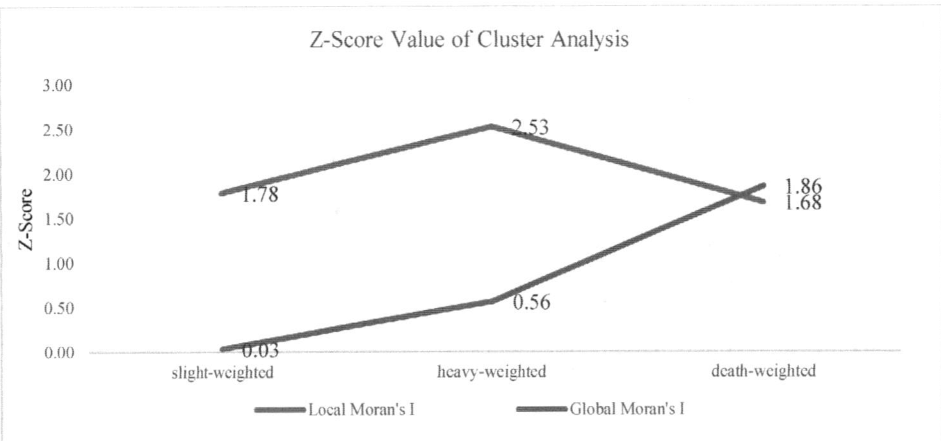

Figure 10. Z-score value of cluster analysis.

In hotspot analysis, the determination of the distance band is based on the average value of the highest z-score obtained from the analysis [34]. In this specific analysis, the average z-score value is calculated to be 1,766. To determine the distance band, this average z-score value is rounded down to the nearest whole number. In this case, the rounded down value of 1.766 is approximately equal to 1.5. This value represents the distance band in which the clustering of accidents is considered significant. These distance units are typically measured in kilometres. The selection of the appropriate distance band is crucial in hotspot analysis as it helps define the spatial scale at which clustering patterns are examined. By considering the average z-score value, the distance band ensures that the analysis captures meaningful spatial relationships and identifies significant clusters of accidents [29].

3.3. Accidents hotspot analysis

Hotspot analysis is a widely used spatial analysis tool available in the ArcGIS Pro application. It offers two primary hot spot analysis tools known for their accuracy and effectiveness. The Local Statistic Analysis: Getis-Ord Gi* tool is used to measure the degree of clustering or dispersion of features within a dataset. This tool provides insights into the presence and significance of spatial clustering patterns, allowing for a deeper understanding of the data. The Kernel Density Estimation tool, on the other hand, is focused on estimating the intensity or concentration of a phenomenon within a dataset. This tool helps identify hotspots or cold spots by visualizing the intensity of the phenomenon across the study area. It provides valuable information about the spatial distribution and density of the features, aiding in the identification and analysis of hotspots within the dataset [29, 32, 35]. Both the Local Statistic Analysis: Gedis-Ord Gi* and Kernel Density Estimation tools are valuable for hotspot analysis as they offer different perspectives on the spatial patterns and clustering of features.

*3.3.1. Local statistical analysis: Getis-Ord Gi**

The results of the Local Statistic Analysis: Getis-Ord Gi* using accident data weighted by death severity are depicted in Figure 11. The visualization in the Figure 11 showcases a gradient of colours ranging from blue to red, which serves to represent the spectrum of hot spots to cold spots. This colour scheme effectively illustrates the intensity and distribution of the analysed data, enabling the identification of areas characterized by high or low levels of clustering or dispersion. Based on the analysis conducted using the Local Statistic Analysis: Getis-Ord Gi* (Figure 11.), it is evident that the occurrence of black spot areas is predominantly concentrated in the national road segment. On the other hand, there is a limited number of black spot areas observed in the provincial collector 3 road segment.

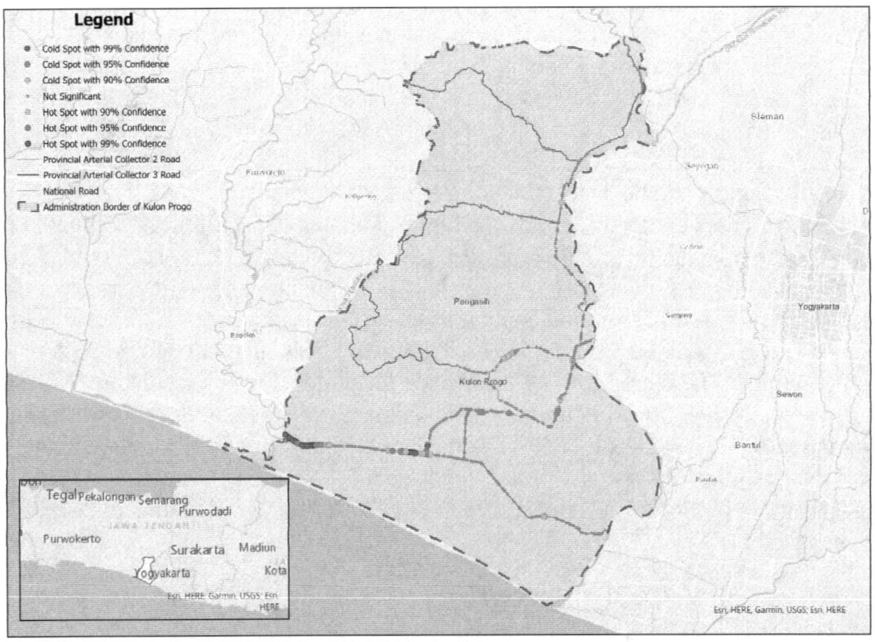

Figure 11. Local statistical analysis: Getis-Ord Gi*.

3.3.2. Kernel density estimation

Figure 12. illustrates the results of the black spot analysis, specifically the analysis conducted using Kernel Density Estimation with a weighting factor for death severity. The visualization in Figure 12. portrays a colour gradient that spans from blue to red, indicating different levels of hot spots. It is evident from Figure 12. that certain areas demonstrate a high concentration of severe accidents resulting in fatalities. These high-concentration areas are primarily observed within the national road segment.

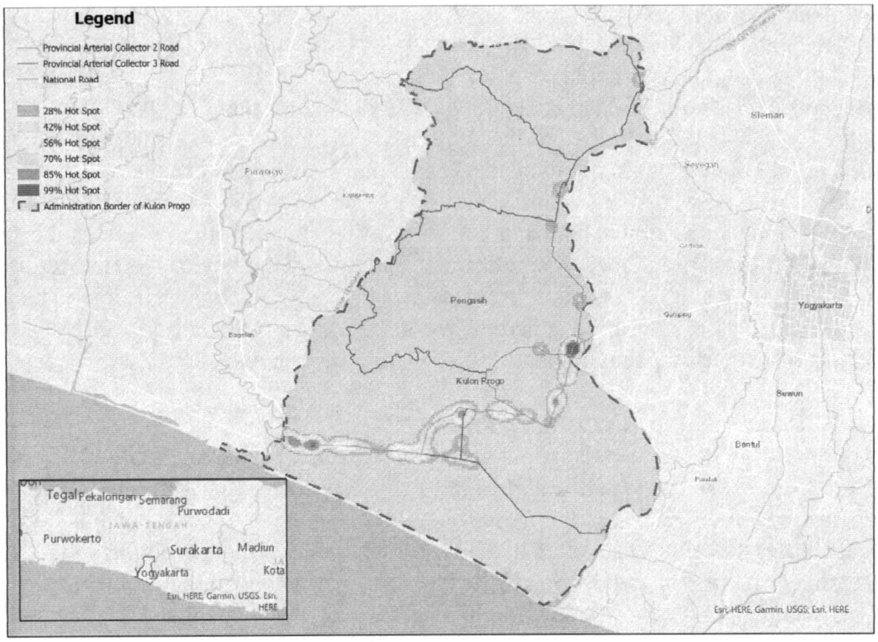

Figure 12. Kernel density estimation with death-weighted analysis.

When comparing the results of the Local Statistic Analysis: Getis-Ord Gi* and Kernel Density Estimation analysis, as depicted in Figure 13., it becomes apparent that the black spot areas identified by both methods exhibit consistency when overlaid. This consistency implies that both analysis methods effectively identify and highlight the same areas as black spots. When multiple analysis techniques produce consistent results, it enhances confidence in the identified areas as reliable black spots.

After analysing the data, it can be deduced that the primary area of focus in Kulon Progo Regency is the National Road Segment. This segment shows a higher number of occurrences compared to other road segments. On the other hand, the provincial Collector 2 Road Segment has relatively fewer instances of black spot areas, while the provincial Collector 3 Road Segment exhibits the smallest extent of such areas. The consistent identification and highlighting of black spot areas by both the Local Statistic Analysis: Getis-Ord Gi* and Kernel Density Estimation analysis methods provide a robust foundation for subsequent actions. These areas can be prioritized for targeted interventions, such as implementing safety measures, improving infrastructure, or increasing enforcement, to address the underlying issues contributing to accidents or incidents. This convergence in findings between different analysis methods demonstrates the reliability and validity of the identified black spot areas, enabling stakeholders to make informed decisions and allocate resources effectively to improve safety and mitigate risks in these specific locations.

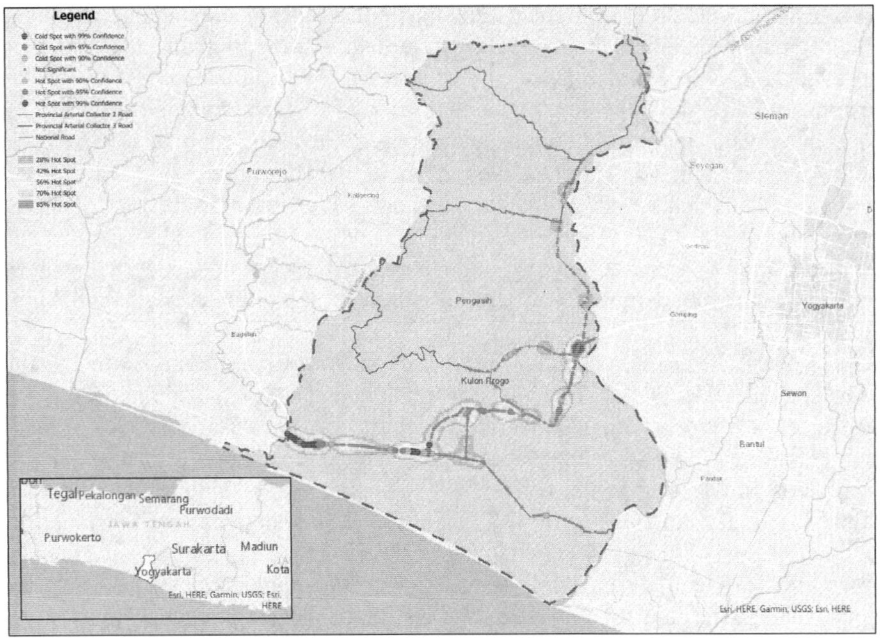

Figure 13. Overlay the result of Local Statistic Analysis: Getis-Ord Gi* and Kernel Density Estimation.

4. Conclusions

The study's analysis has yielded noteworthy findings that significantly contribute to our understanding of accident severity and distribution in Kulon Progo Regency. The application of the Anselin Local Moran's I and Spatial Autocorrelation: Global Moran's I in the Cluster and Outlier Analysis has been instrumental in identifying the severity level of accidents. The z-score values obtained from these methods serve as a reliable indicator for blackspot analysis, providing a quantitative measure of the gravity of accidents in specific areas.

Moreover, the study has determined the distance band value for hot spot analysis by averaging the z-score values obtained from both cluster analyses. The value, rounded down to approximately 1.5 from 1,766, represents a significant distance band within which accident clustering is considered significant. This finding is crucial in identifying and predicting accident-prone areas based on spatial proximity. Our analysis has further revealed that accident concentration in Kulon Progo Regency is primarily found on the National Road Segment. This road segment has been identified as a significant black spot area, indicating a high severity of accidents. In contrast, the Primary Collector 2 Road Segment and the Primary Collector 3 Road Segment have fewer and minimal occurrences, respectively.

In conclusion, these findings provide valuable insights into the spatial distribution and severity of accidents in Kulon Progo Regency. The data underscores the significance of the National Road Segment as a major black spot area, shedding light on the areas that require immediate and targeted intervention to enhance road safety.

5. References

[1] Fan, Z., et al., Research On Black Spot Identification Of Safety In Urban Traffic Accidents Based On Machine Learning Method. Safety science, 2019. 118: p. 607-616.
[2] WHO, Mortality Caused by Road Traffic Injury (per 100000 population), in Annual, T.W. Bank, Editor. 2019, World Health Organization, Global Health Observatory Data Repository.
[3] DIY, K.R.I., Data Kecelakaan Lalu-Lintas Daerah Istimewa Yogyakarta 2015-2023. 2023. p. 1.

[4] Le, K.G., P. Liu, and L.-T. Lin, Determining the Road Traffic Accident Hotspots Using Gis-Based Temporal-Spatial Statistical Analytic Techniques In Hanoi, Vietnam. Geo-spatial Information Science, 2020. 23(2): p. 153-164.

[5] Harirforoush, H. and L. Bellalite, A new integrated GIS-based analysis to detect hotspots: a case study of the city of Sherbrooke. Accident Analysis & Prevention, 2019. 130: p. 62-74.

[6] Dereli, M.A. Erdogan, and Saffet, A New Model For Determining The Traffic Accident Black Spots Using Gis-Aided Spatial Statistical Methods. Transportation Research Part A: Policy and Practice, 2017. 103: p. 106-117.

[7] Tola, A.M. and A. Gebissa, Identifying Black Spot Accident Zones using a Geographical Information System on Kombolcha-Dessie Road in Ethiopia. Int. J. Sci. Basic Appl. Res, 2019. 48: p. 1-12.

[8] Manap, N., et al., Determining spatial patterns of road accidents at expressway by applying Getis-Ord Gi* spatial statistic. Int. J. Recent Technol. Eng, 2019. 8(3S3): p. 345-350.

[9] Jiang, H., et al. Research on identification method for road accident black spots with ordinal clustering method. in 2011 international conference on remote sensing, environment and transportation engineering. 2011. IEEE.

[10] Masron, T., et al., Applying GIS in analysing black spot areas in Penang, Malaysia. The Indonesian Journal of Geography, 2018. 50(2): p. 133-144.

[11] Razzak, J., U.R. Khan, and S. Jalal, Application of geographical information system (GIS) for mapping road traffic injuries using existing source of data in Karachi, Pakistan--a pilot study. Journal of the Pakistan Medical Association, 2011. 61(7): p. 640.

[12] Ouni, F. and M. Belloumi, Pattern of road traffic crash hot zones versus probable hot zones in Tunisia: A geospatial analysis. Accident Analysis & Prevention, 2019. 128: p. 185-196.

[13] Flahaut, B.t., et al., The local spatial autocorrelation and the kernel method for identifying black zones: A comparative approach. Accident Analysis & Prevention, 2003. 35(6): p. 991-1004.

[14] Elvik, R., State-of-the-art approaches to road accident black spot management and safety analysis of road networks. 2007: Transportøkonomisk institutt Oslo, Norway.

[15] Elvik, R., A survey of operational definitions of hazardous road locations in some European countries. Accident Analysis & Prevention, 2008. 40(6): p. 1830-1835.

[16] Lai, P.C. and W.Y. Chan, GIS for road accident analysis in Hong Kong. Geographic Information Sciences, 2004. 10(1): p. 58-67.

[17] Loo, B.P., The identification of hazardous road locations: a comparison of the blacksite and hot zone methodologies in Hong Kong. International Journal of Sustainable Transportation, 2009. 3(3): p. 187-202.

[18] Geurts, K., Ranking and profiling dangerous accident locations using data mining and statistical techniques. 2006: Univ. Hasselt.

[19] Zovak, G., D. Brcic, and Z. Saric. Analysis of road black spots identification method in republic of Croatia. in IX International Conference Road Safety in Local Communities. 2014.

[20] Vistisen, D., Models and methods for hot spot safety work. 2002.

[21] Portal, I.G., RBI Map. Accessed on June 2023

[22] Regency, R.G.o.K.P. Geographic Information of Kulon Progo Regency. 2020 [cited 2023 25 July]; Available from: https://kulonprogokab.go.id/v31/detil/7670/geografis.

[23] Esri, Arcgis Pro. 1969.

[24] Afghari, A.P., M.M. Haque, and S. Washington, Applying a joint model of crash count and crash severity to identify road segments with high risk of fatal and serious injury crashes. Accident Analysis & Prevention, 2020. 144: p. 105615.

[25] Yogyakarta, G.o.D.I., SK Gubernur Daerah istimewa Yogyakarta No 117/KEP/2016

[26] Indonesia, P.o., Law No. 2/2022 on 2nd Revision of Law No. 38/2004. 2022.

[27] Kumar, Y. and G. Sahoo, A two-step artificial bee colony algorithm for clustering. Neural Computing and Applications, 2017. 28: p. 537-551.

[28] Nst, U.R.P., Sutarman, and P. Sirait, Analysis of Representative Value in Clustering using the CURE Algorithm. International Journal of Recent Technolgy and Engineering, 2020. 9(2).

[29] Anselin, L., Local indicators of spatial association—LISA. Geographical analysis, 1995. 27(2): p. 93-115.
[30] Esri. Cluster and Outlier Analysis (Anselin Local Moran's I) (Spatial Statistics). [cited 2023 25 July]; Available from: https://pro.arcgis.com/en/pro-app/latest/tool-reference/spatial-statistics/cluster-and-outlier-analysis-anselin-local-moran-s.htm.
[31] Géza, T., Területi autokorrelációs vizsgálat a Local Moran I módszerével. Tér és Társadalom, 2003. 17(4): p. 39-49.
[32] Getis, A. and J.K. Ord, The analysis of spatial association by use of distance statistics. Geographical analysis, 1992. 24(3): p. 189-206.
[33] Moran, P.A., Notes on continuous stochastic phenomena. Biometrika, 1950. 37(1/2): p. 17-23.
[34] Huang, H., M. Abdel-Aty, and X. Wang, Analysis of vehicle crash injury severity levels at unsignalized intersections. Accidents Analysis & Prevention, 2007. 39(2): p. 10.
[35] Silverman, B.W., Density estimation for statistics and data analysis. 2018: Routledge.

6. Acknowledgement

The authors would like to thank ERASMUS + CBHE PROJECT "ASIASAFE", which is co-funded by EUROPEAN COMMISION with contract number 618325-EPP-1-2020-1-SE-EPPKA2-CBHE-JP and Research and Innovation Institute of Universitas Muhammadiyah Yogyakarta (LRI UMY).

The willingness-to-walk to access public transit in Indonesia

M Hadid[1*], D N Tufail[2], A Hidayat[2] and B S Khofifah[1]
[1] Department of Civil Engineering, Institut Teknologi Kalimantan, Indonesia
[2] Department of Urban and Regional Planning, Institut Teknologi Kalimantan, Indonesia

*Corresponding author's email: hadid@lecturer.itk.ac.id

Abstract. In order to support the implementation of public transportation, an understanding of people's interest in walking is essential. The farther people's walking distance means public transportation will have a larger catchment area. This study aims to get parameters that can be used to improve walking distance in Balikpapan (Indonesia). Data collection in this study was carried out through stated preference surveys to 800 respondents divided into 400 motorcycle users and 400 car users. The data were analysed using the Binary Logistics Regression Analysis method, and the result shows that the respondent's walking distance was 300 meters with a probability of 47% (motorcycle) and 59% (car). The result shows that two aspects significantly influence the willingness to walk, comfort and aesthetics. So, the strategies that are prepared to increase willingness to walk can be started by providing street furniture to support the comfort aspect and planting trees and plants to add aesthetic value.

1. Introduction

Walking is an option for continuing the journey after public transportation [1]. Walking is an essential mode of travel to reach the nearest bus stop from the place of origin [2]. Therefore, bus stops should be within walking distance acceptable to the community. This makes the willingness to walk necessary for transport planners and related agencies in determining the position of bus stops along urban routes. There is a relationship between walking distance and the use of public transportation [3]. The distance of willingness to walk can determine the user base of public transport and the catchment area [4]. Thus, the distance, which is the measure of the catchment area, is often used to make predictions about transit passengers, the basis for regulatory arrangements, as well as a reference for financial planning for the concept of Transit Oriented Development (TOD) in a city [5].

The relationship between walking distance and public transport users makes pedestrian paths an important element in a city [6]. There is a need to increase the characteristics that support the accessibility of pedestrian paths because sidewalks that give the impression of being walkable will increase walking motivation [7]. Various factors can potentially influence the characteristics and variability of people's walking distance, one of which is the built environment in the local area [8]. The same thing was expressed by previous study, which confirmed that one aspect of the built environment that can increase walking activity is the walkability of the sidewalk paths that are passed [7]. The friendliness of the pedestrian path to pedestrians can be assessed from its wider dimensions and higher quality and the provision of supporting facilities such as chairs and bicycle paths. Increasing walkability in pedestrian corridors leading to public transportation stops can increase the distance people are willing to walk [9]. Someone tends to walk farther on a more walkable corridor [10].

On average, people in Kupang Indonesia are willing to walk 115 meters [11]. On the other hand, the walking distance accepted by pedestrians on Penang Island, Malaysia, is 400 to 500 meters [12]. This indicates that the willingness to walk in each city is different [2]. Therefore, even though the standard walking distance to the bus stop has been regulated in the service quality indicators issued by the World Bank (1987), this value does not distinguish the conditions of the city served, even though the characteristics of each city are very different [13]. Therefore, an analysis was carried out regarding the willingness to walk of the community towards the public transport in Balikpapan. This analysis aims to determine the extent of walking distance people accept and identify the factors that most influence this distance. The resulting parameters can be a reference for the development of bus stop locations and the basis for developing strategies to increase people's willingness to walk.

In this study, the analysis was carried out using the binary logistic regression method to know how significant the probability of willingness to walk is for the people of Balikpapan City. Through this, it can also be known what factors influence people's walking decisions, especially private vehicle users. The analytical method used in this study is quite different from the previous research [11] [13], which used the Probability Density Function (PDF) and Cumulative Distribution Function (CDF) methods. This is because this research is expected to produce a probability percentage of the willingness of motorbike and car users for each variation of walking distance. Also, the study's target respondents in Balikpapan City were private vehicle users. This is different from the case studies in Kupang (Indonesia) and Penang Island (Malaysia), which determined pedestrians as respondents. This difference is because by conducting this research, it is hoped that it can simultaneously increase public knowledge about the existence of the Public Transport, which will later be able to participate in growing the people's interest in walking. Along with the increasing distance of willingness to walk, the catchment area will also be more significant and increase transport users' potential.

2. Method

In general, this study was divided into two stages: data collection and analysis. At the data collection stage, interviews were carried out with people active around public transport corridors who use private vehicles (cars and motorcycles) for mandatory activity. The survey technique used was random sampling with the stated preference interview technique. Several alternative situations of imaginary pedestrian facilities will be used, which will be answered by the respondent. The survey was conducted online and offline. Because this study focuses on people who use private vehicles, the number of respondents is divided into two classes, namely motorcycle and car users. The number of respondents used in this study was 800, consisting of 400 motorcycle users and 400 private car users. This aims to obtain a balanced comparison of the analysis results regarding the number of respondents.

2.1. Questioner Design

The questionnaire used was divided into two types, namely the characteristics of the respondents and part of the willingness to walk. The characteristics of the respondents used in this study include age, gender, occupation, the purpose of travel, direction of travel, number of trips, reasons for using private vehicles, number of family members, number of motorcycle and car ownership, the income of respondents, household income, frequency of use public transportation, distance to access the stop point of public transport, and time to the public transport stop point. The second part of the questionnaire is about the willingness to walk, which contains questions about the pedestrian facility scenario.

The pedestrian facility scenario is a combination of several facilities that support walkable, including distance (X1), access (X2), convenience (X3), aesthetics (X4), and safety and comfort (X5) with details shown in Table 1.

Table 1 shows that each facility has several variations. The several facilities in Table 1 are then combined into six scenarios that will be given to the respondents so that each respondent will provide six responses to the pedestrian facilities offered. The various scenarios used in this study are shown in Table 2.

Table 1. Variable for Pedestrian Facilities

Variable	Category
Distance (X1)	1 = 100 m 2 = 200 m 3 = 300 m
Access (X2)	1 = minimalist pedestrian path 2 = wide pedestrian path, visible directions, friendly for elder people and baby stroller
Comfort (X3)	1 = flat surface of pedestrian path 2 = flat surface with canopy, chair, and sparated bike lane
Aesthetics (X4)	1 = pedestrian path without vegatation 2 = pedestrian with bollard, trees, and vegetation
Safe dan Secure (X5)	1 = proper crosswalk 2 = proper crosswalk and pedestrian lighting

Table 2. Scenarios for Pedestrian Facilities

Scenario	Variables				
	X1	X2	X3	X4	X5
1	1	1	1	1	2
2	1	2	1	2	1
3	2	1	1	2	2
4	2	2	2	1	1
5	3	1	2	2	1
6	3	2	2	1	2

2.2. Binary Logit Model

Use the Binary Logit Model to obtain the probability of willingness to walk and the factors influencing it. Binary Logit Model Analysis is used because the variables used only produce outputs, namely willing (1) and not willing (0). Logistic regression is usually used to predict the probability of a phenomenon or event.

3. Result and Discussion

3.1. Characteristic of Respondents

Respondent characteristic data obtained includes respondents' answers to the questions asked. Questionnaire results data on motorbike users and car users are shown in Table 3.

Table 3 informs that respondents who use motorbikes are dominated by people aged 25 years (34.5%), while car driver respondents are dominated by the age group over 26 years (40.5%). Nonetheless, the purpose of travel for motorcyclists and cars is dominated by work (46.5% and 52.5%). Regarding vehicle ownership, respondents who use motorbikes own at least two motorbikes (46%) with more than four family members (37.75%). On the other hand, most car drivers only own one car, with a percentage of 87.75%, and the number of family members is four.

For private vehicles, the motorcycle category prefers it because it saves time (49%), while the car category, as much as 35%, chooses it for convenience. Almost all motorcycle user respondents did not own a car, so they only used motorbikes for daily mobility. This was quite different from car user respondents, who almost owned at least one motorbike but still used a car as their primary vehicle. There is a similarity in the frequency of use of transportation for motorbikes and cars, where almost all respondents did not use public transportation services at all in the past month (87.5% and 97%).

Table 3 also shows that almost half of the motorcycle user respondents (29.5%) have a place to live near the pick-up point (stops or highways). On the other hand, almost half of car user respondents

(35.5%) had to walk for quite a long to access public transportation. This condition was one of the factors that made the percentage of car users' interest in using public transportation lower when compared to motorbikes.

Table 3. Characteristics of Respondents

	Characteristics	Motorcycle user (%)	Car user (%)
1. Age	< 17 years old	4.25	0.50
	17 – 25 years old	34.50	24.00
	26 – 35 years old	33.50	40.50
	36 – 50 years old	22.00	25.00
	> 50 years old	5.75	10.00
2. Gender	Male	47.25	61.75
	Female	52.75	38.25
3. Occupancy	PNS/TNI/POLRI	11.50	15.75
	Pegawai Swasta	30.00	30.75
	Wirausaha	8.50	15.25
	IRT	13.50	975
	Pelajar/Mahasiswa	31.50	23.75
	Lainnya	5.00	4.75
4. Trip Purpose	Work	46.50	52.50
	Education	27.25	23.00
	Recreation	3.75	7.50
	Shopping	15.75	13.25
	Others	6.75	3.75
5. Direction	Balikpapan City	15.50	25.25
	East Balikpapan	10.50	15.50
	South Balikpapan	28.00	20.25
	West Balikpapan	10.50	10.50
	North Balikpapan	22.00	20.00
	Centre Balikpapan	13.50	8.50
6. Number of trips	Twice	50.50	50.75
	3 times	8.25	18.75
	4 times	21.75	20.25
	5 times	2.25	2.75
	> 5 times	17.25	7.50
7. Reason of private mode utilization	Time saving	49.00	30.00
	Multipurpose trip	29.50	23.75
	Secure	2.50	10.25
	Comfort	18.00	35.00
	Others	1.00	1.00
8. Family member	1	1.50	2.00
	2	4.50	5.00
	3	22.50	19.75
	4	33.75	42.00
	> 4	37.75	31.25
9. Motorcycle ownership	0	0.00	3.50
	1	26.75	37.25
	2	46.00	46.75
	3	20.00	11.25
	> 3	7.25	1.25
10. Car ownership	0	75.25	0.00
	1	22.50	87.75
	2	1.50	10.75
	3	0.75	1.25
	> 3	0.00	0.25

Table 3. Characteristics of Respondents (Cont'd)

	Characteristics	Motorcycle user (%)	Car user (%)
11. Individual Income	≤ 3 million	46.00	25.75
	3 - 4 million	23.00	8.00
	4 – 5 million	25.00	27.00
	5 – 6 million	3.75	24.75
	> 6 √	2.25	14.50
12. Household income	≤ 5 million	17.00	8.00
	5 – 7 million	32.25	14.00
	7 – 9 million	16.75	35.25
	> 9 million	17.50	33.00
	Individual income as household income	16.50	9.75
13. Frequency of Public Transit utilization for a month	0	87.50	97.00
	Once	6.50	2.25
	Twice	4.00	0.75
	3 times	1.00	0.00
	> 3 times	1.00	0.00
14. Distance of access public transit	≤ 100 m	26.50	21.50
	101 – 200 m	29.50	19.25
	201 – 300 m	15.75	15.00
	301 – 400 m	6.25	8.75
	> 400 m	22.00	35.50
15. Travel Time to access public transit	≤ 5 minutes	38.00	29.50
	6 – 10 minutes	33.25	24.75
	11 – 15 minutes	15.00	15.75
	> 15 minutes	13.75	30.00
16. Knowing of Public Transport	Yes	26.50	25.25
	No	73.50	74.75
17. Intention to use Public Transport	Yes	56.00	49.75
	No	44.00	50.25

3.2. Validity and Reliability of Questioner Data

Validity and reliability tests were carried out on the respondent characteristic questionnaire at the end of the survey. The r table value for 400 respondents with a significance level of 5% is 0.098, and for a significance level of 1% is 0.128. The results of the validity and reliability tests for the motorcycle and car categories are presented in Table 4.

Table 4. Validity and Reliability of Questioner Data

Respondent class	Number of Question	Reliability
Motorcycle user	12	0.231
Car user	14	0.172

From a total of 17 question items tested for validity and reliability, there were 12 invalid items for motorcycle respondents, and for car respondents there were 14. The conditions mean that there were three invalid items in the MC category, namely age, gender, and frequency of use of public transportation. At the same time, frequency of public transport is invalid in the car user. Furthermore, the reliability test indicates that both motorcycle and car user respondent characteristic questionnaire has low reliability. This condition occurred because the question items such as age, occupation, and income per month, it is possible that the answers will change if the same questions are asked again. So, the questionnaire needs to be more consistent and more reliable.

3.3. Willingness to Walk

The willingness to walk questionnaire from this survey will produce 6 data for each respondent, so 400 respondents will produce 2,400 data. Interpretation of the binary logistic regression analysis results consists of several model tests, namely the overall model test, the goodness-of-fit test, and the coefficient of determination contained in Table 5.

Table 5. Estimated Model

	Motorcycle user	**Car user**
Overall Model Test		
-2 Log Likelihood (block number = 0)	3010.479	3326.625
-2 Log Likelihood (block number = 1)	2846.594	3244.121
Goodness Fit Test (Hosmer and Lemeshow Test)		
Chi-square (χ2)	5.871	7.861
Sig.	0.209	0.097
Nagelkerke R Square	0.165	0.045
Wald test		
Distance (X1)	36.101	23.070
Access (X2)	48.691	17.185
Comfort (X3)	2.199	4.752
Aesthetic (X3)	24.998	13.264
Omnibus test		
Chi-square (χ2)	163.885	82.504
Sig.	0.000	0.000
Odds Ratio		
Distance (X1)	2.079	1.723
Access (X2)	0.437	0.654
Comfort (X3)	0.732	0.643
Aesthetic (X3)	0.552	0.689
Coefficient		
Distance (X1)	0.732	0.544
Access (X2)	-0.828	-0.424
Comfort (X3)	-0.312	-0.441
Aesthetic (X3)	-0.594	-0.373
Constanta (C)	0.304	0.794

When the block number = 0 (the independent variable has not been included in the model), the -2 log-likelihood value for the motorcycle user is 3010.479 and decreases to 2846.594 when the block number = 1 (the independent variable is included). The same applies to car user, the -2 log-likelihood value has decreased from 3326,625 to 3244,121. The decrease in the -2 Log Likelihood value indicates that the regression model is improving, so the hypothesized model fits the data, and H0 is accepted.

Furthermore, the table chi-square value for Degree of Freedom 4 (Sum of independent variables – 1) for the 0.05 significance level is 9.488. The chi-square value on the Hosmer and Lemeshow's Test is low, namely 5.871 (MC) and 7.861 (LV). The significance value of MC (0.209) and LV (0.097) is more than 0.05, so the model is acceptable.

The Nagelkerke R Square MC and LV values obtained in the final survey were 0.165 and 0.045, respectively. So, it was concluded that the ability of the independent variables to explain the dependent variable was only 16.5% for motorbike users and 4.5% for car users. These values experienced a slight decrease compared to the Nagelkerke R Square values obtained in the preliminary survey, which yielded 18.7% for MC and 9.3% for LV.

Furthermore, the results of the Wald test showed that the variables of distance, access, comfort, and aesthetics in the MC and LV categories had a value greater than the chi-square table (1.960955), so they were declared to affect the model partially. Simultaneously, the results of the omnibus test showed that the chi-square value of motorbike and car users was more significant than the chi-square table

(2.375643). This indicated that they had a significant effect after the independent variables were entered into the logistic regression model.

The results of the Odds ratio test on motorcycle and car users show that distance has the highest value, which means that a longer walking distance under certain conditions does not reduce the respondent's interest in walking at all. Furthermore, in the MC category, comfort also contributed to an increased willingness to walk, which was quite large by 0.732 times. Meanwhile, in the LV category, the aesthetic variable gave the second largest contribution after distance, namely being able to increase the willingness to walk by 0.689 times.

Then a probability calculation is also carried out based on the value of the coefficient B presented in Table 5. Then an example of calculating the probability for scenario 1 is carried out.

The calculation of the probability for the MC category for scenario 1 produces a value of 33%. It indicates that only that is the probability of the willingness to walk for the motorcycle community in scenario 1, which has a distance of 100 meters without being supported by wide access and not equipped with other supporting facilities.

A similar method was carried out for the following scenarios for each category of vehicle users, and the recapitulation was obtained as shown in Table 6.

Table 6. Probability

Scenario	Motorcycle User	Car User
1	33%	52%
2	11%	33%
3	36%	57%
4	25%	44%
5	47%	59%
6	41%	58%

The results from Table 6. show that the highest probability for motorbike and car users is the same in scenario 5, with percentage values are 47% and 59%, respectively. Scenario 5 offers a longer walking distance but prioritizes aspects of comfort and aesthetics by providing various supporting facilities such as seats, shade, and bicycle paths, and is accompanied by good environmental management such as planting trees and plants and bollards. Based on scenario 5, the willingness to walk of respondents, both motorbike and car users is sequentially 300 m.

4. Conclusion

In conclusion, the willingness to walk of respondents who are prospective passengers of the Public Transportation and divided into motorbikes and cars is the same, namely 300 meters. Several factors influence the distance, but the results of binary logistic regression analysis show that two factors most influence willingness to walk, namely the comfort and aesthetics of a sidewalk. Thus, improvement strategies formulated to support walking comfort and foster community willingness to walk can be carried out by focusing on these two aspects. In terms of comfort, this can be done by providing flat pedestrian paths equipped with street furniture such as seats, canopies, and bicycle paths. To improve the aesthetic quality of space, this can be done by planting and arranging trees to create shady conditions, adding decorative vegetation, improving pedestrian path designs, and installing bollards.

Reference

[1] Liu, Zhili., Xudong, Jia., dan Wen, Cheng. (2012), "Solving the Last Mile Problem: Ensure the Success of Public Bicycle System in Beijing". Changsa: 8th International Conference on Traffic and Transportation Studies.

[2] Johar, A., Jain, S. S., Garg, P. K., & Gundaliya, P. J. (2015). "A study for Commuter Walk Distance from Bus Stops to Different Destination along Routes in Delhi". European Transport/Transport Europei. Issue 59, Paper no 3, ISSN 1825-3997.

[3] Wu, C. S., & Murray, A.T. (2005). "Optimizing public transit quality and system access: the multiple-route, maximal covering/shortest-path problem". Environment and Planning B, 32(2), 163-178
[4] Andersen, J.L.E. dan A. Landex. (2008), "Catchment Area for Public Transport. Lyngby", WIT Transaction on The Built Environment, Vol. 101, Halaman 175 – 184.
[5] Guerra, E., Cervero, R., & Tischler, D. (2012). "Half-mile circle". Journal of the Transportation Research Board, 2267(1), 101-109.
[6] Carr, Stephen. (1992), "Public Space", Cambridge University Press, Cambridge.
[7] Garcia-Palomares, J. C., Sousa Ribeiro, J., Gutierrez, J., & Sa Marquez, T. (2018). "Analysing proximity to public transport: the role of Street network design". Boletin de la Asociacion de Geografos Espanoles, 76, 102-130. Doi: 10.21138/bage.2517.
[8] Daniels, R., & Mulley, C. (2013), "Explaining Walking Distance to Public Transport: The Dominance of Public Transport Supply". Journal of Transport and Land Use, 6(2), 5-20.
[9] Soest, Dennis van., Rogers, Chris David Foss., & Tight, Miles. (2019), "Exploring the Distance People Walk to Access Public Transport", United Kingdom.
[10] Jiang, Y., Zegras, P.C., & Mehndiratta, S. (2012), "Walk the Line: Station Context, Corridor Type and Bus Rapid Transit Walk Access in Jinan, China". Journal of Transport Geography, Vol.20(1), Hal.1-14.
[11] Jaya, I Putu Agus Merta., Pribadi, Ocky Soelistyo., & Khotimah, Khusnul. (2021), Peningkatan Catchment Area Angkutan Perkotaan Melalui Pendekatan Willingness to Walk (Studi Kasus Trayek Angkutan Perkotaan di Kota Kupang). Sekolah Tinggi Transportasi Darat, Bekasi.
[12] Sukor, Nur Sabahlah Abdul., dan Fisal, Siti Fadhlina Muhammad. (2018), "Factors Influencing the Willingness to Walk to the Bus Stops in Penang Island", Journal of the Malaysian Institute of Planners, Vol. 16, Hal 193 – 204.
[13] Basuki, Imam. (2014), "The Willingness to Walk of Urban Transportation Passenger (A Case Study of Urban Transportation Passenger in Yogyakarta Indonesia)". Journal of Technology and Social Science, Vol.1, No.3.

IOP Conf. Series: Earth and Environmental Science 1294 (2024) 012018 doi:10.1088/1755-1315/1294/1/012018

The development of an intermodal transportation area in Banyuwangi using the transit-oriented development approach

S Sulistyono[1], D J Koesoemawati[2], A J Pramudya[2]
[1] Civil Engineering Study Program, Universitas Jember, Jember-68121, East Java, Indonesia
[2] Urban and Regional Planning Study Program, Universitas Jember, Jember-68121, East Java, Indonesia

Corresponding author: sonya.sulistyono@unej.ac.id

Abstract. Public transportation is one of the means of transport people use to move from one city to another. This makes the integration of transportation in an urban area very necessary. This study discusses the concept of developing an integrated transportation area. This research was conducted in Ketapang Village, Kalipuro District, and Banyuwangi Regency. This area is very strategic if developed into a transportation integration area. This is because this area has a train station and port crossing, and a type A terminal will be built there. However, considering the current condition of the area, this potential has not been appropriately developed. This study aims to determine the suitability of the existing conditions with the concept of transit-oriented development and to determine development priorities that can be applied in this area using an analytical hierarchy process analysis (AHP) involving relevant agencies and stakeholders. The results of this study are the suitability of the existing field conditions with the concept of transit-oriented development and the arrangement of development priorities that can be applied to this area following the concept of transit-oriented development.

1. Introduction

According to data released by Kompas in 2023, Indonesia is the 4th most populous country in the world after China, India, and the United States. According to BPS data, Indonesia's population is 275.7 million people, with 55.9% of the population living on Java Island [1]. With a large population and the vast island of Java, public transportation is needed to connect one city to another. Public transportation is all transportation models used for people's and goods' mobility needs [2]. The provision of transportation services will be an important element of future social, economic, and environmental sustainability in a region [3]. One aspect that supports sustainability is the integration of transportation modes in an area or city. Transportation integration is essential in an area to facilitate the movement of passengers and goods from their origin to their destination [4]. If the transfer of transportation modes can be made more accessible, faster, and more convenient, overall network integration and flexibility will significantly increase [5].

One of the areas planned to become an integration area is the Ketapang intermodal transportation area. An intermodal transportation area is an area of transportation of people and goods from the origin to the destination with at least two modes of transport where the movement is carried out in one area, namely the intermodal terminal [6]. The Ketapang intermodal transportation area is in Ketapang Village, Kalipuro District, Banyuwangi Regency. This area has train stations, Ferry Ports, and type A terminals that are being planned. According to the RTRW of East Java Province 2011–2031, the development of transportation systems is carried out by developing an integrated network supported by facilities and infrastructure. There is also an increase in RTRW in Banyuwangi Regency 2012- 2032, which is said to increase intermodal transit around Ketapang Port and reaffirmed in the Strategic Area Spatial Plan of Ketapang Port and RDTR Part of the Banyuwangi Urban Area 2016-2036, which stipulates this area as an intermodal terminal sub-area.

In addition to this potential, there are several problems found in the Ketapang Intermodal Transportation Area, such as the lack of crossing facilities, the unavailability of signage, the fact that not all areas have pedestrians, and others that make the area comfortable and safe to be used as an integration area. From the potential problems and existing problems, the study was conducted using the transit-oriented development (TOD) approach using the theory of the 2012 Florida guidebook, as evidenced by ITDP 2017. There are three variables to be used: density (building density, BCR, and FAR), diversity (residential and non-residential land use), and design (pedestrian, crossing facilities, bicycle lanes, bicycle parking, motor vehicle parking, shade or shielding, and transit points).

This study aims to determine the suitability of existing conditions with the concept of transit-oriented development using qualitative descriptive analysis and find development priorities that can be applied to this region using analytical hierarchy process (AHP) analysis involving relevant agencies and stakeholders so that this research can make the Ketapang intermodal transportation area a safe and comfortable area for passengers and residents who are active in this region.

2. Method

2.1. Research Location

This research was conducted in Ketapang Village, Kaliporo District, Banyuwangi Regency, East Java. Ketapang Railway Station and Ketapang Ferry Port are within the research area, and the construction of intermodal and intermodal terminals in one research area is being planned. This study used the concept of transit-oriented development in Florida in 2012. The location of TOD according to the Florida guidebook, the travel time of the TOD area is about 10 minutes on foot (about 500–800 meters) from the transit center [7]. Where the TOD center of the Ketapang transportation integration area is located at the intermodal terminal according to the spatial plan of the strategic area of the Ketapang port and RDTR for the Banyuwangi urban area in 2016–2036 in Article 69, Paragraph 3, the radius of the research area used is 500 meters. The study area has temperatures between 20 0 C-36 ^{0}C., the most considerable rainfall between 301-400mm, and the lowest 58.2mm.

2.2. Sampling Techniques and Research Sample

The sampling technique used in this study used purposive sampling techniques. Purposive sampling collects data by selecting respondents most likely to agree with the research conducted [8]. The respondents involved are: 1) Government: Public Works Office, Cipta Karya, Housing and Residental, Transportation Office, ATR/BPN Office, and Ketapang Village Office; 2) Stakeholders: PT. KAI, PT. ASDP Indonesia Ferry.

2.3. Research Variable

The variables in this study are based on the theory of transit-oriented development (TOD) sourced from the 2012 Florida Guidebook, which is the same as the theory of ITDP 2017. The research variables used can be seen in Table 1.

Table 1. Research Variables.

Variable	Indicator	Source
Density	Land use density	Florida guidebook 2012
	Building Coverage Ratio	Florida guidebook 2012
	floor area ratio	Florida guidebook 2012
Diversity	Residential land use	Florida guidebook 2012
	Non-residential land use	Florida guidebook 2012
Design	Pedestrian availability	Florida guidebook 2012

Variable	Indicator	Source
	Bike lane availability	Florida guidebook 2012
	Bicycle parking availability	Florida guidebook 2012
	Vehicle parking availability	Florida guidebook 2012
	Transit point availability	Florida guidebook 2012
	Availability of shade and protector	ITDP 2017
	Availability of crossing facilities	ITDP 2017

2.4. Formatting the title

The analytical method used in this study uses two analyses, namely qualitative descriptive analysis and analytical hierarchy process (AHP) analysis. Qualitative descriptive analysis is an analytical method used to find knowledge or theories of research at a specific time. Meanwhile, AHP Analysis is an analysis procedure using paired comparisons designed for relative assessment and ensuring consistency [9]. The objectives and results of the analysis that will be presented can be seen in Table 2

Table 2. Analysis Methods.

Analysis techniques	Purpose	Analysis results
Qualitative Descriptive Analysis	Knowing the compatibility of existing conditions with the TOD concept in the intermodal transportation area of Ketapang	Classification of the suitability of existing conditions with the concept of TOD, which is divided into four categories (unavailable, not appropriate, close to proper, and appropriate).
Analytical Hierarchy Process (AHP) Analysis	Knowing the most appropriate planning priorities applied to the Ketapang intermodal transportation area	The most appropriate planning ranking is applied to the intermodal transportation area of Ketapang.

3. Results and Discussion

In this study, the division of observation areas based on blocks was carried out to analyze the suitability of TOD theory. The study area is divided into four blocks. Block maps are divided by physical boundaries and most land use. The block division includes Block 1 (Industry and warehousing), Block 2 (High-density residential), Block 3 (Medium-density residential), and Block 4 (trade and services). This block division is intended to make it easier to analyze data [10]. The block division map can be seen in Figure 1.

3.1. Compatibility of existing conditions with the TOD concept

First is variable density. Density is considered important because it directly affects the running of an area or city. After all, higher environmental density has more purposes as well as parks or retail [11]. Variable density is divided into three sub-variables: building density, Building Coverage Ratio (BCR), and floor area ratio (FAR). According to Florida, the intermodal transportation area of Ketapang is included in the type of community center. A community center is the center area of sub-regional or local economic and community activities. It is included in urban centers and cities served by one or more transit centers [12].

According to Florida theory, the building density for the appropriate ketapang intermodal transportation area is 60-80 buildings per hectare. States that building density affects the intensity of the built-up areas, which optimizes land capabilities compared to the land area [8]. Based on the results of field observations and analysis results, it was found that the population density for all blocks in the Ketapang transportation intermodal area received a value of 47 buildings per hectare. For sub-variables, The corresponding Building Coverage Ratio (BCR) within the TOD area is 80-90%. This BCR serves as a tool that regulates the intensity of development, whose function is to control land cover by buildings

IOP Conf. Series: Earth and Environmental Science 1294 (2024) 012018 doi:10.1088/1755-1315/1294/1/012018

[13]. Based on the results of field observations and analysis results, it was found that the BCR found in the intermodal transportation area on average was 80-90%. The TOD area's corresponding floor area ratio (FAR) is > two floors for sub-variables. FAR (floor area ratio) ensures that the carrying capacity of land does not receive excessive loads from human activities [14]. Based on field observations, it was found that the FAR contained in the Ketapang transportation intermodal area was still, on average, one floor. For details of variable density per block can be seen in Table 3.

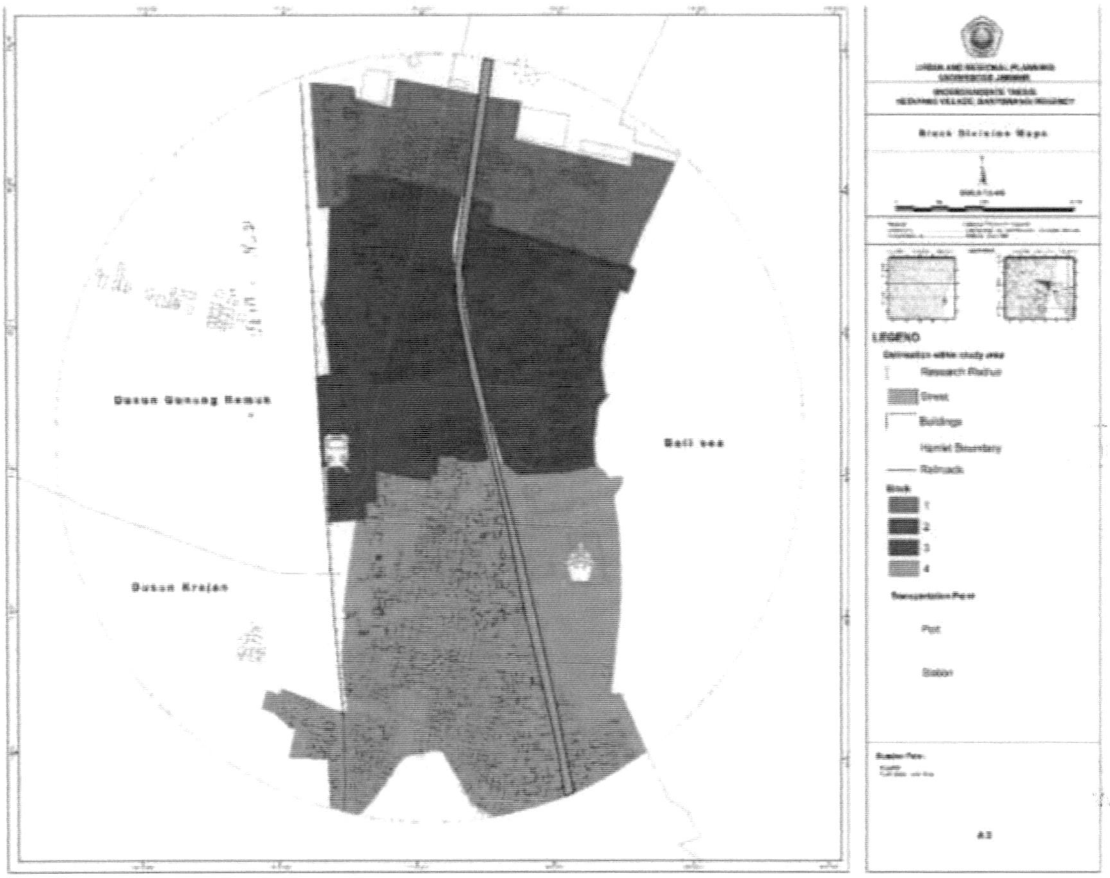

Figure 1. Blok division map.

Table 3. Density variables.

Block	Density (building/ha)	BCR (%)	FAR (%)
1	28	30-100	1-2
2	65	40-100	1-2
3	24	30-100	1-2
4	57	15-100	1-3

The second is the diversity variable; the diversity variable is divided into two sub-variables, namely residential and non-residential land use. In diverse areas, the TOD conformity of residential is 45%, and non-residential is 55%. Diversity or mixed land use is one of the supporting aspects of compact cities in sustainable urban development, focusing on an area development in one unit [15]. The results of field observations and analysis results, it was found that diversity in this area can be seen in Table 4.

Table 4. Diversity variables.

Blok	Residential (%)	Non-residential (%)
1	75,8	24,2
2	86,8	13,2
3	74,6	25,4
4	92,2	7,8

The third variable is design; the design variable is divided into seven sub-variables: pedestrian, crossing facilities, bicycle lanes, bicycle parking, motor vehicle parking, shade or protection, and transit points. The first sub-variable is pedestrian. Pedestrian planning must provide security and be easily accessible and well-connected for pedestrians [16]. According to Florida 2012, suitable pedestrians are available in all areas, or 100% of areas have pedestrians and have a width of 2 meters equipped with vegetation. From the results of field observations, it was found that not all areas have pedestrians. The available pedestrians have a 1.6-1.7 meter width and no vegetation.

The second sub-variable is the crossing facility. Crossing facilities must guarantee the safety of pedestrian crossings by completing the road speed management system, the road system, and supporting facilities for road crossings [17]. According to the ITDP 2017 theory, crossing facilities must have a minimum width of 2 meters, be barrier-free for people with disabilities, and have sufficient lighting [18]. According to field observations, two zebra crossings with a width of 3 meters were found in front of Ketapang port and military area.

The third sub-variable is vehicle parking. According to a 2012 Florida theory, off-street parking facilities should not exceed 50% of any public facility or building. And parking is not allowed facing directly to the road. According to field observations, all blocks already have parking spaces and do not exceed 50% of the facility's area.

Next are sub-variables that are not available at the study location. According to a 2012 Florida theory, bike lanes should consider safety and comfort factors. A good bicycle lane is a bicycle lane that has clear and unfading road markings [19]. Bicycle parking must provide security and comfort for parking space users [20]. According to Florida theory, bicycle parking is provided at each transit facility with at least 12 bicycle parking spaces. According to the theory of ITDP 2017, protective facilities, according to the TOD concept, if the shade or protective area is more than 75%. According to Florida 2012, the distance of the transit point is 500-1000 meters facilitated with a protected waiting room and information boards. For details of design variables per block can be seen in Table 5.

Table 5. Design variables.

Block	Pedestrian (%)	Crossing facilities	Motor vehicle parking (%)	Bicycle parking	Bike lanes	Shading or protector	Transit points
1	0	Available	<50	0	0	0	0
2	44	Not available	<50	0	0	0	0
3	44	Not available	<50	0	0	0	0
4	100	Available	<50	0	0	0	0

After observation and data obtained for each variable, it can be concluded that the suitability of the existing conditions of the Ketapang intermodal transportation area with the transit-oriented development (TOD) concept can be seen in Table 6.

Table 6. TOD conformity.

Variable	Sub variables	Block	Not available	Not Appropriate	Approaching appropriate	Appropriate
Density	Land use density	1		X		
		2				X
		3		X		
		4			X	
	Building Coverage Ratio	1			X	
		2			X	
		3			X	
		4			X	
	floor area ratio	1		X		
		2		X		
		3		X		
		4		X		
Diversity	Residential and non-residential	1		X		
		2		X		
		3		X		
		4		X		
Design	Pedestrian	1	X			
		2		X		
		3		X		
		4				X
	Crossing facilities	1				X
		2	X			
		3	X			
		4				X
	Bike lanes	1	X			
		2	X			
		3	X			
		4	X			
	Bicycle parking	1	X			
		2	X			
		3	X			
		4	X			
	Vehicle parking	1				X
		2				X
		3				X
		4				X
	Shading or protector	1	X			
		2	X			
		3	X			
		4	X			
	Transit points	1	X			
		2	X			
		3	X			
		4	X			

IOP Conf. Series: Earth and Environmental Science 1294 (2024) 012018 doi:10.1088/1755-1315/1294/1/012018

3.2. Prioritization

Prioritization of planning in the Ketapang transportation intermodal area using the analytical hierarchy process (AHP) method with data sorting using the expert choice application. Priority for variables that include density, diversity, and design. The results of the AHP questionnaire for variables can be seen in Figure 2.

Figure 2. AHP variable results.

It can be seen in the priority of planning in the intermodal transportation area Ketapang is more to the Diversity sector (mixed land use) with a yield of 54.1%, then design with 25.1% and density (building density) with 20.9% with 8% inconsistency.

The first thing to be discussed is the sub-variable diversity covering residential and non-settlement land. The results of AHP sub-variables contained in the diversity variable can be seen in Figure 3.

Figure 3. AHP sub-variable diversity results.

It can be seen that the priority of land use in the intermodal transportation area is more towards non-residential land use, with 81.1%. Meanwhile, residential land use scored 18.9% with 0% inconsistency.

The second is the design sub-variables, including pedestrians, crossing facilities, bicycle lanes, bicycle parking, motor vehicle parking, shade or protection, and transit points. The results of AHP sub-variables contained in the design variables can be seen in Figure 4.

Figure 4. Results of AHP sub-variable design.

It can be seen in the design priorities in the intermodal transportation area Ketapang is more to crossing facilities at 26.5%, vehicle parking at 19.5%, pedestrians at 14.1%, shade or protectors at 11.9%, and transit points at 11.6% while the sub-variables of bicycle parking 10.2% and bicycle lanes with 7.5% with inconsistency 4%.

The third is the density sub-variable, which includes building density, building base coefficient, and floor area ratio. The results of AHP sub-variables contained in the density variable can be seen in Figure 5.

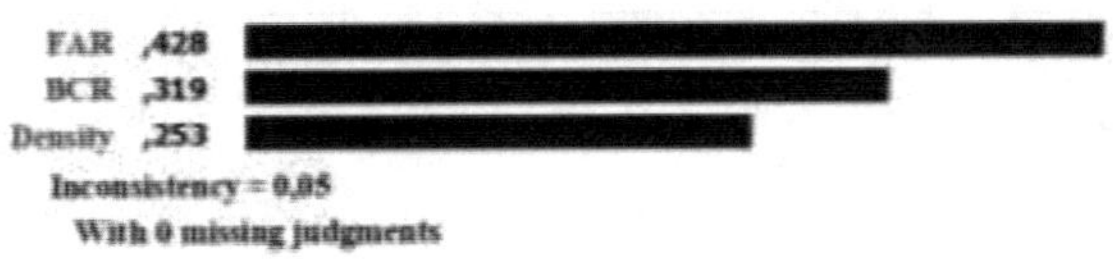

Figure 5. AHP sub-variable density results.

IOP Conf. Series: Earth and Environmental Science 1294 (2024) 012018 doi:10.1088/1755-1315/1294/1/012018

It can be seen that the priority of building density planning in the intermodal transportation area is more to design the floor area ratio with 42.8%. While the basic building coefficient is 31.9%, and the sub-variable of land use density is 25.3% with an inconsistency of 5%.

Next is a discussion of alternative integration area planning. Alternative planning in the Ketapang intermodal transportation area takes the integration at Solo Balapan Station (integration through a pedestrian bridge) and Pasar Senen Station (Integration through pedestrian fireplaces and the addition of shade). Alternative AHP results can be seen in Figure 6.

Figure 6. Result of AHP Alternative Planning.

It can be seen in Figure 8 that alternative planning in the Ketapang intermodal transportation area is more to the integration of the Solo Balapan station with integration through a pedestrian bridge with 86.6% and Pasar Senen station with integration in the form of pedestrian fireplaces and the addition of shade by 13.4%.

4. Conclusion

The conditions in the Ketapang intermodal transportation area still have TOD sub-variables that have not been adjusted, and some follow existing theories. Sub-variables follow the TOD concept found in building density (block 2), pedestrian (block 4), crossing facilities (block 1 and block 4), and availability of vehicle parking (all blocks). Sub-variables that are close to TOD theory are found in land use density (block 4) and building base coefficient (all blocks). Sub-variables that do not follow the TOD concept are found in land use density (blocks 1 and 3), floor area ratio (all blocks), residential and non-residential land use (all blocks), and pedestrian (blocks 2 and 3). In addition, there are also TOD sub-variables that are not available, such as pedestrian (block 1), crossing facilities (blocks 2 and 3), bicycle lanes (all blocks), bicycle parking (all blocks), shade and protection (all blocks), and transit points (all blocks).

Because many areas still have not applied the TOD concept, it is necessary to determine planning priorities in the Ketapang intermodal transportation area with AHP analysis. It is known that the most priority variable is Diversity 54.1% (non-Residental 81.1% and Residental 18.9%), then design with 25.1% (crossing facilities 26.5%, vehicle parking 19.5%, pedestrian 14.1%, shade or protector 11.9%, transit point 11.6%, bicycle parking 10.2%, and bicycle lanes 6.3%) and finally density 20.9% (FAR, 42.8%, BCR 13.9% and building density 25.3%) Then for alternative planning more to integrate planning through skywalk (pedestrian bridge) as available at Solo Balapan station with 86.6% of the results it is known that the intermodal transportation area is more devoted as a non-residential area by paying attention to the floor area ratio, which is a minimum of 2 floors. As well as to become a transportation integration area, an integration system of transportation infrastructure with a skywalk system or flyover with several supporting facilities such as crossing facilities, motorized vehicle parking, bicycle and pedestrian parking must be available in the Ketapang transportation intermodal area.

5. References

[1] BPS 2023 *statistik indonesia 2023* (Jakarta: badan pusat statistik)

[2] Karim A, Lesmani L, Sunarta D A, Suparman A. Yunus A I, Khasanah, Marlita D, Saksono H, Asniar N and Andari 2023 T *Manajemen Transportasi* (Batam: Yayasan Cendikia Mulia Mandiri) 1

[3] Murray A T, Davis R, Stimson R J and Ferreira L 1998 Public Transportation Access *Transp. Res. Part D Transp. Environ* **3** 5

[4] Masiero L, Hrankai R, and Zoltan J 2022 The role of intermodal transport on urban tourist mobility in peripheral areas of Hong Kong *Res. Transp. Bus. Manag* **48**

[5] Malik A, Murtejo T, and Alimuddin A 2022 Analisis Kebutuhan Fasilitas Integrasi Moda Krl Stasiun di Kawasan Kota Bogor *J. Civronlit Unbari* **7** 1

[6] Crainic T G and Kim K H 2007 *Intermodal Transportation* C. Barnhart and G. Laporte (Korea: Pusan National University, Elsavier) 14 467–537

[7] Busha M, Brunot M, DeLaney K, Fett S, Gianniotes A, Hatton W, Hatton C, Little D and Sabol E 2012 *Florida TOD Guidebook* (Florida: Florida Department of Transportation) 256

[8] Campbell S, Greenwood M, Prior S, Shearer T, Walkem K, Young S, Bywaters D and Walker D 2020 Purposive sampling: complex or simple? Research case examples *J. Res. Nurs* **25** 652–661

[9] Chen C F 2006 Applying the Analytical Hierarchy Process (AHP) Approach to Convention Site Selection *J. Travel Res* **4** 2

[10] Glediswandi L and Widartono B S 2013 Pemanfaatan Citra SatelitWorldview dan SIG untuk Evaluasi Pemanfaatan Ruang Terhadap Rencana Detail Tata Ruang Sebagian Kota Wates Tahun 2003 - 2013 Kabupaten Kulon Progo *J. bumi Indonesia* **2** 4

[11] Lek Y, Sangkertadi, and Moniaga I L 2014 "Kadatan Bangunan dan Karakteristik Iklim Mikro Kecamatan Wenang Kota Manado," *J. Lingkung. binaan dan Arsit.*, **6** 3

[12] Lu Y, Xiao Y, and Ye Y 2017 Urban density, diversity and design: Is more always better for walking? A study from Hong Kong *Prev. Med. (Baltim).***103** S99–S103,

[13] Purnomo A B 2021 Koefisien Dasar Bangunan (KDB) dan Koefisien Lantai Bangunan (KLB) Sebagai Dasar Gubahan Massa *Metr. Ser. Hum. DAN SAINs* **2** 2

[14] Joshi K K and Kono T 2009 Optimization of floor area ratio regulation in a growing city *Reg. Sci. Urban Econ.* **39** 4

[15] Sagnita A and Handayani W 2017 Pengukuran Penggunaan Lahan Campuran (Mixed Use) Dengan Indeks Entropy di Kota Semarang *J. riptek*

[16] Gaglione F, Gargiulo C, Zucaro F, and Cottrill C 2022 Urban accessibility in a 15-minute city: a measure in the city of Naples, Italy *Transp. Res. Procedia* **60** 1

[17] Khayesi M and Tiwari G 2023 *Pedestrian safety: a road safety manual for decision-makers and practitioners* (Delhi: World Health Organization),

[18] Institute for Transportation and Development Policy 2017 *TOD Standard 3.0* (New York: Our Cities Ourselves: Principles of Transport in Urban Life) 3

[19] Gumelar A, Dewanti, and Priyanto S 2023 Kinerja dan Rekomendasi Pengembangan Fasilitas Sepeda di Provinsi Daerah Istimewa Yogyakarta *J. Civ. Eng. Plan.* **4** 1

[20] Taufiqurrohim M, Nurhidayat M, and Pujiraharjo Y 2023 Perancangan Tempat Parkir Sepeda Minimalis *e-Proceeding Art Des.* **10** 3

Simulation of the impact of tram operations on road network performances

K H Basuki[1*], D Purwanto[1], A K Indriastuti[1], F R Ananda[1]

[1] Civil Engineering Department, Faculty of Engineering, Diponegoro University, Jalan Prof. Sudharto, Semarang 50275 Indonesia

kamiharibasuki@lecturer.undip.ac.id

Abstract. The City government of Semarang has a plan to build a tram way in Semarang Old City Area as a public transportation mode to accommodate tourism activities. The existence of tram way can impact the performance of existing road network due to being built on existing roads side by side with other vehicles. Street network performances discussed in this paper are on the vehicle-kilometres travelled (VKT), and vehicle-hours travelled (VHT), to determine the aforementioned changes, traffic simulation was done with PTV VISSIM software. Vehicle-kilometres travelled decreased by 12,76%, vehicle-hours travelled increased by 10,54%. To minimize the impacts, two mitigation scenarios are simulated, mitigation 1 which is one-way system and mitigation 2 the combination of one-way system and signal optimizations. In mitigation scenario 1 the value of VKT decreased by 13,12%, VHT decreased by 0,65% compared to existing conditions. In mitigation scenario 2 the value of VKT decreased by 17,54%, VHT decreased by 4,93%.

1. Introduction

Semarang is the capital of Central Java Province in Indonesia. Situated on the economically significant pantura area (Java island northern coast) that is connected by railways and highways to various cities, and was a significant Dutch colonial city pre-independence, Semarang attracts both economic and tourism interests. One of the tourism attraction in Semarang is the Kota Lama area, it was the city center in the 18th to 19th century AD [1].

The City Government of Semarang suggests to build a tram way to accommodate transportation needs. Tram is defined as a light rail system on level with existing roadways [2]. This makes tram existence impacts the performance of existing roadways, with one study in Surabaya, Indonesia concludes that tram existence can raise degree of saturation (DS) value of Darmo street above 0.75, which, according to Indonesian Highway Capacity Manual is considered a traffic jam [3]

This paper examines the effects of tram existence on the performance of road networks, and proposes possible mitigation scenarios to mitigate the effects, with the approach of comparing the change in value of Vehicle-kilometers Traveled (VKT) and Vehicle-hours Traveled (VHT). The changes in VKT and VHT values are acquired from the process of traffic microsimulation using PTV VISSIM software.

2. Literature Review

2.1. Existing Researches on Tram Operations Impacts

There has been a number of studies exploring various aspects of tram railways system, for example a study done by Naznin explores the safety aspects of tram operations[4], however most of the studies were done outside Indonesia, since Indonesia currently does not have any active tram railways system. An example of a study in Surabaya, Indonesia concludes that the existence of tram railways causes potential traffic jam [3], this study was done with the scope of one road only using degree of saturation (DS) value as its performance indicator, which is commonly used in Indonesia to determine the performance of a single road or intersection.

Most studies on tram operations impacts are done with focus only on the impacts on a road link or corridor, with no studies assessing the network-wide effect of the existence of a tram network in Indonesia. An example of a network-wide study through simulation with PTV VISSIM was done in Melbourne, Australia where volume of vehicles reduced by 19.6% and travel time increases by 14.7% [5], however this study did not explore any possible mitigation scenarios.

2.2. Traffic Simulation

For traffic performance studies with the scope of a single road or intersection, determining the performance is possible using manual calculation methods, the existence of computer softwares only serve to simplify the processes, however to determine the performance of a whole road network that consists of multiple roads and intersection, with the whole trip path of all vehicles passing through the road network cannot be determined with current survey methods, the existence of traffic simulation softwares becomes a necessity. Traffic simulation software can also be used to model different alternative scenarios of traffic adjustments before it is implemented, predictions of behavior and performance changes due to changes on the roads and intersections [6]

3. Research Method

3.1. Data Gathering Process

Secondary data used are existing road geometry data, tram rail and car specifications, and tram route plan provided by the City Government of Semarang. Primary data used are signal configurations, vehicle running speed, and traffic volume gathered by drone on observed roads and intersections. Vehicles are classified to three types according to Indonesian Highway Capacity Manual, which are Motorcycle (MC), Light Vehicle (LV), and Heavy Vehicle (HV). Signal configurations on each signalled intersections were counted manually by stopwatch.

3.2. Simulation Modelling

To create the model necessary for simulation in PTV VISSIM, following data have to be inputted:

- Existing road and tram way geometric data
- Traffic volume
- Vehicle routing decisions
- Vehicle speed
- Signal heads and configurations on intersections
- Driving behavior
- Vehicle travel time measurement points

In each run of simulations, PTV VISSIM will simulate the inputted vehicles on the modelled road network, and measure the number of vehicles passing through the assigned vehicle travel time measurement points with the travel length and time taken.

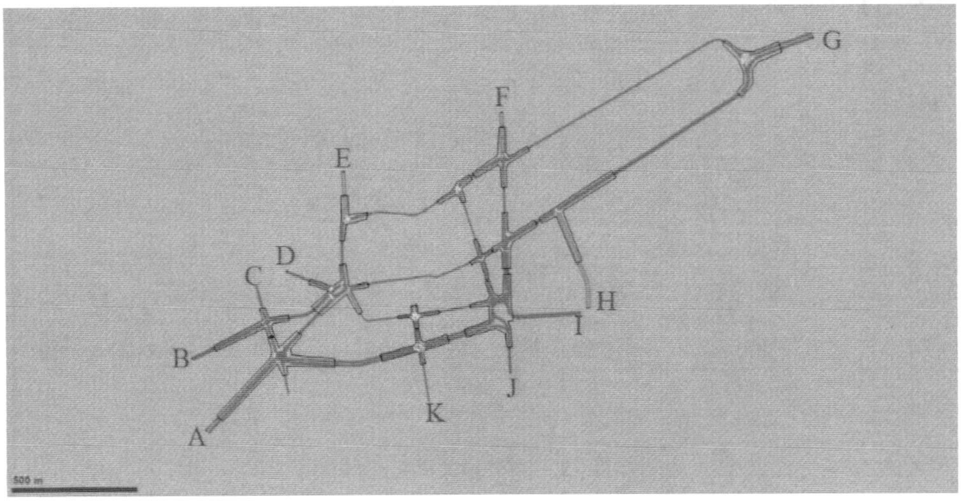

Figure 1. Vehicle Travel Time Measurement Points.

Table 1. Code Assignment of Vehicle Travel Time Measurement Points.

Code	Road Name	Traffic	
		Entry	Exit
A	Jalan Pemuda	v	v
B	Jalan Imam Bonjol	v	v
C	Jalan Kolonel Sugiono (1)	-	v
D	Jalan Kolonel Sugiono (2)	v	-
E	Jalan Empu Tantular	v	v
F	Jalan Ronggowarsito	v	v
G	Jalan Kaligawe Raya	v	v
H	Jalan Widoharjo	v	v
I	Jalan Pattimura	-	v
J	Jalan MT Haryono	v	-
K	Jalan Pekojan	-	v

It is assumed that no transportation mode changes happen with the existence of tram, due to the planned tram being purposed for tourism purposes.

3.2.1. Tram modelling. PTV VISSIM software has a default 3D model for trams which numbers of cars can be configured, dimensions are assumed to be close enough to the planned specifications, gauge width is currently undetermined since Indonesia typically uses 1067 mm cape gauge for metro trains, long-distance trains, and Jakarta Mass Rapid Transit (MRT) system, but implementations in Indonesian High Speed Railways and Jakarta Light Rail Train (LRT) uses 1435 mm standard. For the planned implementations in Semarang, the tram model uses 2 car configuration and takes 3900 mm clearance width on the road [7].

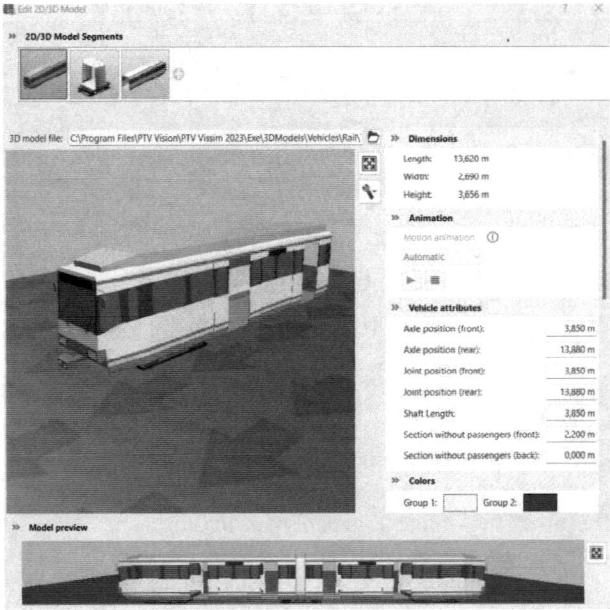

Figure 2. Tram Model Input

The planned tram implementations follows the traffic rules and signals, therefore the tram lane is not exclusive to tram only, except in certain parts where the tram traffic goes against road vehicle traffic. Input of tram line uses the "Public Transport Line input" of PTV VISSIM shown in Figure 3 below.

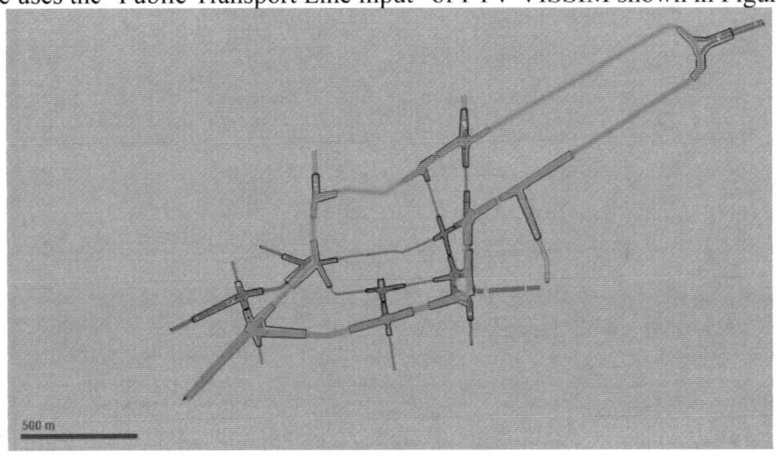

Figure 3. Tram Line Input

3.3. Model Validation

To ensure the accuracy of the simulation model towards the inputted traffic volume data in existing conditions, validation was done with Geoffrey E. Havers (GEH) formula. The GEH equation model is a modified statistical formula of Chi-square by combining the difference between relative and absolute values [8]. The GEH equation is shown in equation 1 and the meaning of the GEH number value is shown in table 2

$$GEH = \sqrt{\frac{(q_{simulated} - q_{observed})^2}{0,5*(q_{simulated} + q_{observed})}} \quad (1)$$

Table 2. GEH Value criteria.

GEH Value Range	Description
GEH < 5,0	Good model, accepted
5,0 ≤ GEH ≤ 10,0	Bad model, possible errors
GEH > 10,0	Model have to be rejected

Validation is done when modelling the existing conditions before modelling the other possible scenarios, by creating model with default driving behavior parameters and calibrating such parameters afterwards. The calibration parameters are based on previous research by Putri et al. [8], with additional changes made by trial and error in the last case.

Table 3. Changed Driving Behavior parameters.

No.	Changed Parameters	Inputted Value		
		Trial 1 (Default)	Trial 2[8]	Trial 3
1	Desired Position at free flow	Middle of Lane	Any	Any
2	Overtake on same lane: on left & on right	Off	On	On
3	Distance standing (at 0 km/h) (m)	1	0,2	0,2
4	Distance driving (at 50 km/h) (m)	1	0,4	0,4
5	Average standstill distance (m)	2	1	0,6
6	Additive part of safety distance (m)	2	1	0,6
7	Multiplicative part of safety distance (m)	3	2	1
8	Observe Adjacent Lane	off	off	on
9	Safety Distance Reduction Factor	0,6	0,6	0,2
10	Consider next turn	off	off	on
11	Overtake reduced speed areas	off	off	on

Table 4. GEH Value results.

Case	Average GEH Value
Trial 1	14,35839
Trial 2	9,09742
Trial 3	6,50187

The last calibrated case is not considered a good model according to criteria in Table 2, this is due to the data gathering process of the traffic volume not done at the exact same time in all data gathering points.

3.4. Scenarios Modelled

The scenarios modelled in this paper are :

- Existing conditions
- With tram conditions
- Mitigation 1 (One way system)
- Mitigation 2 (One way system + signal optimizations)

3.4.1. One way system. One way system is one of the possible ways to mitigate the impacts of tram operations on road network performance, one way system enables more vehicles to enter the road network, or enables vehicles to travel in higher speeds due to less congestions, thus reducing travel times and hence, Vehicle Hours Traveled (VHT).

One way system is applied in Mitigation 1 and Mitigation 2 scenarios on following roads:

- Jalan Widoharjo (1)
- Jalan Ronggowarsito (2)
- Jalan Cendrawasih (3)

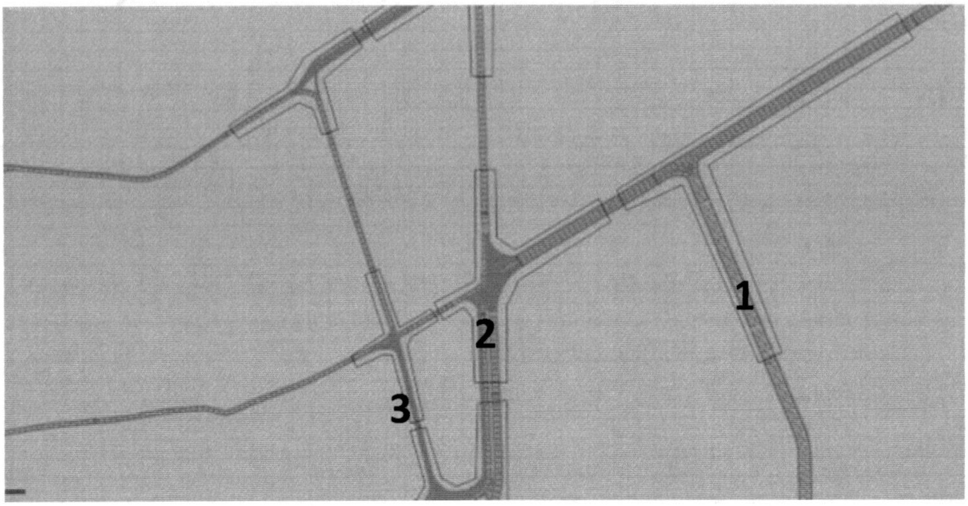

Figure 4. Location of Proposed One Way System.

3.4.2. Signal optimizations. Another way to mitigate the impacts of tram operations is through optimizing signal configurations on signalized intersections. Using the built-in "optimize traffic signals" functions in PTV VISSIM software, the software will determine the new signal configurations by using simulated trial-and-error iterations.

Figure 5 shows the example of pre-optimization and post-optimization signal diagram in KH Agus Salim intersection.

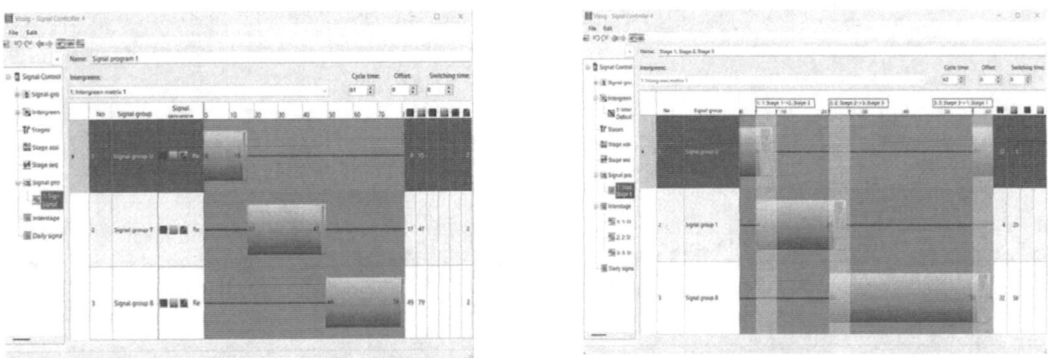

(a) Before optimization **(b) After optimization**

Figure 5. Signal Diagram in KH Agus Salim Intersection.

3.5. Vehicle Kilometers Traveled (VKT) and Vehicle Hours Traveled (VHT)

VKT (or VMT, Vehicle Miles Traveled in United States) is a metric used extensively in transportation planning for a variety of purposes. It is a measurement of the amount of vehicular travel in a geographic region over a given period of time. It refers to the total number of kilometers traveled by all vehicles over a given period of time. VKT data are used primarily by transportation agencies, environmental agencies, and consultants to perform a variety of functions such as allocating resources, estimating vehicle emissions, computing energy consumption, and assessing traffic impacts. VKT may also be used to evaluate conformity assumptions, adjust travel demand forecasts, and identify pavement maintenance needs. It is a key metric in transportation planning because it provides an indication of travel demand and behavior. Additionally, policy decisions and infrastructure investment planning incorporate VKT data [9].

Factors that influence VKT value includes route choice and traffic volume. VKT is calculated using the formula shown in equation 2:

$$VKT\ (vehs.km) = volume * travel\ length \tag{2}$$

Vehicle Hours Traveled (VHT) provides a proxy measure of the overall pressure on a road network due to the congestion impacts of regional travel [10]. VHT can be used to analyze economic impacts due to changes on road networks, specifically on trip time value analysis. VHT is calculated using the formula shown in equation 3:

$$VHT\ (vehs.h) = volume * travel\ time \tag{3}$$

VHT is tied to the travel time needed by vehicles to complete a trip, which is influenced by:

- Delays (due to stops in intersections, traffic jam)
- Vehicle speed
- Driving behavior and interactions between vehicles

4. Results

4.1. VKT Comparison

The VKT calculations results of the scenarios simulated in this paper is shown in table 5 and figure 6 below :

Table 5. VKT Values Comparison.

No	Scenario	VKT (vehs.km)	Change (%)
1	Existing	20331,20439	0,00%
2	With Tram	17736,33179	-12,76%
3	Mitigation 1	17664,47487	-13,12%
4	Mitigation 2	16756,46235	-17,58%

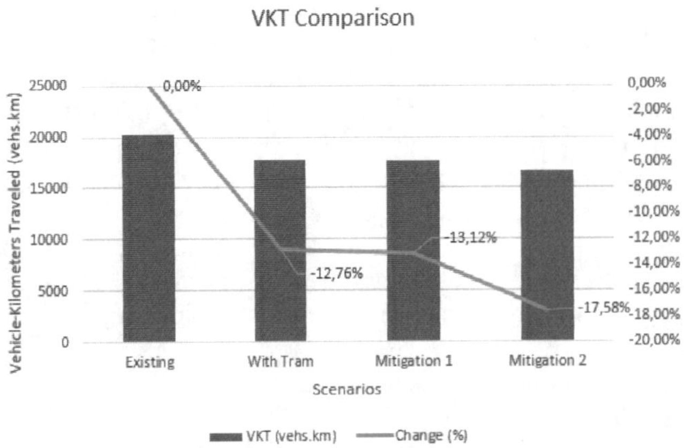

Figure 6. VKT Comparison Graph.

Since traffic volume is defined as the number of vehicles passing over a point at one time, in a traffic jam the measured number of VKT can be lower than in better traffic conditions, due to less vehicle movements passing through the measurement points at one time, however traffic conditions cannot be determined by only looking at the VKT value without considering other indicators.

By only looking at the VKT value changes, in the "with tram" scenario, assuming route choices of the vehicles are unchanged and thus, the travel lengths are also unchanged, this means there are less vehicle able to pass the measurement points at one time in the simulation run, however to determine whether this means the traffic condition is better or worse, VHT value results have to be considered.

4.2. VHT Comparison

The VHT calculations results of the scenarios simulated in this paper is shown in table 6 and figure 7 below :

Table 6. VHT Values Comparison.

No	Scenario	VHT (vehs.h)	Change (%)
1	Existing	1661,015154	0,00%
2	With Tram	1836,016159	10,54%
3	Mitigation 1	1650,144056	-0,65%
4	Mitigation 2	1579,09144	-4,93%

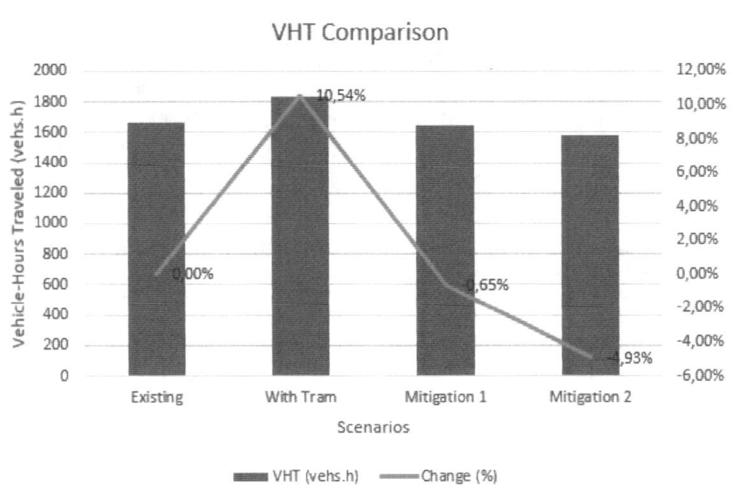

Figure 7. VHT Comparison Graph.

VKT interpretation is tied to the VHT results. In the "with tram" scenario, despite the reduction in VKT value, the VHT value raises up, this proves that the reduction in volume does not happen due to reduction in vehicle use but due to worse traffic conditions, since even with lower traffic volume, the accumulated delay is higher.

5. Conclusions and Suggestions

This paper concludes that the operation of tram on existing roads can negatively impact the performance of road networks. In the proposed Semarang Tram Railways plan, it reduces VKT by 12,76% and increases VHT by 10,54%. Proposed mitigation scenario of one way system successfully reduces the VHT even slightly lower than in existing conditions, however the VKT value does not change significantly. Combination of one way system and signal optimizations further reduces VHT value to be 4,93% lower than existing conditions, however the VKT value decreases even further to be 17,58% below the existing conditions. One way system is recommended to be applied in some roads if the Semarang Tram Railways plan is approved to be built.

It is suggested that to create a more representative modelling, that in further researches, potential mode choice changes have to be considered with the existence of tram as a new mode choice, more suitable survey methods also have to be considered for gathering multiple data at multiple points at the same time to make the simulation model more representative.

Acknowledgements

Usage of PTV VISSIM in the process of writing this paper is licensed by PTV Group Asia Pacific, Pte Ltd., Singapore, as the representative of PTV Plannung Transport Verkehr GmbH Karlsruhe, Germany, the developer of PTV VISSIM, license was provided free-of-charge for academic purposes.

Survey equipments are provided by the Transportation Engineering Lab, Department of Civil Engineering, Faculty of Engineering, Diponegoro University, Semarang, Indonesia.

Data of tram plans are provided by the Transportation Bureau of the City Government of Semarang and CV HRV Consultant as the appointed planner consultant bureau.

IOP Conf. Series: Earth and Environmental Science 1294 (2024) 012019 doi:10.1088/1755-1315/1294/1/012019

References

[1] Ratih Sari S, Rochma Harani A and Werdiningsih H 2017 Pelestarian dan Pengembangan Kawasan Kota Lama sebagai Landasan Budaya Kota Semarang

[2] Chen C L 2018 Tram development and urban transport integration in Chinese cities: A case study of Suzhou Economics of Transportation **15** 16–31

[3] Muhlis Z and Herijanto W 2014 Manajemen Lalu Lintas Akibat Trem Di Jalan Raya Darmo Surabaya Jurnal Teknik POMITS **3** 37–42

[4] Naznin F, Currie G, Logan D and Sarvi M 2016 Safety impacts of platform tram stops on pedestrians in mixed traffic operation: A comparison group before-after crash study Accid Anal Prev **86** 1–8

[5] Nguyen-Phuoc D Q, Currie G, Young W and De Gruyter C 2016 Modelling the Direct Impact of Tram Operations on Traffic

[6] Aghabayk K, Sarvi M, Young W and Kautzsch L 2013 A Novel Methodology for Evolutionary Calibration of VISSIM by Multi-Threading

[7] Dinas Perhubungan Kota Semarang 2021 Studi Perencanaan Pembangunan Trem di Kota Semarang (Semarang: Dinas Perhubungan Kota Semarang)

[8] Putri N H and Irawan M Z 2015 Mikrosimulasi Mixed Traffic pada Simpang Bersinyal dengan Perangkat Lunak VISSIM (Studi Kasus: Simpang Tugu, Yogyakarta)

[9] Williams T A, Chigoy B, Borowiec J B and Glover J D 2016 Methodologies Used to Estimate and Forecast Vehicle Miles Traveled (VMT) (Houston)

[10] Bagloee S, Asadi M and Bozic C 2012 A Sustainability Approach in Road Project Evaluation, Case-Study: Pollutant Emission and Accident Costs in Cost Benefit Analysis pp 295–303

Experience in calibrating the VISSIM microscopic simulation model for an unsignalized intersection in Pekanbaru, Indonesia

M Z Muttaqin[1*], **N Feswiger**[1] , **A K Zaini**[1], **C Kumalasari**[2]

[1]Civil Engineering, Universitas Islam Riau, Jalan Kaharuddin Nasution 113 Pekanbaru, Indonesia
[2]Department of English Education, Universitas Islam Riau, Jalan Kaharuddin Nasution 113 Pekanbaru, Indonesia
*Corresponding author : muchzaenalmuttaqin@eng.uir.ac.id

Abstract. Traffic activity is unique and occurs throughout the region. All moving traffic movements cause a build up of vehicles when there are conflicts between vehicles and result in congestion, one of which is in the intersection area located on Jalan SM Amin, Pekanbaru. Traffic modelling can assist traffic management through accurate simulation results with field conditions based on driving behavior value VISSIM is a simulation tool used to simulate dynamic traffic flow conditions. The research has aim to modelling and calibrating driver behavior value based on traffic flow conditions at unsignalised intersections in Pekanbaru. This research uses simulation modelling with VISSIM as the modeler and GEH and MAPE as validation tools to choose which right model. Simulation process will used calibration process based on driving behavior value for the right model by trial and error method.. The result showed that simulation model based on default parameter of driving Behavior cannot represent for real traffic. Then changes in the combination of Driving Behaviour values affect the accuracy of model conditions with field conditions. The accuracy is measured using the reference volume of vehicles exiting the unsignalled intersection. Therefore, this research could be can assist in smoothing traffic flow model in Pekanbaru which more easily by using live traffic flow simulation that can be utilized for traffic intersection management scenarios.

Keyword : PTV Vissim, Intersection, Calibration, Simulation, Unsignallized Intersection

1. Introduction

The growth of Riau Province vehicles, based on data from the Central Bureau of Statistics of Riau Province in 2017, has increased every year ranging from 8-10% [1] . While the number of motorized vehicles in Pekanbaru has increased yearly, ranging from 7-9%. The figure includes trucks, passenger buses, cars and motorcycles. The high percentage increase in the number of motorized vehicles from year to year, especially in Pekanbaru City, can result in the capacity of road space in Pekanbaru City will be smaller due to the imbalance in the growth of the number of vehicles with the growth of existing road sections in Pekanbaru City. It can certainly be one of the trigger factors for traffic vehicle density during peak hour conditions, especially at intersections near the central area of Pekanbaru City, shopping centers, crowd centers such as city park areas, and places to do other activities. A solution to the current intersection problem is to use simulation technology. Simulation technology can be done through vehicle traffic modeling that looks at vehicle movement patterns during peak hour conditions in the intersection area. One traffic simulation tool that can model traffic flow is VISSIM. VISSIM is a traffic simulation tool or software for traffic engineering, transportation planning, signal time, and public transportation, even at the planning level in a city. Previous research has widely used VISSIM in conducting traffic project analysis, such as at intersections [2]. In detail, previous research in the use of VISSIM in hubs was to calibrate through driver behavior patterns at the study site, such as [3], which simulated VISSIM for roundabout areas in the United States. In addition, research was conducted by [4] for signalized intersection areas, and [5] used for road sections. The entire study used calibration in the modeling results through a comparison of queue length and travel time as validation of the traffic simulation. The difference from the previous study with this study is that this study validated in a simulation using the number of vehicles. This validation follows the Vissim simulation guide, which states that the volume of vehicles exiting the simulation must be equal to the volume input state [6]. In addition, the difference between this study and the previous one is also in the characteristics of traffic. This research was conducted in the realm of knowing the parts of traffic in the territory of Indonesia, especially in the city of Pekanbaru. This is supported by Buck [7], who also states that differences in traffic areas cause different calibration values. Investigating calibration values against simulations will also facilitate the process of traffic modification to provide easier solutions to traffic problems. The ease of use of VISSIM simulation is also given to compatibility with different vehicle types. This program is able to accommodate mixed traffic in the same lane so that the events that occur can be said to be similar to the real world. SM Amin-Naga Sakti intersection is one of the unsignalized intersections located on collector roads in Pekanbaru. Fig 1 showed about traffic condition in this intersection. This intersection has three junctions and serves as one of the main accesses used by the people of Pekanbaru City to the urban center area and the main access to the Riau Main Stadium which is used every day by the people of Pekanbaru City for sports activities. The research has aim to modelling and calibrating driver behavior value based on traffic flow conditions. In details, this research wouldmake model from real traffic and analyze them to make pattern of traffic vehicle movement and intersection performance at the intersection in Pekanbaru.

Fig 1. Traffic condition in Jl.SM Amin-Jl. NagaSakti

2. Literature Review

2.1. Traffic Movement Modeling

According to [3], transportation modeling is a simplification of the similarity of a transportation system in the real world into a formula or visual that has been planned in such a way. A transportation model can be said to be good if it can represent events in the real world. However, a transportation model with a high accuracy level is tough for the movement of transportation, especially in road conditions, is very diverse and tends to be unique [4]. The micro transportation model will consider the driver's behavior in estimating the vehicle's movement while on the road. The results of this behavior will produce estimates of traffic patterns, volume, speed, and changes through distance and time [5]. Therefore, tools are needed in modeling traffic, such as VISSIM. This tool aids the ability to simulate microscopically related traffic flow, such as queues and density levels in a road network, visually, both from two to three-dimensional [6].

2.2. PTV VISSIM

PTV Vissim is a software capable of modeling multimodal traffic flow at the microscopic level. PTV Vissim (hence referred to as VISSIM) can analyze private vehicle modes to public transit, particularly in the construction of the highway network with problems that can be solved, such as input problems of vehicle composition, lane configuration, and crossings [10]. This program was created by VerkehrStadten Simulationsmodel which means traffic simulation modeling [2].which translates to "urban traffic model." This simulation has an edge over others because it can represent traffic conditions graphically in three dimensions by incorporating a variety of vehicles, including cars, motorbikes, trucks, buses, trains, and other vehicles. According to [9], the simulation results obtained by this software can aid in calculating the quality of road traffic services, just as the 1997 MKJI technique does. This requires further consideration for the development of traffic management in Indonesia. However, the standard procedures applied to MKJI have the disadvantage of not being able to detect the development of traffic movements that continue to develop every time. Microscopic simulations such as VISSIM have the main advantage of depicting traffic conditions according to their development. All result in VISSIM is carried out in the microscopic section. The advantage of microscopic parts is that the results obtained are able to describe the detailed characteristics of a traffic movement, such as the characteristics of the driver himself. In addition, Vissim also has an easy interface so there is no need to write complicated programs in the formation of simulations [12].

2.3. Calibration and Validation

Calibration of a model is the process of adjusting each component of a simulation model. The calibration process leads to the condition of simulation findings being able to accurately depict actual conditions in the field, as determined by direct observation and real-time traffic situations [13]. The calibration procedure can determine traffic parameters such as traffic control, flow characteristics, and driving behavior. A process then validates the outcomes of calibration adjustments to the desired model components. This technique involves testing against a calibrated model and then comparing it to field results, such as previously acquired traffic flow data. Validation is closely tied to calibration since calibration results must directly enhance the simulation model's capacity to simulate real-time situations in the physical world. Hellinga [10] explains the sequence of the calibration process in a simulation model, beginning with the definition of goals and objectives, the determination of required data, the decision of the reference performance and criteria of the calibration results to be used, the representation of the road network and routes, and the output of the evaluation model. The rules for the validation and calibration processes are established individually with each researcher. The GEH and MAPE methods are one of the recognized validation techniques for assessing VISSIM traffic simulations. The GEH test (Geoffrey E. Havers) is utilized to compare two sets of real-world vehicle volume parameters. Validated parameters pertain to the flow of traffic. This validation test is backed by [12], which demonstrates that GEH is one of the most effective approaches for comparing the input and output data of VISSIM simulations. GEH is described as a modified form of theChi–squared statistic that combines the difference between relative and absolute values [14]. The GEH formula is depicted in the following equation:

$$GEH = \sqrt{2\,(M - C)^2/(M + C)} \tag{1}$$

With M was value in the model and C as value calibrated. It means this formula which compare for modelled value and calibrated value and lowets score would be the best score and represented as the best driving behavior value for the model.References in simulation validation using GEH have specific provisions of the error value generated as follows, lees than 5 as accepted, between 5 and 10 as error also more than 1- as not accepted.

In addition, validation is also performed using the MAPE parameter. It was a statistical measure used to perform an accuracy of the predicted value in the forecasting method [14]. This measurement uses the principle that the smaller the value of a percentage error from the difference in the expected value, the more accurate the value will be. The formula of MAPE is as follows,

$$MAPE = \frac{\sum_{t=1}^{n}\left|\left(\frac{A_t - F_t}{A_t}\right)100\right|}{n} \quad (2)$$

With At as Actual request to t, Ft as forecasting result to t, and N as magnitude of forecasting data Then, criteria for the MAPE value can be interpreted that less than 10% as Very Good, between 10-20% as Good, between 20 – 50% as Poor and more than 50% as Reject.

3. Research Methodology

Data collection in the field through direct observation by placing several surveyors on each junction of the intersection at the research site. What can be obtained is geometric intersection data through walking measurement for basic modeling. In addition, the primary data collected is the volume of traffic flow calculated using the traffic counting method during the afternoon peak hours at 05.00 pm - 06:00 pm by classifying the types of vehicles into 4 types, namely cars, buses , motorcycles, and trucks. This time has already was selected after conducting a survey for 6 hours, namely 3 hours in the morning, namely from 06.00 to 09.00 WIB, and in the afternoon, which is from 15.00 to 16.00 WIB. The highest hours are in the afternoon, namely 17.00 – 18.00 WIB. Next The speed of each type of vehicle is also calculated using the journey speed method. Data simulation is run through VISSIM tools by entering input data according to field data . The input process in this simulation is briefly shown in the following sequence,

1. Road Network creation
2. Vehicle volume for each vehicle type
3. Vehicle speed
4. Input for vehicle composition
5. Input for vehicle route composition
6. Managing for conflict area
7. Managing for priority rules
8. Managing for Calibration and Validation rules

Furthermore, the calibration and validation process was carried out using the GEH and MAPE methods. The best results in this simulation are based on the minimum value obtained from both GEH and MAPE values. The calibration process is carried out by trial and error methods by changing the Driving behavior parameters contained in the VISSIM simulation tool.

4. Results and Discussion

4.1. Traffic Characteristics

The geometric conditions to be discussed concern the widths of lanes, sidewalks, and medians. All of themwould concern for batasan dari wilayah simulasi lalu lintas yang terjadi. Sharma [16] lebih lanjut menyatakan bahwa geometri ini menjadi parameter utama yang menunjukkan dimana pergerakan lalu lintas akan terjadi. Direct measurements will be taken with a meter or length measuring equipment. This crossroads has three junctions because it serves as the primary entrance to Riau Main Stadium. This intersection falls under the classification of collector roads. The intersection's geometric conditions are depicted in figure 1 below. The North Junction of the intersection has a two-way six-lane design with a median, the South Junction has a two-way five-lane configuration with a median, and the West Junction has a four-lane, two-way configuration with a median.

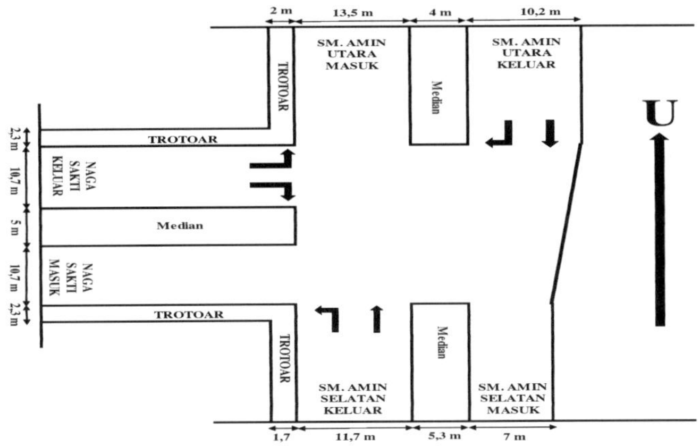

Fig 2. Research location

From the results of daily vehicle traffic surveys conducted at the intersection of Jl. SM. Amen-Jl. Naga Sakti from August 13 to 16, 2022 resulted in the number of vehicles passing through the intersection of Jl. SM. Amen-Jl. Naga Sakti on Saturday, August 13, 2022, for the most types of motorcycle vehicles comes from Jl. SM. South Amin at 07.00-08.00 WIB with a total of 3133 vehicles, the most types of car vehicles come from Jl. SM. North Amin at 16.00-17.00 WIB with a total of 1443 vehicles, the most types of bus vehicles come from Jl. SM. North Amin at 15.00-16.00 WIB with a total of 15 vehicles and for the most types of truck vehicles come from Jl. SM. South Amin at 17.00-18.00 WIB with a total of 128 vehicles. In addition, the number of vehicles passing through the intersection of Jl. SM. Amen-Jl. Naga Sakti on Sunday, August 14, 2022, for the most types of motorcycle vehicles comes from Jl. SM. South Amin at 16.00-17.00 WIB with a total of 2262 vehicles, the most types of car vehicles come from Jl. SM. South Amin at 16.00-17.00 WIB with a total of 1099 vehicles, the most types of bus vehicles come from Jl. SM. North Amin at 15.00-16.00 WIB with a total of 18 vehicles and for the most types of truck vehicles come from Jl. SM. South Amin at 15.00-16.00 WIB with a total of 88 vehicles. the number of vehicles passing through the intersection of Jl. SM. Amen-Jl. Naga Sakti on Monday, August 15, 2022 for the most types of motorcycle vehicles comes from Jl. SM. South Amin at 07.00-08.00 WIB with a total of 3994 vehicles, the most types of car vehicles come from Jl. SM. North Amin at 17.00-18.00 WIB with a total of 1509 vehicles, the most types of bus vehicles come from Jl. SM. North Amin at 07.00-08.00 WIB with a total of 14 vehicles and for the most types of truck vehicles come from Jl. SM. South Amin at 16.00-17.00 WIB with a total of 188 vehicles. the number of vehicles passing through the intersection of Jl. SM. Amen-Jl. Naga Sakti on August 16, 2022 for the most types of motorcycle vehicles comes from Jl. SM. North Amin at 17.00-18.00 WIB with a total of 3990 vehicles, the most types of car vehicles come from Jl. SM. North Amin at 17.00-18.00 WIB with a total of 1632 vehicles, the most types of bus vehicles come from Jl. SM. South Amin at 16.00-17.00 WIB with a total of 16 vehicles and for the most types of truck vehicles come from Jl. SM. South Amin at 15.00-16.00 WIB with a total of 156 vehicles.

4.2. Calibration and Validation

Based on the results of modeling the intersection of Jl. SM. Amen-Jl. Naga Sakti with PTV Vissim 20 software using the default parameters of Vissim, it can be concluded that the modeling results carried out are not yet valid, because after modeling the validation test process was carried out with the GEH (Geoffrey E. Havers) and MAPE (Mean Absolute Percentage error) methods were not met, namely GEH > 5 and MAPE > 10%. Next, the calibration process is carried out using trial and error by referring to previous research on calibration and validation using VISSIM. Calibration is carried out by following the car following model which is technically through the driving behavior parameter available in VISSIM. Driving behavior is human nature in responding to events that occur during driving [17]. Car following behavior is the behavior of drivers in following vehicles from each other, such as maintaining a safe distance from the front and rear, choosing empty lanes to minimize distance and time, as well as maintaining a safe distance on the side of the vehicle. In addition, it also accommodates the driver's behavior while in the signaling area, namely whether the driver will continue to drive or stop. All of these parameters in VISSIM can be shown, among others,

1. Look Ahead Distance (Min) means the minimum visibility of the rider towards the front.
2. Look Ahead Distance (Max) means the maximum visibility of the rider towards the front.
3. Observed vehicle means the number of vehicles observed by the rider.
4. Look Back Distance (Min) means the minimum visibility of the rider towards the rear.
5. Look Back Distance (Max) means the maximum visibility of the driver towards the rear.
6. Average Standstill Distance means the average stopping distance of the vehicle.
7. Additive Part of Safety Distance means the value used for the calculation of safe distance, the default value of 2.
8. Multiplication Part of Safety Distance means the value used for the calculation of the safety distance is greater the default value of 3.
9. Min Cleareance means a minimum of empty/clean area.
10. Safety Distance Reduction means the safe distance between vehicles in front and behind or the gap and clearing distance between vehicles.
11. Desired Position at Free Flow means the desired position at Free Flow
12. Overtake at Same Lane (Right) means overtaking in the same lane (Right)
13. Overtake at Same Lane (Left) means overtaking in the same lane (Left)
14. Minimum Lateral Distancing (Driving) means the minimum distance between the rider's sides when driving
15. Minimum Lateral Distancing (Standing) means the minimum distance between the rider's sides when stopping
16. Behavior at Red/ Amber Signal means behavior at the time of a red or yellow light

The geometric conditions to be discussed concern the widths of lanes, sidewalks, and medians. Direct measurements will be taken with a meter or length measuring instrument. This intersection has three arms because it serves as the primary entrance to Riau Main Stadium. This intersection falls under the classification of collector roads. The intersection's geometric conditions are depicted in figure 1 below. The North Arm of the intersection has a two-way six-lane configuration with a median, the South Arm has a two-way five-lane configuration with a median, and the West Arm has a four-lane, two-way configuration with a median,

$$d = ax + bx \tag{3}$$

$$bx = (bx_{aad} + bx_{mult} * z)*\sqrt{v} \tag{4}$$

This formula describe with αx as Idle distance, V as Vehicle speed [m/s], and z as the range value (0.1), which is normally distributed around 0.5 with a standard deviation of 0.15.

Table 1. Calibration Parameter List

Calibration Identity	List Parameter of Calibration															
	1	2	3	4	5	6	7	8	9	10	11	12	13	14	15	16
Default	0	250	4	0	150	2	2	3	0.5	0.6	Middle	Off	Off	1	0.2	Go
Cal 1	0	250	4	0	150	2	2	3	0.5	0.6	Any	On	On	1	1	Stop
Cal 2	0	250	4	0	150	1.5	1.5	2.5	0.5	0.6	Any	On	On	0.8	1	Stop
Cal 3	0	250	4	0	150	1	1	2	0.5	0.6	Any	On	On	0.7	0.7	Stop
Cal 4	0	100	3	0	100	1	1	2	0.3	0.5	Any	On	On	0.7	0.5	Stop
Cal 5	0	100	3	0	100	0.5	0.5	1	0.3	0.5	Any	On	On	0.5	0.5	Stop
Cal 6	0	100	3	0	100	1	1	2	0.2	0.5	Any	On	On	0.8	0.6	Stop
Cal 7	0	100	4	0	100	0.6	0.6	1	0.3	0.5	Any	On	On	0.7	0.5	Stop
Cal 8	0	100	4	0	100	0.4	0.4	0.8	0.3	0.5	Any	On	On	0.7	0.5	Stop
Cal 9	0	100	4	0	100	0.3	0.3	0.6	0.3	0.5	Any	On	On	0.7	0.5	Stop
Cal 10	0	100	4	0	100	0.2	0.2	0.4	0.3	0.5	Any	On	On	0.7	0.5	Stop

All results of calibration in the form of traffic volume are collected, and the difference at each junction is depicted as shown below. The graph shows that the second calibration has the most considerable difference in simulation volume, where the second calibration has the type of change in the value of the Weidman 74 model driving behavior parameter, namely Average Standstill Distance, Additive Part of Safety Distance, and Multiplic Part of Safety Distance. However, after the second calibration, the value of the simulated traffic volume began to rise again until, at the 10th calibration, the volume value of the simulated vehicle was very close to the observation results, according to the graph. The best calibration results provide simulation results close to field results or observations. Furthermore, Manjunatha et al [18] explained that the simulation results should be the same as the field results. This will provide simulation characteristics that make it easy for the accuracy of engineering strategies when needed. The calibration process is carried out by trial and error method until it reaches the best validation value in the validation performed. Figure 3 below shows the difference in vehicle volume results issued by simulation results and vehicle volume results in the field. In detail, Figure 3 shows that the change in look back distance value of 1.5 and average standstill balance of 2.5 in the second calibration gives the largest difference between simulation results and field results. However, subsequent calibration developments gave better results, where the simulation value was close to the observation value. In the end, the vehicle volume value in calibration result 10 shows the approach to the difference in vehicle volume value in the simulation and the smallest observation, which is less than 10 vehicles / hour.

The validation process will be carried out with statistical tests of GEH (Geoffrey E. Havers) and MAPE (Mean Absolute Percentage Error) validation. The results of the calculation of statistical tests validated by GEH (Geoffrey E. Havers) and MAPE (Mean Absolute Percentage Error) will produce which calibration is closest to the behaviour of drivers in the field. Support for this validation process is obtained from Feldman's statement that the accuracy of simulation models in traffic uses GEH and MAPE validation values where the concept of validation is based on a statistical approach that concentrates simulation data close to observational results.

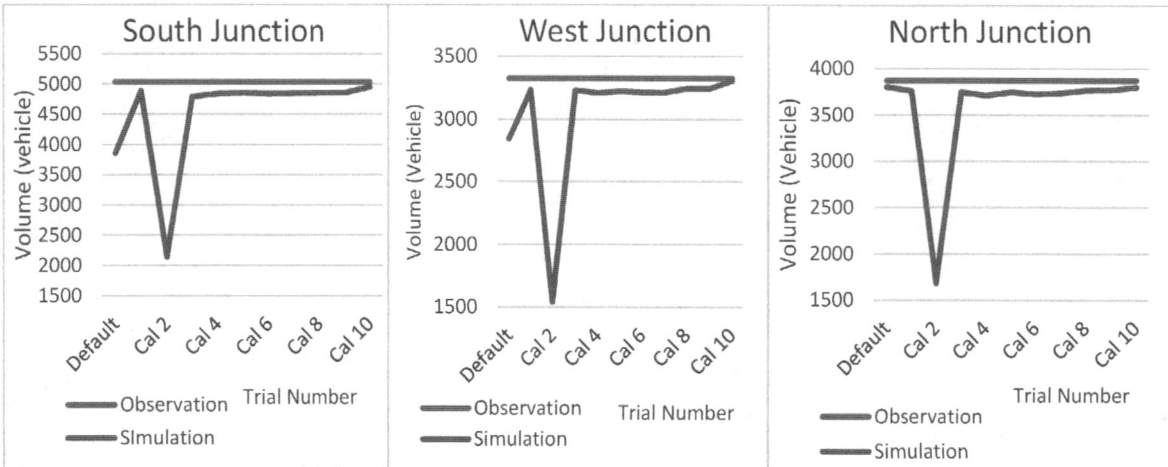

Fig 3. Calibration result for each junction

The validation results can be seen in Table 2. Results of statistical validation tests GEH (Geoffrey E. Havers) and MAPE (Mean Absolute Percentage Error) on intersection modelling Jl. SM. Amen-Jl. Naga Sakti was using peak clock volume data. According to Table 2 below, the results of calibration parameters with statistical tests of GEH and MAPE validation are best at the 10th calibration, which is called cal 10. The result of calibration parameter 10 produces the same value as calibration parameter 10 marked in GEH 0.36 and MAPE 0.63 in the western segment, GEH values 1.17 and MAPE 1.65 on the northern section and GEH values of 1.15 and MAPE 1.83 on the southern section.

Table 2. Validation Result

Condition	GEHValue			Conclusion	MAPE Value			Conclusion
	West	South	North		West	South	North	
Default	8.64	17.73	1.11	Error	16.87	30.69	1.81	Good
Cal 1	1.61	2.12	1.81	Accepted	2.77	2.96	2.89	Very Good
Cal 2	36.19	48.23	41.57	Not Accepted	53.68	57.40	56.57	Reject
Cal 3	1.66	3.41	1.99	Accepted	2.86	4.75	3.18	Very Good
Cal 4	2.03	2.75	2.58	Accepted	3.49	3.83	4.11	Very Good
Cal 5	1.78	2.57	2.02	Accepted	3.07	3.60	3.23	Very Good
Cal 6	1.94	2.73	2.35	Accepted	3.34	3.81	3.74	Very Good
Cal 7	1.98	2.66	2.14	Accepted	3.40	3.72	3.41	Very Good
Cal 8	1.43	2.57	1.70	Accepted	2.47	3.60	2.71	Very Good
Cal 9	1.41	2.60	1.63	Accepted	2.44	3.64	2.61	Very Good
Cal 10	0.36	1.17	1.15	Accepted	0.63	1.65	1.83	Very Good

Visually, what can be observed about modeling before calibration and after calibration is the difference in driver behavior in modeling with driver behavior at the intersection of Jl. SM. Amen-Jl. Naga Sakti, before calibration was carried out, it could be seen that the distance between vehicles in the modeling was still too tenuous, and it did not match the space of motorists who were at the intersection of Jl. SM. Amen-Jl. Naga Sakti, as shown below, while after modeling the calibration process is carried out, the distance between vehicles is appropriate and resembles the driver's behavioral conditions. Based on the calibration results, it was also found that changes to the modified Wiedmann 74 type calibration parameters resulted in a safe distance formed in the simulation model of d, which is 2 meters, with the average vehicle speed at each junction of the junction being 36 km/h. This figure is generated from VISSIM, and It shows that each rider has a relatively close perceptual distance when approaching an intersection. Empirically, this indicates that the vehicle has a span of 2 meters with the vehicle in front of it as the vehicle's safe limit in driving at the intersection. This is consistent with Buck's statement[7] that vehicles have the characteristic of getting closer to the intersection when stopped. Fig 4 shows significant differences in VISSIM simulation results before and after calibration. The simulation results after calibration can provide clarification that the phenomenon of queuing vehicles at intersections does tend to approach and the queue pattern is narrower than when on a continuous road. In addition, the simulation results after calibration also show the characteristics of vehicles in Indonesia when they are at intersections. Vehicles in Indonesia are expected to follow traffic rules and signals, but it is not uncommon to see drivers ignoring traffic signals or driving recklessly. This often happens especially during crowded conditions. This was agreed by Prasetjo et al who stated that the speed of the vehicle when at the intersection reached 12 km per hour. This shows that vehicles in Indonesia tend to want to immediately release congestion so that they can drive according to the desired route [20].

(a) Before calibration (b) After Calibration

Fig 4. Visualization model three dimension VISSIM

5. Conclusion

Traffic modeling generated from VISSIM software can run very well, by simulating a road network in mixed traffic conditions. The calibration process is also needed to adjust the simulation conditions to field conditions. The calibration process is carried out with the concept of Car Following Model with the Wiedemann 74 type. The calibration parameters that affect the VISSIM simulation results are average standstill distance, additive part of safety distance and multiplicative part of safety distance. All of these meters provide major changes in the calibration process which is carried out 10 times through trial and error methods until the average error difference between observation data and pre-calibrated data reaches the smallest MAPE value of 1.83 for traffic flow volume.

Acknowledgments

The authors wish to thanks to Universitas Islam Riau as its support in this study.

6. References

[1] BPS, *Pedoman Pencacahan Survey angkutan penumpang dan barang*. 2015.

[2] N. Wu, "Total approach capacity at signalized intersections with shared and short lanes; generalized model based on a simulation study," *Transp. Res. Rec.*, no. 2027, pp. 19–26, 2007.

[3] Z. Li, M. DeAmico, M. V Chitturi, A. R. Bill, and D. A. Noyce, "Calibration of VISSIM roundabout Model: A critical gap and follow-up headway approach," *92nd Annu. Meet. Transporation Res. Board*, no. January, pp. 1–22, 2013.

[4] U. Shahdah, F. Saccomanno, and B. Persaud, "Application of traffic microsimulation for evaluating safety performance of urban signalized intersections," *Transp. Res. Part C Emerg.*

Technol., vol. 60, pp. 96–104, 2015.

[5] J. Kima, B. Park, J. Leeb, and J. Won, "Determining optimal sensor locations in freeway using genetic algorithm-based optimization," *Eng. Appl. Artif. Intell.*, vol. 24, no. 2, pp. 318–324, 2011.

[6] M. Fellendorf and P. Vortisch, "Validation of the Microscopic Traffic Flow Model VISSIM in Different Real-World Situations," *Transp. Res. Board 80th Annu. Meet.*, no. January 2001, pp. 1–9, 2001.

[7] H. S. Buck, N. Mallig, and P. Vortisch, "Calibrating vissim to analyze delay at signalized intersections," *Transp. Res. Rec.*, vol. 2615, no. 1, pp. 73–81, 2017.

[8] Z. F. M. Li, "A generic characterization of equilibrium speed-flow curves," *Transp. Sci.*, vol. 42, no. 2, pp. 220–235, 2008.

[9] Z. Lu, T. Fu, L. Fu, S. Shiravi, and C. Jiang, "A video-based approach to calibrating car-following parameters in VISSIM for urban traffic," *Int. J. Transp. Sci. Technol.*, vol. 5, no. 1, pp. 1–9, 2016.

[10] P. Maheshwary, K. Bhattacharyya, B. Maitra, and M. Boltze, "A methodology for calibration of traffic micro-simulator for urban heterogeneous traffic operations," *J. Traffic Transp. Eng. (English Ed.*, vol. 7, no. 4, pp. 507–519, 2020.

[11] T. Kretz, S. Hengst, and P. Vortisch, "Pedestrian Flow at Bottlenecks - Validation and Calibration of Vissim's Social Force Model of Pedestrian Traffic and its Empirical Foundations," 2008.

[12] Z. Li, M. DeAmico, M. V Chitturi, A. R. Bill, and D. A. Noyce, "6th Road Safety on Four Continents Conference CALIBRATING VISSIM ROUNDABOUT MODEL USING A CRITICAL GAP AND FOLLOW-UP HEADWAY APPROACH," *16th Road Saf. Four Cont. Conf. 15-17 May 2013*, no. May, pp. 0–15, 2013.

[13] B. (Brian) Park and J. D. Schneeberger, "Microscopic Simulation Model Calibration and Validation: Case Study of VISSIM Simulation Model for a Coordinated Actuated Signal System," *Transp. Res. Rec. J. Transp. Res. Board*, vol. 1856, no. 1, pp. 185–192, Jan. 2003.

[14] M. R. Muslim, M. Hustim, and M. Pasra, “Analisis Kinerja Simpang Bersinyal Haji Bau – Jl . Cendrawasih – Jl . Arif Rate Di Makassar,” Makassar, 2018.

[15] H. D. Golakiya and A. Dhamaniya, "Modeling Speed and Capacity Estimation at Urban Midblock Sections under the Influence of Crossing Pedestrians," *J. Transp. Eng. Part A Syst.*, vol. 145, no. 9, pp. 1–14, 2019.

[16] H. K. Sharma and M. Swami, "Effect of turning lane at busy signalized at-grade intersection under mixed traffic in India," *Eur. Transp. - Trasp. Eur.*, no. 52, 2012.

[17] M. Y. Madi, "Investigating and Calibrating the Dynamics of Vehicles in Traffic Micro-simulations Models," *Transp. Res. Procedia*, vol. 14, pp. 1782–1791, 2016.

[18] P. Manjunatha, P. Vortisch, and T. Mathew, "Methodology for the Calibration of VISSIM in Mixed Traffic," *Transp. Res. Board 92nd Annu. Meet.*, no. August, p. 11, 2013.

[19] O. Feldman, "The geh measure and quality of the highway assignment models," no. October 2012, 2014.

[20] J. Prasetijo and H. Ahmad, "Capacity Analysis of Unsignalized Intersection Under Mixed Traffic Conditions," vol. 43, pp. 135–147, 2012.

IOP Conf. Series: Earth and Environmental Science 1294 (2024) 012021 doi:10.1088/1755-1315/1294/1/012021

Study of Integration of Greater Bandung BRT Service in Bandung City

A C Putra[1,2], **S S Wibowo**[1,3], **F Zukhruf**[1,4] and **K N S Nawassa**[1,5]
[1] Master's Program in Civil Engineering
Faculty of Civil and Environmental Engineering, Institut Teknologi Bandung
Jl. Ganesha No. 10 Bandung, 40132
[2] E-mail: alvincputra@gmail.com
[3] E-mail: sonyssw@hotmail.com
[4] E-mail: febri.zukhruf@ftsl.itb.ac.id
[5] E-mail: kardina@ftsl.itb.ac.id

Abstract. The Trans Metro Pasundan (BRT) requires a comprehensive review for network integration with existing modes like *angkot*, Damri Bus, Trans Metro Bandung (TMB), and commuter trains (KRD). This study aims to employ a GIS model to assess BRT integration in Bandung City through spatial and trip analysis. Key network integration indicators include: (1) overlap level; (2) area coverage within catchment areas; (3) the number of transfers between corridors; and (4) route availability. Analysis shows that the unclear intermodal hierarchy has caused high overlap level and low coverage areas. Low intermodal connectivity also leads to high number of transfers. These issues do not meet the concept of equity set out in the regulations. BRT overlap exceeds 60% with *angkot* and 90% with Damri, failing to enhance coverage in certain areas, particularly four eastward sub-districts, with catchment ratios of just 13% - 27%. Utilizing ArcGIS, 26 out of 900 trips between sub-districts, originating at BRT stops, require more than the regulation's designated two corridor transfers. Route availability is measured by the walking distance needed to access a stop between routes, which remains below the 500-meter regulatory limit.

1. Introduction

The BRT Trans Metro Pasundan, a public mass transport program developed by the Ministry of Transportation, GIZ, and local governments, will be integrated with other modes of transportation to reduce congestion in the area. The BRT program in Greater Bandung is planned to start operating in stages until 2023. Trans Metro Pasundan is present as a BRT transportation to support community mobilization which is integrated with other mass transportation services. In addition, it is proposed that the BRT development will use direct services to be more flexible and have a wider coverage, so that passengers are not forced to transfer.

Currently, the Bandung City Transportation Agency is also designing an integrated transportation system that is assisted by the British Embassy through the Future Cities program, where there will be an evaluation of the public transportation system that has been operating in Bandung which is divided into feeder, main, and branch routes. The main routes will maximize bus modes, namely BRT, Trans Metro Bandung, Damri. The public transport modes that are currently operating in Bandung City consist of Trans Metro Bandung (TMB), Damri, commuter train (KRD), and *angkot*, within the city and across cities. Branch networks can be filled by *angkot* because it must be integrated with mass transportation.

The integration of existing public transportation service areas in Bandung City as of 2020 has been analysed, namely *angkot*, Damri Bus, Trans Metro Bandung, and KRD by Putra [1]. Several areas in Bandung City are still not covered by the existing public transport catchment areas, namely in the eastern part of Bandung in Mandalajati, Ujung Berung, Cibiru, Panyileukan; to the middle areas, namely Arcamanik and Cinambo; the southern part, namely Gedebage and Rancasari; and the southwest, namely Babakan Ciparay and Bojongloa Kidul. Therefore, it is necessary to review whether BRT services can provide additional coverage in areas in Bandung City that have not been served well.

In this sense, the main objective of this study is: "To conduct a study of integration of Greater Bandung BRT Service in Bandung City" to answer the research question: "Is Greater Bandung BRT service integrated in Bandung City?". The sub-objectives are to identify the catchment areas of public transport services in Bandung City, to develop GIS models for evaluation of the integration of public transport services, and to evaluate the new public transport services in terms of catchment areas and integration (case study: proposed BRT Trans Metro Pasundan). Due to limited data and conditions, the scope of this study is limited to Bandung City only. The modes studied are angkutan kota (*angkot*), Damri Bus, Trans Metro Bandung, KRD (Bandung local train), and BRT Trans Metro Pasundan. The transit routes are also based on existing legal documents. This research focuses on network integration, from the spatial and trip perspective. The analysis of this study was primarily conducted in 2021, hence the data of public transport network routes is using the 2021 version, while the O-D projection data available and used in this study is for 2031.

2. Literature review

One of the indicators that is most often developed in studies on the integration between modes of public transport is the overlap level. A study conducted on the optimization of integration between metro and bus modes mentions that competition and cooperation between the metro and public transportation are very important in an integrated public transportation network [2]. Another study also uses an indicator of the overlap level in assessing public transportation connectivity, which can be quantified by the ratio of overlapping route lengths [3]. The overlap analysis between routes can also be based on the coverage area along the route where it can trigger a reduction in passengers and competition between fleets [4].

Another research on intermodal integration conducted in other cities analysed integration with the identification of catchment areas at railbus stops, as well as a potentially multimodal travel approach from determining the origin to destination zone (O-D matrix) [5]. Analysis of the characteristics and patterns of user travel demand, characteristics of origin-destination within and outside the city, is also studied, using a quantitative approach, namely spatial analysis (overlap) [6]. An analysis related to network or route integration consisted of an analysis of route overlap and the availability of routes that accommodate direct trips between nodes, along with an analysis of the number of mode changes [7].

Regarding spatial analysis in the study of public transport, a study evaluated the relationship of the spatial distribution of the population to the public transportation network to ensure that the population's needs are met, with components of bus coverage area, population density, and origin-destination (O-D) data [8]. GIS was utilized as a mapping in evaluating the effectiveness and coverage of public transport areas. The developed indicator or parameter system is spatially correlated with the public transportation network and population distribution [9]. Another research that applies GIS in spatial analysis is the modelling of the accessibility of public facilities with an analysis of the coverage area of public transport supported by data in the form of population density and land use in public facilities, especially residential areas [10].

However, there are still several aspects in the evaluation of the public transport network that have not been investigated further. In the technical study of the Trans Metro Pasundan BRT network planning conducted by Bandung City Transportation Agency [11], the actual overlap level has been studied, but only between BRT corridors, not yet on modes in Bandung City that have the potential to become a competition, including *angkot*. Research that evaluates public transport networks based on population density in the region is not commonly found in Indonesia, even though the population is a component that needs to be supported by public transportation to ensure accessibility [10]. Analysis of the availability of feeder routes and the hierarchy of public transport routes is the minimum service standard stipulated in the guidelines in Indonesia, namely Minister of Transportation No. 10 of 2012 [12], the Decree of the Director General of Transportation No. 687 of 2002 [13], and Guidelines for Assessment of the Level of Integrated Intermodal Transportation [14]. However, these factors have not been widely analysed in Indonesia.

2.1. Catchment area
One aspect that needs to be considered to conduct a public transit analysis is the catchment area, which is related to the potential users who are interested in public transit, depending on the location of bus stops and routes of public transit. The catchment area is an area that is still quite comfortable to walk to transit routes or stops [15]. The size of the service area of a route is very dependent on how far the walking distance still feels comfortable. The walking distance tolerance of each location varies, depending on geographical conditions, climate, land use characteristics, walking habits, and walking goals [16].

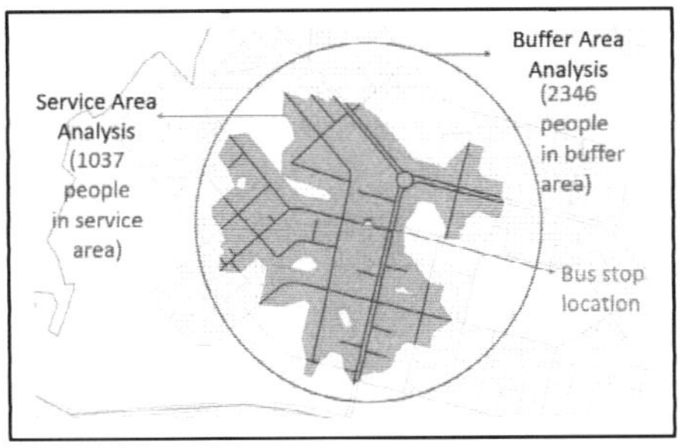

Figure 1. Comparison of analysis of service areas and buffer areas.
Source: Aslan & Kocaman, 2018

Analysis of service area instead of buffer area will lead to better results for spatial or service area analysis [17]. Figure 1 shows a comparison of the analysis of service areas and buffer areas. Analysis of the buffer area of the image seems to include more space, so more area is affordable. However, given the fact that people will not be able to reach the bus stop directly from a bird's-eye view, a better analysis is through the analysis of service areas by assuming people will use existing roads in the catchment area.

2.2. Indicators of public transit network evaluation
In Law No.22 of 2009 concerning Road Traffic and Transportation [18], it is explained that in the urban scope, the Regency/City Regional Government is obliged to ensure the availability of public transportation for the transportation of people and/or goods within the Regency/City area within its administrative area. From this statement, accessibility of public transportation is the aspect that is focused on in this case.

Various guidelines or previous studies indicate that the transfer or trip of modes or corridors in the public transport network should be minimized. A good public transportation route follows the trip pattern of transportation passengers to create a more efficient trip so that the mode of transfer that occurs when passengers travel by public transport can be minimized [13]. Likewise, locations that have the potential to become travel destinations are sought to become service priorities. Public transportation is required to ensure the ease of passenger trips between corridors from origin to destination to ensure accessibility. The number of corridor transfers is set to a maximum of 2 (two) times [12].

The public transit integration model consists of the main mode and the feeder mode. The main characteristic of multimodal networks is to have a network that is connected between modes and recognizes the different levels of the network [19]. The main modes are designed to provide services to residential areas, trade, and activity centres as well as transportation nodes. This planned service is on the main corridors to ensure high travel speed and reliability of operation. For Bandung City, the modes of public transit with greater capacity are the railway, the BRT, Damri Bus and Trans Metro Bandung, making it suitable to focus on becoming the main modes. *Angkot* that has a lower capacity, can be focused on operating in the feeder route, by reaching users in smaller areas and roads, then feeding it to the main modes. *Angkot* routes can be rerouted or restructured to support their function as feeder modes for city buses. The criteria for the integration of public transport service networks, both intermodal and the same mode, include the availability of public transport routes that serve trips between transportation nodes [14]. In addition, the hierarchy of the route network that serves the trip is also reviewed, whether it is the main route or the feeder route.

Competition and cooperation between the metro and public transportation are very important in an integrated public transportation network [2]. Therefore, the integration of the public transport network can be achieved if there is minimum competition between modes, namely between BRT and other modes. Bus modes (BRT, Damri, and TMB) and train modes (KRD) as the main modes should be supported by *angkot* as feeder mode. If the BRT route still overlaps with routes from other modes, they would compete over travel demands on the road, and integration of the public transport network will not be achieved. In terms of network efficiency and operation, route overlap should also be minimized. If the city transportation route is diverted to other areas or areas, that can cover a wider area so that it can feed passengers for city buses, the public transportation network can be said to be better integrated. In addition, a wider area coverage certainly increases the distribution of public transport areas as well.

The main factor in determining the network of public transport routes is that the existing public transport routes are sought as close as possible to reach areas with high potential demand, namely those with high population density [13]. Area with a high intensity of activity has high trip generation and attraction, so it should be accessible by public transportation. However, apart from potential service areas, public transportation should also cover all existing urban areas. This follows the concept of equal distribution of services to the provision of public transportation facilities. The more public transit routes are distributed in a city, the more roads are reachable by the catchment area of public transit.

3. Methodology

The data used are secondary data from official documents issued by the Regional Development Planning Agency (Bappeda) in Bandung City, and various literatures that can support research. The proposed BRT will be reviewed against four other modes of development plans in Bandung City, namely *angkot*, Damri buses, TMB, and local trains or KRD. Therefore, information on the corridors of each mode is needed, accompanied by the location of the bus stops/terminals/stations scattered throughout Bandung City. The network and region data are superimposed or stacked into a map via ArcGIS. The O-D matrix between sub-districts is collected from the Department of Transportation of Bandung City to identify patterns of passenger trips in Bandung City. Sub-district density data to review public transport network services in meeting demand in Bandung City is obtained from Statistics Indonesia (BPS) 2020.

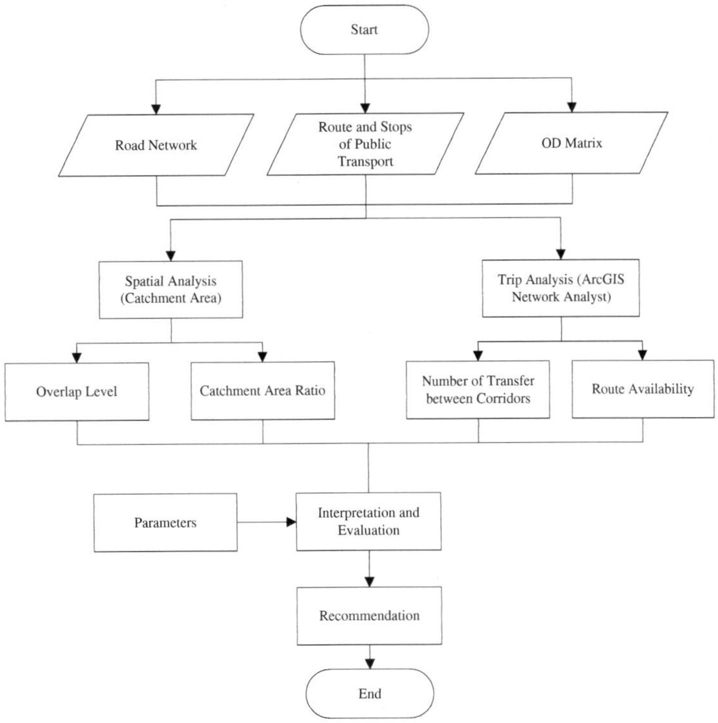

Figure 2. Analysis flowchart.

3.1. Trip analysis with Network Analyst

The integration of the BRT network with other modes is analysed by reviewing the trip of users. To analyse the trip taken in a trip in the transportation network system, including the number of transfers and the availability of routes in the public transport network, a modelling is needed that describes the network (link) and stops (nodes) containing information in the form of distance, time, obstacles, direction, and other attributes. The Network Analyst feature in ArcGIS can help find the best route (quickest path), create an O-D cost matrix, and create a model for route analysis, to help analyse the trip in Bandung City with public transportation to obtain predetermined integration indicators, namely the number of mode transfers and the route availability. Network Analyst on ArcGIS software version 10.3 is used.

A network dataset containing public transportation networks along with points or stops is built and equipped with information on distance, speed, direction, and route number. The public transport routes that pass-through Bandung City are 39 city *angkot* routes, 80 intercity *angkot* routes across Bandung City, 11 Damri Bus corridors, 6 Bandung Trans Metro corridors, and 1 KRD corridor, accompanied by plans for 12 BRT corridors. To avoid overlap in the model, the network dataset is made in three dimensions (3D) so that each route has a different elevation. The stops on each route are then connected to pseudo stops on the road network that have zero elevation, with connectors at each stop. The connector will also be used in the number of transfer attributes.

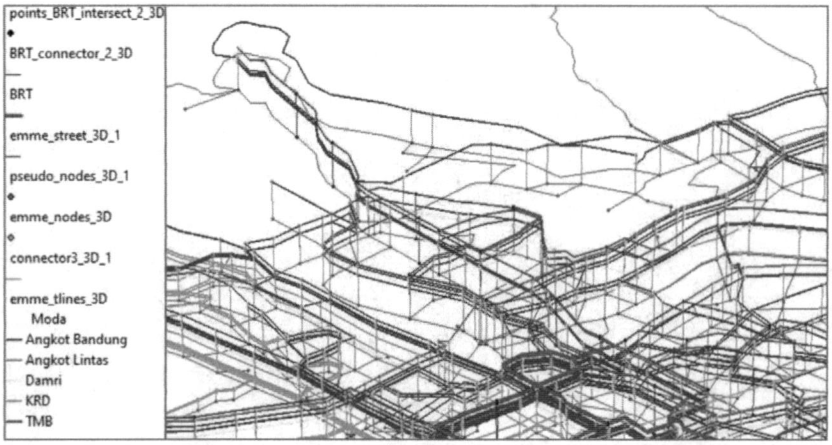

Figure 3. The 3D view of the network dataset of the public transport network in Bandung City.

The impedance used is travel time, by assuming the travel speed of each mode from various sources. Travel time in a trip by public transportation consists of travel time, access/egress time and transfer time. The standard transfer time for public transportation in Indonesia is set at 5 – 10 minutes [16]. Access/egress time by foot uses the assumption of walking speed. Travel time in a trip depends on the planned travel speed of the mode used. The attributes are calculating the distance travelled from each mode, calculating the number of transfers by counting the number of connectors used, calculating the walking distance, and setting the direction of each route.

After that, the start/end locations of the trip to issue the quickest route are put at the 30 BRT stops in the sub-district in Bandung City, or *angkot*/bus stops if there are no BRT stops available. The point is placed at one of the pseudo-stops on the road network, then the Network Analyst will decide to move through the connector to get to a stop on a route that is already at a different elevation or start the journey on foot on the road network to get to a stop on another route.

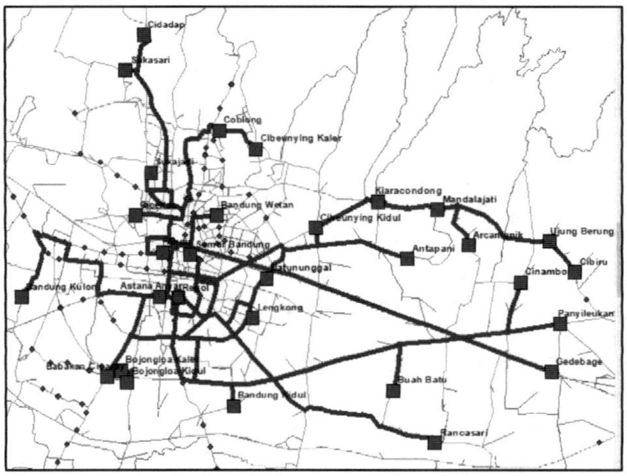

Figure 4. Example of the generation of the quickest route example from sub-districts in Bandung City to Regol.

IOP Conf. Series: Earth and Environmental Science 1294 (2024) 012021 doi:10.1088/1755-1315/1294/1/012021

With the user-based indicator method, the following two trip indicators can be identified.

1. The number of intermodal transfers to public transport corridors that must be carried out by users when travelling to a destination in Bandung City.
2. Availability of routes in serving trips in Bandung City

The results of the analysis of the number of transfers and the route availability are then recapitulated in an O-D matrix to review the trips that have not been served properly, as well as to evaluate the Bandung City public transportation network.

3.2. Spatial analysis

The service catchment area is within walking distance of the route or transit stop. The total catchment area of each mode (*angkot*/city bus/train) can be calculated using ArcGIS software. Meanwhile, the overlap level between two public transport routes can use the following ratio [3].

$$\mathrm{OP}_{r1,r2} = \mathrm{ol}_{r1,r2}/\mathrm{l}_{r1} \quad (1)$$

where r1 and r2 are the notation for each route, olr1,r2 is the length of the overlapping route between the two routes; and lr is the total length of the route. In this study, the ratio was based on catchment area. Overlap analysis is performed per BRT route on each of the other modes' routes to see the proportion of bus routes that overlap with other modes. On the other hand, if a city bus route overlaps with many *angkot* routes, but the length of each *angkot* route is low, then the bus route is not interrupted by those *angkot* routes.

To review the distribution of public transportation routes in Bandung City in responding to demand, the ratio of the catchment area of public transportation available in each sub-district is reviewed on the population density indicating residential, and on the generation and attraction numbers related to trip and destination centres in Bandung City. The catchment area in each sub-district is calculated based on the road [20], while the area that represents the public transportation catchment area is the total catchment area which has considered overlap. It is analysed whether areas with high density or becoming centres of departure and destinations are priorities for public transport services, especially with the presence of BRT in Bandung City.

$$\text{Catchment area ratio in sub-district} = \frac{\text{Catchment area of public transport}}{\text{Catchment area of network}} \quad (2)$$

The performance of the integration of BRT network services with other modes, namely Damri buses, TMB, KRD local trains, and *angkot* in Bandung will be discussed further. The analysis of each indicator that is presented in maps/tables/graphics, namely the number of transfers, route availability, overlap level, and catchment area ratio, is reviewed to identify routes or locations for BRT stops or other modes that still do not provide good connectivity and answer the needs in a certain area. Then, recommendations are developed based on the indicators in this study, which include the proposed addition or rerouting of public transport routes and the addition of bus stops in certain areas.

4. Analysis

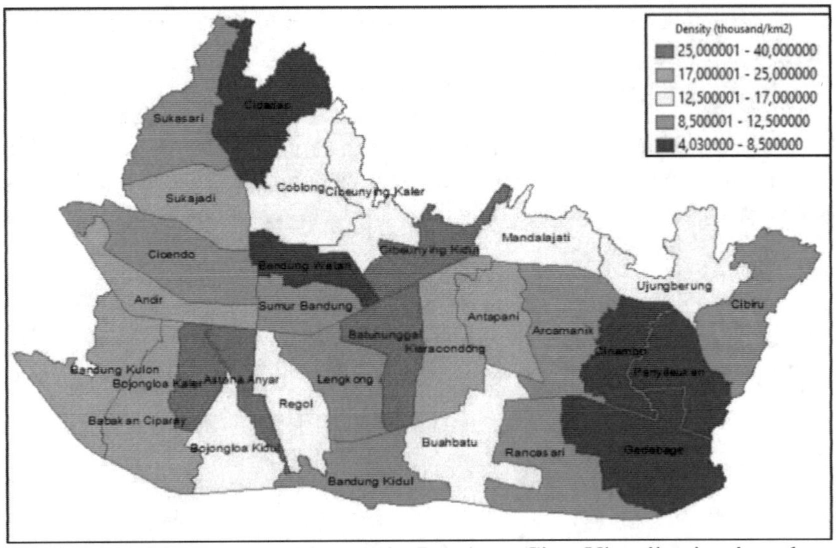

Figure 5. Population density map in Bandung City. Visualization based on data from Bandung City in Numbers (2020).

The most populous sub-districts in Bandung City are marked in red in Figure 5, namely Bojongloa Kaler in the southwest part of Bandung City, followed by Astana Anyar, just to the east. The other dense areas are Cibeunying Kidul and Batununggal which are in the middle of Bandung City. The sub-districts with the lowest density are marked in dark green, dominating the eastern and southeastern areas of Bandung, namely Gedebage, Panyileukan, and Cinambo, as well as Cidadap in northern Bandung.

For public transit modes that will be reviewed, based on Bandung Mayor Decree in 2008, officially there are 39 *angkot* routes within Bandung City. There are also 80 intercity *angkot* routes across Bandung City. There are 11 Damri bus routes, with five of them connecting points within the administrative boundaries of Bandung City, while the remaining six connect across regions outside of Bandung City, namely Jatinangor, Tanjung Sari, and West Bandung. In addition to Damri, there are currently five Trans Metro Bandung (TMB) bus routes in the city. There is also the local Greater Bandung train or KRD, a rail-based mass transportation currently available in Bandung, which is a series of commuter local trains. With a length of 51.7 km, KRD serves stations from Ciroyom, Bandung Station, Andir, Cimindi, Cikudapateuh, Gedebage, and Kiaracondong.

Five regencies/cities in Greater Bandung are part of the development of road-based public transportation (BRT), according to the Minister of Transportation Regulation No. 9 of 2020. The BRT program in Greater Bandung under the name Trans Metro Pasundan is planned to start operating in stages until 2023. BRT will be integrated with other modes of transportation to reduce congestion in the area. Trans Metro Pasundan is present as a BRT to support community mobilization which is integrated with other mass transportation services. It is also proposed that BRT development will use direct service services to be more flexible and have a wider coverage so that passengers are not forced to change buses. 12 corridors that will be developed in the Greater Bandung City that connect Bandung City with the surrounding area and stretch for 23 km. BRT connects points within Bandung City as well as with surrounding areas such as Cimahi City, Baleendah and Soreang in Bandung Regency, and Jatinangor in Sumedang Regency. Several areas in Bandung City which are the centres of departures and arrivals for BRT routes are Cicaheum Terminal, BEC Mall, and Dipatiukur.

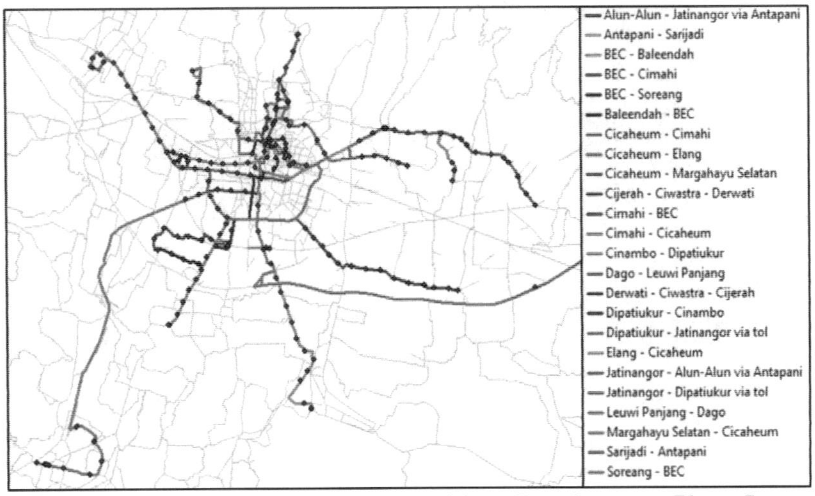

Figure 6. Greater Bandung BRT Corridors Development Plan. Source: Identification based on Directorate General of Land Transportation's Webinar (2021).

In conducting a spatial analysis of public transportation routes in Bandung, the radius of the catchment area is determined in advance based on the walking distance. The catchment area of public transportation is measured from transfer points, namely stops/terminals, because passengers are allowed to get on and off only at stops, such as bus modes (BRT, Damri, and TMB) and KRD. However, users of *angkot* in Bandung City go up and down the road, so the catchment area is assumed along the road. Therefore, the catchment area of an *angkot* is calculated from the roads traversed by each route.

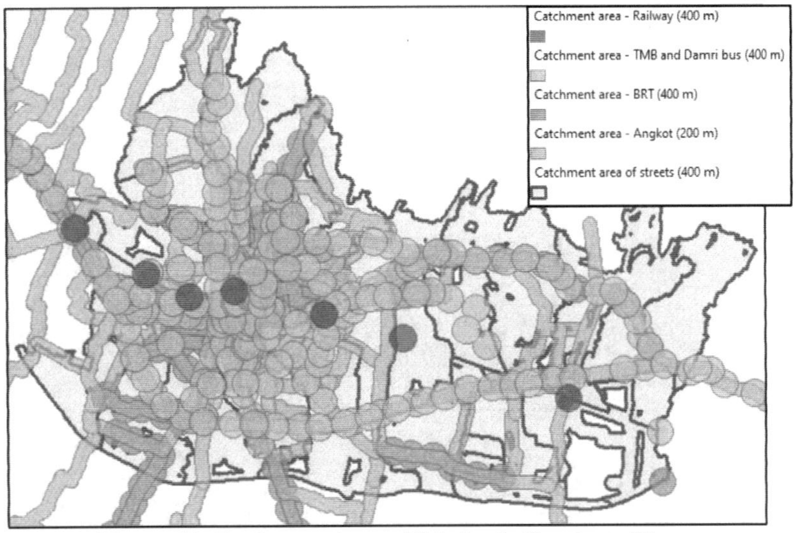

Figure 7. Catchment Area of Modes in Bandung City.

Assuming the catchment area of *angkot* is linear along the route, and its role as a feeder, it is different from bus and rail modes which act as the main mode which is larger and has a catchment area drawn from the bus stop, so the radius or walking distance for *angkot* modes is also different. Therefore, considering the characteristics of Bandung and the trip of its citizens, a walking distance of 200 m is used for public transit in the city, while for trains is 400 m. 400 m to city bus stops (BRT,

Damri, and TMB), as well as stations that are passed by local trains or KRD. The road network also uses a 400 m radius [16].

Figure 7 shows the catchment area of each mode depicted through ArcGIS. The entire catchment area of the BRT (coloured orange) is covered by the catchment area of other modes. This phenomenon can be an early indication that BRT is not yet integrated with other modes because it serves the same area in Bandung City and does not provide additional accessible coverage areas.

4.1. Catchment area ratio

Table 1. Catchment area ratio compared to population density and traffic generation and attraction in Bandung City.

No	Sub-district	Catchment Area Total Ratio	2020 Data: Density (thousand/km^2)	2020 Data: Covered Density (thousand/km^2)	2031 Transit O-D Matrix: Generation (PCU/hours)	2031 Transit O-D Matrix: Attraction (PCU/hours)
1	Sumur Bandung	100,0%	10,62	10,62	313,5	429,9
2	Bandung Wetan	100,0%	8,41	8,40	729,8	800,5
3	Astana Anyar	97,4%	27,64	26,93	187,6	147,1
4	Regol	88,7%	16,91	14,99	716,0	770,1
5	Bojongloa Kaler	88,1%	39,57	34,88	482,1	280,3
6	Lengkong	85,2%	12,06	10,28	408,2	400,7
7	Andir	84,2%	23,49	19,78	505,1	662,6
8	Coblong	75,6%	15,62	11,80	805,2	1013,0
9	Babakan Ciparay	74,1%	19,63	14,54	224,8	283,9
10	Batununggal	73,8%	25,08	18,52	713,3	793,8
11	Cicendo	72,5%	12,32	8,94	245,1	501,1
12	Sukajadi	71,9%	19,19	13,79	308,1	232,4
13	Cinambo	71,6%	5,91	4,23	25,5	215,0
14	Cibeunying Kaler	65,7%	15,04	9,87	244,9	106,8
15	Bandung Kulon	61,2%	19,11	11,70	389,5	400,3
16	Bojongloa Kidul	56,8%	16,47	9,36	138,8	234,1
17	Antapani	56,7%	18,62	10,55	203,9	193,5
18	Buahbatu	56,1%	13,54	7,60	114,3	667,7
19	Cibeunying Kidul	55,9%	27,27	15,23	757,6	373,7
20	Arcamanik	52,4%	10,05	5,27	237,9	436,5
21	Sukasari	49,1%	12,10	5,94	173,8	129,2
22	Kiaracondong	46,8%	22,43	10,49	753,8	176,6
23	Cidadap	46,5%	6,37	2,96	46,5	54,3
24	Rancasari	44,4%	11,90	5,29	774,9	206,4
25	Panyileukan	43,7%	7,40	3,23	426,3	522,2
26	Bandung Kidul	35,0%	11,01	3,86	50,1	254,0
27	Mandalajati	26,6%	14,78	3,93	326,3	158,0
28	Ujungberung	23,2%	13,82	3,20	283,5	267,3
29	Gedebage	21,0%	4,03	0,85	382,2	186,2
30	Cibiru	13,6%	10,64	1,45	76,0	147,3
	Total	**57,4%**	**14,63**	**8,40**	**11045**	**11045**

To evaluate the distribution of public transportation routes in the sub-districts of Bandung City, the ratio of the available public transportation service areas in Bandung City to the catchment area of the road network in each sub-district is calculated, as shown in equation (2) in Chapter III. The catchment area ratio shows the area in the sub-district that can be reached by public transportation. The ratio can be seen in Table 1, which shows the colour gradation for each sub-district against other sub-districts in that mode. Almost the entire road network in Sumur Bandung and Bandung Wetan sub-districts is served by public transportation, which is 98.8% and 92.3% respectively of the catchment area of the road network. When combined with other modes (the "Total" column is red), the entire road network in the two sub-districts is served by public transport (100%). For BRT mode, the catchment area ratio is highest in Andir and Sumur Bandung Subdistricts, which are busy areas in the centre of Bandung City. Overall, the sub-districts that have the lowest catchment area coverage of public transport are Cibiru (13.6%), Gedebage (21.0%), and Ujungberung (23.2%), an area in the eastern part of Bandung City. This means that more than 75% of the road network in the sub-districts is not served by any mode of public transportation.

If the catchment area ratio of public transportation is reviewed further on population density and the value of generation and attraction in each sub-district in Bandung City, it can be analysed how far the fulfilment of the needs of public transportation trips can be. The population density figure indicates residence, while the generation and attraction rates are related to the trips and destination centres in Bandung City.

The population density value in each sub-district is symbolized by a green circle in Figure 8, while the ratio of the total catchment area is indicated by the background colour in each sub-district. The worse the service area ratio, the darker the colour, which indicates a critical condition. Bojongloa Kaler sub-district as the most populous sub-district has a good catchment area ratio, i.e., 88% of the road network is covered by public transportation (large circles with light background colours). Therefore, the value of population density covered is still high, only decreasing from 39.57 thousand/km^2 to 34.88 thousand/km^2. Other sub-districts that have a high population density and good public transport coverage are Andir and Astana Anyar.

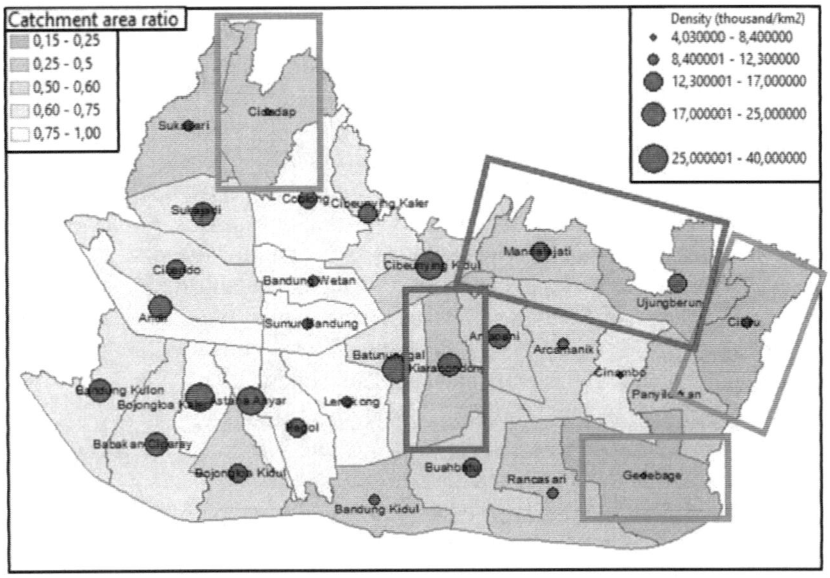

Figure 8. Visualization of population density towards catchment area ratio in Bandung City.

The sub-districts that have a relatively high population density but have a low catchment area ratio are Kiaracondong, Mandalajati, and Ujungberung Districts – indicated by the red box – which are in the north-to-east part of Bandung City. This does not meet the regional integration criteria that public transport routes are sought to reach areas with high demand potential, namely those with high population density [13]. This phenomenon is related to the next spatial indicator, namely the overlap level in the area, namely BRT for *angkot* which reaches 60% and for Damri Buses with 92%.

On the other hand, the less densely populated sub-districts (small circles) tend to have a low catchment area ratio (dark background), as shown by the blue box in Figure 8, namely Cibiru and Gedebage sub-districts in the east and Cidadap in the north. This phenomenon shows that the service area of public transportation in Bandung City is uneven and does not meet the criteria [13] so it can hinder regional development. One of the goals of BRT mode planning is to reduce the gap in infrastructure services that can hinder economic growth, but this is not achieved because public transportation remains difficult to access for some people in Bandung City in some areas, even BRT does not provide additional accessible areas, as shown in Figure 7.

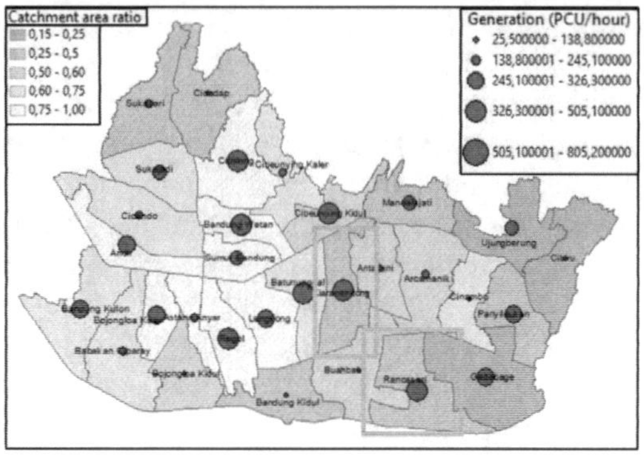

Figure 9. Visualization of traffic generation in sub-districts towards catchment area ratio in Bandung City.

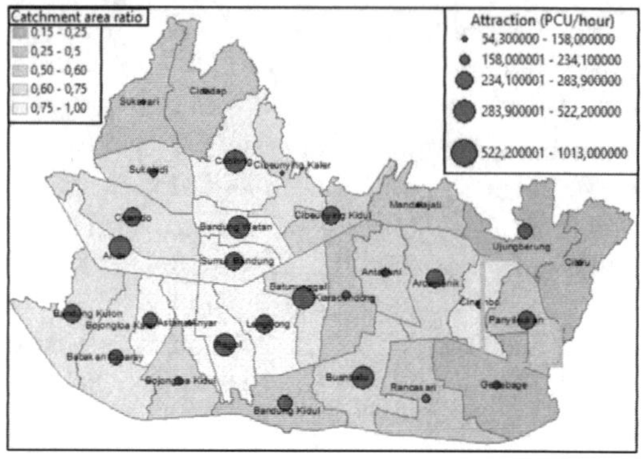

Figure 10. Visualization of attraction in sub-districts towards catchment area ratio in Bandung City.

Apart from population or residential density, the catchment ratio of public transport areas in Bandung City is also reviewed against the generation and attraction values in each sub-district. Table 1 shows the comparison of generation and attraction currents in each sub-district to the catchment area ratio. The sub-districts that have good public transportation coverage are those with high traffic demand, such as Bandung Wetan, Batununggal, Coblong, and Regol, especially Regol which is the location of various terminals and stations in Bandung City. However, some areas have a relatively low public transport catchment area ratio compared to the generation and/or attraction demands, as shown by the yellow boxes in Figure 9 and Figure 10, namely Kiaracondong and Rancasari which have high generation demand for public transport, and Panyileukan for the attraction value.

4.2. Overlap level

The first overlap level analysis is to compare the BRT mode as the main mode to *angkot* as the feeder mode in Bandung City. The overlap level can be related to the previous spatial indicator, the catchment area ratio, as shown in Figure 11. Catchment areas that are darker in colour indicate a higher overlap level between the two routes. The red box captions indicate BRT routes in the northeastern part of Bandung City (Mandalajati, Ujungberung, and Antapani, Cibiru), and Rancasari in the south, which have a high overlap level with *angkot*. This condition reduces the coverage in the area, which can be seen in the dark colour background of the sub-district, as well as explains the previous phenomenon that the sub-district has a relatively high population density but a low catchment area ratio. Both BRT and *angkot* serve major roads, thus creating competition. This condition reduces coverage in the area because local roads are underserved, whereas *angkot* should instead cover small roads to increase the coverage area that is accessible and reduce overlap.

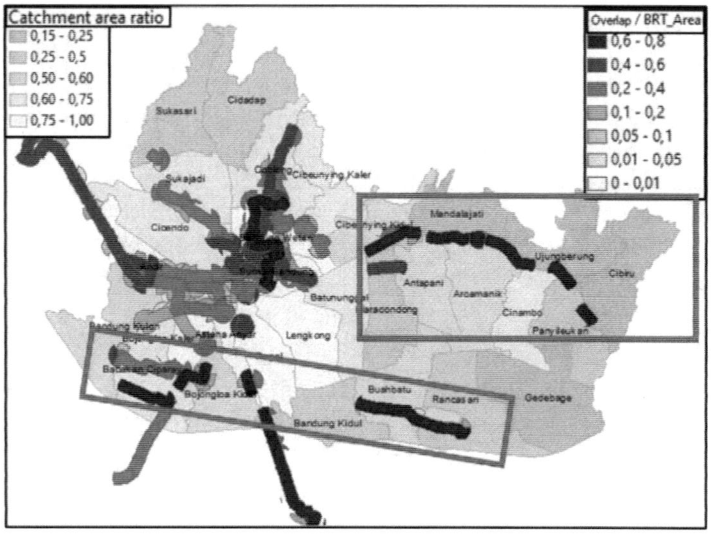

Figure 11. Visualization of overlap level between BRT corridors vs. *Angkot* in Bandung City.

Apart from *angkot*, the BRT corridors have even worse overlap level with other bus modes, namely Damri and TMB. And like *angkot*, the BRT corridor that overlaps with Damri is in the northeastern area, accompanied by corridors to and from Cimahi. The unclear division of the role of BRT with other main modes has resulted in high regional competition in some areas and corridors. Meanwhile, shown in Figure 13, the overlap level of the main modes other than BRT, namely Damri and TMB buses, and KRD to *angkot* is lower than BRT to *angkot*, where no corridor has an overlap level of more than 60 %, as happened in the BRT corridors.

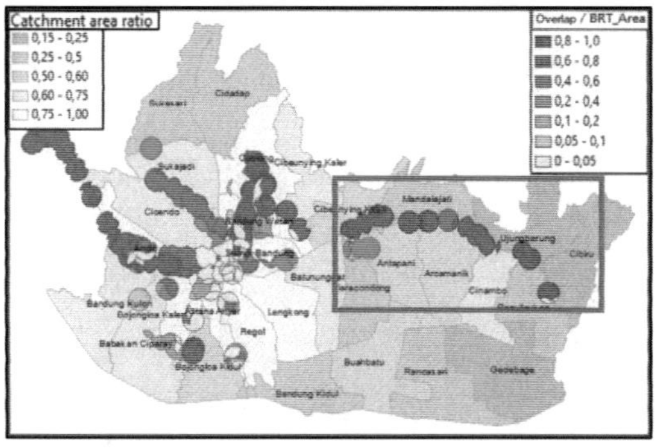

Figure 12. Visualization of overlap level between BRT corridors vs. Damri, TMB, and KRD in Bandung City.

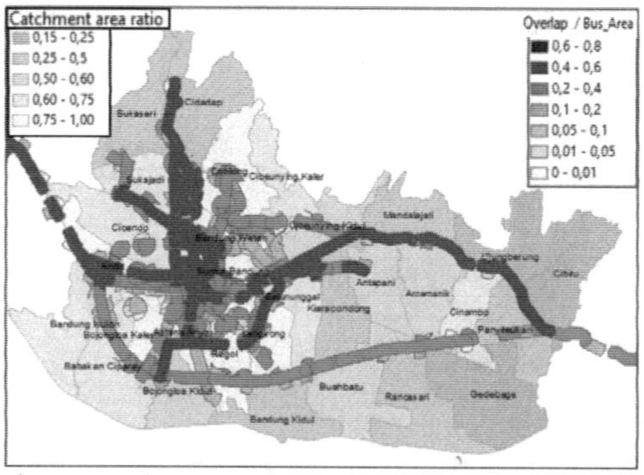

Figure 13. Visualization of overlap level between Damri, TMB, and KRD corridors vs. *Angkot* in Bandung City.

The phenomenon of BRT overlap level which is worse than the existing main modes is also shown by the graph of the percentage of route pairs that overlap at a certain level in Figure 14. The overlap level between routes dominates at 20% and below because the catchment area between routes must overlap at a low level to be connected and integrated with other modes of transportation nodes such as bus stops/stations. As many as 12.2% of the total pairs of BRT routes to *angkot* in Bandung City have an overlap level in the range of 20% to 40%, and when added with an overlap level of 40% and above, the value becomes 13.8% (12.2% + 1.2% + 0.4%). The overlap level for *angkot* across Bandung City at the level of 40% and above is also above this figure, to be precise at 15.6%. When compared with the percentage of pairs of existing main modes (Damri, TMB, and KRD) vs. urban *angkot*, the figures are not much different (13.0%), but there are no overlapping routes at the level of 60% and above. The percentage for overlapping levels of intercity *angkot* also decreased considerably to 8.7%. The conclusion is that the existing mode of overlap with *angkot* is better than the BRT overlap. The overlap level of BRT to the existing main modes (Damri, TMB, and KRD) is higher than that of *angkot*, which when added up is 28.7% at an overlap level of above 20%, indicated by grey bars. This phenomenon needs to be a concern to reduce competition among bus modes in Bandung City so that integration runs better.

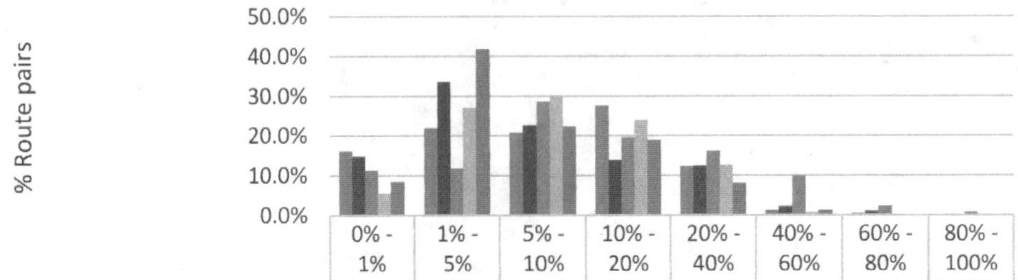

	0% - 1%	1% - 5%	5% - 10%	10% - 20%	20% - 40%	40% - 60%	60% - 80%	80% - 100%
BRT vs Angkot	16.1%	21.9%	20.6%	27.6%	12.2%	1.2%	0.4%	0.0%
BRT vs Intercity Angkot	14.7%	33.5%	22.6%	13.8%	12.4%	2.2%	1.0%	0.0%
BRT vs Damri, TMB, KRD	11.2%	11.8%	28.6%	19.4%	16.1%	9.9%	2.2%	0.7%
Damri, TMB, KRD vs Angkot	5.5%	27.1%	30.2%	24.0%	12.5%	0.7%	0.0%	0.0%
Damri, TMB, KRD vs Intercity Angkot	8.3%	41.6%	22.2%	18.8%	7.9%	1.1%	0.0%	0.0%

Overlap level

Figure 14. Percentage of intermodal route pairs based on overlap level.

4.3. Trip indicators

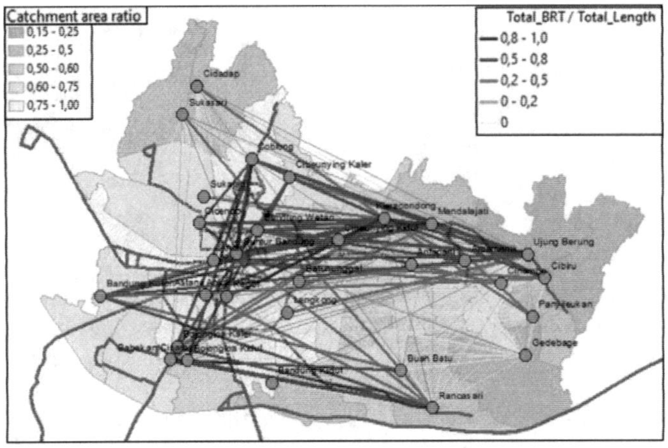

Figure 15. Visualization of the percentage of BRT usage in quickest routes with start/end points at BRT stops.

After the quickest route between sub-districts was issued, the use of the BRT mode was relatively minimal, namely only 150 out of 900 trips using the BRT mode more than 20% of the total distance traveled. This phenomenon can be caused by the BRT route which does not follow the trip pattern in Bandung City and has a high overlap level with the existing modes, so that the Damri, TMB, or *angkot* routes have been served, and the Network Analyst uses the existing modes.

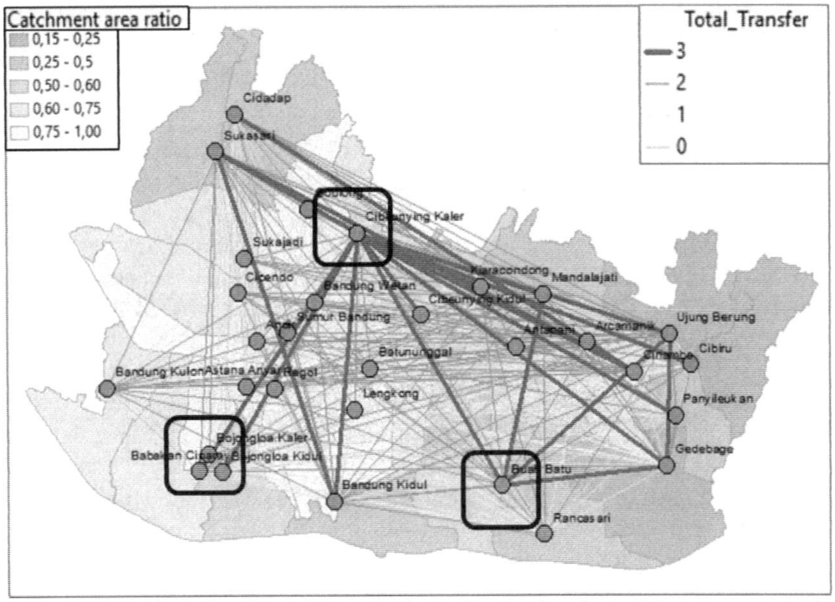

Figure 16. Visualization of the number of transfers in quickest routes with start/end points at BRT Stops

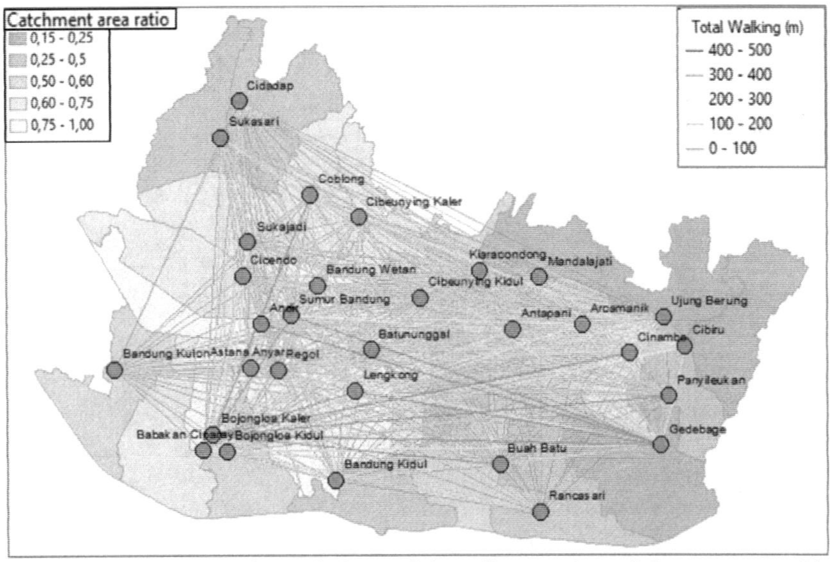

Figure 17. Visualization of the walking distance in quickest routes with start/end points at BRT Stops

The first trip indicator analysed is the number of transfers between corridors, which are set a maximum of 2 (two) times in [15] and [16]. However, based on the results of the analysis by Network Analyst for the combination of BRT, Damri, TMB, KRD, and *angkot* modes in Bandung, there are still 26 out of 900 trips (~3%) that require users to transfer three times, namely trip to Cibeunying Kaler and Sukasari sub-districts, as shown in Figure 16. This phenomenon can also be seen in the trip to and from the South Bandung area (blue box), which includes the areas of Bojongloa Kaler and Bojongloa Kidul, as well as Buah Batu.

The next indicator in the integration of the public transport network is the route availability in a trip with public transport. To quantify the definition of route availability by Network Analyst, the distance to reach the stop was used, which was set at 300–500 m for the urban area and 500–1000 m for the rural area [13]. In this study, the figure of 500 m is used to accommodate trips in the urban and rural areas. The distance to a stop is calculated using the walking attribute of the Network Analyst, which indicates the distance it takes to move from one corridor to another. Figure 17 shows a visualization of the walking distance indicator the user has travelled in the trip, and several trips require walking to transfer, but all of them are still under 500 m. Other trips are dominated by green lines which indicate the journey does not need to be taken on foot at all. Overall, users do not need to walk far to reach a stop to transfer in travelling in Bandung City by public transport, which means that at least the public transport routes are connected and available for other routes and meet the indicators of route availability.

26 inter-sub-district trips in Bandung City do not meet the indicators for the maximum number of transfers or route changes, which requires three transfers in one trip. The trip was dominated by the destination of Cibeunying Kaler. It was found that the 26 trips had a relatively low demand for transit, and even had no demand at all. Table 2 shows 20 inter-district trips with the highest transit demand in Bandung City. The trip only requires one transfer, some even do not need to transfer. The walking distance is also under 200 m, which means that every route taken is connected to other routes, so it can be said that public transportation in Bandung City meets the indicators of network integration in terms of the maximum number of transfers and the availability of routes for a trip with the highest flow.

However, the use of BRT in these trips should be noted, because only 2 of the 20 trips use BRT in their quickest routes, even though start/end points are located at BRT stops in each sub-district. The previous analysis of spatial indicators can explain this phenomenon, that the BRT overlap level is very high for *angkot* and bus modes, namely Damri and TMB, as well as Figure 7 that shows the catchment area of the BRT already served by other modes. In addition, the planned BRT route does not follow the trip pattern of users in Bandung City, which requires more travel time due to potential transfers if BRT is still used on the route by the Network Analyst. In other words, the existence of the BRT plan does not have a significant impact on Bandung City.

Table 2. Modes used in quickest routes for trips with high traffic demand.

Origin	Destination	Traffic (PCU/hour)	Total Transfer	Walking (m)	*Angkot* (km)	BRT (km)	Damri/TMB (km)
Panyileukan	Coblong	92,94	1	146	11,8	0,0	4,9
Bandung Wetan	Coblong	88,07	1	0	3,2	0,0	0,0
Coblong	Bandung Wetan	87	1	0	3,3	0,0	0,0
Andir	Coblong	83,5	0	146	0,0	0,0	4,4
Rancasari	Bandung Wetan	77,72	1	0	8,4	0,0	3,5
Bandung Kulon	Andir	77,7	1	0	2,8	3,1	0,0
Batununggal	Coblong	76,28	0	146	0,0	0,0	6,0
Cibeunying Kidul	Bandung Wetan	75,84	1	0	0,0	0,0	4,1
Regol	Batununggal	75,47	1	1	0,0	0,0	4,5
Regol	Bandung Wetan	75,22	1	83	0,0	0,0	5,9
Cibeunying Kidul	Regol	74,54	1	1	2,2	0,0	4,3

Origin	Destination	Traffic (PCU/hour)	Total Transfer	Walking (m)	*Angkot* (km)	BRT (km)	Damri/TMB (km)
Bojongloa Kaler	Bandung Kulon	74,22	1	180	5,6	0,0	0,0
Panyileukan	Regol	73,45	0	1	0,0	0,0	13,3
Rancasari	Regol	73,44	1	1	11,4	0,0	0,0
Cibeunying Kidul	Coblong	71,3	1	146	0,0	0,0	6,4
Rancasari	Batununggal	70,5	1	0	8,4	0,0	1,9
Bojongloa Kaler	Andir	69,36	0	180	0,0	0,0	4,7
Cibeunying Kidul	Batununggal	69,13	0	0	2,9	0,0	0,0
Bandung Kulon	Cicendo	66,23	1	0	6,2	0,0	0,0
Coblong	Andir	66,05	1	0	0,0	3,4	4,0

From the results of the analysis, several issues were found related to efforts to integrate the public transport network in Bandung City, namely regional competition due to overlapping, low distribution of area coverage in certain areas, and the high number of transfers in several trips due to lack of intermodal connectivity, even after the BRT mode plan. On the other hand, the indicator of the availability of routes in trips is fulfilled.

4.4. Discussion: High BRT overlap level due to unclear intermodal hierarchy

The BRT corridors have a high overlap level, which has the potential to create competition with other modes. The Alun-Alun - Cibiru Corridor has the worst overlap level because the overlap is very high, both for *angkot* as feeders and for other main bus modes, namely Damri. The overlap level with the Alun-Alun - Jatinangor corridor via Antapani even reached 92.5%, completely replacing the role of the other corridor. These results indicate the lack of clarity in the hierarchy of the public transport network in Bandung, namely the role of BRT against other main modes, namely city buses, as well as for feeder transport, namely *angkot*, so that they fight over the passenger area. In addition, the Alun-Alun – Cibiru BRT corridor also overlaps with *angkot*, namely the A10 Cicaheum – Cileunyi route. The two corridors overlap along routes within Bandung City so that BRT has the potential to decrease the role of the *angkot*.

Table 3. Recapitulation of BRT corridors overlap level and number of usages in quickest route.

		vs. *Angkot*		vs. Damri/TMB/KRD		Used in quickest route	
Fleet Size	BRT Corridors	Max	>40%	Max	>40%	Sum	%
medium	Alun-Alun - Cibiru	61,8%	2	92,5%	6	88	9,8%
medium	Cibiru - Alun-Alun	61,8%	2	92,5%	6		
large	Elang - Cicaheum	29,6%	0	67,6%	8	14	1,6%
large	Cicaheum - Elang	29,8%	0	72,7%	5		
medium	BEC - Cimahi	48,0%	3	68,5%	3	24	2,7%
medium	Cimahi - BEC	42,4%	2	55,4%	2		
medium	Baleendah - BEC	62,8%	8	15,8%	0	7	0,8%
medium	BEC - Baleendah	44,3%	5	20,9%	0		
large	Jatinangor - DU	24,9%	0	71,9%	4	0	0,0%
large	DU - Jatinangor	24,6%	0	66,5%	2		
small	DU - Cinambo	62,0%	2	44,7%	2	154	17,1%
small	Cinambo - DU	60,1%	2	44,1%	2		
small	Derwati - Ciwastra - Cijerah	63,7%	2	15,9%	0	80	8,9%
small	Cijerah - Ciwastra - Derwati	40,1%	2	15,9%	0		
medium	Leuwi Panjang - Dago	47,8%	4	51,3%	3	124	13,8%

		vs. *Angkot*		vs. Damri/TMB/KRD		Used in quickest route	
Fleet Size	BRT Corridors	Max	>40%	Max	>40%	Sum	%
medium	Dago - Leuwi Panjang	37,0%	0	61,4%	3		
large	Cicaheum - Cimahi	35,5%	0	56,7%	4	58	6,4%
large	Cimahi - Cicaheum	37,9%	0	47,7%	6		
large	Cicaheum - Margahayu Selatan	32,9%	0	44,7%	6	23	2,6%
large	Margahayu Selatan - Cicaheum	30,5%	0	48,3%	6		
medium	Antapani - Sarijadi	45,4%	1	43,3%	1	70	7,8%
medium	Sarijadi - Antapani	44,8%	2	44,1%	1		
medium	BEC - Soreang	25,3%	0	7,9%	0	0	0,0%
medium	Soreang - BEC	19,3%	0	21,2%	0		

Good to bad

Besides, Table 2 shows the minimal use of BRT mode in the quickest route with Network Analyst, for the trip with the highest demand for public transport use. This phenomenon can be further explained through Table 3 below, that some underused BRT corridors have a high degree of overlap with other main modes and/or *angkot*, as shown by the Elang – Cicaheum corridor. The corridor has an overlap rate of ~70% with the Damri Bus and TMB which also share the Cibeureum (Elang) – Cicaheum corridor. In addition to the two bus corridors, the Elang – Cicaheum BRT corridor also has an overlap rate of above 40% with many other corridors. This condition causes the BRT corridor to be used only 14 times in inter-sub-district trips in Bandung City for both directions, a very small number, as well as shows the existence of this corridor is less significant.

Other BRT corridors that are less used in the quickest routes are connecting Greater Bandung, such as Baleendah – BEC, Cimahi – BEC, and Soreang – BEC which ends at BEC, so it is natural that it is not used for trip within Bandung City, even though the overlap level with *angkot* and/or Damri buses is also very high. The Jatinangor – Dipatiukur BRT Corridor via the toll road is not being used at all because there is also a Damri Jatinangor – Dipatiukur bus corridor.

This high overlap level harms public transport operations because they compete for passengers in the same service area. In addition, overlap causes headway irregularities or the time between fleet arrivals, where there are times with a busy number of fleets, but there are also hours that have irregular fleet arrival schedules. This is detrimental to public transport users.

4.5. Discussion: Lack of equitable coverage of public transport areas

In addition to causing regional competition between modes, the high overlap level also causes a decrease in regional coverage because they serve the same area. The role of the feeder, which should feed passengers to the main mode by serving smaller travel-generating areas that are not served by the main mode, is less visible due to the overlap. The issue of the lack of clarity in the hierarchy is again the cause of the lack of feeder roles in increasing the coverage area in Bandung City. Some areas in the BRT corridor that have high overlap with *angkot* have a low catchment area ratio, and this phenomenon is found from the results of spatial analysis.

The Alun-Alun – Cibubur BRT Corridor serves some of the same areas as the Gedebage – Simpang Dago *angkot* in the eastern area of Bandung City, a collection of sub-districts that have relatively high residential densities such as the Mandalajati, Ujung Berung, and Cibiru. It can be seen in Figure 18 that the area has a low catchment area ratio, which is in the range of 15% - 25% only. This means that more than 75% of the road network in these sub-districts is not served by public transportation at all. There should be an opportunity for *angkot* to feed users from residential areas to use public transportation.

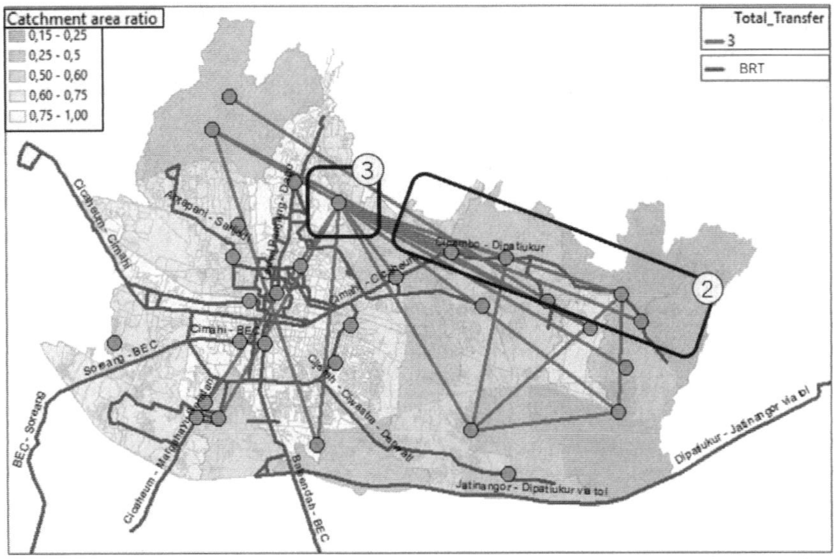

Figure 18. Examples of low catchment area ratio issues (2) and high number of transfers for trips to and from Cibeunying Kaler (3).

One of the reasons for this low demand is the lack of public transportation in the area, making it difficult to develop the area if mobility to the area is not supported by mass transportation. This is also not following the concept of equity adopted by public transport, especially BRT which can be accessed by all classes of society to reduce the gap in infrastructure services that can hamper economic growth. This phenomenon also does not meet the concept of equal distribution of services to the provision of public transportation facilities that in addition to paying attention to potential service areas, public transportation also covers all existing urban areas [13]. It is also stated that the availability of public transportation for people's transportation services must be guaranteed within the district/city area which is in the administrative area [18].

4.6. Discussion: Fulfilment of route availability indicator

An indicator of the availability of routes or other modes at the node for each trip is quantified by walking distance in the quickest route generated in the Network Analyst. From the analysis results visualized in Figure 17, all trips do not require a walking distance of more than 500 meters to reach a stop. Thus, every inter-sub-district trip in Bandung City has a route that can continue the journey from the previous route, and the integration of the public transportation network in Bandung City meets the indicators of route availability in the trip regulated in [12], [13], and [14].

4.7. Discussion: Number of transfers more than two due to lack of intermodal connectivity

Public transportation needs to minimize the number of transfers between corridors in trips in urban areas, as stipulated in the guidelines, in an effort to follow the pattern of trips in Bandung City. The number of corridor changes in urban areas is set at a maximum of 2 (two) times. Based on the quickest route shown in Table 2, indicators of the number of transfers and availability of routes have been met for trips with the highest demand for transit use, which indicates the public transportation network in Bandung City has followed the existing trip pattern. However, with the combination of BRT modes with *angkot*, Damri Bus, TMB, and KRD, the use of BRT for this trip is still very minimal.

In addition, there are still 26 trips in Bandung City that require 3 (three) transfers, especially towards Cibeunying Kaler from the south and east areas of Bandung City, which shows a lack of connectivity or intermodal integration in the Cibeunying Kaler area, especially towards BRT as a new mode plan that is expected to increase coverage in Bandung City. However, from another perspective,

IOP Conf. Series: Earth and Environmental Science 1294 (2024) 012021 doi:10.1088/1755-1315/1294/1/012021

these 26 trips have a relatively low demand for transit usage or even no demand at all. The low demand for trip by public transport for these destinations can cause the trips to become less of a priority in planning public transport networks, including BRT and other modes. But on the other hand, public transport routes that are less integrated into facilitating the trip, on the contrary, can trigger low demand for the use of public transport. This does not meet the initial public transport planning objectives, such as one of the objectives of the BRT plan is to have a wider coverage so that passengers do not change buses.

4.8. Discussion: The BRT Trans Metro Pasundan corridor hierarchy is not in aligned with the hierarchy of the public transport network in Bandung City

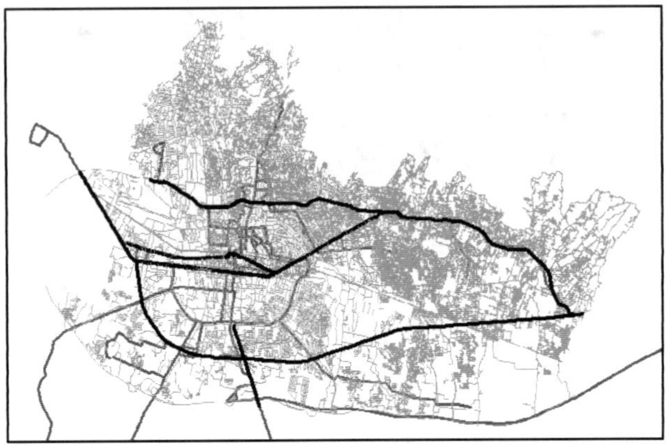

Figure 19. Comparison of the BRT Trans Metro Pasundan corridor to the main corridors in the Bandung City public transport network hierarchy (black lines).

In the spatial and trip analysis, the BRT corridor plan has a high overlap level and low usage in the quickest route for trips between sub-districts in Bandung City. This indicates the lack of clarity in the hierarchy of the public transportation network in Bandung City, where BRT as the main mode should serve the main transportation corridor in Bandung City, which includes the trips from west to east in three areas in Bandung City, namely the north, centre and south, as well as the trips from the north (Bandung City) to the south (Soreang), as shown in Figure 19. There is also a need for the role of supporting corridors, both from BRT and other bus modes, to serve places that are the centre of trip in Bandung City, such as shopping centres, markets, and educational or office areas. Finally, *angkot* as feeder transport can fill gaps in areas that have not been served by main transport and supporting transport, as well as feed users to the corridor.

The following is a summary of the evaluation of network integration in Bandung City:

- The overlap rate of BRT for *angkot* is 62%, and even 92% for Damri Buses. BRT dominates in overlapping levels above 40%, compared to overlapping existing modes vs. *angkot*.
- Due to the overlap, BRT does not increase coverage in some areas, such as 4 sub-districts in the eastern part of Bandung City, with a catchment area ratio of only 13% - 27%. The role of *angkot* as a feeder is also insignificant, indicating the unclear hierarchy of the public transport network in Bandung.
- From the quickest route between sub-districts with start/end points at BRT stops, there are 26 out of 900 trips that require transfer between corridors more than 2 times [14]. These trips dominate the destination of Cibeunying Kaler, which has low transit demand.
- The indicator of route availability is met through quantification with walking distance to reach a stop to transfer, which is still below the maximum limit of 500 m [13].

Of the three issues of integration of the public transport network in Bandung City that have been discussed based on the analysis results, an example of scenario recommendations for each issue is compiled, namely removing overlapping BRT routes to reduce intermodal competition, rerouting *angkot* routes to reduce overlap while increasing regional coverage, and rerouting *angkot* to improve connectivity with BRT corridors.

4.9. Recommendation Scenario: BRT corridor removal to reduce competition due to overlap

Two BRT corridors have a very high overlap level with *angkot* and/or Damri and TMB buses. Therefore, recommended scenarios are prepared to reduce intermodal competition while reducing network inefficiency.

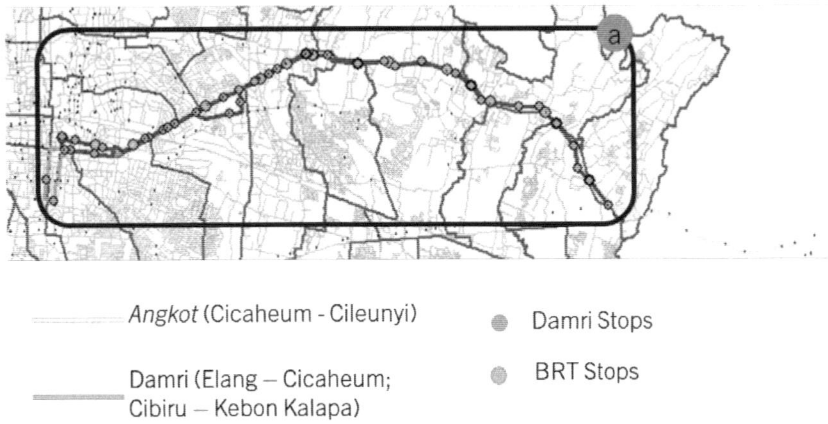

Figure 20. Recommendation to remove BRT Alun-Alun Cibiru corridor (a) that overlaps with *angkot* Cicaheum - Cileunyi and Damri Cibiru – Kebon Kalapa.

a) Removing the BRT Alun-Alun – Cibiru corridor which has an overlap of 92% of its bus stop area with the Damri corridor (Cibiru – Kebon Kalapa), so that the role of the planned corridor is no longer significant. The opportunity to reduce overlap by shifting the bus stops in the corridor is also relatively small because the position of the Damri corridor stops is also crowded. The BRT route also overlaps by 62% with the Cicaheum – Cileunyi (A10) *angkot* route, which serves the same route as the BRT within Bandung City, starting from Cicaheum Terminal.

b) Removing the Elang – Cicaheum BRT corridor which is less used in the quickest route between sub-districts in Bandung City, even though the start/end points are at the BRT bus stop, which is caused by the overlap with the Damri (Cibeureum - Cicaheum) and TMB (Cibeureum - Cicaheum) corridors at overlap level ~70%.

   In addition, because the TMB and Damri corridors serve the same route, it is recommended to remove one of the two corridors as well.

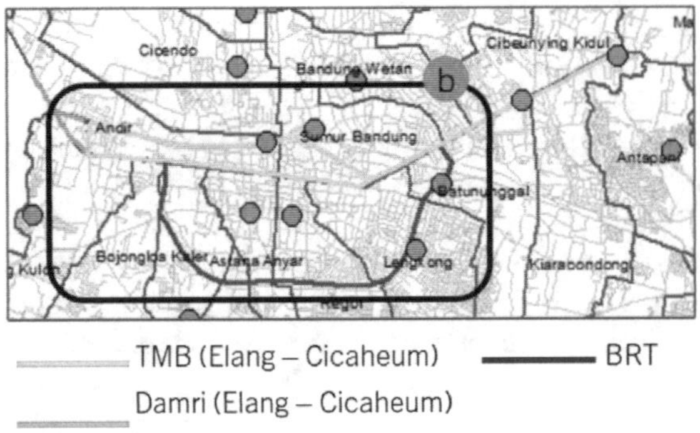

Figure 21. Recommendation to remove BRT Elang – Cicaheum corridor and either of the bus corridors, TMB's and Damri's Elang – Cicaheum (b).

4.10. Recommendation Scenario: Catchment area ratio improvement in the eastern area of Bandung City

This region of Mandalajati, Ujung Berung, and Cibiru sub-districts in the eastern area has a low catchment area ratio, even though it has a decent population density and low generation and attraction demand, so the integration of the public transport network does not fulfil the concept of equity. Therefore, *angkot* as a feeder needs to play its role in reaching users in areas that are not served by the main modes. The following is a recommended scenario for the proposed *angkot* rerouting.

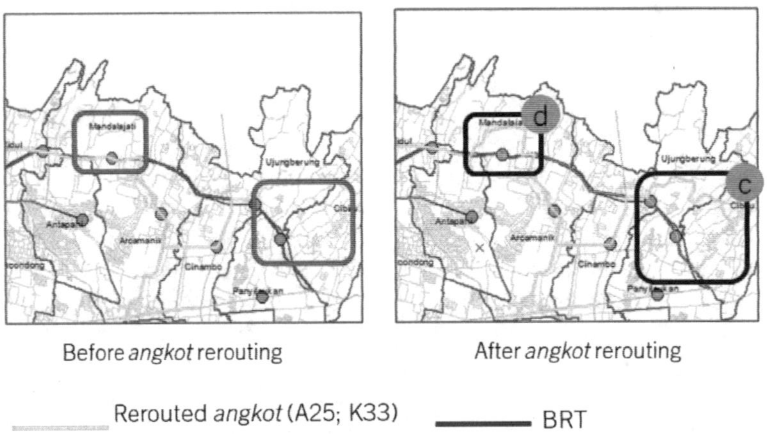

Figure 22. Recommendation to extend *angkot* route A25 Ciwastra – Ujung Berung (c) and reroute *angkot* route K33 Gedebage – Simpang Dago (d), to improve catchment area ratio.

c) Extending the Ciwastra – Ujung Berung (A25) route which initially ends at the Ujung Berung Sub-Terminal, to reach Cibiru District. The route is extended until it reaches the main mode again on collector road A.H. Nasution Street so users who are fed by the A25 route can continue their journey by bus. After the extension of the route, the catchment area ratio in Ujung Berung District increased from 23.2% to 38.5%, and in Cibiru District from 13.6% to 26.9%.

d) Rerouting Gedebage – Simpang Dago (K33) to a smaller road network to increase area coverage in Mandalajati while reducing overlap with BRT and other bus modes on main roads. After rerouting, the catchment area ratio in Mandalajati District increased from 26.6% to 42.4%.

4.11. Recommendation Scenario: Intermodal connectivity improvement at Cibeunying Kaler to reduce the number of transfers

The number of transfers for trips to and from Cibeunying Kaler exceeds the maximum limit of 2 [11]. There are no bus modes that serve in the Cibeunying Kaler area, including the BRT plan. Planning that does not pay attention to the area can be caused by one of the reasons for the low demand for public transportation to and from Cibeunying Kaler. Therefore, the importance of the role of *angkot* as a feeder that can feed users to and from the area without having to transfer too often.

Connectivity between corridors and between modes requires that there are routes that overlap or coincide with each other, or adjacent stops, to ensure the availability of routes for trips and minimize walking. There is only one *angkot* route to Cibeunying Kaler that crosses the BRT corridor in the Dago area, namely the K34 Caringin – Sadang Serang route. Other *angkot* routes also have the potential to cross the Cibeunying Kaler area and improve *angkot* connectivity with BRT, namely the K33 Gedebage – Simpang Dago corridor. Therefore, the following is a proposed transfer of the K33 route to improve the connectivity of *angkot* with BRT.

f) Rerouting Gedebage – Simpang Dago (K33) from the Surapati (south) route which does not intersect with the BRT stop to the Tubagus Ismail route which can intersect with the BRT bus stop in Dago. Thus, users to and from Cibeunying Kaler have other options and do not rely entirely on the Caringin – Sadang Serang route to continue their journey by BRT.

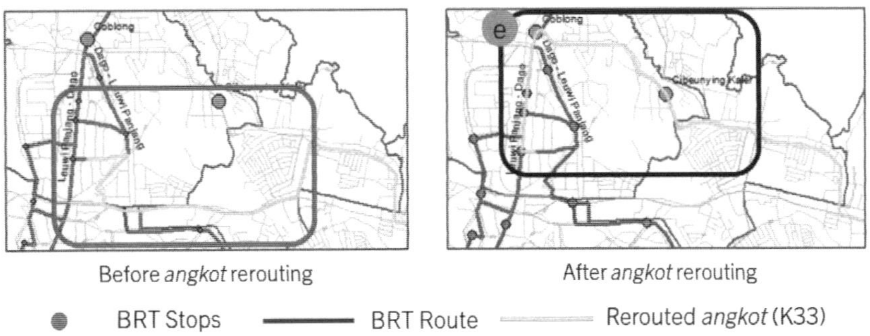

Figure 23. Recommendation to reroute *angkot* route K33 Gedebage – Simpang Dago (e), to improve connectivity in Cibeunying Kaler.

After testing with the Network Analyst feature, it was proven that after the K33 *angkot* rerouting, the number of corridor changes for trips from 7 sub-districts to Cibeunying Kaler decreased from 3 times to 2 times, namely trips from the eastern area (Cibeunying Kidul, Mandalajati, Ujung Berung, Cibiru, and Penyileukan) due to the increase in *angkot* options to Cibeunying Kaler, and from the South Bandung area (Bojongloa Kidul and Babakan Ciparay) due to the increasing use of the Dago – Leuwipanjang BRT corridor.

5. Conclusions

Based on the evaluation and analysis of the public transit network, the conclusion can be drawn from this study.

- Based on spatial analysis (service area) and O-D Matrix trip analysis to assess the integration of the Greater Bandung BRT Trans Metro Pasundan network to the existing modes of *Angkot*, Damri Bus, Trans Metro Bandung, and KRD or commuter train in Bandung City, several conclusions can be drawn that the unclear intermodal hierarchy causes 1) high overlap level and 2) low area coverage; and low connectivity causes 3) high intermodal transfers and results in low demand with

IOP Conf. Series: Earth and Environmental Science 1294 (2024) 012021 doi:10.1088/1755-1315/1294/1/012021

transit. These issues do not meet the concept of equity set out in the guidelines and objectives of public transport itself.

- With a catchment area radius of 200 m along the *angkot* routes and 400 m at BRT, Damri, TMB, and KRD stops, the overlap level of the Derwati – Ciwastra – Cijerah BRT corridor towards the same *angkot* route (Derwati – Ciwastra – Cijerah) is 63%. Concerning other major bus modes, the overlap level of the catchment area of the BRT stops for the Alun-Alun – Cibiru corridor even reaches 92.5% over the Cibiru Damri – Kebon Kalapa corridor. When compared to the existing overlap of modes vs. *angkot*, BRT dominates at an overlap level of >40%, where there is 12.8% of the total combination of BRT route pairs vs. other main modes (Damri, TMB, and KRD), 3.2% for BRT vs. intercity *angkot*, and 1.6% for route pairs vs. *angkot*, compared to overlapping of existing modes vs. *angkot* which only reached 0.7% for *angkot* and 1.1% for intercity *angkot*. The results of this analysis show the unclear hierarchy of public transportation modes in Bandung City, especially the BRT plan towards *angkot* modes as feeders and Damri, TMB, and KRD Bus modes as the main modes.
- Due to the overlap, BRT does not increase the coverage area in some areas that have a low catchment area ratio, such as 3 sub-districts in eastern Bandung, namely Mandalajati, Ujung Berung, and Cibiru, which only have public transport coverage of 13% - 27% of the available road network, despite having a relatively low population density (~10,000 – 15,000 people/km^2) which needs to be supported in order to thrive. Likewise, the southern part of Bandung, namely Panyileukan and Rancasari with a catchment area ratio of 44%, even though they have a relatively high demand for generation/attraction demand (~500 – 700 PCU/hour), so they need to be supported so as not to trigger congestion. This shows the lack of role of *angkot* as feeder transport in the area.
- With Network Analyst in ArcGIS, the quickest route between sub-districts is issued with a start/end point at the BRT stop to review trip indicators. After the existence of BRT to complement the existing modes of *angkot*, Damri, TMB, and KRD, there are still 26 out of 900 trips (~3%) that require more than 2 transfers. The 26 trips that require transfers 3 times are dominated by the destinations of Cibeunying Kaler. Other trips are from the southern part of Bandung, which includes the areas of Bojongloa Kaler and Bojongloa Kidul, as well as Buah Batu. When related to the O-D projection, the demand for transit to Cibeunying Kaler and this trip is low or zero. Twenty (20) trips with the highest demand for transit in Bandung City meet the maximum number of transfers and route availability, but only two (2) use the BRT corridor in the quickest route.
- Indicators of route availability are met through quantification by walking distance to reach stops in changing corridors. Of the 900 quickest routes for 30 sub-districts in Bandung City, the distance is still below the maximum limit of 500 m [16].
- On the other hand, a clear hierarchy of public transportation networks is needed in Bandung City, where the Trans Metro Pasundan BRT as the main mode should serve the main transportation corridor in Bandung City, namely the trip from west to east in three areas in Bandung City, namely the north, central, and south, as well as the trip from the north (Bandung City) to the south (Soreang). Then, the role of supporting corridors from BRT and other bus modes is also needed for the trip centre in Bandung City. *Angkot* as feeder transportation should be a gap filler in areas that have not been served by main corridors and supporting corridors.
- In this study, sample scenario recommendations have been developed to resolve the issues identified in the integration of the Bandung Raya BRT network in Bandung City: 1) Removing the Alun-Alun – Cibiru and Elang – Cicaheum BRT corridors to reduce the high BRT overlap with *angkot* and Damri buses that cause regional competition and some BRT corridors are underused in the quickest route; 2) Rerouting the *angkot* routes from Ciwastra – Ujung Berung (A25) and Gedebage – Simpang Dago (K33) to reduce the overlap level while increasing the catchment area ratio in the eastern part of Bandung City which has a low population density, generation and attraction demands; and 3) Rerouting the *angkot* route Gedebage – Simpang Dago (K33) to

increase the option of feeder *angkot* to Cibeunying Kaler and connecting with the BRT bus stop in Dago, thereby reducing the number of transfers from seven sub-districts to Cibeunying Kaler, from 3 times to 2 times.

6. References

[1] Putra A C 2020 *Catchment Area Evaluation of Public Transport in Bandung City with Geographic Information System (GIS)* (Final Paper of Bachelor's Program, Institut Teknologi Bandung)
[2] Wei J, Long K, Gu J, Ju Q and Zhu P 2020 *Optimizing Bus Line Based on Metro-Bus Integration*
[3] Hadas Y 2013 *Assessing Public Transport Systems Connectivity Based on Google Transit Data* (Journal of Transport Geography) vol 33, pp 105-116
[4] Fitriyani A A and Judiantono T 2017 *Evaluasi Tingkat Pelayanan Angkutan Umum Penumpang di Kota Cimahi*
[5] Chairi, Yossyafra and Putri 2017 *Perencanaan Integrasi Layanan Operasional Antar Moda Railbus dan Angkutan Umum di Kota Padang* (Jurnal Rekayasa Sipil) vol 13 no 1
[6] Manullang O R and Sitorus P A 2020 *Kajian Kebutuhan Integrasi Layanan Angkutan Umum Massal di Kota Semarang dan Sekitarnya* (Jurnal Pengembangan Kota) vol 8 no 1
[7] Gusleni Y 2016 *Keterpaduan Pelayanan Angkutan Umum di Kota Cirebon* (Jurnal Transportasi Multimoda) vol 14 no 04
[8] Rudke A P, Martins J A, Santos A M, Silva W P, Caldana N F, Souza V A and Albuquerque T T 2021 *Spatial and Socio-Economic Analysis of Public Transport Systems in Large Cities: A Case Study for Belo Horizonte, Brazil* (Journal of Transport Geography) vol 9 ISSN 0966-6923
[9] Domènech A and Gutiérrez A 2017 *A GIS-Based Evaluation of the Effectiveness and Spatial Coverage of Public Transport Networks in Tourist Destinations* (International Journal of Geo-Information)
[10] Taki H M and Lubis M Z 2017 *GIS Modelling Accessibility of Community Facilities: A Study Case of Depok City, Indonesia* (Journal of Applied Geospatial Information) vol 1 no 2
[11] Directorate General of Land Transportation 2020 *Review Dokumen Teknis Dalam Rangka Penyusunan Dokumen Perencanaan Pengembangan Angkutan Massal Berbasis Jalan Dengan Skema Pembelian Layanan (Buy the Service) di Kawasan Metropolitan Bandung Raya*
[12] Minister of Transportation Regulation No. 10 of 2012 *Minimum Service Standards for Road -Based Mass Transportation*
[13] Directorate General of Land Transportation 2002 *Technical Guidelines for the Implementation of Public Passenger Transport in Urban Areas in Fixed and Regular Routes* (SK.687/AJ.206/DRJD/2002)
[14] Multimodal Research and Development Centre 2013 *Guidelines for Assessment of the Level of Integrated Intermodal Transportation*
[15] LPKM-ITB 1997 *Modul Pelatihan: Perencanaan Sistem Angkutan Umum, Lembaga Pengabdian Masya* (Institut Teknologi Bandung)
[16] Wibowo S S, Tanan N and Tinumbia N 2015 *Walkability Measures for City Area in Indonesia (Case Study of Bandung)*
[17] Aslan H and Kocaman H 2018 *GIS Based Bus Stop Optimisation for Sakarya Public transit System* (Sakarya University Journal of Science)
[18] Law No.22 of 2009 *Road Traffic and Transportation*
[19] Nes R 2002 *Design of Multimodal Transport Network, a hierarchical approach*

Acknowledgments
Most of the data and analysis contained in this paper have been presented in the first author's final paper in the Master's Program of Civil Engineering in Bandung Institute of Technology under second, third, and fourth author supervision. More detailed work can be obtained in the university's electronic library.

Addressing the occupancy problem of Soekarno-Hatta Airport rail link during its first five years of operation

A T Syarif[1], **I Syabri**[2,*]

[1] Master Student, Graduate Program of Transportation, School of Architecture, Planning, and Policy Development, Bandung Institute of Technology, Bandung, Indonesia
[2] Associate Professor, Graduate Program of Transportation, School of Architecture, Planning, and Policy Development, Bandung Institute of Technology, Bandung, Indonesia
* Corresponding author

Emails: 24221008@mahasiswa.itb.ac.id, syabri@itb.ac.id

Abstract. Airport rail links have long been acknowledged as a crucial component of the modern transportation system. Their attractiveness among air travellers stems from their convenience, speed, and environmental benefits. Despite their advantages, these services often face several occupancy problems that hinder their effectiveness and efficiency. Soekarno-Hatta airport rail link's operation in 2018-2022 has been marked by low occupancy. This study aims to assess the realisations of several critical attributes of a successful airport rail link service according to TRB studies in the context of the Soekarno-Hatta airport rail link. In this context, it was shown that only four of the seven key attributes have met the minimal level to attain such goals. Despite numerous strategies undertaken during the period, which coincides with the COVID-19 pandemic, most fall under the operating or service adjustments category. The minimal realisations of the seven key attributes confirm the low occupancy levels, yet efforts to increase the ridership were limited during the pandemic. These findings provide avenues for a more successful transfer of operations from PT Railink to PT Kereta Commuter Indonesia with the clear intention of boosting the rail link's ridership. Alongside several expansion plans by the Directorate General of Railways, the findings shall help public policymakers promote a modal shift and pave the way towards a broader integration with other modes of public transportation in the Jabodetabek region.

1. Introduction

Since its launch in early 2018, the Soekarno-Hatta Airport rail link (SHARL) has enabled high connectivity between the airport and the surrounding area of Banten and Jakarta provinces. Being only Indonesia's second airport rail link, the SHARL was intended as an alternative ground access mode to relieve road congestion to Soekarno-Hatta International Airport (SHIA). The SHIA is the primary airport serving the Greater Jakarta Metropolitan Area, also known as Jabodetabek (Jakarta, Bogor, Depok, Tangerang, Bekasi). The airport ranks among the world's busiest airports: in 2018, SHIA was used by more than 66.9 million passengers, making it the 18th busiest airport in the world [1]. The 37.3-km journey between the Airport station (Bandara Soekarno-Hatta station) and Manggarai station takes about 56 minutes, stopping at three stations in between, namely the Batu Ceper, Duri, and Sudirman Baru (BNI City) stations. From the four stations outside SHIA, passengers may transfer to Jabodetabek

Commuter Line (KRL Commuter Line) trains that call at each of the four stations, as well as TransJakarta buses and MRT Jakarta trains, which are available outside BNI City station. The SHARL terminus at Manggarai, currently linked to nearby TransJakarta BRT stops, is projected to become Jakarta's future main central station, providing further connections with several KRL lines and long-distance trains [2].

The development, construction, and operation of SHARL by the Indonesian Railways Co. (the PT Kereta Api Indonesia, PT KAI) was mandated in the Regulation of the President of the Republic of Indonesia No. 83 Year 2011 (Perpres No. 83/2011) [3]. It was followed by a feasibility study which forecasted an annual demand of 2.8-3.2 million passengers in the planned opening year of 2016, gradually increasing to 7.6-8.7 million in 2025 [4]. However, a problematic first-year period resulted in an annual ridership of 840,988 passengers by the end of 2018, then peaked at 1,492,410 passengers in 2019, a condition only worsened by the COVID-19 pandemic a year afterwards [4]. The pandemic caused a drop in SHARL ridership for two years, such that the services were ground to a halt twice for several months. In 2022, ridership numbers eventually rose to 926,608 passengers.

This paper thus aims to assess how SHARL operators have mitigated low ridership levels during the 2018-2022 years compared to some key attributes of a successful airport rail link proposed by several studies. A successful airport rail link—or transit system in general—is indicated by a high number of passengers, which is further linked to a high modal share of trips coming into and out of the airport using public modes. By "operators", we also mean to analyse how the change of operating companies in early 2023—from PT Railink to PT Kereta Commuter Indonesia (PT KCI), the company operating KRL Commuter Line—as well as how expansion plans, might have impacts to the future of SHARL. The approach used in this study is mainly qualitative and descriptive, based on seven key attributes proposed by TRB [5] and additional reports highlighting the main problems of SHARL [6, 7], which will be discussed in parallel. Some precedents of airport rail link services in other countries will also be presented as a benchmark. In doing so, secondary sources will be utilised, such as statistical data published by the regulators and operators, other data from official statistical publications, announcements or publications by the operators, and news reporting.

2. Literature review

Several studies have explained that ground access to airports is an essential aspect of urban public transportation planning, most notably to serve a high number of trips generated by airline passengers and airport workers [8, 9]. Research by TCRP in the United States, as quoted by Mandle et al. [8], mentioned eleven types of airport ground access modes with railway, scheduled (express and multi-stop) buses, and shared-ride or door-to-door vans categorised as public transportation modes. In contrast, private vehicles, rental cars, courtesy vehicles, airline crew vehicles, taxicabs, town cars, prearranged limousines, and chartered buses and vans are categorised as private transportation modes. *Public transportation modes* are defined as the services available to the general public and aim to transport multiple passengers travelling at one time [8].

Furthermore, Schabas [9] grouped airport rail link services into five broad categories, namely:

1. a metro or rapid transit line to the CBD;
2. a dedicated express railway to the CBD;
3. a major station within the regional rail network;
4. a station on a commuter rail line/network; and
5. a hybrid commuter and express rail line to the CBD.

Schabas' study also differentiated the characteristics of two primary user groups of airport rail link (or airport access modes in general): the airline passengers and airport workers, listed in Table 1. The airline passengers' market was further categorised based on their trip purposes and residential status [5], as illustrated in Figure 1.

A study by de Neufville [10], which explained the effect of LCCs on the planning of airport ground access transportation, reveals that the operating model of LCCs is aimed at serving demands from small-sized feeder airports within a metropolitan area. The LCCs' model differs from full-service legacy carriers as they mainly operate from larger main airports, the so-called "fortress hubs" in the US. Thus, airlines operating out of smaller airports can reduce their airport-related expenses and, in turn, minimise the tariff imposed on their passengers. An airport primarily serving a price-sensitive market (i.e., preferring cheaper air tickets) means the viability of building an expensive, dedicated airport express railway towards such an airport is low, considering it will not attract high levels of ridership. In all, de Neufville suggests building an affordable airport access mode in parallel to facilitating door-to-door travel throughout the urban area by integrating it with existing transit modes and opting for more cost-effective alternatives [10].

Table 1. Characteristics of air passengers and airport workers using airport rail link. (Source: [9])

Air Passengers	Airport Workers/Employees
• Air passengers who are also residents have a relatively dispersed residential origin/destination within the region (hence the need for more connections), while non-residents primarily head for the CBD.	• Airport employees may form a captive market because they carry no luggage, travel to their workplaces at fixed times, and are more price-sensitive than air passengers.
• The type of air traveller (business or leisure) may indicate his or her willingness to pay and desired quality of service besides the origin/destination location at the city-end of the trip.	• The employees' usage of airport rail links might be low since their workplaces are not concentrated in one place, and their working shifts usually extend beyond public transport running hours.
• The type of airline operating from the airport (low-cost carriers/LCCs or full-service airlines) determines a passenger's willingness to pay for more premium rail services and the potential of using public transport at all if travelling in groups.	• As they are more price sensitive, employees of the airport and surrounding areas might be attracted to using commuter rail line that extends to the airport as a viable alternative to premium rail services.

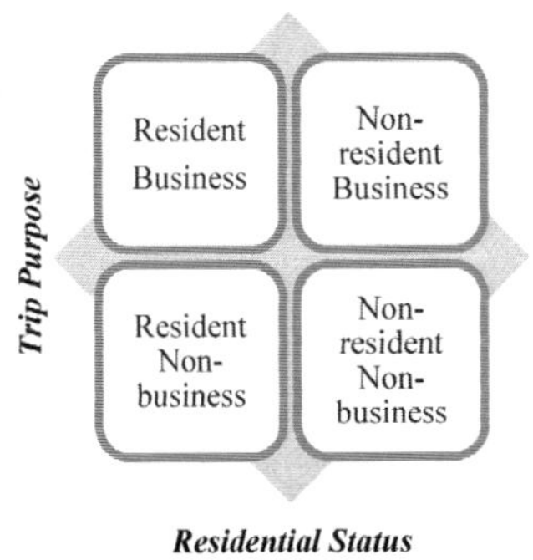

Figure 1. The "Four Cell Matrix" of the four market segments of air passengers using the airport rail link. (Source: [5])

In the context of the differing passengers' and workers' demands, it is noteworthy to distinguish dedicated and shared public transportation services to airports, as defined in [5]. When airport link services and vehicles are designed specifically to meet the needs of air travellers, they are categorised as dedicated services. On the other hand, air travellers using shared services are provided with the same vehicle types run in the same corridor as other public transportation passengers. Moreover, while most dedicated services offer direct, high-speed service to only one terminal in the downtown area, most shared services call at several stations upon entering the urban or downtown area, thus making for a relatively slower service but many intermodal transferring points [5]. As for key factors of successful rail-based airport access modes, our study refers to the seven desired attributes summarised in [5], which are explained in Table 2.

Table 2. Seven key attributes of a successful airport rail link. (Source: [5])

Attribute	Explanation
1. The proportion of air travellers with trip ends in downtown or the transit-rich core areas	At most airports, less than 15% of all travellers end their trips in the downtown area. Hence, the coverage area of a direct airport rail link to the downtown or CBD only serves a relatively small percentage of the total air passenger market.
2. Characteristics of the air traveller market	While air travellers with little or no checked baggage tend to use rail link service, large family groups (often including well-wishers) refrain from using rail. The passengers' familiarity with local public transit systems also creates a potentially high rail link ridership.
3. Regional travel time	The availability of a direct service to avoid the need for transfers or multiple stops is essential. Among the concerns of air travellers that drive them towards using rail service are: i) uncertain travel times on airport access roads and ii) the lack of or the difficulty of finding a convenient parking space at the airport.
4. Ability to walk between station and destination	Rail service is more attractive to air passengers if it is within walking distance from their final destination at the airport, hence completing door-to-door travel. The need to transfer to a second mode once they arrive at the station may shift travellers away from using the airport rail link.
5. Extent of regional coverage	A large catchment area means more accessibility for passengers, resulting from the interconnectivity between the airport rail link and the wider railway network, thus serving a more significant portion of the market.
6. On-airport travel time	Similar to factor no. 4, a more convenient airport rail service may yield when air passengers are only required to travel between the station and the aeroplane gate within a short time and distance. Another supporting factor is the airport's terminal layout (single terminal vs. multiple terminal buildings).
7. Frequency of service	Waiting times of 10 minutes or less are preferred, and they need to be adjusted with the headway of the rail service. The availability of late-night and weekend services is also essential for passengers (and workers) who need to access the airport during early or late hours.

Another TRB study lists several strategies, actions, or initiatives that transport operators may seek to improve the ridership of public transit systems. **Table** 3 below summarises the four categories of such strategies; further details on the actions/initiatives can be referred to [11].

Table 3. General strategies to improve public transportation system ridership. (Source: [11])

Category	Subcategory	
Operating/service adjustments	• Routing/coverage adjustments • Scheduling/frequency adjustments	• New types of service • Improved amenities
Partnerships/coordination	• Partnerships	• Coordination
Marketing and information initiatives	• Marketing/promotional initiatives	• Information improvements
Fare collection/structure initiatives	• Fare collection improvements	• Fare structure changes

3. Methodology

We use a qualitative approach owing to the availability of data at the time of writing, which is mainly a secondary one. We then apply nonreactive techniques (see [12]), which are content analysis and existing statistics research, to extract information from the available data. The information will be discussed and compared with the theoretical definitions, guidelines and existing precedents.

As for the data concerned, some important datasets have been obtained, such as the ridership data of SHARL for the years 2018-2023 (monthly) from the Directorate General of Railways (DGR), Ministry of Transportation, and for January-April 2023 (daily) from PT KCI; as well as the annual ridership data of Damri Soekarno-Hatta airport feeder buses for the years 2017-2023 (aggregated data coverage for year 2023 is only for January-May) the from the Perum Damri. To obtain the most recent data related to SHARL's operations, we held a public seminar session on July 4, 2023, to which several stakeholders were invited, including the DGR and PT Railink. Other data sources will also be used, mainly the annual Air Traffic Report published by the Port Authority of New York and New Jersey (PANYNJ), which provides extensive statistical reporting of world's busiest airports; the annual Air Transportation Statistics published by the Statistics Indonesia (*Badan Pusat Statistik*, BPS); and the JABODETABEK Urban Transportation Policy Integration Project Phase 2 or JUTPI 2 report published by the Japan International Cooperation Agency (JICA).

4. An overview of transportation conditions

4.1. The Soekarno-Hatta Airport and its rail link

Soekarno-Hatta International Airport (IATA code: CGK, ICAO code: WIII), located in Tangerang Municipality, Banten, is one of two airports serving Jabodetabek region; the other airport is Halim Perdanakusuma Airport located within the province of Jakarta Special Capital Region (DKI Jakarta). In 2010-2022, the number of passengers travelling through SHIA is shown in Figure 2 [1, 13, 14]. The number of airlines operating are 30 international and 10 Indonesian airlines, of which eight national airlines operate from SHIA as their main hub [15]. Based on BPS's data from 2010-2021, around 94-98% of the passengers are travellers originating or terminating at SHIA, indicating the potentially heavy land-based traffic coming into and from the airport [14].

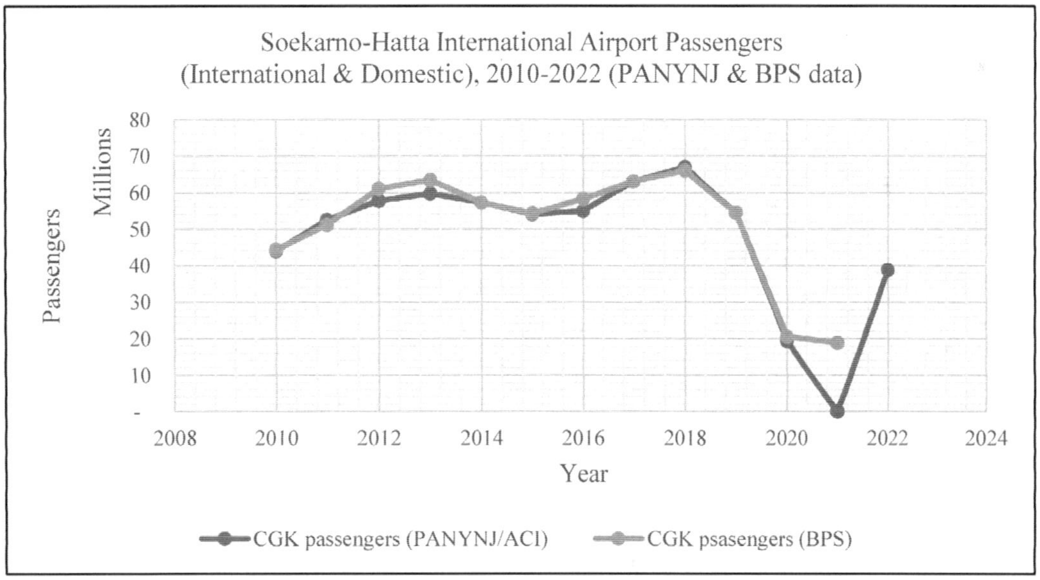

Figure 2. Number of international and domestic passengers travelling through Soekarno-Hatta International Airport, 2010–2022[a]. (Source: [1, 13, 14])

[a] No available data from PANYNJ/ACI for CGK airport in 2021.

As shown in Figure 3 [4], the 37.3-km Soekarno-Hatta Airport rail link connects five stations: Bandara Soekarno-Hatta (BST), Batu Ceper (BPR), Duri (DU), Sudirman Baru or BNI City (SDB), and Manggarai (MRI). The first two stations are in Tangerang Municipality, and the remaining three are in DKI Jakarta. The journey takes about 52 minutes with a 30-minute headway, and the service operates between 05:00-22:37 daily [4]. The ticket for an "executive" class ride costs Rp10,000-Rp50,000, whereas the "premium" class—trialled in mid-2021 and launched in early 2022—costs between Rp10,000-Rp30,000. For 15 months in 2018-2019, the service was also extended to Bekasi station before passenger facilities at Manggarai were completed.

The SHARL's ridership since launched in early January 2018 has fluctuated, as indicated in Figure 4 [4]; it peaked at 179,134 monthly passengers in November 2019. Another important metric is the load factor levels. Data obtained from the DGR, Ministry of Transportation had calculated a 28-day average of SHARL load factor: 42% on weekdays and 45% on weekends, with an average of 113 passengers/day on weekdays and 95 passengers/day on weekends [4].

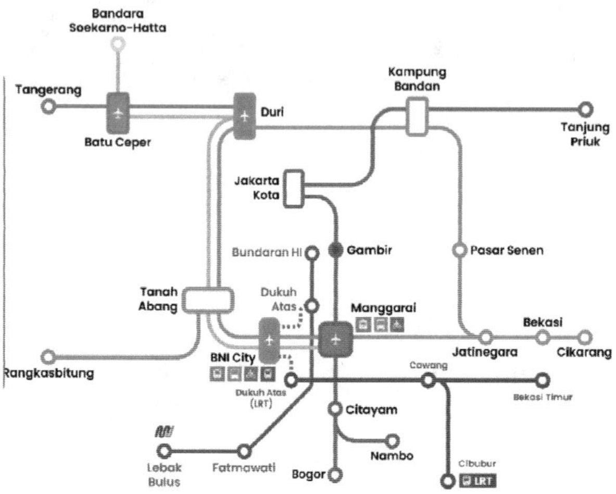

Figure 3. Schematic map of Soekarno-Hatta airport rail link and other rail-based transportation lines in Jabodetabek region. The SHARL line is coloured orange on the map (Source: [4]).

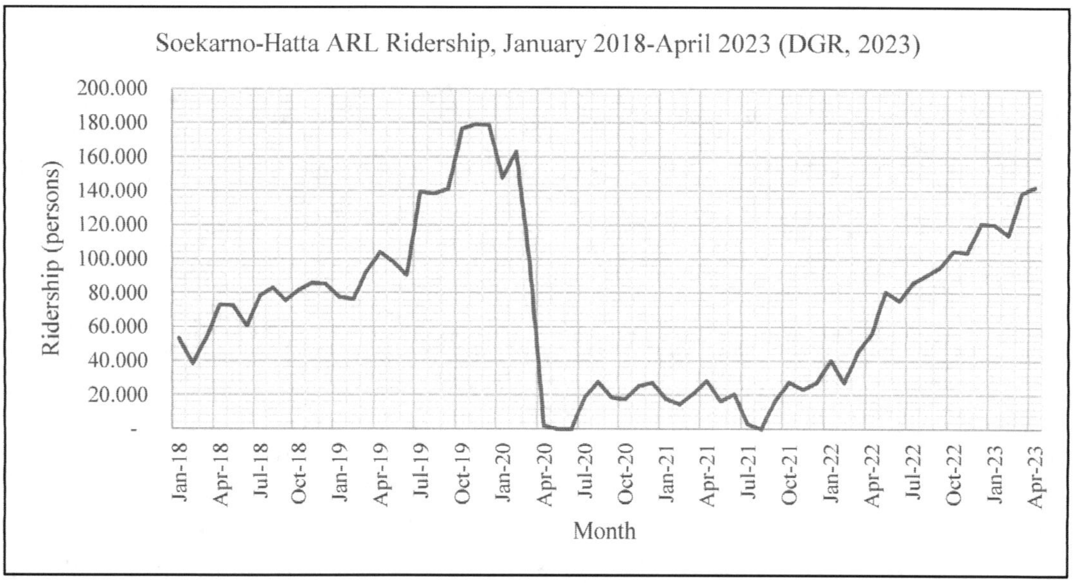

Figure 4. Passenger ridership of Soekarno-Hatta airport rail link, January 2018–April 2023. (Source: [4])

4.2. Other modes of transportation

Some other public transport options for travelling to SHIA offer connections to the Jabodetabek region and beyond, such as Jabodetabek Airport Connexion (JA Connexion) large buses, other Jabodetabek and intercity large buses, and intercity shuttle vans (locally known as "travel") mainly travelling between Bandung (West Java)-SHIA routes [16]. An overview of modal share to SHIA is obtained from the 2019 JUTPI 2 report (see Figure 5) [17], whose surveys were done in the year prior. Since the JUTPI 2 survey did not yet include SHARL among alternative modes going into and from SHIA, the public transportation modes were only represented by the large, medium, and small buses.

The Damri Soekarno-Hatta airport buses are one of the most popular alternative modes. The company operates large-bus services under the JA Connexion name and independently under its name from within and outside Jabodetabek. A Damri airport bus ticket costs Rp80,000-Rp130,000 (Jabodetabek) and Rp140,000-Rp200,000 (outside Jabodetabek). Figure 6 shows the region-aggregated share of Damri bus ridership in 2018 (referring to JUTPI 2 data obtained in that same year), totalling 7,118,811 passengers across the six regions [18].

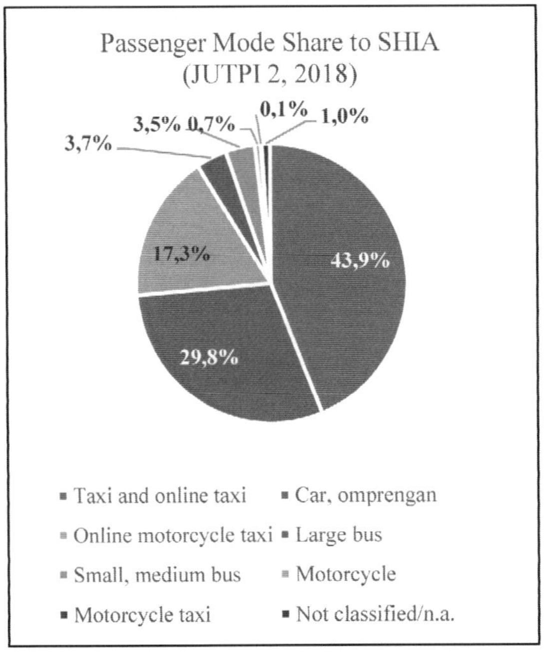

Figure 5. Access mode share of Soekarno-Hatta International Airport passengers, 2018. (Source: [17])

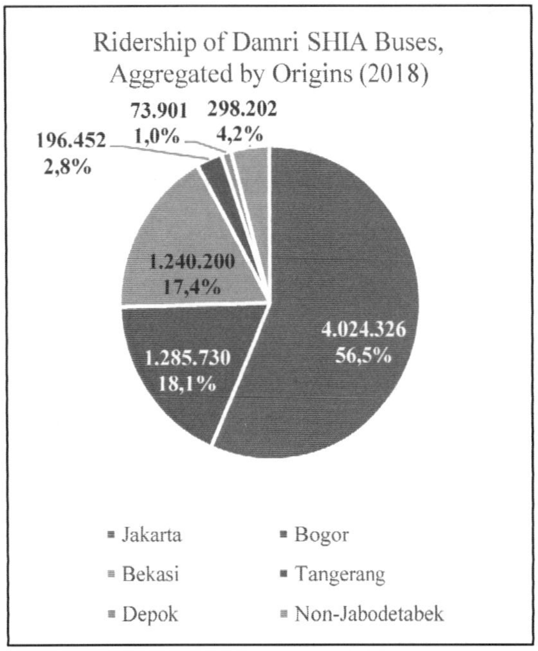

Figure 6. Ridership of Damri Soekarno-Hatta airport buses from six regions of origin, 2018. (Source: [18])

5. Results and discussion

5.1. SHARL problems, as identified in 2019 reports

In early 2019, two articles emerged detailing the problems of SHARL after just one year of operation. The reports agreed that SHARL's low occupancy must be addressed from both the city-end and airport-end sides [6, 7]. The following points are the summary of the two reports.

- Assuming a projected "70% occupancy per trip", the modal split of SHARL was still low among all trips to or from SHIA, which numbered "around 5-6% of the estimated 16%" (as quoted). The daily passenger numbers in November 2018 never exceeded 5,000. In addition, SHARL's capacity (in pphpd, passengers per hour per direction) is relatively low compared to similar services in other countries.

- SHARL fares from Jakarta city centre were more expensive than other public transport modes, such as Damri shuttle buses (Rp75,000 for SHARL compared to Rp40,000 for Damri buses when the reports were published).
- The Airport train station (BST) is not directly connected to any of the terminals at SHIA. Passengers travelling to terminals 1, 2, or 3 must take the automated people mover system ("Skytrain") or the airport shuttle buses. Furthermore, the Skytrain stops at each terminal are not located on the same level as the departure or arrival levels or within the terminal building. Thus, interconnectivity at the airport side is poor.
- Passenger access or transfer to SHARL stations in the city centre (SDB and MRI) was poor. When the reports were published, various passenger facilities were still in construction: the Jakarta MRT, which was due to have one station in the Dukuh Atas-Sudirman Baru area, the connecting corridor between SHARL's SDB and Commuter Line's Sudirman stations, and SHARL facilities at MRI. Moreover, SHARL does not serve Gambir or Jatinegara station, which enables better connectivity with intercity/long-distance train passengers who may opt to continue their journey by plane. On the other hand, there is a Damri bus service from Gambir to SHIA.
- An "odd-even" traffic management policy on Sudirman and MH Thamrin streets during morning and evening rush hours has forced passengers using private cars to make a long detour to access SDB station.
- There was (or is) a lack of coherent and coordinated policy among transport stakeholders to support the modal shift to SHARL. For example, expanding parking spaces at SHIA will encourage airline passengers to use private cars.

In conclusion, SHARL occupancy problems arose from competition with other modes (Damri buses), poor interconnectivity with other modes, station capacity constraints, and lack of coordination among the stakeholders, which posed challenges to its successful operation early on.

5.2. Airport rail link precedents in Asia

Some precedents of successful and unsuccessful airport rail links across Asia are presented in Table A1 (see Appendix). Among those services, the dedicated *Express Line* of Bangkok's Airport Rail Link is an example of an unsuccessful one. It was closed after just three years of operation, yet its counterpart, the commuter-type *City Line*, still operates today.

5.3. The realisations of the desired key attributes in the SHARL service

5.3.1. The proportion of air travellers with trips ending in downtown areas. More detailed data is needed to make a conclusive statement of the share of air passengers' trips ending in Jakarta's CBD due to the ridership data being aggregated from all five stations. However, since the SHARL was first conceived in the 1980s, the planners had already considered two alternative routes, each with several stops beyond or near the borders of Jakarta's urban area, to mitigate such a possibility of low passenger share should a direct service be operated [19]. Currently, the operational route follows the "most economical" alignment to build, although it deviated from the alignment mandated by the 2013 regulation [20], which links SHIA and Halim airport through a northerly route instead of a westerly one. The lessons from Bangkok's *City Line* and *Express Line*, and other surviving airport rail link operating with a commuter service, might also become noteworthy evidence to the statement.

5.3.2. The characteristics of the air traveller market. From the JUTPI 2 survey results, about 22.0% and 12.0% of the SHIA air passengers stated "family matters" and "shopping/leisure/tourism" as their travel purposes, respectively [17]. This information alone may indicate a high proportion of air passengers who travel in groups or carry heavier/more luggage. On the other hand, the highest share of trip purposes flying through SHIA comprised work-related travels, numbered 48.3% [17]. Nonetheless, a relatively high modal share of combined car-based modes (taxis, online taxis, and cars) to SHIA accounts for about 73.7% of all trips (see Figure 5). In all, there is an opportunity for SHARL to serve the remaining market of individual passengers currently travelling by private and non-public vehicles.

5.3.3. Regional travel time. Although it does not provide a direct service between SHIA and Jakarta's CBD, the SHARL completes its 37.3 km journey within one hour (52 minutes). Compared with other available road-based access modes (cars, vans, and medium to large buses), SHARL shall maintain a reliable travel time even during peak hours. However, a benchmark of a similar service, that is, Hong Kong's MTR Airport Express, would not justify the travel time that SHARL currently attains from a comparable total distance and number of stops (SHARL: 37.3 km distance, 52 minutes, five stations; MTR: 35.3 km distance, 28 minutes, five stations). Indeed, faster train service is needed for SHARL to gain significant advantages over the road-based modes.

5.3.4. The ability to walk between the final station and destination, and on-airport travel time. One particular problem noted in the 2019 reports was the distance between the Airport station and the terminals. Passengers arriving by train must board the Skytrain or shuttle buses from the "Integrated Building" to finally get to their intended terminals. A 2019 study reflects the problem of integrating the time between SHARL and Skytrain arrivals, specifically for flights arriving after 23:00, whose passengers cannot continue their journey by SHARL, although the Skytrain is still operating by that time [21]. In-airport connection difficulties are exacerbated because none of the Skytrain stops are located *within* any of the three terminal buildings. Terminal 3, the largest terminal, is the farthest terminal from the Integrated Building, despite having a Skytrain stop relatively closer to its doorsteps than the other two terminals.

5.3.5. The extent of regional coverage. Stopping at major transit stations—Duri, Manggarai, Sudirman Baru/BNI City—nominally shows that the SHARL's regional coverage is extensive when all the connected rail-based modes are considered, notwithstanding the other public transit modes (for example, TransJakarta BRTs). The Bekasi extension during 2018-2019 should have attracted more passengers; its closure was likely due to incomplete station and track facilities. Hence, the SHARL competed with more frequent commuter trains on the same corridor. However, such interconnectivity with other high-speed, high-capacity transport modes in the Jabodetabek region has yet to materialise in a significant increase in SHARL's ridership levels compared to Damri buses (see Figure 6).

5.3.6. Frequency of service. The latest iteration of service adjustments now enables the SHARL to have 56 daily trips (and vice versa) with a 30-minute headway [4]. This frequency marks a drop from as many as 82 daily trips allocated for both directions in early 2018 [22]. Nevertheless, a recovery from a near-total termination of services during the pandemic era should be appreciated. In addition, the current frequency and headway can be maintained throughout the week. However, as marked by the DGR load factor data, the weekday occupancy level is higher than the weekends. The operating hours of SHARL should also be reconsidered as they hinder the late night and early morning passengers (and airport workers) from using the service.

5.3.7. Other aspects of service. Since early 2022, the SHARL introduced "premium class" trains on selected schedules. Initially trialled for just one month during the Covid-19 restrictions in April 2021, the premium class is still available with reduced fares on ten daily trips, and vice versa [4]. A review of the trial period noted an increase in ridership due to a larger capacity seating arrangement being introduced [23]. Another notable observation is the SHARL's higher ridership on weekdays than at weekends. Some possible explanations might be concluded: the business air passengers mostly travel on Mondays through Fridays, and the workers choose to ride SHARL whether or not they work at the airport. Hence, defining the appropriate market segment to be served by SHARL adds up to another task for its operator. Finally, the most considerable change during the first five years of SHARL operations is the transfer of the operating company from PT Railink to PT KCI [24]. Now under the same management as the KRL Commuter Line, it follows Hong Kong's MTR precedent, which operates both the subway and the Airport Express, thus supplanting the airport rail link with an extensive network of commuter or mass rapid transit system.

5.4. The future of SHARL

The Directorate General of Railways is planning to reopen the Bekasi extension of SHARL, yet with an additional stop at Jatinegara station [4]. Another plan is to establish a new route which branches towards the north at Duri and heads to Kampung Bandan and Pasar Senen stations (refer to Figure 3). The resumption of service towards Bekasi via a southern route may be considered due to the still ongoing double-double track improvements between Manggarai-Jatinegara-Bekasi and the upgrading project of Manggarai station, which are expected to complete as early as 2024 [2]. Nonetheless, if the new northern route is to be constructed, a thorough planning of the development of railway facilities throughout the corridor is required since the "Eastern line" of Jakarta railway loop also serves as the freight train route towards the Tanjung Priok harbour. A far more important consideration is that operating the two planned SHARL routes in the future would make them completely overlap with the current Bekasi/Cikarang KRL line. Analysing how the Commuter Line demand will respond to service adjustments after such decisions shall be prioritised. Despite the necessary operational adjustments and physical improvements to enable a seamless connectivity between SHARL and other modes, it is imperative that other aspects of intermodal integration—such as legal/regulatory, institutional, informational, and economical/tariff aspects—need to be taken into account, which is a subject for further research. Ultimately, addressing these challenges will not only improve the efficiency and utilization of SHARL, but also contribute to a more sustainable and seamless travel experience for passengers.

6. Conclusions

In the analysis above, an assessment of SHARL's operational conditions and problems in the first five years of operation has shown that only four of the seven attributes have met what is minimally prescribed in the TRB study [5], which are the characteristics of the air traveller market, regional travel time, the extent of regional coverage, and frequency of service. However, room for improvement in the four attributes is necessary to boost the SHARL's ridership levels further. Furthermore, based on another TRB study [11], the operator's strategies to mitigate a low occupancy mostly fall under the operating or service adjustments category. In all fairness, a more comprehensive range of strategies, actions, or initiatives was limited because of the global pandemic during most of the five years. Despite plans for future service expansion by the DGR [4], which are also related to the same type of strategy, they lack a vision and clear strategy to fully integrate the SHARL service with other modes of public transportation in Jabodetabek. Such integration shall, in turn, pave the way towards higher ridership levels due to a more seamless transfer between the modes. The strategy to combine the SHARL and KRL Commuter Line operations under a single company in early 2023 has enabled "integration" in a business sense. Nonetheless, since the first day of SHARL operation, physical connectivity of the two modes has materialised with four of the five stops located at Jakarta's major rail interchanges. While it is still too early to conclude what PT KCI's takeover of the operation and the DGR's planning may bring about, the progress to an increase in SHARL's ridership is underway, along with the general transition from the pandemic to the endemic era.

7. Appendix

Table A1. Some precedents of airport rail links in Asian cities

Name of ARL	*KLIA Ekspres, KLIA Transit*	*Airport Rail Link (ARL)*	*MTR Airport Express*	*AREX*
Location	Kuala Lumpur, Malaysia	Bangkok, Thailand	Hong Kong, China	Seoul, South Korea
Type of service	dedicated express (*Ekspres*); commuter-express hybrid (*Transit*)	commuter (*City Line*); dedicated express (*Express Line*)	dedicated express	dedicated express (*Express*); commuter (*All Stop*)
Year of operation	2002-present	2010-present (CL); 2011-2014 (EL)	1998-present	2007-present

Total distance	57 km	28.6 km	35.3 km (Hong Kong–Expo)	63.8 km[a]
Travel time	28 min. (E); 36 min. (T)	26 min. (CL); 15 min. (EL)	24 min. (to Airport); 28 min. (to Expo)	43 min. (E)[a]; 59 min. (AS)[a]
No. of stations	3 (E); 6 (T)	8 (CL); 2 (EL)	5	3 (E); 11 (AS)
Headway	15/30 min. (E); 20/30 min. (T)	10/15 min. (CL); 60 min. (EL)	10-15 min.	40-50 min. (E); 10-20 min. (AS)
Tariff from CBD	RM55.00[b]	THB15-45[c] (to Bangkok CBD)	HK$110[d] (Octopus card)	₩9,500[e] (E); ₩4,750[e] (AS)
Type of intermodal integration at city	KTM trains, LRT, MRT, monorail, city buses, City Air Terminal (KL Sentral)	MRT, BTS, SRT, BMTA buses	MTR, buses, airport shuttle, in-town check-in (Hong Kong & Kowloon)	Korail trains, KTX, Seoul Subway, City Air Terminal (Seoul station); Incheon Subway, LRT
Type of intermodal integration at airport	Stations within terminals 1 & 2	Station within the airport terminal	Station within the airport terminal	Incheon Airport Maglev, stations within airport terminals

[a] Distance and time to Incheon Airport terminal 1
[b] RM10.00 = Rp32,125 = US$2.15
[c] THB100 = Rp43,084 = US$2.88
[d] HK$10.00 = Rp19,091 = US$1.28
[e] ₩1,000 = Rp11,713 = US$0.78
(All exchange rates as of July 12th, 2023)
Sources: [25–28]

8. References

[1] The Port Authority of NY & NJ. *2018 AIRPORT TRAFFIC REPORT*. The Port Authority of NY & NJ, 9 April 2019.

[2] Putri RS. Proyek Revitalisasi Manggarai Jadi Stasiun Sentral Terhambat Pembebasan Lahan, Ini Respons Kemenhub. *Tempo*, https://bisnis.tempo.co/read/1730860/proyek-revitalisasi-manggarai-jadi-stasiun-sentral-terhambat-pembebasan-lahan-ini-respons-kemenhub (2023, accessed 14 July 2023).

[3] Peraturan Presiden Republik Indonesia Nomor 83 Tahun 2011 tentang Penugasan kepada PT Kereta Api Indonesia (Persero) untuk Menyelenggarakan Prasarana dan Sarana Kereta Api Bandar Udara Soekarno-Hatta dan Jalur Lingkar Jakarta-Bogor-Depok-Tangerang-Bekasi. 114, 2011.

[4] Direktorat Jenderal Perkeretaapian, Kementerian Perhubungan. INTEGRASI TRANSPORTASI ANTARMODA PADA KA BANDARA SOEKARNO-HATTA.

[5] Airport Cooperative Research Program. *Ground Access to Major Airports by Public Transportation*. Washington, D.C.: Transportation Research Board.

[6] Abrar RY. Instran Beberkan 10 Kekurangan KA Bandara - TIMES Indonesia, https://timesindonesia.co.id/peristiwa-nasional/195898/instran-beberkan-10-kekurangan-ka-bandara (2019, accessed 13 July 2023).

[7] Hariyono WP. After one year: Improving international airport train. *The Jakarta Post*, https://www.thejakartapost.com/academia/2019/01/12/after-one-year-improving-international-airport-train.html (2019, accessed 13 July 2023).

[8] Mandle PB, Mansel DM, Coogan MA. Use of Public Transportation by Airport Passengers. *Transportation Research Record* 2000; **1703**: 83–89.

[9] Schabas M. Who Rides Airport Railways?: The Case of London City Airport. *Transportation Research Record* 2013; **2350**: 80–90.

[10] Neufville RD. Planning Airport Access in an Era of Low-Cost Airlines. *J. American Plan. Assoc.* 2006; **72**: 347–356.

[11] Transit Cooperative Research Program. *Elements Needed to Create High-Ridership Transit Systems*. Washington, D.C.: Transportation Research Board.

[12] Neuman WL. *Social Research Methods: Qualitative and Quantitative Approaches*. 2014.
[13] The Port Authority of NY & NJ. 2010-2023 AIRPORT TRAFFIC REPORT, https://www.panynj.gov/airports/en/statistics-general-info.html (2023).
[14] Badan Pusat Statistik. Statistik Transportasi Udara 2010-2021, https://www.bps.go.id/ (2022).
[15] Info Maskapai. *Soekarno-Hatta International Airport*, https://soekarnohatta-airport.co.id/maskapai (accessed 14 July 2023).
[16] Direktorat Jenderal Perhubungan Udara, Kementerian Perhubungan. KEBIJAKAN INTEGRASI TRANSPORTASI STUDI KASUS BANDAR UDARA SOEKARNO HATTA.
[17] Japan International Cooperation Agency (JICA), ALMEC Corp. *JABODETABEK Urban Transportation Policy Integration Project Phase 2 in the Republic of Indonesia - Annex 02: JABODETABEK Urban Transportation Master Plan (Detailed RITJ)*. October 2019.
[18] Perum Damri. Data Penumpang & Jadwal Keberangkatan Bus Bandara Soekarno-Hatta.
[19] Japan International Cooperation Agency (JICA). *NEW RAILWAY LINE JAKARTA-CENGKARENG AIRPORT*. Tokyo, 1983.
[20] Keputusan Menteri Perhubungan Republik Indonesia Nomor KP. 1264 Tahun 2013 tentang Penetapan Trase Jalur Kereta Api Umum Nasional dari Bandar Udara Soekarno-Hatta ke Halim Melalui Manggarai. 2013.
[21] Viarsyah D, Sumabrata J. Time integration analysis of Soekarno-Hatta International Airport (SHIA) Train with Skytrain. *IOP Conf Ser: Earth Environ Sci* 2019; **248**: 012015.
[22] ray. Operasional KA Bandara Soetta Bikin Kacau KRL Duri-Tangerang. *CNBC Indonesia*, https://www.cnbcindonesia.com/news/20180406112705-4-9933/operasional-ka-bandara-soetta-bikin-kacau-krl-duri-tangerang (2018, accessed 14 July 2023).
[23] KAI Bandara Railink. Kenapa Premium? *@kabandararailink*, https://www.instagram.com/p/CXvsAy5hNN8/?utm_source=ig_web_copy_link&igshid=MzRlODBiNWFlZA== (2021, accessed 14 July 2023).
[24] KAI Bandara Railink. Siaran Pers 3 Maret 2023 PT. Railink (KAI Bandara) Resmi Tak Lagi Menjadi Operator KA Bandara Soekarno-Hatta. *@kabandararailink*, https://www.instagram.com/p/CpZVq8uvhub/?utm_source=ig_web_copy_link&igshid=MzRlODBiNWFlZA== (2023, accessed 14 July 2023).
[25] Express Rail Link Sdn Bhd. KLIA Ekspres - The Fastest Way to KLIA. *KLIA Ekspres*, https://www.kliaekspres.com/http:www.kliaekspres.com/ (accessed 14 July 2023).
[26] Bangkok Airport Train Guide. Bangkok Airport Train Guide - Suvarnabhumi Airport Rail Link. *Bangkok Airport Train Guide - Suvarnabhumi Airport Rail Link Passenger Information*, https://bangkokairporttrain.com/ (2023, accessed 13 July 2023).
[27] MTR Corporation Limited. MTR > Airport Express Services. *MTR*, https://www.mtr.com.hk/en/customer/services/airport_express_index.html (accessed 14 July 2023).
[28] AIRPORT RAILROAD Co., Ltd. Welcome to Incheon Airport Railroad. *AREX Airport Railroad*, https://www.arex.or.kr/main.do (accessed 14 July 2023).

Acknowledgments
The authors would express their gratitude to the stakeholders who had participated in the SAPPD Webinar #3 held on July 4th, 2023: Directorate General of Civil Aviation, Ministry of Transportation; Directorate General of Railways, Ministry of Transportation; PT Railink; Jakarta Provincial Agency for Transportation; Professor Harun A. S. Lubis of ITB's Faculty of Civil and Environmental Engineering; and Mr. Harya S. Dillon, Ph.D. of The Indonesian Society for Transportation (MTI). We also thank PT KCI and Perum Damri for the data provided. Some data and analysis presented in this paper are obtained from the first author's coursework (unpublished) of the TR5004 Public Transportation System course taught in the Graduate Program of Transportation in SAPPD, and his master thesis (unpublished) under the second author's supervision. More detailed work of the thesis will be made available through the university's electronic library by the time the paper is presented in the Symposium.

IOP Conf. Series: Earth and Environmental Science 1294 (2024) 012023 doi:10.1088/1755-1315/1294/1/012023

Perspective of senior citizens regarding intersectional equity on walkability

A K Adisaputro*, T B Joewono and P Hartieni

Civil Engineering Department, Parahyangan Catholic University, Ciumbuleuit 94 Bandung, Indonesia

*Correspondence: aurelia.krishna@gmail.com

Abstract. With the current trend of exploring walkability measures using objective approach, it is essential not to overlook subjective perspectives. Intersectionality examines how heterogenous members of a specific group might have different experiences depending on several aspects of their identities, including gender, age, sexuality, ethnicity, religion, and other social factors. This study explores the influence of age on walkability perceptions, revealing that senior citizens have distinct perspectives on this matter compared to the average population. The study involved 392 senior citizens who rated various variables affecting their walkability. These ratings were then analyzed using Random Forest Classifier to identify key variables that predict their willingness to walk. The findings indicate that walkability is positively associated with senior citizens' willingness to walk when certain variables are present, such as adequate lighting, evenness of pedestrian walks, short travel time to recreational facilities, availability of benches or resting spaces, the presence of family or friends during walks, the presence of security officers, availability of pedestrian walks, and the availability of social activities and gatherings. This study contributes to the understanding of intersectional equity in pedestrian facilities and provides considerations to its future development by including the perspective of senior age individuals.

1. Introduction

Walkability measures the capabilities of a built environment to enable walking condition by being easily traversable, compact, physically-enticing, or safe [1]. Many variables can affect a neighborhood's walkability, such as residential density, street connectivity, traffic conditions, and aesthetics. Increased walkability can encourage people to use public transportation, which may become a solution to high traffic issues, whilst also bringing a positive trend to public health and economy [2].

The measure of walkability can be explored from an objective and subjective standpoint. The trend of current study of walkability measures tend to use a more objective standpoint. However, the subjective standpoint of the pedestrian is just as important in determining walkability [3]. Environments that have high walkability score does not necessarily have high pedestrian density. The walking behavior and neighborhood characteristics also promote public's motivation to walk. Measures of walkability needs to consider more socioeconomic factors so that environments with high walkability will be able to promote more walking trips [4].

Intersectionality is a concept that examines how heterogenous members of a specific group might have different experiences depending on several aspects of their identities [5]. Intersectionality includes gender, age, sexuality, ethnicity, religion, and other social locations. These aspects are interrelated and influence one another to create unique social experiences for each individual members of society [6].

Pedestrians' perspective on walkability can be affected by their intersectionality. People of different gender, age, and cultures can have different perspectives on what they consider as a walkable environment [7]. Intersectional equity for walkability is achieved by considering the different walking experiences of the pedestrians of a built environment. An ideal walkable environment should be designed for diversity so that it becomes beneficial for all [8].

World Health Organization (WHO) defines individuals aged 60 and above as senior citizens [9]. This age group often faces specific health and mobility challenges that can limit their ability to engage in physical activities like walking [10]. Therefore, it is essential to focus on the development of infrastructure and services tailored to support senior citizens in maintaining a healthy and active lifestyle [3]. To achieve this goal effectively, it becomes essential to consider the perspectives of senior citizens regarding walkability variables [8]. By doing so, walking trips among senior citizens can be better promoted and encouraged.

Departing from that idea, this study is developed to fulfill two aims. Firstly, it seeks to comprehensively analyze the variables that influence the perspective of senior citizens regarding walkability in urban environments. By examining these factors, we aim to gain a deeper understanding of the specific elements that are significant to senior citizen considering their walking experiences. Secondly, this study aims to prioritize these variables, shedding light on which factors carry greater influence in shaping the walkability perceptions of senior citizens. By achieving these objectives, this study aims to provide valuable insights into the unique needs and priorities of senior citizens in urban planning, hence contributing to the development of more age-friendly and inclusive communities.

2. Methods

2.1. Sample

Senior citizens of Bandung will be the primary source of this study. The study is focused on the perspectives of senior citizens towards walkability on Braga Street, who have taken walks on Braga Street for the last 30 days of the data collection period. The whole length of Braga Street is applicable to this study. The surveyor is required to first identify senior citizens who fit into the criteria before approaching them to fill in the survey questionnaire.

Braga Street is chosen due to its abundance of recreational facilities such as shops and restaurants, also for its close vicinity to shopping malls and parks. Braga Street is also known as one of the historic sites in Bandung City [11]. Its status as a recreation spot provides higher population of pedestrians [12]. Higher population of pedestrians also means that there is higher chance of finding senior citizens who walk in the area. It is important to select a location with many pedestrians since the number of senior citizens who travels by walking can be quite low.

2.2. Data Collection

Data collection of this study is done using an online survey via Google Forms. The survey questionnaire is separated into four parts. The first part of the questionnaire asks about the respondents' characteristics, such as their gender, age, and travel behavior. The second part of the questionnaire consists of filter questions that made sure that the respondents are senior citizens who have walked on Braga Street within 30 days and have not filled in the questionnaire before. The third part of the survey asks the respondents to rate 16 different variables of walkability on a scale of 1 to 5. The final part of the survey consists of a statement question used for the Random Forest Classifier model used for this study.

The variables of walkability for this study are taken from walkability variables listed by previous studies. The studies referenced are Global Walkability Index [13], Field Walkability Survey Parameters adjusted to Asian cities [14], Walkability Audit Elements [15], and Facilitators of Neighborhood Activity [16]. The walkability variables and its sources can be seen in Table 1.

2.3. Data Analysis

The survey result will be split into training data and test data for the random forest classifier. The training data will be used to build and train the random forest classification model. The test data will be used to test the accuracy and precision of the model.

The training data will be separated into 8 bootstrapped datasets in order to build 8 decision trees that contains 2 variables each. Then, the Gini Index of each variable can be calculated to decide the root node and branch node of each decision tree. The decision tree will then be used to make a prediction on senior citizens' willingness to walk based of majority voting between 8 decision trees. The accuracy and precision of the model is measured using a Confusion Matrix. Accuracy is the ratio between the number of correct predictions to the total number of samples, whereas precision is the ratio between the number of correctly classified positive sample to the total number of samples classified as positive [17].

Table 1. Walkability variables for senior citizens.

Parameter	Variable	Source
Availability of walking paths	Availability of pedestrian walks	Field Walkability Survey Parameters [14]
Path size	Width of pedestrian walks	Walkability Audit Elements [15]
Quality and maintenance of walking path surface	Evenness of pedestrian walks	Global Walkability Index [13]
Availability of crossings	Availability of crossings	Global Walkability Index [13]
Security	Presence of security officer	Global Walkability Index [13]
Security	Adequate lighting	Global Walkability Index [13]
Cleanliness	Cleanliness of pedestrian facilities	Global Walkability Index [13]
Aesthetics	Aesthetic of pedestrian facilities	Walkability Audit Elements [15]
Amenities	Verdancy and adequate shading	Walkability Audit Elements [15]
Connectivity	Travel distance to nearest recreational facilities	Global Walkability Index [13]
Connectivity	Travel time to nearest recreational facilities	Global Walkability Index [13]
Connectivity	Availability of public transportation	Global Walkability Index [13]
Connectivity, disability infrastructure	Accessibility for bicycles and mobility aids	Global Walkability Index [13]
Amenities	Availability of benches or other resting space	Field Walkability Survey Parameters [14]
Incentive of social interaction	Availability of social activities and gatherings	Facilitators of Neighborhood Activity [16]
Incentive of social interaction	Presence of family or friends during walks	Facilitators of Neighborhood Activity [16]

3. Results

The survey was conducted on June 17th to June 29th 2023. From this survey, a total of 392 responses can be used for data analysis. The respondents were comprised of majority females (58.67%) gender-wise, and 60-64 years old (54.08%) in age group. Moreover, all of the senior citizen respondents walk at least once a month, with more than 70% of them actually revealed to walk at least once a week. The most popular destinations among them are around their neighborhood (31.38%), parks (29.85%), and shopping districts (28.57%). Finally, this study unveils that the most preferred mode of transportation of the senior citizens are by car (34.95%) and motorcycle (28.57%), in contrast to non-motorized mode such as walking (12.76%) and bicycle (4.08%). Surprisingly, 14.54% of the respondents prefer using local public transportation, *angkot*, which has various available routes around the city and relatively cheaper. These descriptions reflect typical characteristics of senior age individuals who are enjoying their retirement and spending their time doing leisure activities. More details on the respondents' characteristics can be seen in Table 2.

In the section about walkability, there are 16 variables that has been rated by the respondents in the survey. The average rating of each variable is then counted and sorted from highest to lowest. The variable with the highest mean is availability of pedestrian walks at 4,2347 and the variable with the lowest mean is availability of social activities and gatherings at 3,7959. The overall mean for all 16 variables is at 4,0886. The list of variables, the distribution of the rating and the average/mean of each variable can be seen in Table 3.

Table 2. Respondent characteristics.

Respondent Characteristics		Quantity	Percentage (%)
Gender	Male	159	40,5612
	Female	230	58,6735
	Would rather not specify	3	0,7653
Age	60-64 years old	212	54,0816
	65-69 years old	144	36,7347
	70-74 years old	32	8,1633
	>= 75 years old	4	1,0204
Walking frequency	Every day	57	14,5408
	A few times a week	109	27,8061
	Once a week	125	31,8878
	Every other week	81	20,6633
	Once a month	20	5,1020
	Very rarely	0	0,0000
Most frequent walking destination	To work	13	3,3163
	To school	9	2,2959
	To a park	117	29,8469
	To a shopping district	112	28,5714
	Around the neighborhood	123	31,3776
	Others	18	4,5918
Most preferred mode of transportation	Walking	50	12,7551
	Bicycle	16	4,0816
	Motorcycle	112	28,5714
	Car	137	34,9490
	Angkutan kota (*angkot*)	57	14,5408
	Bus	14	3,5714
	Trains, MRT, or LRT	6	1,5306

The respondents were also asked to rate the current walkability of Braga Street on a scale of 1 to 5, with 1 being very poor and 5 being excellent. Out of 392 respondents, there were no respondents that rated the walkability of Braga Street at 1. There were 7 respondents who gave a rating of 2, 54 respondents gave a rating of 3, 106 respondents gave a rating of 4, and 222 respondents gave a rating of 5, bringing the mean of the rating to 4,40.

The data is then separated into 292 responses for the training data and 100 responses for the test data. The training data is then bootstrapped into 8 datasets to build the decision trees. The decision trees with its variables can be seen in Table 4, which is later used as a foundation to predict senior citizens' willingness to walk more often. The prediction of training data resulted in 237 true positive (TP), 0 true negative (TN), 31 false positive (FP), and 24 false negative (FN), which leads to 81,2% accuracy and 88,4% precision. Moreover, model estimation on the test data generates 83 TP, 0 TN, 11 FP, and 6 FN, with 83% accuracy and 88,3% precision produced.

The higher the model's accuracy, the better it is at making a prediction. Hence, with the test data result, the model is deemed to be accurate with satisfactory precision, meaning the result is ideal. The accuracy and precision of the model can be improved further, but it is not recommended since it can cause overfitting.

Table 3. Walkability variable rating.

Walkability Variable	Rating Distribution					Mean
	1	2	3	4	5	
Availability of pedestrian walks	1	4	67	150	170	4,2347
Cleanliness of pedestrian facilities	0	1	71	172	148	4,1913
Availability of benches and other resting space	1	3	77	160	151	4,1658
Adequate lighting	0	2	70	183	137	4,1607
Availability of public transportation	0	0	79	184	129	4,1276
Availability of crossings	1	2	80	177	132	4,1148
Accessibility for bicycles and mobility aids	0	1	83	180	128	4,1097
Verdancy and adequate shading	0	1	80	187	124	4,1071
Aesthetic of pedestrian facilities	0	1	86	179	126	4,0969
Width of pedestrian walks	0	2	87	179	124	4,0842
Evenness of pedestrian walks	0	4	90	169	129	4,0791
Presence of family or friends during walks	0	2	87	183	120	4,0740
Presence of security officer	1	2	91	179	119	4,0536
Travel time to nearest recreational facilities	0	7	90	180	115	4,0281
Travel distance to nearest recreational facilities	0	3	106	173	110	3,9949
Availability of social activities and gatherings	8	27	90	179	88	3,7959

Table 4. Decision Trees with Root Nodes, Branch Nodes and Gini Index.

Decision tree	Variable	Gini index	Root node/ branch node
1	Width of pedestrian walk	0,1954	Branch node
	Adequate lighting	0,1929	Root node
2	Evenness of pedestrian walks	0,1622	Root node
	Cleanliness of pedestrian facilities	0,1637	Branch node
3	Availability of crossings	0,1293	Branch node
	Travel time to nearest recreational facilities	0,1274	Root node
4	Availability of public transportation	0,1301	Branch node
	Availability of benches or other resting space	0,1300	Root node
5	Accessibility for bicycles and mobility aids	0,2262	Branch node
	Presence of family or friends during walks	0,2220	Root node
6	Presence of security officer	0,0761	Root node
	Aesthetic of pedestrian facilities	0,0763	Branch node
7	Availability of pedestrian walks	0,1451	Root node
	Verdancy and adequate shading	0,1468	Branch node
8	Travel distance to nearest recreational facilities	0,1126	Branch node
	Availability of social activities and gatherings	0,1124	Root node

4. Discussion

The most popular walking destinations amongst the respondents of this study are parks at 117 responses (29,8%), shopping districts at 112 responses (28,6%), and there are 123 responses (31,4%) that stated that they often walk around their neighborhood. This result is consistent with the findings of previous studies that stated that senior citizens tend to walk to destinations that are close to their neighborhood [10]. This finding suggests that walkable environments need to be close to parks and shops to appeal more towards senior citizens. Shopping areas and neighborhood parks need to provide accessible walking paths for senior citizens to encourage more walking trips amongst senior citizens within the area. This suggestion takes into consideration that age-friendly infrastructures can encourage walking as a physical activity amongst senior citizens [3].

The walkability of Braga Street was rated quite high by the respondents, with the average rating at 4,40. This means that the respondents consider the pedestrian facilities of Braga Street to have satisfactory performance in support their walking activities. However, the high rating does not guarantee that the senior citizens are willing to walk. Amongst 392 respondents, 362 stated that they will walk more often if their walkability preferences are satisfied and 30 respondents stated that they would not walk more often even if their preferences are satisfied. This finding shows that high rating towards walkability does not guarantee high pedestrian density [4].

In the questionnaire, the respondents were asked to rate 16 variables by how much each variable affects walkability based on their personal experiences. Availability of pedestrian walk is rated the highest by the respondents of this study. This shows that availability of pedestrian walk has high priority as a variable of walkability. This finding is consistent with Field Walkability Survey Parameters that set this variable with the highest weight for walkability measure in Asian cities [14]. However, some variables that were highly rated, such as: cleanliness of pedestrian facilities that has second highest rating, followed by availability of benches and other resting space, adequate lighting, and availability of public transportation are not put at high priority by previous studies [14,15]. This shows the difference between walkability measures with objective standpoint with the subjective views of pedestrians. This supports the notion that subjective measures must also be considered for walkability measurement since the priorities of objective measures and subjective measures can be different [1].

Random Forest Classifier is utilized to differentiate variables that are better at identifying senior citizens who are willing to walk more often from senior citizens who are not. The model suggests that adequate lighting, evenness of pedestrian walks, travel time to nearest recreational facilities, availability of benches or other resting space, presence of family or friends during walks, presence of security officer, availability of pedestrian walks, and availability of social activities and gatherings are variables that are better at predicting senior citizens' willingness to walk compared to the other variables. Adequate lighting, evenness of pedestrian walks, and availability of benches or other resting space can also help with physical limitations and mobility issues faced by senior citizens. The benefits of these variables may have affected their ability to predict senior citizens' willingness to walk. This finding suggests that minimizing barriers related to physical limitations and mobility issues can motivate senior citizens to do more physical activities, including walking [16].

5. Conclusion

The variables that are considered to affect walkability typically concerns safety, comfort, and aesthetic parameters of pedestrian facilities. The significance of each parameter to walkability varies with pedestrians' perspective and affected by their intersectionality. Age as factor of intersectionality can affect perspective on walkability, thus senior citizens can have unique perspectives on walkability variables compared to the average pedestrians.

Most senior citizens do not see walking as their main mode of transportation and tend to take walking trips for recreational purposes or for its health benefits. Walkability variables that are highly rated by senior citizens are typically beneficial to help with their health and mobility issues common with old age, such as availability of benches and other resting space can help individuals with mobility disadvantages and adequate lighting can help individuals with eyesight issues.

The findings from the model stated that variables that are rated higher by senior citizens are not necessarily better at predicting their motivation to walk. For example, evenness of pedestrian walks is better at predicting senior citizens' motivation to walk compared to cleanliness of pedestrian facilities despite being rated lower by the respondents. Furthermore, this study reveals that the variables that are better at predicting senior citizens' motivation to walk are adequate lighting, evenness of pedestrian walks, travel time to nearest recreational facilities, availability of benches or other resting space, presence of family or friends during walks, presence of security officer, availability of pedestrian walks, and availability of social activities and gatherings.

This study explores how age influences perceptions of walkability, revealing that senior citizens have distinct perspectives on this matter compared to the average population. It emphasizes the need for urban planning and walkability assessments to consider age-related factors, advocating for a more inclusive and age-sensitive approach. Ultimately, the study aims to contribute to the development of healthier, more supportive, and inclusive communities for senior citizens and all residents.

6. Recommendations for future works

Considering the findings and limitations of this study, several recommendations can be proposed for future research in the field of walkability assessment. Firstly, there is an opportunity to delve deeper into the intersectionality of age with other factors, such as gender, ethnicity, or socioeconomic status, to comprehensively understand how these intersections influence pedestrians' perceptions of walkability. This could provide a more nuanced view of how various aspects of identity interact with age to shape individual perspectives.

Secondly, to enhance the understanding of pedestrian density among senior citizens, future studies could consider the implementation of pedestrian counting stations in the study area. This data collection method could provide valuable insights into the concentration of senior pedestrians in specific areas. Subsequently, researchers could explore whether there is a correlation between areas with high senior pedestrian density and the travel behavior of senior citizens, as well as how this density relates to their unique perspectives on walkability. Such analyses could yield important information for urban planners and policymakers aiming to create age-friendly and walkable communities.

References

[1] Forsyth, Ann 2015 What is a walkable place? The walkability debate in urban design *Urban Design International* **20**, 4: 274–92

[2] Wang H and Yang Y 2019 Neighbourhood walkability: A review and bibliometric analysis *Cities* **93** 43–61

[3] Lee E and Dean J 2018 Perceptions of walkability and determinants of walking behaviour among urban seniors in Toronto, Canada *Journal of Transport & Health* **9** 309–20

[4] Herrmann T, Boisjoly G, Ross NA and El-Geneidy AM 2017 The Missing Middle: Filling the Gap Between Walkability and Observed Walking Behavior *Transportation Research Record* **2661(1)** 103–10

[5] Atelowogun D 2018 Intersectionality Theory and Practice [Internet] [cited 2023 Jun 30] Available from: https://doi.org/10.1093/acrefore/9780190224851.013.48

[6] Collins PH and Bilge S 2020 *Intersectionality* (John Wiley & Sons)

[7] Delgado-Ron, JA and Erdman C 2020 *Supporting the creation of walkable complete communities* [Internet] (Vancouver: The University of British Columbia) Available from: https://sustain.ubc.ca/about/resources/supporting-creation-walkable-complete-communities-understanding-walkability-through

[8] Stafford L and Baldwin C 2018 Planning Walkable Neighborhoods: Are We Overlooking Diversity in Abilities and Ages? *Journal of Planning Literature* **33(1)** 17–30

[9] Dyussenbayev A 2017 Age Periods Of Human Life *Advances in Social Sciences Research Journal* **4(6)** 258–63

[10] Van Holle V, Van Cauwenberg J, Van Dyck D, Deforche B, Van de Weghe N and De Bourdeaudhuij I 2014 Relationship between neighborhood walkability and older adults' physical activity: results from the Belgian Environmental Physical Activity Study in Seniors (BEPAS Seniors) *International journal of behavioral nutrition and physical activity* **11** 1–9

[11] Abdillah D, Ardiwidjaja R, Marhanani Tri Astuti, Putranto D 2022 Development of Eco-Friendly Travel Patterns based on Cultural Heritage Attractions in the City of Bandung *EAJMR* **1(11)** 2507–20

[12] Natalia TW, Rohmawati T 2019 The Relationships between the Characteristics of Pedestrian and the Increase of Facilitation of Sidewalk *IOP Conf Ser: Mater Sci Eng* **662(4)** 042030

[13] Krambeck HV 2006 THE GLOBAL WALKABILITY INDEX

[14] Gota S, Fabian HG, Mejia AA and Punte SS 2010 Walkability surveys in Asian cities *Clean Air Initiative for Asian Cities (CAI-Asia)* **20** 2017–21

[15] Dannenberg AL, Cramer TW and Gibson CJ 2005 Assessing the Walkability of the Workplace: A New Audit Tool *American Journal of Health Promotion* **20(1)** 39–44.

[16] Stathi A, Gilbert H, Fox KR, Coulson J, Davis M and Thompson JL 2012 Determinants of Neighborhood Activity of Adults Age 70 and Over: A Mixed-Methods Study *Journal of Aging and Physical Activity* **20(2)** 148–70

[17] Gad AF 2020 Evaluating Deep Learning Models: The Confusion Matrix, Accuracy, Precision, and Recall [Internet] [cited 2023 Jul 21] Available from: https://blog.paperspace.com/deep-learning-metrics-precision-recall-accuracy/

Travel pattern analysis and freight trip generation modeling of textile commodities. Case study: Thamrin City Shopping Center, Jakarta

Halisa Khaerani[1], Nahry[2], and Andyka Kusuma[3]

Department of Civil Engineering, Universitas Indonesia, Kampus Baru UI Depok, 16424, Indonesia

halisa.khaerani@ui.ac.id [1], nahry@eng.ui.ac.id [2], andyka.k@eng.ui.ac.id [3]

Abstract. One of the sectors supporting the country's economic development is the commercial sector. With the increasing population, the demand for consumer goods will increase, impacting trip generation. Thus, it is necessary to conduct a study to predict the trip generation of freight transport in shopping centers, especially in DKI Jakarta, with textile stores as trip generators. This research aims to determine the travel pattern of freight transport and freight trip generation models produced and attracted at the Thamrin City shopping center and to analyze the factors influencing freight trip generation. The research was conducted using a quantitative research method with a cross-sectional. Data collection used a questionnaire given to 105 merchants. Data analysis used linear regression methods with the number of freight trip production and attraction as the dependent variable. The independent variables used are the number of employees and gross floor area. The study results show that the delivery of goods using expedition services is sent to the distribution center before being sent to the destination. While shipping goods using private vehicles, sent directly to the destination location. The gross floor area significantly influences the trips generation. Transportation planners can use this research to estimate the number of trip generation in shopping center areas in the future.

1. Introduction

Technology has become a key driver of economic, regional, and urban growth in this modern era. Economic growth is closely related to the ability to produce and distribute goods and services. One of the sectors supporting economic development in a country is the commercial sector [1]. Over time the number of people in urban areas will increase. The larger the population, the greater the demand for consumer goods [2]. This growing number of people encourages the development of a transportation system to handle the entire flow of goods. However, through the development of this transportation system, problems such as congestion can arise.

One of the areas in Jakarta that often faces congestion problems is the Tanah Abang area. This area is known as the largest textile trading center in Southeast Asia. Using land transportation such as trucks as the main mode of transportation for goods movement also causes this congestion. Based on the results of the Indonesia Infrastructure Initiative (2014) study shows that the average number of freight transportation trips is 75.3% using road mode. This congestion causes longer travel times and additional transportation costs that can reduce the productivity of trade activities [4]. It is necessary to plan the freight transportation system to solve this problem by estimating the freight trip generation. Through this estimation, the number of freight trips generated and attracted from the Thamrin City shopping center and the model of the number of freight trips generated or attracted from the Thamrin

City shopping center can be estimated. This study's model estimation focuses on a commercial area in a dense urban area with textile commodities as the study object. This research is expected to contribute to the planning of supporting facilities and infrastructure for freight transportation.

2. Literature Review

Trip generation is one of the stages of transportation planning to estimate the number of movements generated and attracted from an area. Holguin-Veras et al. (2013) explained that in freight transportation planning, there are two approaches that can be taken to estimate the generation that occurs. The two approaches are Freight Generation modeling and Freight Trip Generation modeling. Freight Generation (FG) is the generation of commodities transported by vehicles. FG modeling refers to the production and attraction of commodities measured in tonnage or volume (m^3). In comparison, Freight Trip Generation (FTG) generates vehicle trips generated in transporting commodities in freight generation. FG is related to the business size; the more significant the industry scale, the more commodity moves into or out of the business location. Meanwhile, FTG is not only influenced by the size of the item transported but also the size of the mode of transportation that transports these commodities. Thus, an increase in FG does not show a linear relationship with an increase in FTG. Commodity generation can increase without necessarily increasing the amount of trip generation. Industry can increase FG/FTG by increasing the size of commodities to be shipped and changes in the size, type, and mode of transportation [5].

In modeling freight generation (FG) and freight trip generation (FTG), the relationship between the number of vehicle trips and the amount of cargo with land use factors is analyzed. Based on previous research, several factors affect FG and FTG are raw materials, production goods [6], land area [7], number of employees [7]–[14], distance to market [9], [10], and type of industry[10], [12], [13]. In addition, several methods are used for this modeling, such as linear regression analysis [8], [10], [11], [15], ross-classification analysis [12], and multiple classification analysis [9], [12]. Linear regression and multiple classification analyses are suggested for FG/FTG modeling due to their simplicity in modeling and presentation of results [12]. In addition, linear regression analysis is the most frequently used method in estimating freight trip generation [8]. However, in conducting regression analysis or OLS regression modeling, it is necessary to evaluate the assumptions used because it will affect the quality and accuracy of the modeling.

3. Study methodology and data collection

The research was conducted on July 18, 2022, with textile commodity merchants at the Thamrin City shopping center as an object study. Sampling was carried out using the simple random sampling method. Two types of variables are used, namely, independent, and dependent variables. The independent variables used in this study are store area and number of employees. Meanwhile, the dependent variables are freight trip production (FTP) and freight trip attraction (FTA). From the sampling results, 105 textile commodity merchants and information related to company characteristics and freight shipping were obtained as respondents. Statistical calculations of the survey results are shown in **Table 1**. Before the data is subjected to regression analysis, it is necessary to identify outliers. The outlier identification process is carried out using the boxplot method in the SPSS application. **Table 2** shows the comparison results before and after outlier removal. The calculation results show a change in the average value and standard deviation. The standard deviation becomes smaller than the mean for NE and GFA variables. However, for the FTP and FTA variables, after outlier removal, the standard deviation is larger than the means; this indicates that individual data points are far from the mean value or the data is widely dispersed [16].

Table 1. Summary statistics of dependent and independent variables.

Variable	FTP	FTA	NE	GFA
Unit	Trips/week	Trips/week	No.	m^2

Thamrin City Shopping Center (n = 105)				
Mean	0.0706	0.0123	2.5619	7.6667
Median	0.0214	0.0040	2	6
Std. Deviation	0.1717	0.0312	3.1953	4.9022

Table 2. Comparison of statistical calculations before and after outlier removal.

Variable	FTP	FTA	NE	GFA
Unit	Trips/week	Trips/week	No.	m^2
Outlier not being removed (n = 105)				
Mean	0.0706	0.0123	2.5619	7.6667
Median	0.0214	0.0040	2	6
Std. Deviation	0.1717	0.0312	3.1953	4.9022
Outlier being removed (n = 92)				
Mean	0.0321	0.0054	1.9565	6.7174
Median	0.0126	0.0030	2	6
Std. Deviation	0.0441	0.0061	1.0154	2.6698

4. Model specifications and result analysis

4.1. Freight trip generation modeling

This study used the linear regression method with the model form shown in equation (1) for simple linear regression and equation (2) for multiple linear regression. According to [11], in modeling, three possible cases occur, namely (a) statistically significant constants and coefficients, (b) only statistically significant coefficients, and (c) only statistically significant constants. It is also mentioned that modeling with constants is recommended except for statistically insignificant constants.

$$FTG_i = \beta_0 + \beta_i X_i + \varepsilon_i \quad (1)$$

$$FTG_i = \beta_0 + \beta_1 X_1 + \beta_2 X_2 + \varepsilon_i \quad (2)$$

Where, FG_i = Weekly FTG in trips predicted using variabel NE dan GFA; β_0= intercept; β_1, β_2= coefficients, and ε_i= error.

The normality test results show that the data is not normally distributed, and if used for linear regression, the results can be inaccurate. Data transformation is needed into SQRT (x) or square root so the data becomes normally distributed. Before testing linear regression analysis, a correlation test is carried out to measure the relationship between the dependent and independent variables. The calculation results show that both NE and GFA variables are positively correlated for FTP and FTA models, as shown in **Table 3**.

Table 3. Correlations Test Results.

Variable	FTP	FTA

IOP Conf. Series: Earth and Environmental Science 1294 (2024) 012024 doi:10.1088/1755-1315/1294/1/012024

	Sig	Pearson Correlation	Sig	Pearson Correlation
NE	0.001	0.354	0.047	0.207
GFA	0.000	0.460	0.004	0.295

Linear regression modeling was performed with a combination of two and one variable. The best model is determined after all models are tested for linear regression assumptions. One of the assumption tests that need to be fulfilled is the t-test to partially assess the level of significance between the dependent and independent variables. The calculation results shown in **Table 4**, show that the multiple variable models with intercept do not significantly affect FTP and FTA. The model without intercept shows a strong correlation between variables, so it cannot be used together in modeling. In the NE-based model, it is found that the model with intercept does not have a significant effect on the FTP and FTA models, as shown in **Table 5**. Thus, since the model's results with intercept have no significant impact, a model without intercept can be used [11]. Both without intercept models have a substantial effect on FTG and FTA. In the calculation of the GFA-based model shown in **Table 6**, it is found that the model with intercept does not fulfill the t-test for the FTA model. However, for the FTG, the model with intercept meets the t-test. Thus, for the GFA-based models, the FTG equation uses a model with intercept, and FTA uses a model without intercept. Based on the t-test results, the equation model with multiple variables does not meet the t-test requirements, so it is unnecessary to do the f-test since the model that meets the f-test is only one variable.

Table 4. Multiple variable FTG models.

FTP Models						FTA Models					
Description	With Intercept			Without Intercept		Description	With Intercept			Without Intercept	
	β_o	NE	GFA	NE	GFA		β_o	NE	GFA	NE	GFA
n = 92	-0.135	0.038	0.087	0.017	0.047	n = 92	0.003	0.004	0.032	0.005	0.022
t-stat	-2.431	0.951	3.283	0.432	2.202	t-stat	0.121	0.290	2.092	0.323	2.783
R VIF	-	1.612	1.612	25.44	25.44	VIF	-	1.643	1.643	26.474	26.474
^{2}, S.E.	0.220, 0.103			0.661, 0.105		R^2, S.E.	0.088, 0.038			0.739, 0.037	

Notes: R^2 are not comparable for models with and without intercept.

Table 5. Employment based FTG models.

FTP Models				FTA Models			
Description	With Intercept		Without Intercept	Description	With Intercept		Without Intercept
	β_o	NE	NE		β_o	NE	NE
n = 92	-0.021	0.118	0.103	n = 92	0.030	0.024	0.044
t-stat	-0.467	3.587	12.797	t-stat	1.782	2.012	15.155
R^2	0.125		0.643	R^2	0.043		0.716
S.E.	0.108		0.107	S.E.	0.038		0.039

Notes: 1) Highlightes cells represents the preferred models; 2) R^2 are not comparable for models with and without intercept.

Table 6. Area based FTG models.

FTP Models	FTA Models

Description	With Intercept		Without Intercept	Description	With Intercept		Without Intercept
	β_o	GFA	GFA		β_o	GFA	GFA
n = 92	-0.124	0.102	0.056	n = 92	0.004	0.023	0.024
t-stat	-2.282	4.916	13.301	t-stat	0.185	2.929	16.027
R^2	0.212		0.660	R^2	0.087		0.738
S.E.	0.102		0.105	S.E.	0.037		0.037

Notes: 1) Highlightes cells represents the preferred models; 2) R^2 are not comparable for models with and without intercept.

Of the four equations that meet the requirements of the t-test, heteroscedasticity testing is carried out to test the occurrence of inequality of variance from the residuals of one observation to another. The conclusion reached is invalid if heteroscedasticity occurs in the equation [17]. From the calculation results shown in **Table 7**, it is known that for the FTP model, both equations fulfill the heteroscedasticity test. Meanwhile, for the FTA equation model, the heteroscedasticity test requirement is only met by the equation with the GFA variable.

Table 7. Heteroscedasticity Test Result.

Variable	FTP	FTA
	Sig	
NE	0.189	0.023
GFA	0.966	0.671

Notes: Highlighted cells represent models that pass the heteroscedasticity test.

In modeling freight trip production (FTP), two models were obtained. The best model is selected based on the coefficient of determination and the number of independent variables in the equation. The greater the coefficient of determination and fewer independent variables in the equation, the better the modeling. However, in FTP modeling, it is impossible to compare the coefficient of determination for selecting the best model due to differences in the type of equation, namely the first equation without intercept and the second with intercept. The best modeling is chosen based on the presence or absence of constants in the equation. In the FTP model, the equation with intercept is chosen because if the constant value of the equation is still significant, it is more advisable to use the equation [11]. From the equation, it is known that the GFA variable is used to explain the relationship between land use factors and the number of freight trips generated from Thamrin City. The GFA variable is also obtained for FTA modeling to explain the relationship between land use factors and the number of freight trips generated from Thamrin City. The equation obtained in **Table 8** is the result of linear regression from the data transformation of the SQRT (x) variable or square root so that in its use, it is necessary to include GFA data in the form of square roots and the amount of FTP or FTA in the form of square roots.

Table 8. FTP and FTA regression model.

FG Type	Model Type	Model	R^2	RMSE
FTP	Without Intercept	$FTP^{1/2} = 0.103NE^{1/2}$	0.641	0.366
FTP	With Intercept	$FTP^{1/2} = 0.102GFA^{1/2} - 0.124$	0.212	0.374
FTA	Without Intercept	$FTA^{1/2} = 0.024GFA^{1/2}$	0.738	0.183

Notes: 1) Highlightes cells represents the preferred models; 2) R^2 are not comparable for models with and without intercept.

In the FTP model, 0.102GFA indicates the relationship between the GFA and FTP variables has a linear component with a coefficient of 0.102, or every one unit increase in GFA will lead to a 0.102 increase in FTP. The constant -0.124 indicates the value added to the relationship between FTP and GFA. This negative constant value of 0.124 estimates the amount of FTP when the value of GFA is 0. However, since the amount of storage area value (GFA) will never be zero, this negative constant value can be ignored. The constant value can provide a meaningful interpretation if the value of the dependent variable is equal to zero within the range in the sample [18]. The R^2 value in the FTP equation is 0.212, which means that the GFA variable affects FTP by 21.2%, and other factors influence the rest. In the FTA equation, the R2 value is 0.738, which means that the GFA variable affects FTA by 73.8%, and other factors influence the rest. In addition, the RMSE value in the FTP and FTA equations are 0.374 and 0.183, respectively. Based on these values, it is known that the average prediction error of the model against the actual data is 0.374 for FTP and 0.183 for FTA. In addition, it can also be seen that the GFA variable is a better predictor variable for modeling FTA because the RMSE value is smaller than the FTP RMSE.

4.2. Freight travel pattern at Thamrin City shopping center

The travel pattern of freight transport at Thamrin City Shopping Center is divided into trip production and trip attraction. Most of the delivery of goods from Thamrin City is done using expedition services. The types of vehicles used by expedition services are vans and trucks, as shown in **Figure 1.** Delivery from Thamrin City to the distribution center generally uses a van with an average load occupancy of 70%, then for delivery from the distribution center to another distribution center or the final location using a truck with an average load occupancy of 100%. Delivery to the final location can also use other transportation, such as trains, ships, and planes, depending on the sellers' demand. There are expedition companies that pass through two distribution centers in delivering goods. The first distribution center collects all goods from the area around Tanah Abang. After that, the goods are sent to the second distribution center to be sorted to the destination. For expedition companies that only pass through one distribution center, the goods are sorted directly at the distribution center and grouped according to the destination of the goods. Most goods are sent to the Jakarta, Bogor, Depok, Tangerang, and Bekasi areas. The time required for the goods to arrive at their destination is around 2-3 days.

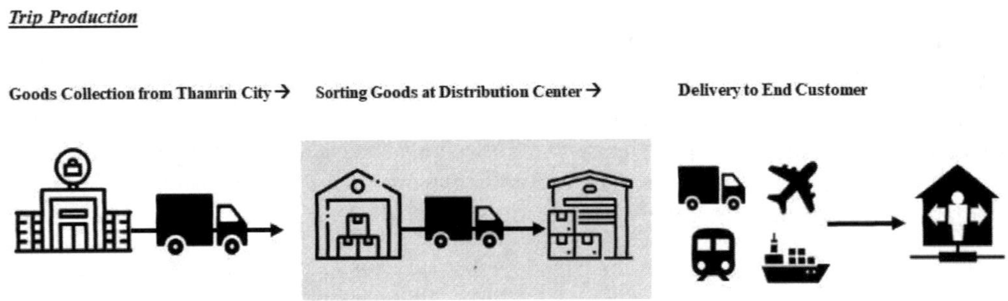

Figure 1. Schematic of trip production chain.

The process of delivering goods to Thamrin City is divided into two types: delivery using expedition services and privately owned vehicles, as shown in **Figure 2**. Delivery of goods using expedition services uses several kinds of vehicles from the initial location to the distribution center, such as trucks, planes, trains, and ships. The trip from the distribution center to Thamrin City usually uses a van-type car. Expedition companies are also divided into shipments that pass through two or one distribution center. Generally, the first distribution center is to collect all goods that come from the same area. After that, the goods are sent to the second distribution center for sorting goods in the Tanah Abang or Thamrin City area. Delivery of goods using privately owned vehicles is divided into three types of vehicles, namely small, medium, and large trucks. Generally, delivery of goods using privately owned vehicles is sent directly to Thamrin City without stopping elsewhere. Most goods are delivered from the Jakarta, Bogor, Depok, Tangerang, and Bekasi areas.

Figure 2. Schematic of trip attraction chain.

5. Conclusions

Based on the research that has been done, the area of the store is the land use factor that affects the freight trip generation in the Thamrin City shopping center. The best multiple linear regression model to predict freight trip generation in the Thamrin City shopping center uses the store area variable (GFA). The model equation for freight trip production is $FTP^{1/2} = 0.102GFA^{1/2}-0.124$ with a coefficient of determination of 0.212. Then, the model equation for freight trip attraction is $FTA^{1/2} = 0.024GFA^{1/2}$ with a coefficient of determination 0.738.

The pattern of freight transport in the Thamrin City shopping center is divided into two, namely, trip production and trip attraction. In shipping goods, most goods are sent using expedition services using vans to the distribution center before being sent to the final consumer using trucks, trains, ships, or planes. In the process of receiving goods, goods can be delivered from the location of origin using private vehicles or expedition service vehicles. Delivery by personal vehicles uses small, medium, and large truck vehicle types. Meanwhile, delivery by expedition services uses truck vehicles from the location of origin to the distribution center and vans from the distribution center to Thamrin City. Most delivery destinations and origin of goods received are Jabodetabek, with the time required to arrive at their destination around 2-3 days.

Acknowledgement

This research was supported by POLAR - UP2M, part of the Faculty of Engineering at the Department of Civil Engineering, University of Indonesia.

References

[1] G. D. Yulianto, A. Jumiati, and Santoso Edy, "Analisis Peranan Sektor Perdagangan Terhadap Perekonomian Wilayah di Kabupaten Jember," *Artikel Ilmiah Mahasiswa*, 2015.

[2] J. Li, "On the relationship between population and means of subsistence," *Renkou Yanjiu*, no. 1, pp. 8–14, Jan. 1984.

[3] Indonesia Infrastructure Initiative, "Naskah Teknokratik Penyusunan RPJMN 2015- 2019 Sektor Transportasi," 2014.

[4] T. Erdelić, T. Carić, M. Erdelić, L. Tišljarić, A. Turković, and N. Jelušić, "Estimating congestion zones and travel time indexes based on the floating car data," *Comput Environ Urban Syst*, vol. 87, May 2021, doi: 10.1016/j.compenvurbsys.2021.101604.

[5] J. Holguín-Veras, M. Jaller, I. Sánchez-Díaz, S. Campbell, and C. T. Lawson, "Freight Generation and Freight Trip Generation Models," in *Modelling Freight Transport*, Elsevier Inc., 2013, pp. 43–63. doi: 10.1016/B978-0-12-410400-6.00003-3.

[6] A. M. C. Rahayu, E. P. Rahajo, A. D. Dwipayana, and I. M. Suraharta, "Identifikasi Faktor Pengungkit Bangkitan danTarikan Perjalanan Kegiatan Industri (Kawasan Industri Bekasi)," *Jurnal Teknologi Transportasi dan Logistik*, vol. 3, pp. 23–28, 2022.

[7] B. Puente-Mejia, L. Palacios-Argüello, C. Suárez-Núñez, and J. Gonzalez-Feliu, "Freight trip generation modeling and data collection processes in Latin American cities. Modeling framework for Quito and generalization issues," *Transp Res Part A Policy Pract*, vol. 132, pp. 226–241, Feb. 2020, doi: 10.1016/j.tra.2019.10.013.

[8] L. K. de Oliveira, G. G. F. de Araújo, B. V. Bertoncini, C. D. Pedrosa, and F. G. F. da Silva, "Modelling Freight Trip Generation Based on Deliveries for Brazilian Municipalities," *Sustainability (Switzerland)*, vol. 14, no. 16, Aug. 2022, doi: 10.3390/su141610300.

[9] J. Chaudhari, R. N. Shukla, and C. R. Patel, "IMPACT ANALYSIS OF URBAN FREIGHT TRIPS GENERATED FROM WHOLESALE MARKET, AHMEDABAD," *Acta Logistica*, vol. 9, no. 3, pp. 325–336, Sep. 2022, doi: 10.22306/al.v9i3.316.

[10] B. N. Dhonde and C. R. Patel, "Estimating urban freight trips using light commercial vehicles in the Indian textile industry," *Transp Res Interdiscip Perspect*, vol. 11, Sep. 2021, doi: 10.1016/j.trip.2021.100411.

[11] A. Pani, P. K. Sahu, G. R. Patil, and A. K. Sarkar, "Modelling urban freight generation: A case study of seven cities in Kerala, India," *Transp Policy (Oxf)*, vol. 69, pp. 49–64, Oct. 2018, doi: 10.1016/j.tranpol.2018.05.013.

[12] A. R. Alho and J. de Abreu e Silva, "Modeling retail establishments' freight trip generation: a comparison of methodologies to predict total weekly deliveries," *Transportation (Amst)*, vol. 44, no. 5, pp. 1195–1212, Sep. 2017, doi: 10.1007/s11116-016-9700-z.

[13] I. Sánchez-Díaz, "Modeling urban freight generation: A study of commercial establishments' freight needs," *Transp Res Part A Policy Pract*, vol. 102, pp. 3–17, Aug. 2017, doi: 10.1016/j.tra.2016.06.035.

[14] G. Günay, G. Ergün, and I. Gökaşar, "Conditional Freight Trip Generation modelling," *J Transp Geogr*, vol. 54, pp. 102–111, Jun. 2016, doi: 10.1016/j.jtrangeo.2016.05.013.

[15] L. Cheah, R. M. Mepparambath, and G. M. Ricart Surribas, "Freight trips generated at retail malls in dense urban areas," *Transp Res Part A Policy Pract*, vol. 145, pp. 118–131, Mar. 2021, doi: 10.1016/j.tra.2021.01.015.

[16] R. N. Hidayat, L. M. Sabri, and Awaluddin Moehammad, "Analisis Desain Jaring GNSS Berdasarkan Fungsi Presisi (Studi Kasus: Titik Geoid Geometrri Kota Semarang)," *Jurnal Geodesi Undip*, vol. 8, no. 1, 2019.

[17] A. Sukriyah, "Analisis Heteroskedastisitas Pada Regresi Linier Berganda," Universitas Islam Negeri Maulana Malik Ibrahim, Malang, 2011.

[18] W. Mendenhall and T. T. Sincich, *A Second Course in Statistics: Regression Analysis*, 7th ed. Upper Saddle River: Prentice Hall, 2011.

T-UFF (Tracker stuff): application development for warehouse tracking

Wahyu Padma Baskara[1], Thomas More Elang Eucharisto[1], Ni Kadek Rania Utari[1], Ahmad Soimun[1] and Riz Rifai Oktavianus Sasue[1]

[1]Departement of Logistic Management, Bali's Land Transportation Polytechnic, Cempaka Putih Street, Bali, Indonesia

riz@poltradabali.ac.id

Abstract. Era Society 5.0, requires the logistics agents to adapt and follow technology development to create supply chain efficiency. One of the ways is to apply the IoT to warehousing activities, creating a system design Warehouse Tracking app that works to increase order picking time efficiency. This system design application method is using the System Development Live Cycle (SDLC) and follows Object-oriented programming (OOP). This final result study is a system design application based on the IoT and QR codes for operating order-tracking goods in a warehouse. With this Tracker Stuff (T-UFF) app, the time it takes for order picking will be decreased so that it can create activity-efficient logistics for the company.

Keywords: Logistics, System Development Live Cycle (SDLC), Object-oriented programming (OOP), Tracking, Warehouse.

1. Introduction

In Indonesia, great progress has been made in the company industry in the era of Society 5.0. Refers to the concept of development technology such as the Internet of Things (IoT), big data, and artificial intelligence, used to significantly increase the quality of human life. In the industrial sector operational processes, will never miss from the existing role-effective warehouse to use carry out process control shopping and production goods [1]. Activities inside warehouses must consider related planning, organizing, and control logistics in a technical or administrative manner; therefore, the application concept of management warehousing is urgently needed [2].

Warehouse management can be defined as governing science-in-matter practice storage and output stuff in warehouses [3]. This application concept of warehouse management aims to avoid the increased time of order picking which is time for the taking process of an item from storage in a warehouse for fulfilling customer orders with a fast time and will get high satisfaction from the customer [4]. The inefficiency of order picking time is known to cause difficulty for the operator in finding location storage for goods to be sent as a result of a bad inventory.

Order picking is the highest activity cost in warehousing and can reach up to 55% of the total cost of operational warehousing, and is therefore considered a priority in enhancing productivity [5]. Order picking is also important in matters influencing the service level of a customer order, which is determined by the efficiency of order picking and accuracy. Efficiency shows how fast an order can be taken and how quickly that order is available for delivery to customers. Therefore, development systems capable of tracking goods push order-picking time will increase efficiency and effectiveness in operational warehousing [6].

Internet of Things (IoT) implementation in the era of Society 5.0 especially in the warehousing sector can be utilized optimally through system tracking. Logistics tracking is defined as the process of recording goods trip interval from place origin to place goals through company transport [7]. In this research, tracking refers to an activity of item location searching in a warehouse. Based on that, the author offers Tracker Stuff (T-UFF), an application that is capable of giving information related to position-stored items through the use of a QR code to show the route location of the goods by lighting the lights connected to the microcontroller. Creating the system via Android Studio, which uses a method of the System Development Live Cycle (SDLC) and object-oriented programming (OOP). SDLC is a structured and systematic approach related to the development of software products [8].

SDLC's normally used for designing and developing software with high quality. This method can streamline the cost and time required for making a piece of software but still can maintain quality products to be created. Stages the beginning that was done on the SDLC, do planning system consisting from define the problem and determine the objective system. Then, the stage furthermore is to do an analysis system consisting of, identification problem, and defining need information. The third stage is a design system with a flowchart system, use case diagrams, and class diagrams [9].

The warehouse tracking system was developed several times, but the results finally reached a maximum. One study was previously carried out by [9], entitled ' System Design Information for Tracking Tracing in the Warehouse with Use Technology QR Code, where development study developed with the method approach System Development Live Cycle (SDLC) gets to the stage design system, with initial stages, with requirements analysis, followed by the design to use method object-oriented programming. The tracking tracing process is carried out to help determine the position from inventory in the warehouse and to ensure that the inventory issued follows the FIFO rule. However, the system built its own weakness, that is, the system Existing ERP, which there is still not yet integrated with system information-based IoT, besides it, the system still needs to be developed for software engineering.

Another reviewed study, is research conducted by [10], with the title ' Research on Intelligent Warehouse Management System Based on RFID, Research on Intelligent Warehouse Management System Based on RFID.' This research is known as a system intelligent management warehousing RFID based. Radio-frequency identification (RFID) technology consists of a transmitter, receiver, microprocessor, antenna, and label. RFID can be divided into three types: readers, antennas, and tags. First, readers send an electronic signal through antennas, the transmit tag stores identifiable information internally after the signal is received, and the reader accepts and recognizes the information and sent back by tags through antennas. Finally, the readers send the identification results to the hosts. Based on the study, the RFID-based application described thus far has limited development-related information goods in warehouse management only.

Despite the limitations of previous studies regarding the development of tracking systems, this study introduces the application of tracking stuff (T-UFF) as a display-capable solution route to the location of goods based on the information obtained from the database by scanning QR Code. This application is also equipped with visual instructions in the form of light, which can guide the user to the intended rack. The system app is connected to a microcontroller that sends a signal to a side-mounted LED light which guide you to specified shelf. The microcontroller uses Arduino because of it's suitability for projects that require high capacity for large circuits [11]. This study aims to design and test the performance of the application. The hope is that this research will produce an output that can be applied in warehousing management, whether for the storage of goods on shelves or off the shelf.

2. Methodology

In conducting this research, the methodology used was the process of applying applications and simulations that will be applied.

2.1. Flowchart System

The methodology described in the system flowchart is illustrated in Figure 1.

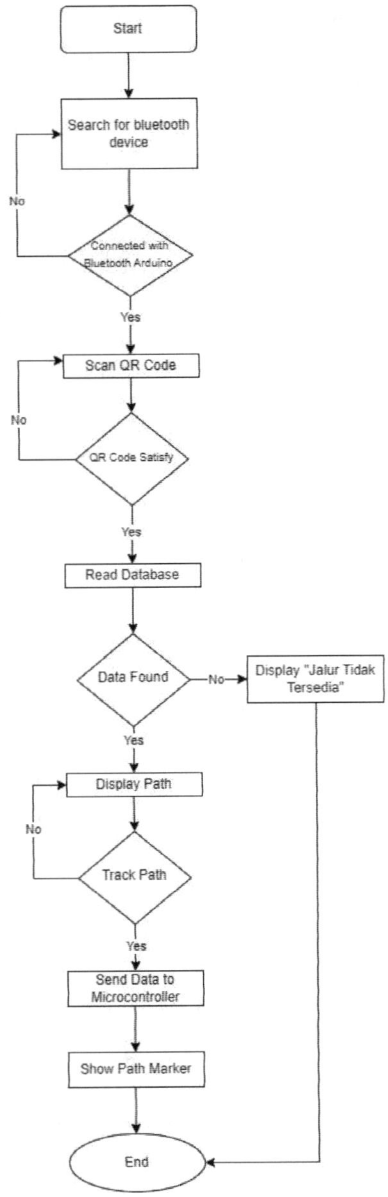

Figure 1. Application System Flowchart.

The system can be explained as follows.

a. The user opens the system, and the screen display appears to search for bluetooth devices.
b. When successfully connected to bluetooth, the user can enter the application. If the connection fails, the button to enter the application cannot be pressed and the user must return to search for bluetooth devices.
c. Then, the application will display a, route layout image, QR Code scan option“start” button, and “stop” button
d. When the scanned barcode is read by the system, it displays the route layout image according

to the QR code.

e. If the scanned barcode is not read by the database system, the information track is not available, and the system ended up appearing early.
f. The successful system display of the route layout will send data to the microcontroller for turning on the light marker route according to the QR Code by "Mulai" button, and the system will be ends with "stop" button or the light will be off.

2.2. Usecase Diagram

To describe the interaction between actors (actor) and the system in this application can be illustrated through the use case diagram that has been shown in figure 2.

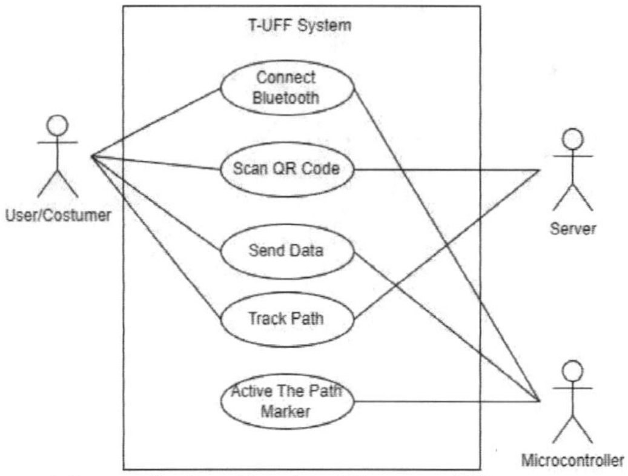

Figure 2. Application Usecase Diagram.

In the picture above, from the use-case diagram owned by the system, three entities are involved: the the user, server, and microcontroller, respectively. First, the microcontroller is connected to the user via bluetooth. When the QR Code is successfully read and appropriate with the database, the server will produce a layout image at once sending data to the microcontroller to turn on the LED according to the specific rack route based on the scanned barcode

2.3. Class Diagram

To find out about the description of the coding structure in this application, it can be seen in Figure 3.

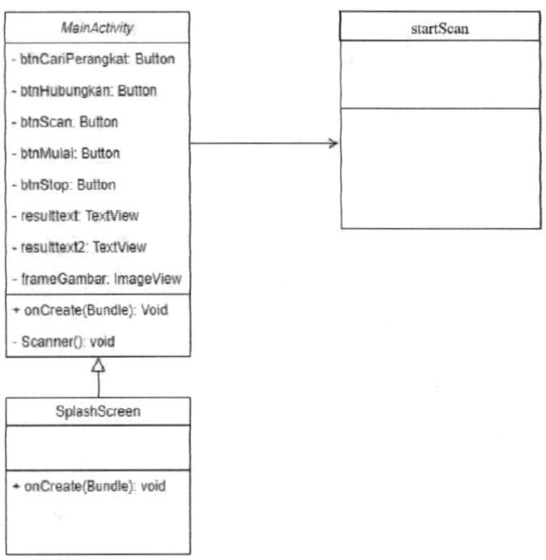

Figure 3. Application Class Diagram.

a. MainActivity:
   1. Contains variable member private: btnCariPerangkat (button), btnHubungkan (button), btnMulai (button), btnStop (button), btnScan (Button), resulttext (TextView), resulttext2 (TextView), and imageframe (ImageView).
   2. Define the ` onCreate` method to initialize the layout and set the ` OnClickListener` to the ` btnScan` button , 'btnCariPerangkat' button, 'btnHubungkan' button, 'btnmulai' button, 'btnStop' Button.
   3. Contain method private `Scanner` in charge answer For configure and launch barcode scanner.
b. Start Scan:
   1. Inherit class ` CaptureActivity` from package ` com.journeyapps. barcodescanner`.
   2. Represent activity custom: Start the scanning process.
c. Splash Screen:
   1. Inherit class ` AppCompatActivity.'
   2. Represent splash screen activity.
   3. Arrange appearance content and define the ` onCreate` method for starts `MainActivity` after pause time.

2.4. Development Method Prototype

The Development Method Prototype is a research method applied in the manufacturing system writing end of prototypes. This prototype aims to be able to know about the schema implementation T-UFF application in warehousing. The application of the prototype method is as follows:

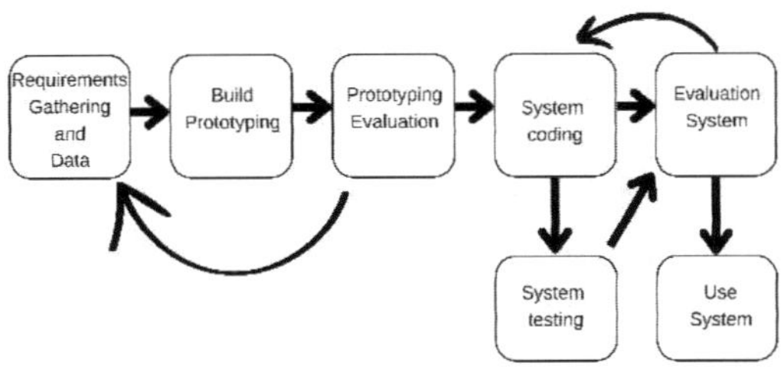

Figure 4. Development Model System Prototype.

The prototype method is used for the development of device software, which represents a new paradigm in the manufacturing or development of software [12]. The prototype method is evolving in the world of development or manufacturing device software, which has also revolutionized the method of development or manufacturing device software, the sequential system known as the waterfall method [13].

3. Analysis and Discussion

3.1. T-UFF Application Start Page

Fig. 5 shows the initial display when the application was opened. To operate this application, the tablet device must be connected to bluetooth, which sends a signal to the Arduino. At the start of the application, the bluetooth connection settings display will appear, and the user presses the search device option; if the device is found, they can press the connect option; therefore, the application is ready to use by pressing the MASUK APP button.

IOP Conf. Series: Earth and Environmental Science 1294 (2024) 012025 doi:10.1088/1755-1315/1294/1/012025

Figure 5. The Initial appearance of the T-UFF application.

3.2. Main Menu Display of T-UFF Application

On the page, the main menu display is shown in relation to the route layout image warehouse used as a whole. This page also displays two captions: scan data and routes/paths. Information scans will provide information about the identity rack you want to know, and the route track contains route-related information about going to the rack you search for. If we want to know the route to the rack, click the SCAN QR button to open the camera, and then it will work to find the identity rack by the way they scan the QR code rack. After identifying the identity rack, the available START button sends a signal to the microcontroller that turns on the LED light to the location of the goods according to the QR Code. The type of microcontroller for this case is Arduino Mega 2560 R3. Viewing this initial scan data as well as the track will provide no information on is data available because it hasn't been done QR scan that has registered in the database application. Meanwhile, the STOP button can be used if the user wants to stop and turn off the lights of a particular shelf and search for a route to another shelf.

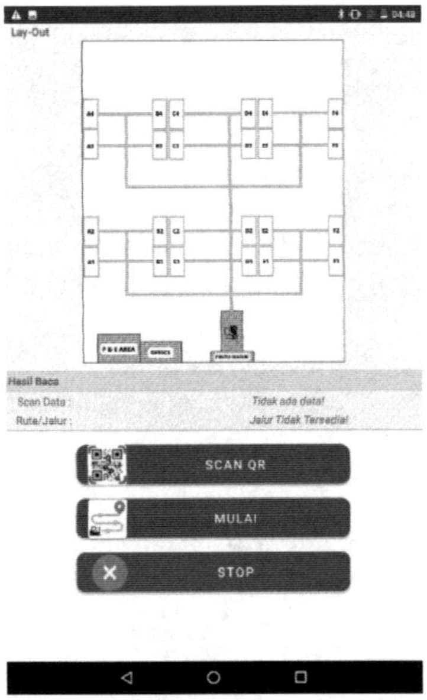

Figure 6. Display of the Main Menu of the T-UFF Application.

3.3. Appearance QR Code use Showing Identity Rack

If it is easy to search for an item, then each shelf in the warehouse is given a number identity from A1-A4, B1-B4, C1-C4, D1-D4, E1-E4, and F1-F4 which can be read via the QR Code.

Figure 7. Qr Code Identity Rack.

3.4. App Display When Reading Shelf Identity and Showing Routes

If the scanned QR Code can be read by the database, the application will display layout images, and the results will show the location of the goods and the route according to the QR Code. Subsequently, the user can push the start button to send data to the microcontroller to power the LED light route. From this view, the user can easily follow the route and finally get to the rack.

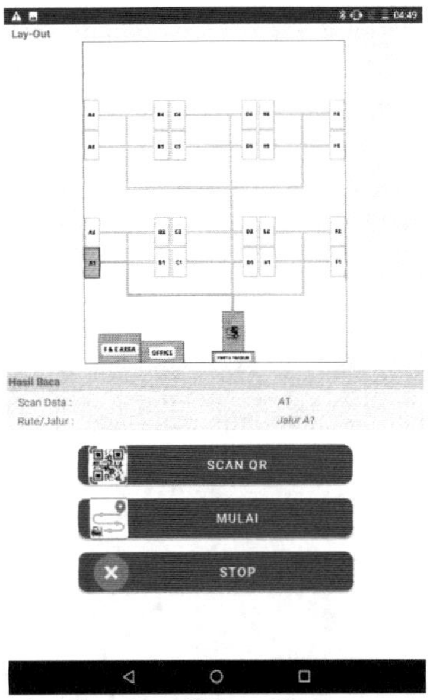

Figure 8. Display Applications When Shows route Goods Positions.

3.5. Appearance Application When Reading No QR Code Registered

If one's own ability displays the route going to shelves that have been identified in the manufacturing process, the application will enter related to the route on each rack that is entered into the database. Thus, when the application reads a QR Code that is not registered, the results on the scanned data can only read the QR Code text and display images, and routes/paths are not available.

IOP Conf. Series: Earth and Environmental Science 1294 (2024) 012025 doi:10.1088/1755-1315/1294/1/012025

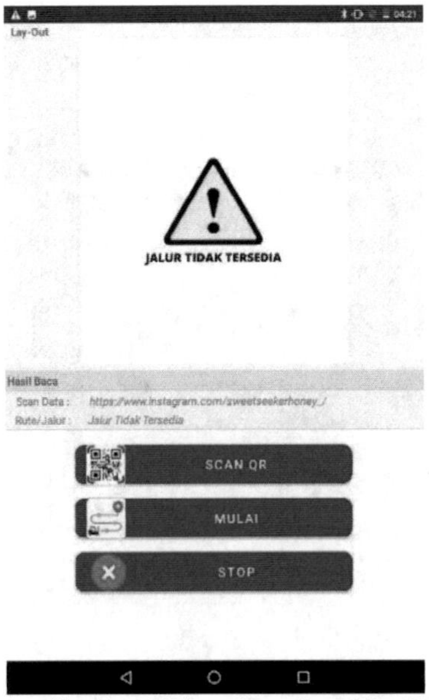

Figure 9. Display Application When Reading No QR Code Registered.

3.6. Display Prototype Implementation The T-UFF Application

When the QR code is successfully scanned, the application can read according to its database. Using the START button, the application sends a signal to the microcontroller, which turns on the LED light route according to the QR code location, as shown in Fig. 8, through a bluetooth signal. The LED light is turned on because it receives an order from the Arduino as a microcontroller or giver command. The flow that occurs when the LED lights up, that is, the Arduino receives the signal over the bluetooth from the T-UFF application; then, the Arduino issues power by 5 V to the relay module with a capacity of 5V to be connected to each lamp and the route ordered. The LED lights have a capacity of 12V and are divided into segments. Relays will get help power from the power supply to turn on the light route. The total number of segments in the prototype is 42. The segment is every path in the warehouse layout. When the user has finished turning on the light as a guide to determine the location of the shelf, they can press the STOP button via the application, which sends a signal to the Arduino to turn off the light that was previously turned on.

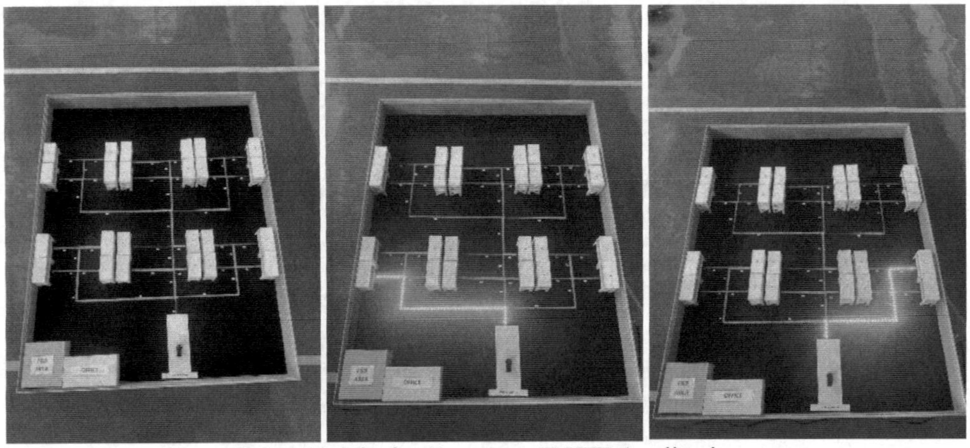

Figure 10. Display Prototype T-UFF Application.

3.7 Benefit of T-UFF Application

The T-UFF application has the benefit of being able to directly show the forklift route to the shelf where the desired goods are stored. In addition, by implementing this application, you can avoid placing items incorrectly, thereby preventing the loss of things. In this way, it will reduce order picking time, thereby speeding up the movement of goods in the warehouse to create efficiency and improve the quality of customer service.

4. Conclusion

After designing and manufacturing the T-UFF application, this system can identify related storage locations for goods and show the route to that location using the Android application. Utilization of the camera on the smartphone serves to support the operation Scan QR feature to read the identity shelf via QR Code. Through the identification of the identity shelf, the application sends a signal to the microcontroller via Bluetooth and turns on the LED light route according to the identified QR code location. In use, the user can know the position placement of goods and routes to that location instantly. In addition, the user can avoid placing items incorrectly, thereby preventing the loss of things. In this way, it will reduce order picking time, thereby speeding up the movement of goods in the warehouse to create efficiency and improve the quality of customer service. This research can continue to develop, one of which is the implementation of a real-time system and an upgrade of the best route guide. Subsequently, the application can show the current location of the forklift on the best route to further speed up the order-picking process by minimizing route-selection inefficiencies.

5. Reference

[1] Oktaviana D, Barizi A, Zubad M, Yaqin N. The Challenges of The Society 5.0 Era: The Evaluation of Learning in Primary Education. Jurnal Ilmiah Pendidikan Guru Madrasah Ibtidaiyah. 2022;12(1):10–7.
[2] Ratnawati S, Oktarina N. Manajemen Pengendalian Logistik Berbasis Sistem Informasi Gudang. Business and Accounting Education Journal. 2022 Apr 30;3(1):27–35.
[3] Indriyani S. Analyzing The Warehouse Management System at PT. Pos Manado Analisa Sistem Manajemen Gudang di PT. Pos Manado. Jurnal EMBA. 2020;8(4):503–11.
[4] Herliani. Perbaikan Produktivitas Picking Order dengan Metode Routing Heuristic di Gudang Pusat Suku Cadang Otomotif [Internet]. Vol. 10. 2018. Available from: https://www.otosia.com.
[5] Chandra. Estimasi Jarak Tempuh Order Picking Manual dengan Metode Analitik di PT GMS. Jurnal Teknik Industri. 2014;15(2):155.
[6] Santoso S, Wawolumaja R, Yudiantyo W, Hidayat K, Rustandi JO. Usulan metode heuristik untuk optimasi proses order picking gudang. Journal Industrial Servicess. 2022 Mar 19;7(2):226.
[7] Alfatonah Si. Perancangan Sistem Order dan Tracking Pengiriman Kendaraan Secara Online Berbasis PHP dan MYSQL pada PT Wira Granada Lestari. 2020.
[8] Adlak Praveen Kumar Singh S. A Comparative study of Various Models in Software Development Life Cycle. International Journal of Research and Analytical Reviews [Internet]. 2018;5(4). Available from: www.ijrar.org
[9] Taufan I, Santoso A, Rungkut JR, Rungkut K, Rungkut K, Sby K, et al. Desain Sistem Informasi untuk Tracking dan Tracing pada Warehouse dengan Menggunakan Teknologi Qr Code. Vol. 16, Jurnal Teknik Industri. 2021.
[10] Junhong M. Research on Intelligent Warehouse Management System Based on RFID. 2020.
[11] Sasue R. Development of a Safe Driving Distance Detection Lidar System. Jurnal Rekayasa Sipil dan Lingkungan. 2022;6(2):193.
[12] Mulyono S. Sistem Monitor Barang Logistik dalam Ruang Gudang dengan Teknik Presence Detection menggunakan Bluetooth Beacon di Perusahaan Pengiriman

Logistik Abdul Rohman Sholeh, Imam Much Ibnu Subroto, Sri Mulyono. Jurnal Transistor Elektro dan Informatika (TRANSISTOR EI. 2022;4(3):155.

[13] P. M. Ogedebe "Software Prototyping : A Strategy to Use When User Lacks Data Processing Experience. Software Prototyping : A Strategy to Use When User Lacks Data Processing Experience. 2012;2(6):219–24.

Analysis of traffic flow due to lane changes by heavy vehicles

Zuniar Ayu Permata Sari[1], Nahry[2], Gari Mauramdha[3]

Department of Civil Engineering, Universitas Indonesia, Kampus Baru UI Depok, 16424, Indonesia

zuniar.ayu@ui.ac.id [1], nahry@eng.ui.ac.id [2], gari.mauramdha01@ui.ac.id [3]

Abstract. The transportation of goods via heavy vehicles (HV) plays a crucial role in urban economics. However, the operational and physical characteristics of these vehicles negatively impact traffic flow, leading to reduced speeds and increased congestion. This study examines the influence of HV lane changes on average traffic speed by implementing diverse lane change intensities and alternative strategies that restrict HV to specific lanes, narrowing their lane-changing space. Five HV classes based on axle count were analyzed using Vissim traffic simulation software on a 600 m JORR toll road segment. Three strategies were assessed across various scenarios, with differing HV composition percentages. Results show that HV lane changes, simulated using Vissim, significantly affect average traffic speed. Comparing high and low-intensity lane changes revealed an approximately 12% speed increase. Additionally, every 5% increment in HV composition led to an average 3% traffic speed reduction. In high-density conditions for every 5% increase in the composition of HV from high-intensity compared to medium and low-intensity lane changes, the average speed of traffic flow increases by 4% and 8% in low-density conditions. This research highlights the importance of enforcing designated lane usage for HV on toll roads to enhance traffic flow performance.

Keywords : *highways, heavy vehicles, lane changes, speed average, vissim*

1. Introduction

Cities face various challenges due to rapid population growth and increasing lifestyles which have an impact on the distribution of goods [1]. Data from the Central Statistics Agency 2022 shows that Indonesia's population reaches 275.77 million people. In fulfilling its needs for goods, Supply Chain Indonesia 2022 said that HV or trucks dominate the mode of transportation of goods. This can be seen from the high growth in the transportation and warehousing sector in early 2022 which was accompanied by a decrease in the number of goods transported by sea and rail transportation modes. This statement is supported by data from the Central Statistics Agency 2021 regarding the contribution of road transportation modes to the GDP of the transportation subsector, which reached 69.8%. In addition, the Indonesian Transportation Society 2022 also predicts that heavy vehicles will dominate up to 90% of freight transport in Indonesia by 2022.

In its operations on the Indonesia HV toll road it has been regulated to be in lane 1 or the left lane which is intended for low-speed vehicles. This is contained in Government Regulation No. 15 of 2005. However, the phenomenon that occurs is the Jakarta Outer Ring Road (JORR) experiencing a decrease in the average speed of traffic flow because heavy vehicles pass not on their lanes during peak hours.

This is exacerbated by the movement of lane changes (lane changing) carried out by these heavy vehicles.

Many researchers are interested in studying traffic flow from various aspects, one of which is lane changing. In the last 60 years, a number of studies have been conducted to investigate lane-shifting behavior [2]. Lane change maneuvers have negative implications, namely 5% of traffic accidents in 2015 in China were caused by inappropriate lane changes. Then, the higher the frequency of lane changing behavior increases the probability and rate of traffic accidents [14]. Previous research has also provided macroscopic evidence that congestion can occur and cause a decrease in capacity due to the high intensity of lane change rates. One study found that the frequent act of changing lanes by heavy vehicles can have a negative impact on traffic performance [3]. An increased frequency of lane changes can also increase density and reduce the average speed of traffic flow [4].

Understanding lane change behavior is important in several application areas such as capacity analysis and safety studies [11]. Approximately 539,000 two-vehicle lane change collisions occurred in the US in 1999 [7]. Therefore at this time, research will be carried out with a microscopic model using the help of the PTV Vissim application to analyze the effect of lane changing by heavy vehicles on the average speed of traffic flow on the Jakarta Outer Ring Road segment. Previously there had been research that conducted lane change modeling, but no one had conducted further research on the impact of lane changes by heavy vehicles specifically on toll road traffic flow. This research can also be evidence of the need for vehicle compliance in using the appropriate lanes on the toll road. Seeing these conditions, research is needed to analyze traffic flow due to lane changes by heavy vehicles.

2. Literature Review

The number of heavy vehicles and their proportion in traffic flow has increased in the last decade which has had a significant impact on traffic flow. This is one of the causes of traffic congestion due to the size and slower operation of heavy vehicles compared to light vehicles [5]. Especially if heavy vehicles change lanes, this will have an impact on the surrounding traffic conditions as well. The existence of traffic congestion can cause negative impacts such as increasing vehicle queues, reducing the average speed of vehicles, longer travel times, economic losses, emissions etc. [6].

Lane change is one of the most common driver habits in traffic flow. In the last 60 years, a number of studies have been conducted to investigate this lane-shifting behavior [2]. Previous studies have been conducted with different focuses. [9] analyzed the duration of lane changes in heavy vehicles and passenger vehicles. While [7] analyzed the differences between the behavior of heavy and light vehicle drivers in changing lanes on toll and arterial roads. Then [10] investigated the effect of microscopic characteristics and analyzed the decision-making model by changing the lanes of heavy vehicles. [8], [11], [12], [13] conducted several studies on lane changes aiming to investigate the decision-making process of lane change manoeuvres for various drivers based on a two-stage test-drive, investigate the impact of a lane change strategy on expressway traffic operations, and analyze the behavior of new follower vehicles changing lanes in the target lane when facing anticipation (pre-insertion process).

The research conducted has its own variables. Toledo and Zohar (2007) focus on the duration of lane changes for heavy and light vehicles. Then [7] observed the width of the gap and the length of the vehicle. [5] focused on the motivation to change lanes, the selection of the target lane, and the decision to change lanes. Whereas [8], [11], [12], [13] focuses on strategies for changing lanes, time, distance, speed, density, and gap width.

Research locations regarding lane changes also vary. [9] located on the California highway. Other studies were conducted on the Berkeley and Hollywood motorways [10],the motorway between Delft and Rotterdam [11], and urban roads in Colombo Sri Lanka [6]. The types of vehicles observed are not the same in all studies. Research by [12] does not include heavy vehicles as its object. Meanwhile [9], [7], [10] made heavy vehicles the object of their research.

In conducting their research, there are researchers who use software assistance to model the simulation. using the help of the MOTUS software [12] and used PTV Vissim software [10]. In this study, the author will use PTV Vissim software. The other researchers used the help of PTV Vissim software to model 2, 3, and 4 lane freeway traffic flow with the assumption of 10% heavy vehicle volume to analyze the parameters that affect the distribution of lane flow on multilane toll roads.

With the history of previous research regarding lane changes, of course there are still many areas that have not been focused. Therefore, this study will analyze traffic flow due to lane changes by heavy vehicles with direct observation and microscopic simulation on PTV Vissim software. This research also focuses more on heavy vehicles passing on the Jakarta Outer Ring Road toll road segment to determine the average speed of traffic flow caused by lane changes (LC) by heavy vehicles. Modeling lane changes by heavy vehicles is very important for safety studies and capacity analysis. This is done so that toll roads can increase their capacity and traffic flow safety.

3. Study Methodology

The main goal of this paper is to investigate the impact of the intensity of lane change on traffic flow characteristics. To this end, the strategy-based lane change model has been implemented in a microscopic simulator. Three different simulation scenarios have been defined. In scenarios (1) with highest intensity lane changes by HV, which is no HV restrictions on all lanes. In scenarios (2) with moderate intensity lane changes, which there is one lane that limits HV from passing the lane. Scenario (3) with lowest intensity lane changes, which there are two lanes that limits HV for passing the lanes. Lane change by HV scenario strategy in Vissim. the data validation method in this modeling uses the GEH method (Geoffrey E. Havers).

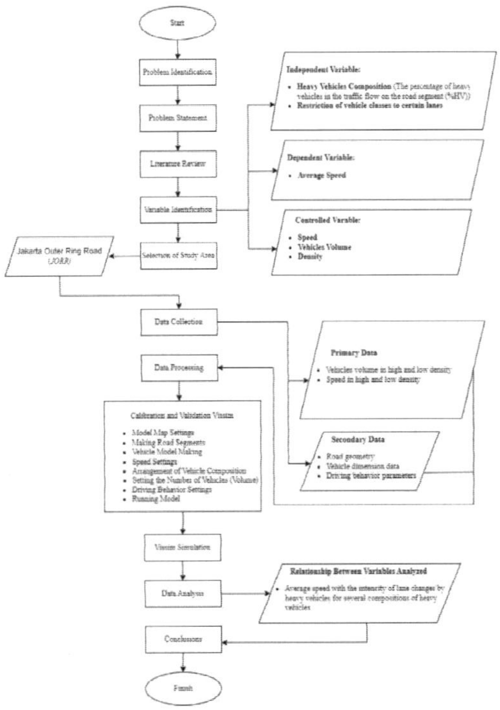

Figure 1. Research flow.

4. Data Collection

4.1. Study Area for Model Development

This research was conducted on the Jakarta Outer Ring Road section of the TB Simatupang toll road. Video recording of the road segment with a length of 600 m through the AD Premier Office building. The survey was conducted for 24 hours by placing cameras at two adjacent points. Data processing is carried out under two conditions, namely during high and low-density traffic flows. This aims to obtain a better analysis.

Figure 2. Overall situation of study area.

Figure 3. Situation at point 1 observation.

Figure 4. Situation at point 2 observation.

4.2. Primary and Secondary Data

Table 1. Data collecting in high density traffic flow.

High Density Traffic Flow									
Time Interval	JORR TB Simatupang Toll Road Segment - 1 Lane Direction to TMII								
	Total Vehicles						Traffic Volume (vehicles/hour)	Average Speed (km/hour)	Density (vehicles/km)
	LV	2 Axle HV	3 Axle HV	4 Axle HV	5 Axle HV	6 Axle HV			
03.00 - 03.05 PM	331	72	21	4	1	2		39.623	125
03.05 - 03.10 PM	282	53	22	6	0	2		34.700	143
03.10 - 03.15 PM	304	57	10	3	3	5		30.939	161
03.15 - 03.20 PM	282	50	14	3	4	4		24.483	203
03.20 - 03.25 PM	337	68	21	6	1	1		28.014	177
03.25 - 03.30 PM	333	51	17	2	0	2	4972	30.686	162
03.30 - 03.35 PM	330	67	14	2	1	2		33.914	147
03.35 - 03.40 PM	319	38	15	4	0	0		26.510	188
03.40 - 03.45 PM	346	52	22	6	3	1		23.204	214
03.45 - 03.50 PM	366	75	16	4	2	3		21.597	230
03.50 - 03.55 PM	381	44	21	6	3	9		24.797	201
03.55 - 04.00 PM	376	45	17	4	2	2		23.975	207
Average								28.537	180

Table 2. Data collecting in low-density traffic flow.

Low-Density Traffic Flow										
Time Interval	JORR TB Simatupang Toll Road Segment - 1 Lane Direction to TMII									
	Total Vehicles						Traffic Volume (vehicles/hour)	Average Speed (km/hour)	Density (vehicles/km)
	LV	2 Axle HV	3 Axle HV	4 Axle HV	5 Axle HV	6 Axle HV			
06.00 - 06.05 AM	257	45	11	4	1	0	4537	46.065	108
06.05 - 06.10 AM	301	57	4	2	1	2		48.261	103
06.10 - 06.15 AM	289	46	7	4	0	1		44.468	112
06.15 - 06.20 AM	295	38	9	1	0	1		43.950	113
06.20 - 06.25 AM	373	34	10	1	0	1		42.889	116
06.25 - 06.30 AM	359	49	7	0	0	1		47.845	104
06.30 - 06.35 AM	326	40	5	0	0	2		43.923	113
06.35 - 06.40 AM	343	47	7	0	0	1		45.862	108
06.40 - 06.45 AM	350	41	7	0	0	1		49.936	100
06.45 - 06.50 AM	321	53	7	0	0	1		49.986	99
06.50 - 06.55 AM	322	39	8	1	1	1		48.012	104
06.55 - 07.00 AM	348	44	6	2	0	2		45.977	108
Average								46.431	107

Table 3. JORR toll road geometric specifications.

Location	JORR Toll Road
(KM 27 – 28 TB Simatupang)	
Lane Width	12.5 m
Number of Lanes	4
Lane 4	2 m
Lane 1	3.5 m
Lane 2	3.5 m
Lane 3	3.5 m
Gradients	0.3%

Table 4. Category class, type, 3D model, and length of the vehicle in the simulation.

Vehicle Class	Vehicle Category	Vehicle Type	3D Model
LV (Light Vehicle)	Car	LV	Ford Explorer 2008
HV (Heavy Vehicle)	HGV	2 Axle HV	EU 5 Tractor
		3 Axle HV	EU 4 Tractor
		4 Axle HV	EU 1 Tractor + US AASTHO WB-40 Trailer
		5 Axle HV	EU 1 Tractor + EU 1 Trailer
		6 Axle HV	US AASHTO WB-50 Tractor + US AASTHO WB-50 Trailer

Table 5. Driving behavior parameters.

Parameter Driving Behavior	**Value (m)**
Average standstill distance	0.6
Additive part of a safety distance	0.6
Multiplicative part of a safety distance	1

4.3. Vissim Model Development

This research was conducted by microscopic simulation with the help of Vissim software. Before running the model, it is necessary to have input parameters so that the simulation results can run properly. In short, the parameters that need to be set to run the simulation model are as follows.

- Model map settings
- Making road segments
- Vehicle model making
- Speed settings
- Setting the composition of vehicles
- Setting the number of vehicles (volume)
- Driving behavior settings
- Running model
- Calibration and validation

4.4. Data Input Preparation for Calibration

The data used in the calibration model input is when high-density traffic flows around 03.00 PM and low density around 06.00 AM. The data used are vehicle volume, vehicle relative flow, and vehicle speed distribution. The data is provided according to the grouping of vehicles. The following is an example of the Vissim model input data tabulation.

Table 6. Example of vehicle volume data input on the Vissim model.

High-Density Traffic Flow						
JORR TB Simatupang Toll Road Segment - 1 Lane Direction to TMII						
Parameters	Total Vehicles					
	LV	2 Axle HV	3 Axle HV	4 Axle HV	5 Axle HV	6 Axle HV
Relative flow	0.802	0.135	0.042	0.01	0.004	0.007
Traffic Volume (veh/hour)	4972					

4.5. Trial and Error Model Calibration and Process Validation

The calibration model that has been prepared according to the input data for the calibration model is then run to get the output data. Running is carried out for a duration of 1 hour according to the data used as input. For each data generated by the Vissim running program, a suitability test was conducted to test the suitability of the Vissim output data with the actual data at the research location. The test was carried out using the Geoffrey E. Havers method. Previously, driving behavior adjustments were made by referring to existing research on vehicle characteristics in Indonesia.

Table 7. Trial and error experiment.

Trial Number	Car Following Model			Lateral				
	Average standstill distance	Additive part of a safety distance	Multiplicative part of a safety distance	Desired position at free flow	Observe adjacent lane	Distance driving	Overtake on the same lane	
	(m)	(m)	(m)	-	-	(m)	right	left
Default	2	2	3	middle of lane	no	1	no	no
1	0.6	0.6	1	any	yes	0.4	yes	yes

Table 8. Results of the Vissim program simulation model validation.

Traffic Flow Conditions	Traffic Volume (Veh/hour)		GEH Value	Data Validation Code
	Survey Results	Vissim Simulation		
High-Density Traffic Flow	4972	4785	2.677	Accepted
Low-Density Traffic Flow	4537	4521	0.238	Accepted

The results of this validation become a consideration for model acceptance for feasibility in the next simulation scenario process. The smaller the GEH value and below the number five, the more accurate the model is with the real conditions at the research location.

5. Result Analysis

5.1. Simulation Results

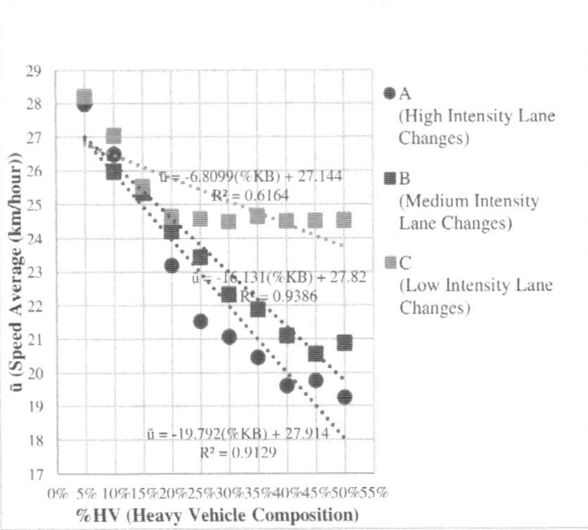

Figure 5. Effect from intensity lane changes by HV and HV composition plots for the three simulations scenarios in high density traffic flow.

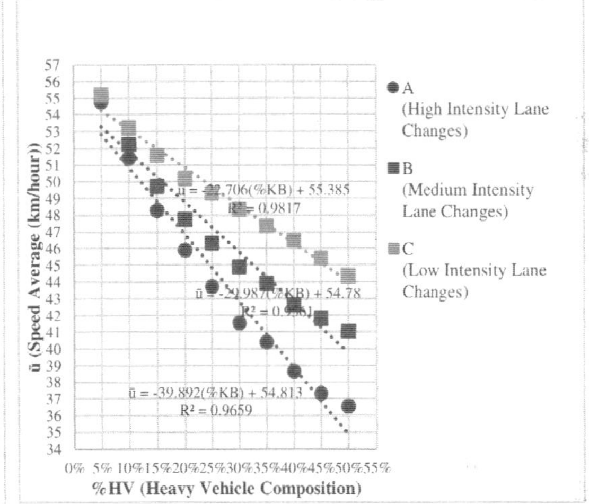

Figure 6. Effect from intensity lane changes by HV and HV composition plots for the three simulations scenarios in low density traffic flow.

From the results of processing simulated data with scenarios that have been determined in Vissim, an analysis was carried out using the linear regression method. For the distribution of the average speed of traffic flow in Figures 7 and 8, where the graph of the relationship between average speed and lane change by HV and the composition of HV in high and low-density conditions, the simulation model has a linear trendline with an R^2 value close to 1.0 indicating that the calibration results are suitable. The R^2 values (coefficient of determination) for scenario groups A, B, and C for high-density traffic flow are 0.9129, 0.9386 and 0.6164, respectively. Meanwhile, the R^2 values for scenario groups A, B, and C for low-density traffic flow were 0.9659, 0.9561, and 0.9817, respectively. This linear line can be used as a reference to find out how much deviation there is between the observed results and the existing simulation model.

From the linear regression line, the trend line equation is obtained. Where there are parameters that represent the slope of the line. The three lines of the high-density graph have the same slope direction, namely from the top left to the bottom right which indicates that the greater the composition of heavy vehicles in the traffic flow, the lower the average speed of all vehicles. In scenario group A, the parameter value or slope is -19,792. while in scenario group B the slope value is -16,131. then for scenario C the slope value formed is -6.8099. The minus sign on this accompaniment value indicates the direction of the trendline descending. The three trend lines have different slope values, this indicates that the different lane change intensity scenarios affect the average speed of traffic flow. The highest slope value of the three lines is shown in scenario group A. This occurs because the intensity of lane changes by HV is low. Then by adding a little bit of the composition of heavy vehicles to the traffic flow, the average speed of the traffic flow becomes very sensitive to decrease.

Meanwhile, from the low-density graph. In scenario group A, the parameter value or slope is -39,892. while in scenario group B the slope value is -29,987. then for scenario C the slope value formed is -22,706. The minus sign on this accompaniment value indicates the direction of the trendline descending. The three trend lines have different slope values, this indicates that the different lane change intensity scenarios affect the average speed of traffic flow. The highest slope value of the three lines is shown in scenario group A. This occurs because the intensity of lane changes by HV is low. Then by adding a little bit of the composition of heavy vehicles to the traffic flow, the average speed of the traffic flow becomes very sensitive to decrease.

From the graph of high-density traffic flow conditions, the gradient value can also be analyzed that in conditions of high-intensity lane changes, the intercept value is 27,914. In a lane change condition with moderate intensity the intercept value is 27.82, while in a lane change condition with low intensity, the intercept value is 27.144. This value indicates that in the condition that the %HV is zero, the average speed of the traffic flow corresponds to the intercept, which in this condition is an average of 27 km/hour.

Whereas in conditions of low-density traffic flow, the gradient value can also be analyzed that in conditions of high-intensity lane changes the intercept value is 54,813. In a lane change condition with moderate intensity the intercept value is 54.78, while in a lane change condition with low intensity, the intercept value is 55.385. This value indicates that in the condition that the %HV is zero, the average speed of the traffic flow corresponds to the intercept, which in this condition is an average of 54 km/hour. In high-density traffic flow conditions scenario A, the average vehicle speed reaches 22 km/hour, then in scenario B, the average vehicle speed is 23 km/hour, and in scenario C, the average vehicle speed increases to 25 km/hour. The average traffic speed increases by 6% in each lane change scenario with less intensity. From the table data, it can also be analyzed that for every 5% increase in KB composition from high-intensity lane changes compared to medium and low-intensity lane changes, in high-density conditions the average speed of traffic flow increases by 4%.

In low-density traffic flow conditions scenario A, the average vehicle speed reaches 43 km/hour, then in scenario B, the average vehicle speed is 46 km/hour, and in scenario C, the average vehicle speed increases to 49 km/hour. The average traffic speed increases by 6% in each lane change scenario with less intensity. From the table data, it can also be analyzed that for every 5% increase in the composition of KB from high-intensity lane changes compared to medium and low-intensity lane changes, in high-density conditions the average speed of traffic flow increases by 8%.

5.2. Contextualization of Research Results

Research that has been conducted to determine the impact of lane changing by heavy vehicles using various strategies for limiting heavy vehicle classes and the composition of heavy vehicles on traffic flows shows that in conditions of high-density traffic flow, the average speed increases by 4% in each lane changing scenario which is getting lower. Then for the addition of the composition of heavy vehicles, the speed of traffic flow decreases by an average of 2% for every 5% addition of heavy vehicles. Meanwhile, in conditions of low-density traffic flow, the average speed has increased by 8% also for each lane-changing scenario which is getting lower. Then for the addition of the composition of heavy vehicles, the speed of traffic flow decreases by an average of 3% for every 5% addition of heavy vehicles.

The results of this study are in line with research [3] where the behavior of changing lanes by heavy vehicles frequently or at high intensity can have a negative impact on traffic performance, one of which is the average speed of traffic flow. Research [4] also supports the results of this study, where the difference in average speed of traffic flow between vehicles changing lanes and vehicles without changing lanes decreased from 51 km/h to 35 km/h due to congestion worsening. A study [15] in Iran showed that increasing the composition of heavy vehicles reduced the average traffic speed by 15% at a composition of 40% HV. Meanwhile, in this study, an increase in the composition of heavy vehicles reduced the average traffic speed by up to 25% at a composition of 40% HV. This can happen because of the different driving behavior between countries. It was proven in research [15] that the average desired speed was 164.5 km/hour, whereas in this study only the maximum value only reached 80 km/hour. Then the distance vehicle in research [15] averaged 1.39 m. While in this study it was 0.6 m.

6. Conclusions

Lane changing by heavy vehicles has an impact on traffic flow performance, namely the average vehicle speed. in traffic flow with high and low density, the average speed of traffic flow increases by ±12%. The presence of an increased composition of heavy vehicles in the traffic flow can make the impact of changing lanes by heavy vehicles worse because the average speed of vehicles decreases compared to the composition of small heavy vehicles. The addition of the HV composition causes a significant reduction in traffic speed with an average of ±3% for every 5% addition of the HV composition. In high and low-density conditions, the difference is that for every 5% increase in the composition of HV from high-intensity lane changes compared to medium and low-intensity lane changes, the average speed of traffic flow increases by 4% in high-density conditions and 8% in low-density conditions.

Acknowledgments

This research was supported by POLAR - UP2M, part of the Faculty of Engineering at the Department of Civil Engineering, University of Indonesia.

References

[1] Hezorg B. O. 2010. *Angkutan Barang Perkotaan di Kota-kota Negara Berkembang.* http://www.sutp.org
[2] Gipps, P. G. 1986. *A Model for The Structure of Lane-Changing Decisions* (Vol. 20, Nomor 8).
[3] Farooq, D., Moslem, S., Jamal, A., Butt, F. M., Almarhabi, Y., Tufail, R. F., & Almoshaogeh, M. 2021. *Assessment of significant factors affecting frequent lane-changing related to road safety: An integrated approach of the ahp–bwm model.* (International Journal of Environmental Research and Public Health, 18(20))
[4] Ahmed, I., Karr, A. F., Rouphail, N. M., Chase, R. T., & Tanvir, S. 2022. *Characterizing lane-changing behavior and identifying extreme lane-changing traits.* Transportation Letters.
[5] Moridpour, S., Rose, G., Sarvi, M., & Mazloumi, E. 2012. *Influence of the Surrounding Traffic Characteristics on Lane Changing Decision of Heavy Vehicle Drivers Article in Road and Transport Research.*
[6] Jayatilleke, S., Wickramasinghe, V., & Amarasingha, N. 2022. *Introduction of a Simple Estimation Method for Lane-Based Queue Lengths with Lane-changing Movements.* (Journal of The Institution of Engineers (India): Series A).
[7] Aghabayk, K., Moridpour, S., Young, W., Sarvi, M., & Wang, Y. B. 2011. *Comparing heavy vehicle and passenger car lane-changing maneuvers on arterial roads and freeways.* (Transportation Research Record, 2260, 94–101).
[8] Chen, K., Knoop, V. L., Liu, P., Li, Z., & Wang, Y. 2022. *How gaps are created during anticipation of lane changes.* (Transportmetrica B).
[9] Toledo, T., & Zohar, D. (2007). *Modeling duration of lane changes.* (Transportation Research Record, 1999, 71–78).
[10] Moridpour, S., Sarvi, M., Rose, G., & Mazloumi, E. 2012. *Lane-changing decision model for heavy vehicle drivers.* (Journal of Intelligent Transportation Systems: Technology, Planning, and Operations, 16(1), 24–35).

[11] Knoop, V. L., & Buisson, C. 2015. *Calibration and validation of probabilistic discretionary lane-change models.*

[12] Keyvan-Ekbatani, M., Knoop, V. L., & Daamen, W. 2016. *Categorization of the lane change decision process on freeways.* (Transportation Research Part C: Emerging Technologies, 69, 515–526).

[13] Keyvan-Ekbatani, M., Knoop, V. L., Grébert, V., & Daamen, W. 2016. *Lane Change Strategies on Freeways: A Microscopic Simulation Study.* (Springer International Publishing).

[14] Kong, D., Guo, X., Yang, B., & Wu, D. 2016. *Analyzing the Impact of Trucks on Traffic Flow Based on an Improved Cellular Automaton Model.* (Discrete Dynamics in Nature and Society, 2016).

[15] Afandizadeh, S., Nasiri, N., Mirzahossein, H. 2021. *Evaluating the Efect of Heavy Vehicles on Trafc of Two-Way Two-Lane Road in Mountainous Area of Iran as a Developing Country.* (Iranian Journal of Science and Technology, Transactions of Civil Engineering (2022) 46:3387–3398).

Identifying heterogeneity in university students' transport mode choice

Willy Kriswardhana[1,2], Domokos Esztergár-Kiss[1]

[1]Departement of Transport Technology and Economics, Budapest University of Technology and Economics, Budapest, Hungary
[2]Department of Civil Engineering, Universitas Jember, Jember, Indonesia

Email: willy.kriswardhana@edu.bme.hu

Abstract. College students are argued to have some interest in adopting alternative mobility solutions. Even though previous works have examined the transport mode choice of college students, the inclusion of shared mobility services is scarcely found. Thus, this study aims to examine the transport mode choice of university students by revealing the heterogeneity in transport mode usage frequency incorporating socio-demographic and travel characteristics as covariates in the latent class cluster analysis (LCCA) framework. The analysis results in four clusters. Clusters of public transport users and walkers are mostly characterized by lower-income students who can reach the university in less than 11 minutes. Private car users mostly fall into a cluster where students generally come from higher-income families. Policies supporting sustainable transportation among college students are discussed, including the mobility points around the university and public transport stops.

1. Introduction

Travel behavior of younger generations, including university students, is different from the preceding generations, where young individuals are most likely to be more flexible in transport mode choices [1,2]. They are proven to be the early adopters of alternative mobility solutions like shared mobility services [3,4]. This condition might be caused by the fast adoption of recent technologies among young individuals and consequently affect their travel behavior [1].

Previous studies have investigated the transport mode choice among university students considering several factors (e.g., socio-demographic, travel characteristics, attitudes, and built environment) using logit models [3,5–7]. These studies examine conventional transport modes, such as private cars, bikes, walking, and public transportation (PT), but little attention has been paid to new mobility services. Information technology is continuously developing, and it brings effects to transportation sectors, where travel options are more varied [8]. Shared transport modes are more available in cities worldwide [9]. Furthermore, by examining the heterogeneity in transport mode choice, we could differentiate the students' travel behaviour. Once the travel behaviour is clustered, we could draw specific policies based on the identified clusters to raise the use of environmentally friendly modes, instead of providing general policies.

Considering the preceding background, this study aims to cover the gaps by examining the heterogeneity of transport mode choice among college students using the latent class cluster analysis (LCCA). The extra layer of heterogeneity is presented by including a wide range of covariates in the modeling framework.

2. Literature review

Several studies have examined the essential factors affecting college students' mode choices, including socio-demographic characteristics. Male students are argued to be motorcycle users, while their counterparts are more likely to travel to school by walking [6]. Different findings state that male students are more likely to be active travelers [7] and use PT than shared taxis and private cars [5,10]. Regarding income, students from high-income families are more inclined to use private cars [5] and are not interested in traveling by active modes and PT [6]. In the case of travel characteristics, increasing travel time leads to a higher likelihood of using bicycles and cars [3]. In this study, socio-demographic and travel characteristics are included in the model as covariates.

Meanwhile, previous works have examined the heterogeneity in transport mode choice preferences. A study argues that educated young people with high technology literacy are more likely to be in a group of mobility-on-demand early adopters [11]. Another study states that the multimodal travel group is related to highly educated young individuals and small households [12]. Individuals who live in urban areas are more likely to be active travelers and PT users [1]. Mentioned studies prove that several factors significantly influence travel mode choice, but mostly include conventional transport modes (e.g., PT, car, active modes).

Even though previous works have investigated transport mode choices among college students, most studies neglect shared transport modes. Thus, this study includes shared mobility services as indicators in the model. Furthermore, socio-demographic and travel characteristics are incorporated as covariates, giving an extra layer of heterogeneity.

3. Method

This study was realized in Budapest, Hungary's capital and most populated city. The metropolitan area of Budapest covers 525.5 km^2, with around 1.7 million inhabitants. The PT services in Budapest integrate metro, tram, bus, and trolley buses under one system and can be accessed through a single travel pass. Emerging mobility services are present in Budapest such as car-sharing, bike-sharing, and e-scooter-sharing.

An online survey was conducted among the Budapest University of Technology and Economics (BME) students from April 2022 to May 2022. The survey consists of three parts. The first part aims to capture the socio-demographics of the participants. The second part is about travel behavior, including their frequency of using the transport modes, ranging from (1) "never" to (8) "several times per day". The number of valid responses is 687. It is worth noting that BME can be easily accessed by all means of urban transport modes, with a metro station, tram stops, and bus stops around the university. Collection points for bike-sharing and e-scooter-sharing are present in the surrounding area of the university.

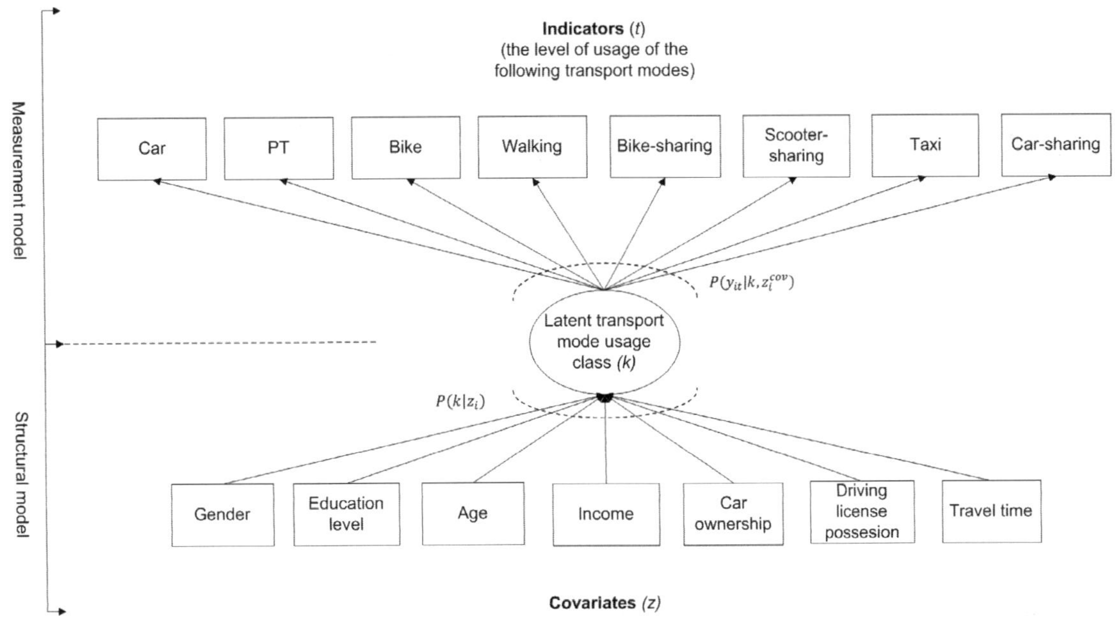

Figure 1. Modeling framework.

LCCA is performed to identify the heterogeneity of travel patterns among students. LCCA is a statistical method following a probabilistic approach to reveal different unobserved subgroups of populations that share specific observed characteristics [13]. LCCA is chosen since the method has advantages, such as the opportunity to include mixed types of variables and statistical criteria to identify the optimal number of clusters [14]. The probability of observing response patterns is formulated as follows [15]:

$$P(y_i|z_i^{cov}) = \sum_{k=1}^{K} P(k|z_i^{cov}) \prod_{t=1}^{8} P(y_{it}|k, z_i^{cov}) \tag{1}$$

Figure 1 represents the modeling framework of this study, inspired by [12], with different indicators and covariates. Let y_{it} denote the response of participant i on indicator t. The latent class k has K categories and intervenes between indicators y and covariates z_i^{cov}. Two kinds of probabilities are present in LCCA: the probability of the membership model $P(k|z_i)$ and the probability of the measurement model $P(y_{it}|k, z_i^{cov})$. The optimal number of clusters is examined using Bayesian Information Criterion (BIC), where the low values are associated with better model fit [16].

4. Results

The descriptive statistics of the participants are given in Table 1. Most participants are male (68%), undergraduate students (65%), and aged under or equal to 23 (73%). The sample is close to being representative since around 69% and 61% of BME students are male and undergraduate students [17]. The majority of respondents come from higher-income households (69%) and have at least one car in their households.

Table 1. Descriptive statistics of respondents.

Category	%	Category	%
Gender		**Car-ownership**	
Male	68	0	19
Female	32	1	33
Education		2	32

Undergraduate student	65	> 2	16
Postgraduate student	35	**Driving license possession**	
Age		No	19
≤ 23	73	Yes	75
> 23	27	No, but planning to have one	6
Household income		**Travel time to the university**	
≤ €700	31	< 11 minutes	20
> €700	69	11 – 30 minutes	34
		> 30 minutes	46

Regarding transport mode usage, Figure 2 shows that more than half of the participants use PT several times per day. Private cars are used occasionally, as 25% use them several times per week. The figure indicates that shared transport modes are not popular among respondents.

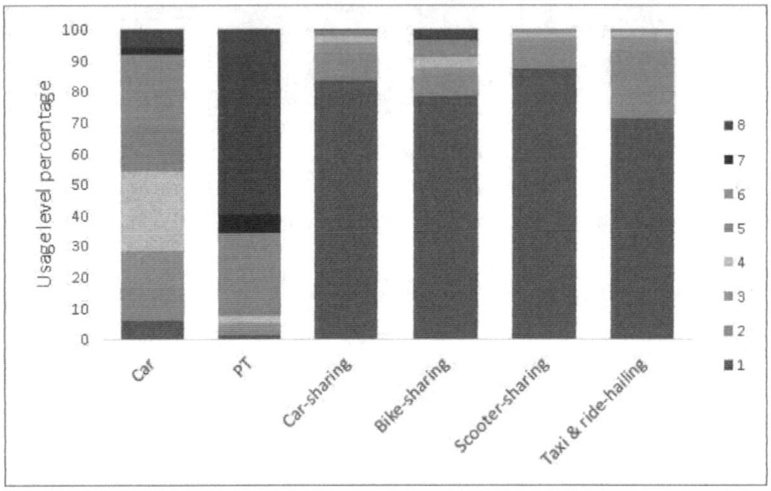

Figure 2. Distribution of participants' usage level of transport modes.

We tested ten models to examine the most suitable number of clusters. The BIC values continue to decline until the 4-cluster model. Thus, the 4-cluster model is selected for further analysis.

Table 2. Model fit of the LCCA models.

N classes	Npar	LL	BIC(LL)
1	52	−5932	12204
2	72	−5820	12110
3	92	−5732	12066
4	112	−5654	12041
5	132	−5606	12074
6	152	−5567	12127
7	172	−5514	12151
8	192	−5483	12220
9	212	−5461	12307
10	232	−5449	12413

Table 3 presents the parameters of the final model. The class-specific parameters indicating the (un)attractiveness of the indicators are given in the measurement model. The most significant positive sign for cluster-1 is walking. For cluster-2, PT is mostly preferred by the respondents. Meanwhile, participants in cluster-3 tend to have higher shared mobility usage frequency. Finally, college students in cluster-4 are private car users.

Table 3. Parameters of the LCCA model with 4 clusters.

	Cluster-1	Cluster-2	Cluster-3	Cluster-4	Wald	z-value
Prediction of the indicators (the measurement model)						
Car_freq	0.275	−1.158	−0.617	1.500	64.728*	7.423
PT_freq	0.189	0.449	−0.128	−0.510	48.059*	6.426
Bike_freq	0.197	−0.448	0.478	−0.228	43.797*	5.911
Walking_freq	0.501	0.118	−0.128	−0.492	38.950*	5.509
Car_sharing_freq	−0.452	−0.469	0.498	0.423	20.441*	3.633
Bike_sharing_freq	−0.008	−0.190	0.453	−0.255	43.018*	5.854
Scooter_sharing_freq	−0.061	−1.146	0.878	0.330	24.065*	4.065
Taxi_RS_freq	−0.415	−0.251	0.367	0.299	22.406*	3.871
Prediction of latent class membership (the structural model)						
Intercept	1.293	1.9828	0.1465	-3.4222	16.5922	
Gender						
Male	−0.052	−0.392	0.425	0.020	12.778*	2.568
Female	0.052	0.392	−0.425	−0.020		
Education						
Undergraduate student	−0.013	0.044	−0.002	−0.030	0.126	−2.326
Postgraduate student	0.013	−0.044	0.002	0.030		
Age						
≤ 23	0.332	0.113	−0.074	−0.371	11.357*	2.330
> 23	−0.332	−0.113	0.074	0.371		
Income						
≤ €700	0.223	0.139	−0.152	−0.211	5.821	1.170
> €700	−0.223	−0.139	0.152	0.211		
Car−ownership						
0	−1.819	2.686	2.048	−2.916	27.896*	3.093
1	0.143	−0.390	−0.242	0.489		
2	0.809	−0.760	−0.892	0.844		
> 2	0.867	−1.536	−0.914	1.583		
Driving license possession						
No	0.922	0.670	−0.392	−1.200	27.788*	3.719
Yes	−0.571	−1.666	0.049	2.188		

No, but planning to have one	−0.351	0.995	0.343	−0.988		
Travel time						
> 11 minutes	0.449	0.209	0.183	−0.841	23.224*	3.186
11 – 30 minutes	−0.412	−0.432	0.543	0.301		
< 30 minutes	−0.037	0.223	−0.727	0.540		

*significant at the 5% level

Moving on to the structural model, female, younger, and lower-income students are more likely related to cluster-1 and cluster-3, while male, older, and higher income students are positively associated with cluster-3 and cluster-4. More cars lead to higher positive parameters in cluster-4. Similarly, students having a driving license are generally related to cluster-4. Additionally, college students who spend less than 11 minutes going to the university are primarily associated with cluster-1. In comparison, cluster-4 is more related to college students who spend more than 30 minutes to reach the university.

Table 4. The within-cluster distribution of indicators.

		Clusters				Total
		WPC	PW	BS	C	
Indicators						
Car_freq	Mean	5.07	2.73	3.63	6.81	4.40
PT_freq	Mean	7.24	7.52	6.60	5.04	6.99
Bike_freq	Mean	3.28	1.96	4.45	2.25	3.03
Walking_freq	Mean	7.80	7.58	7.25	6.07	7.48
Car_sharing_freq	Mean	1.13	1.13	1.87	1.73	1.31
Bike_sharing_freq	Mean	1.48	1.27	3.40	1.22	1.71
Scooter_sharing_freq	Mean	1.13	1.03	1.81	1.24	1.23
Taxi_RS_freq	Mean	1.26	1.32	1.73	1.66	1.39

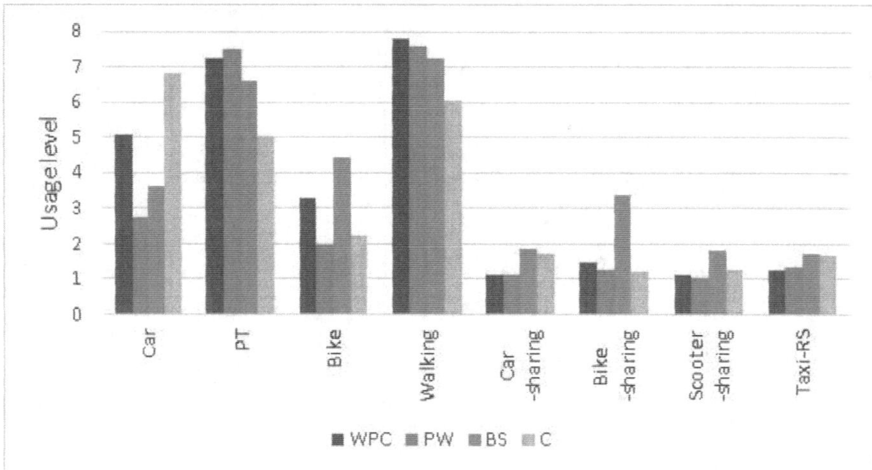

Figure 3. Distribution of the usage level of transport mode within clusters.

The profiles for clusters are given in Table 4 and Figure 3, where clusters are named after the level usage of transport modes. The clusters are Walking + PT + Car (WPC), PT + Walking (PW), + Bike +

Shared mobility services (BS), and private car users (C). The WPC cluster comprises nearly half of the respondents (48%), while 26% fall into the PW cluster. Smaller shares of participants belong to BS and C clusters (16% and 10%).

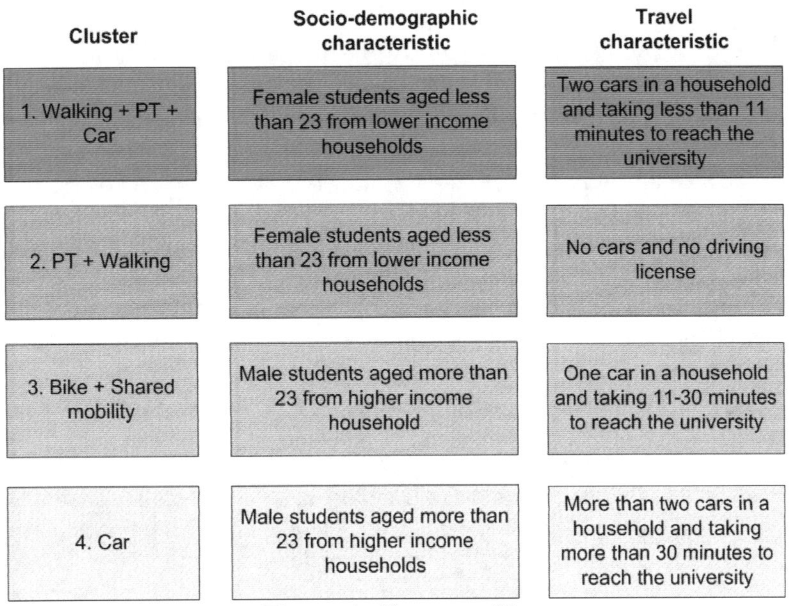

Figure 4. Cluster profile.

The socio-demographic and travel characteristics of clusters are presented in Figure 4. The WPC cluster and PW cluster are represented by female students (31% and 42%) who are aged under 23 (86% and 68%) and come from lower-income households (36% and 35%). Meanwhile, the BS cluster and C cluster are characterized by male students (82% and 71%) who are aged more than 23 (45% and 47%) and come from higher-income households (79% and 84%). In the case of travel characteristics, students in the WPC cluster have two cars in their households (45%) and spend less than 11 minutes going to the university (26%). The most prominent characteristics of students falling into the PW cluster are those who do not have cars (47%) and a driving license (32%). College students in the BS cluster have one car in their household (39%), and take around 11-30 minutes to reach the university (63%). Finally, the C cluster is represented by students who have more than two cars in their households (32%), and spend more than 30 minutes going to the university (64%).

5. Discussion

This study investigates the heterogeneity in transport mode choice among university students, where the LCCA is performed to reveal cluster membership. Four clusters are identified, including WPC (group of students who use PT, walking, and car), PW (group of students who use PT and walking), BS (groups of students who use the bike and shared mobility services), and C (group of students who use private cars). The WPC and PW clusters mostly consist of female students, which aligns with studies stating that female students are more likely to walk [6] and dominate the PT users class [18]. The majority of lower-income students belong to the WPC and PW clusters, which aligns with the study implying low-income millennials are related to the most frequent PT users [1]. The BS cluster is dominated by male students, which resonates with a study arguing that male students are likelier to bike to the university [7]. A significant percentage of college students from higher-income households is present in the BS and C clusters. This is aligned with a study in Australia revealing that individuals with relatively high incomes increase the probability of bike-sharing memberships [19]. Similar findings are reported by a Dutch study confirming that higher-income people are more inclined to use car-sharing [20].

In terms of travel characteristics, college students in the WPC cluster can reach the university in less than 11 minutes. Most probably, they live near the university. College students in this cluster are occasional users of private cars. The PW cluster consists of college students who have no cars in their households. Conversely, the C cluster is characterized by students who have more than two cars in their households. It resonates with a study arguing that students with cars available to their families are more likely to commute to the university by car [5]. Meanwhile, university students who spend around 11-30 minutes to reach the university are more likely to fall into the BS cluster. This is understandable since most probably they do not live near the university, while their families only have one car. Thus, besides PT, using bikes or shared mobility services is the most reasonable choice for them.

As for policies to encourage the shift to more environmentally friendly transport modes, we focus more on the C cluster. While it is clear that college students using private cars are mostly present in this cluster, they also show some interest in car-sharing and e-scooter-sharing services. Thus, placing more shared mobility points around the university and PT stops could be a good way to decrease private car use. The multimodal classes (WPC, PW, BS) already show sustainable travel behavior. Thus, a breakthrough mobility solution, such as Mobility as a Service (MaaS) could target college students as its potential users since the service promises more accessible ways to use a wider range of transport modes through one single system [21], and college students show a high interest in adopting the service [22]. Since college students in WPC and PW classes come from lower-income households, the mobility companies could support them in favor of enhancing the use of shared sustainable transport modes by giving discounts.

The first limitation of this study is related to the use of self-reported travel pattern data. While the self-reported travel pattern data is generally faster and more affordable, the reliability is questionable [23]. Future studies could use travel data from the government or mobile applications with GPS trackers. Since this study focuses on college students, the results are not necessarily generalizable to other groups of individuals.

6. Conclusion

This study investigates the heterogeneity in transport modes of university students, including shared mobility services. The LCCA method is used, where the level of transport mode usage serves as indicators, and socio-demographic and travel characteristics are included as covariates. Four clusters are identified: WPC, PW, BS, and C. The WPC cluster captures students who mostly walk to the university and use PT and a car occasionally. The PW cluster consists of lower-income college students who use PT several times per day. The BS and C clusters consist primarily of higher-income students. College students who use bikes and shared mobility services are in the BS cluster, while those who use private cars mostly fall into the C cluster. Responsible institutions could utilize the current study to set the policies encouraging the shift to more environmentally friendly transport modes.

Acknowledgements

Project no. TKP2021-NVA-02 has been implemented with the support provided by the Ministry of Culture and Innovation of Hungary from the National Research, Development and Innovation Fund, financed under the TKP2021-NVA funding scheme.

References

[1] Lee Y, Circella G, Mokhtarian PL, Guhathakurta S. Are millennials more multimodal? A latent-class cluster analysis with attitudes and preferences among millennial and Generation X commuters in California. Transportation (Amst) 2020;47:2505–28. https://doi.org/10.1007/s11116-019-10026-6.

[2] Limanond T, Butsingkorn T, Chermkhunthod C. Travel behavior of university students who live on campus: A case study of a rural university in Asia. Transp Policy 2011;18:163–71. https://doi.org/10.1016/j.tranpol.2010.07.006.

[3] Whalen KE, Páez A, Carrasco JA. Mode choice of university students commuting to school and the role of active travel. J Transp Geogr 2013;31:132–42. https://doi.org/https://doi.org/10.1016/j.jtrangeo.2013.06.008.

[4] Narayanan S, Antoniou C. Shared mobility services towards Mobility as a Service (MaaS): What,

who and when? Transp Res Part A Policy Pract 2023;168:103581. https://doi.org/https://doi.org/10.1016/j.tra.2023.103581.

[5] Danaf M, Abou-Zeid M, Kaysi I. Modeling travel choices of students at a private, urban university: Insights and policy implications. Case Stud Transp Policy 2014;2:142–52. https://doi.org/10.1016/j.cstp.2014.08.006.

[6] Nguyen-Phuoc DQ, Amoh-Gyimah R, Tran ATP, Phan CT. Mode choice among university students to school in Danang, Vietnam. Travel Behav Soc 2018;13:1–10. https://doi.org/10.1016/j.tbs.2018.05.003.

[7] Zhou J, Wang Y, Wu J. Mode choice of commuter students in a college town: An exploratory study from the United States. Sustain 2018;10. https://doi.org/10.3390/su10093316.

[8] Yan X, Levine J, Zhao X. Integrating ridesourcing services with public transit: An evaluation of traveler responses combining revealed and stated preference data. Transp Res Part C Emerg Technol 2019;105. https://doi.org/10.1016/j.trc.2018.07.029.

[9] Acheampong RA, Siiba A, Okyere DK, Tuffour JP. Mobility-on-demand: An empirical study of internet-based ride-hailing adoption factors, travel characteristics and mode substitution effects. Transp Res Part C Emerg Technol 2020;115. https://doi.org/10.1016/j.trc.2020.102638.

[10] Cattaneo M, Malighetti P, Morlotti C, Paleari S. Students' mobility attitudes and sustainable transport mode choice. Int J Sustain High Educ 2018;19:942–62. https://doi.org/10.1108/IJSHE-08-2017-0134.

[11] Wang X, Yan X, Zhao X, Cao Z. Identifying latent shared mobility preference segments in low-income communities: Ride-hailing, fixed-route bus, and mobility-on-demand transit. Travel Behav Soc 2022;26:134–42. https://doi.org/10.1016/j.tbs.2021.09.011.

[12] Molin E, Mokhtarian P, Kroesen M. Multimodal travel groups and attitudes: A latent class cluster analysis of Dutch travelers. Transp Res Part A Policy Pract 2016;83:14–29. https://doi.org/10.1016/j.tra.2015.11.001.

[13] Hagenaars JA, McCutcheon AL. Applied latent class analysis. Cambridge University Press; 2002.

[14] Vermunt JK, Magidson J. Latent class cluster analysis. Appl Latent Cl Anal 2002;11:60.

[15] Vermunt JK, Magidson J. Technical guide for Latent GOLD 5.0: Basic, advanced, and syntax. Belmont, MA Stat Innov Inc 2013.

[16] Nylund KL, Asparouhov T, Muthén BO. Deciding on the number of classes in latent class analysis and growth mixture modeling: A Monte Carlo simulation study. Struct Equ Model 2007;14:535–69. https://doi.org/10.1080/10705510701575396.

[17] BME. Tények és adatok a Műegyetemről; Facts and Figures on BME. 2021.

[18] Ton D, Zomer LB, Schneider F, Hoogendoorn-Lanser S, Duives D, Cats O, et al. Latent classes of daily mobility patterns: the relationship with attitudes towards modes. Transportation (Amst) 2020;47:1843–66. https://doi.org/10.1007/s11116-019-09975-9.

[19] Fishman E, Washington S, Haworth N, Watson A. Factors influencing bike share membership: An analysis of Melbourne and Brisbane. Transp Res Part A Policy Pract 2015;71:17–30. https://doi.org/https://doi.org/10.1016/j.tra.2014.10.021.

[20] Kim J, Rasouli S, Timmermans H. Satisfaction and uncertainty in car-sharing decisions: An integration of hybrid choice and random regret-based models. Transp Res Part A Policy Pract 2017;95:13–33. https://doi.org/https://doi.org/10.1016/j.tra.2016.11.005.

[21] Ho CQ, Mulley C, Hensher DA. Public preferences for mobility as a service: Insights from stated preference surveys. Transp Res Part A Policy Pract 2020;131:70–90. https://doi.org/10.1016/j.tra.2019.09.031.

[22] Kriswardhana W, Esztergár-Kiss D. Exploring the aspects of MaaS adoption based on college students' preferences. Transp Policy 2023;136:113–25. https://doi.org/https://doi.org/10.1016/j.tranpol.2023.03.018.

[23] Kelly P, Doherty A, Mizdrak A, Marshall S, Kerr J, Legge A, et al. High group level validity but high random error of a self-report travel diary, as assessed by wearable cameras. J Transp Heal 2014;1:190–201. https://doi.org/10.1016/j.jth.2014.04.003

IOP Conf. Series: Earth and Environmental Science 1294 (2024) 012028 doi:10.1088/1755-1315/1294/1/012028

Evaluation Of The Implementation Of Traffic Separation Scheme (TSS) In The Sunda Strait

A A Priadi[1], **R Ivan**[2],**Darsani**[3], and **C Anindhyta**[4]

[1]Nautical Lecturer of STIP Jakarta
[2]Head of Tanjung Priok Type B Navigation District
[3]Head of VTS Merak
[4]Universitas Negeri jakarta

Corresponding author: antoni.kemenhub@gmail.com

Abstract. The sea traffic separation chart or what is commonly called the Traffic Separation Scheme (TSS) is generally used to assist international shipping navigation and is also part of a country's maritime zone as mandated in UNCLOS 1982. In this case, Indonesia has implemented TSS in the Sunda Strait and in the Lombok Strait on 1 July 2020. The Directorate General of Sea Transportation, the Ministry of Transportation is the leading agency responsible for enforcement of the scheme through the Directorate of Navigation and the Directorate has established the Merak Vessel Traffic Service (VTS) system in the Province of Banten to monitor the Sunda Strait and VTS Benoa in the Province of Bali to monitor the Lombok Strait. Since the implementation of the TSS in the Sunda Strait until now, it turns out that there are still many ships that do not comply with the traffic procedures in the TSS as regulated in such a way. The research suggestion that violations in addition to administrative sanctions for those ships who commit violations, also implement a violation reporting system application to speed up the process of reports from VTS Operators to related officials.

1. Introduction

The sea traffic separation chart or commonly called the Traffic Separation Scheme (TSS) is generally used to assist the safety of international shipping navigation and is also part of a country's maritime zone as mandated in UNCLOS 1982. In this case, Indonesia implemented TSS in the Sunda Strait and the Lombok Strait on 1 July 2020. The Directorate General of Sea Transportation, the Ministry of Transportation is the leading agency responsible for enforcement of the scheme through the Directorate of Navigation and the Directorate has established the Merak Vessel Traffic Service (VTS) system in the Province of Banten to monitor the Sunda Strait and VTS Benoa in the Province of Bali to monitor the Lombok Strait.

This TSS is the first of its kind to be adopted by an archipelagic nation. Legally, the Sunda Strait and Lombok Strait are part of Indonesia's territorial waters, and no part of these areas is included in international maritime territory or borders with other countries. In comparison, the Malacca Strait has four states as its littoral, namely Indonesia, Malaysia, Singapore and Thailand. For Indonesia, managing TSS itself is a new experience. This has sparked euphoria among the local community and authorities such as port operators as well as parties involved in shipping, they are excited and see it as an opportunity to organize commercial activities such as bunkering services, ship goods/food supply, pilotage, alternative shipping routes, etc.

The legal basis for the application of the Sunda Strait TSS, namely:

1. IMO Circular Number COLREG.2/Circ.74 dated 14 June 2019 concerning New Traffic Separation Scheme which contains an international circular regarding the announcement of the implementation of TSS in the Sunda and Lombok Straits.

2. IMO Circular Number SN.1/Circ.337 Routeing Measures Other Than Traffic Separation Schemes IMO which contains further explanation regarding areas of caution in the Lombok and Sunda Strait TSS areas;

3. Government Regulation Number 5 of 2010 concerning Navigation, this regulation covers all matters concerning sea navigation, establishing shipping traffics, route systems, traffic procedures, and ship berthing areas according to their interests.

All ships sailing in the Sunda Strait TSS must comply with the provisions of rule 10 COLREGs 1972 related to TSS, consist of sailing in the direction of a suitable carriageway, stay away from the separation line/ zone, sntering/leaving the traffic at the end of the traffic, avoid cutting traffic traffics, avoid anchoring, avoid catching fish.

The Sunda Strait TSS is located in the Sunda Strait, the Province of Lampung and the Province of Banten, Republic of Indonesia. In addition, the Sunda Strait TSS is also right on the Indonesian Archipelagic Sea Traffics (ALKI - I) or SeaLine - I.

Some of the functions of implementing the Sunda Strait TSS among others reducing the number of head-on situations to improve shipping safety, by separating opposing traffic flows. reducing the risk of ship accidents caused by ships running aground on rocks, reduce the danger of collisions at intersections by establishing alert areas. protecting the Marine Environment from the effects of pollution in the event of a shipping accident or disaster.

It is known that the Sunda Strait TSS consists of

a) Northern Part of TSS
   - Overall Width : 2.67 Nm
   - Long Separation Zone : 0.30 Nm
   - Width : 2.60 Nm

b) Alert Area
   - North Side Width : 2,60 Nm
   - South Side Width : 1.20 Nm
   - Length : 2.80 Nm

c) TSS South Area
   - North Side Width : 1.20 Nm
   - North Segment Width : 1.25 Nm
   - South Side Width : 2.27 Nm
   - Traffic Length : 2.73 Nm
   - Total Length of TSS : 9.38 Nm

d) North and South Side Report Lines

e) East and West Side Report Lines

f) North and South Anchorage Areas

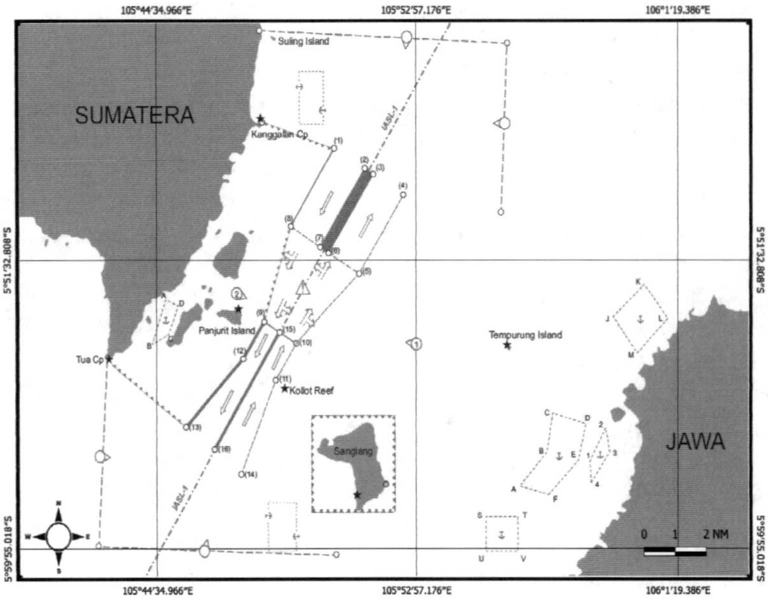

Figure 1. The Chart of Sunda Strait TSS

The ships crossing the Sunda Strait both entering the TSS from the North and from the South, Ferry/ Roro/ Merak – Bakauheni Ferries when entering the Alert Area and the ships going to leaving and entering ports in Banten, especially in Ciwandan, Cigading, Merak to Bojonegara, Banten are currently using the reporting and management system of the Merak VTS, Class I Navigation District, Tanjung Priok.

Technical instructions for ship management in the Sunda Strait TSS are contained in PM 130 of 2020 concerning traffic procedures in the Sunda Strait Traffic Separation Scheme (TSS). Meanwhile, for Ferry/ Roro/Merak-Bakauheni ferries, schedule arrangements and entry and exit are regulated by the Local Port Service (LPS) Merak and LPS Bakauheni belonging to BPTD Banten and BPTD Bangkulu - Lampung the Directorate General of Land Transportation. For ships that will enter the wharf at TERSUS and TUKS in the Banten KSOP area, it is regulated by the 3 (three) existing pilot stations.

In reality in the implementation of TSS in the Sunda Strait since July 2022 until now there are still many ships that do not comply with traffic rules in the TSS. This is under the supervision of the State Patrol Ship belonging to the Sea and Coast Guard Unit (KPLP) Directorate General of Sea Transportation, Ministry of Transportation.

The KPLP Patrol State Vessel conducted observations of vessels suspected of committing violations in the TSS traffic based on information from VTS. After that, the KPLP State Patrol Ship communicated via radio with the ship suspected of committing a violation on the TSS traffic to inform of the violation, while at the same time observations were made with Radar, AIS, Marine Maps and GPS on the State Patrol Ship. If the ship suspected of having committed a violation answers the summons, then the Master of the State Vessel Patrol PLP will inspect the ship suspected of having committed a violation. Meanwhile, if the ship does not answer the call, then the Captain of the State Vessel Patrol KPLP will make a Report and Minutes of the Event to the local Syahbandar. For ships that commit violations related to Traffic procedures on the TSS Traffic which are dangerous and can cause accidents in the TSS area, law enforcement will be carried out by imposing sanctions in accordance with the applicable Laws and Regulations.

In the Director General's Decree specifically for the Sunda Strait TSS, the person in charge of Patrol in the context of Law Enforcement in the Sector of Traffic Safety in the Sunda Strait TSS is carried out by the Tanjung Priok Class I PLP Base at any time, including from the Sunda Strait from the North of the Java Sea to the South of Indian Ocean and vice versa for Indonesian-flagged ships, from Merak to Bakauheni and vice versa, and from the north direction of the Java Sea to the south of the Indian Ocean and vice versa for foreign-flagged ships. Based on the background above, the formulation of the problem in this discussion is (1) What influences shipping on TSS; (2) How is the evaluation of the implementation of the Sunda Strait TSS?

2. Literature Review

VTS is a maritime coast-based service to assist ship traffic in certain areas, the VTS area. This service is provided by a VTS operator (VTSO) using Very High Frequency (VHF) radio to provide information essential for safe navigation within the area. Having been introduced after World War II, the VTS was little more than a radar chain. When ships are delayed during periods of poor visibility, port operations come to a halt, spreading ship delays throughout the inshore transport chain, regularly causing disruptions and problems in the storage of goods. To minimize this interference, a coast-based radar chain for traffic monitoring was implemented to ensure traffic flows efficiently in the port area even during periods of reduced visibility. The first radar chain was deployed at Douglas, Isle of Man, in 1948, but other major ports in northern Europe followed soon after, namely Amsterdam (IJmuiden) in 1952 and Rotterdam in 1956 [1].

Currently VTS is on the verge of change and several simultaneous developments, such as e-Navigation [2], chain planning [3], expanded ship-shore route planning [4], and encouragement for advances in VTS system organization and technology. Originally introduced as a RADAR chain system to assist secure port entry during periods of poor visibility [1], the VTS system is now a complex socio-technical system realized in various ways around the world. Depending on local conditions, VTS can offer services ranging from information services, containing information critical to the safety of conducting, to complex traffic organization systems, where traffic is planned and deconflicted prior to its arrival in the area.

VTS is a system with the goal of promoting safety and efficiency, Resilience Engineering (RE), has been adopted as a theoretical framework to analyze and evaluate how VTS systems balance system goals during day-to-day operations. The concept of resilience originally came from ecology and is concerned with how a system maintains a stable state over time. In the JCS setting, resilience describes how a system is able to maintain or regain dynamic stability which means producing stable performance output based on dynamic input-status to be able to continue to function in the presence of continuous stress in daily operations [5].

As in the VTS domain, system goals, such as efficiency and security, are often conflicting and require trade-offs because the two goals cannot be achieved concurrently. As a result, human operators in high-risk industries are often forced to improvise and find solutions to be able to cope with limited resources [6]. In RE, resilience is the intrinsic ability of a system to maintain its function under various operational conditions. As mentioned above, there are four cornerstones to endurance engineering; monitor, respond, anticipate and learn. These four capabilities are essential for a system to be able to recognize challenging conditions, respond to them, evaluate responses, and prepare for future events. The four capabilities are interdependent, and each represents one facet of system function. By analyzing day-to-day operations with the help of capabilities, the analyst is able to do so by identifying ways in which the capacity of the system to know what to do (respond to), what to look for (monitor), what to expect (anticipate) and what has been occur (learning) can be strengthened [6].

Due to several major oil spills in the 1970s, public awareness of accidents in the maritime area increased. This is followed by the need for harmonization and cooperation between radar chain operators and other actors, such as pilots. As a result, the existing radar chain that was originally implemented to support the efficiency of traffic movement was changed to a shore-based assistant service, VTS, which in addition to efficient operation, also strives to ensure that traffic flows safely and reduce its environmental impact. In 1985, the International Maritime Organization (IMO) adopted Resolution A.857, Ship Traffic Services Guidelines. This resolution was superseded in 1997 by IMO Assembly Resolution VTS A.857(20). There are currently more than 500 operational services worldwide offering VTS, with the main objective of promoting the efficient, safe and environmentally friendly movement of maritime traffic in areas with high traffic density, such as the Hook of Holland, and sensitive sea areas, such as islands or coral reefs [1].

The definition of Traffic Separation Scheme or in Indonesian language a shipping traffic separation chart or a sea traffic separation scheme by quoting from IMCO Resolution A.284 (VIII) adopted on 20th November 1973 concerning Routeing Systems explains the meaning of a Sea Traffic Separation Chart or Sea Traffic Separation Scheme commonly referred to as Traffic Separation Scheme is a scheme which separates traffic proceeding in opposite or nearly opposite directions by the use of a separation zone or line, traffic lanes or by other means which, when interpreted, can mean a scheme by separation sea traffic lanes in the opposite or nearly opposite direction by using an area or separation line for sea traffic lanes or other navigational aids.

The explanation regarding Local Port Service is based on the Decree of the Director General of Land Transportation number KP – DRJD 4937 of 2022 concerning Local Port Services (LPS) Equipment and Maintenance Standards at Ferry Ports. The definition of Local Port Services (LPS) is ship traffic services at internal ferry ports. the framework of regulating and controlling the operations of ships, ports, harbour and terminals at a crossing. Meanwhile, according to the Regulation of the Minister of Transportation number PM 26 of 2011 concerning Telecommunications - Shipping explains the meaning of Local Port Service as a ship traffic service which is limited to providing information regarding data relating to the needs and operations of ports and terminals that are not responsive to shipping traffic in related station coverage area.

The definition of Vessel Traffic Service or abbreviated as VTS according to the Minister of Transportation Regulation number PM 26 of 2011 concerning Telecommunications - Shipping explains the meaning of Vessel Traffic Service is ship traffic services in designated areas that are mutually integrated and carried out by the authorized party (Ministry of Transportation) and designed to improve ship safety, navigation efficiency and protect the environment, which has the ability to interact and respond to ship traffic development situations in the VTS area by means of radio and navigation electronic devices.

The explanation regarding the Directorate of Navigation is one of the Directorates under the auspices of the Directorate General of Sea Transportation with the main task of carrying out the formulation and implementation of policies, preparation of norms, standards, procedures and criteria, provision of technical guidance and supervision as well as evaluation and reporting in the field of navigation.

The Directorate of Sea and Coast Guard Unit is one of the Directorates under the auspices of the Directorate General of Sea Transportation with the main task of carrying out the formulation and implementation of policies, preparation of norms, standards, procedures and criteria, provision of technical guidance and supervision as well as evaluation and reporting in the field of patrols and security, enforcement law and advocacy, orderly shipping, disaster management and underwater work, facilities and infrastructure. The explanation regarding state ships regulated in Law number 17 of 2008 concerning Shipping explains the meaning of state ships are state-owned ships used by certain Government agencies that are given functions and authorities in accordance with statutory provisions to enforce the law and other Government tasks.

The Office of the Land Transportation Management Center or hereinafter abbreviated as the BPTD Office is the Technical Implementation Unit within the Ministry of Transportation which is under and responsible to the Director General of Land Transportation. BPTD which carries out the management of land transportation in areas with land characteristics that have road transportation services, as well as river, lake and commercial and pioneer crossing. Explanation regarding Terminals for Own Interests/ TUKS according to Minister of Transportation Regulation Number PM 51 of 2011 concerning Special Terminals and Terminals for Self Interests explains Terminals for Self Interests are terminals located within the Work Environment Area and Port Interest Environment Area which serve their own interests in accordance with main business.The definition of a Special Terminal/TERSUS according to the Regulation of the Minister of Transportation Number PM 51 of 2011 concerning

Special Terminals and Terminals for Own Interests explains that a Special Terminal is a Terminal that is located outside the Work Environment Area and Port Interest Environment Area which is part of the nearest Port to serve its own interests according to the main business.

The Port Authority and Port Authority Office is a Technical Implementation Unit under the auspices of the Ministry of Transportation with the task of carrying out supervision and law enforcement in the field of shipping safety and security, coordinating government activities at ports as well as regulating, controlling and supervising port activities at commercially operated ports.

3. Methodology

This type of research is Descriptive Quantitative Observational which will be carried out from 1 July 2020 to the end of 2022 in the Sunda Strait TSS. Descriptive research is research that seeks to answer existing problems based on data. The process of analysis in descriptive research is presenting, analyzing, and interpreting [7]. In descriptive quantitative observational research, descriptive research presents the frequency distribution of the data presented (known as univariate data analysis). This descriptive research is considered lower than analytic research, of course, there are times and times when a research needs to be done descriptively or analytically.

Case Reports and Case Series are part of a descriptive quantitative observational research design. The sample size is about 2,600 merchant ships that commit violations. In this study, systematic random sampling was used, which is a simple sampling technique that was carried out systematically at random. The sampling technique for systematic random sampling according to [8] is a sampling technique based on the sequence of population members who have been given serial numbers. The steps of this technique are: (1) Determine the target population, in this study the respondents that are the target population; (2) Determination of checkpoints; (3) Determine the time to be used to determine sampling; (4) Organizing the field, especially at checkpoints. This orientation will be used as the basis for determining the basic interval for the number of respondents. (5) Determine the sample size (n) to be surveyed. The type of data collected is primary data which is carried out by structured observation using data collected by VTS Merak.

4. Disussion

4.1. Factors Influencing Shipping On The TSS in The Sunda Strait

There are 4 (four) important things in the navigation and reporting system in the Sunda Strait TSS which affect the creation of sailing safety in the Sunda Strait TSS, there are all ships sailing in the TSS of both straits must comply with the provisions of rule 10 COLREGs 1972 related to the TSS, namely: (i) Sailing in the direction of an appropriate carriageway; (ii) Stay away from the separation line/zone; (iii) Entering/leaving the traffic at the end of the traffic; (iv) Avoid cutting traffic traffics; (v) Avoid anchoring; (vi) Avoid catching fish.

Then, communication between ships on voyages in the Sunda Strait is carried out through easy-to-understand and short conversations. Communication in the Sunda Strait TSS via VHF Radio on channel 22 or 68, with the nickname Merak VTS. All ships passing through the TSS must fully carry out listening watch duties.

The Reporting and Navigational Information System in the Sunda Strait is mandatory for all Indonesian-flagged ships passing, crossing or crossing the TSS in the alert area. As for foreign ships, it is highly recommended to participate in the Navigation Information and Reporting System

About report format, in the interests of safety of navigation and protection of the marine environment, it is highly recommended for all ships that wish to pass through the TSS in the Sunda Strait to provide information on vessel size, both in ballast and loaded conditions; the type of cargo, whether it is toxic and dangerous as defined in the relevant international conventions.

4.2. Analysis of Ship Trafic Discrepancies in The Sunda Strait

Since the implementation of the Sunda Strait TSS on July 1 2020 until December 2022 there have been no accidents or ship accidents in the Sunda Strait TSS. This cannot be separated from the role of VTS Merak in its task of monitoring and serving ship traffic in the Sunda Strait, especially in the TSS according to the type of VTS service on VTS Merak that has been implemented, namely Information Navigation Service (INS) and Navigational Assistance Service (NAS).

The role of the VTS is very vital in the implementation of the Navigation Reporting and Information System, considering that ships will communicate with the VTS, related to the ship reporting function, as well as related to the INS (Information Navigation Service) and NAS (Navigational Assistance Service) services, where all ships sailing in both straits is recommended to use the information broadcast by VTS Merak.

The number of ship traffic crossing the Sunda Strait for the period July 2020 to November 2022 was 52,484 (this number does not include roro ferries crossing from the Merak ferry port to Bakauheni and vice versa)

Table 1. Table of Ship Traffic in The Sunda Strait

	Jan	Feb	Mar	Apr	May	June	July	Aug	Sept	Okt	Nov	Dec
2020							1598	1711	1708	1731	1616	1563
2021	1492	1261	1577	1600	1804	1987	2019	1964	1831	1910	1851	1942
2022	2050	1754	1958	1961	1846	1928	1860	2067	1880	2071	1948	

In Table 1. which displays Ship Traffic in the Sunda Strait, it can be seen that from year to year, from January 2020 to December 2022, the trend of ship traffic in the Sunda Strait is always increasing. The highest increase occurred in October 2022 where it touched 2071 ship traffic that occurred.

In the same timeframe since the enactment of the Sunda Strait TSS, not all ships have sailed according to the applicable provisions according to PM 130 of 2022 concerning procedures for traffic in the Sunda Strait TSS. With a percentage of 4.5% (2,401 ships) that violated it and 95.5% followed the rules that had been determined.

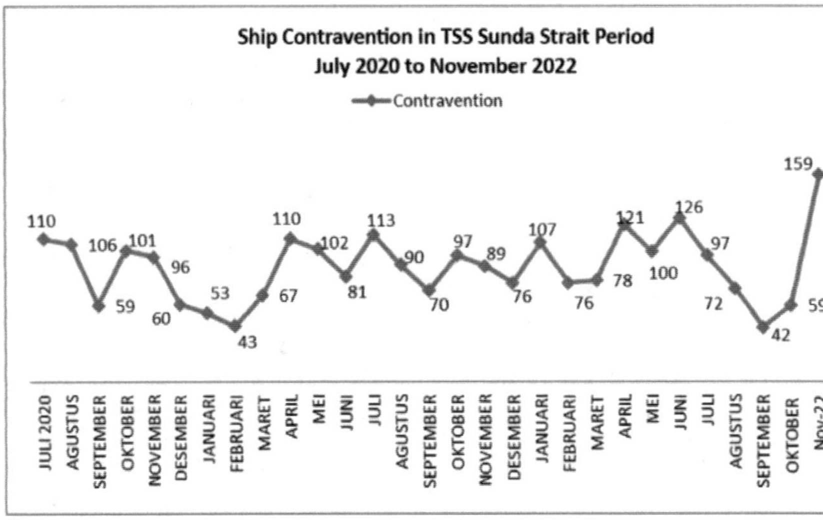

Figure 2. Chart of Ship Contravention in The Sunda Strait TSS

From Figure 2, which displays the trend of ship violations (contravention) in the Sunda Strait TSS from July 2020 to November 2022, it is noted that the smallest number of violations was in September

2022 with 42 violations, but there was a spike in violations in November 2022 of 159 violations, this figure reached 3 times the number of violations (contravention) in the previous month.

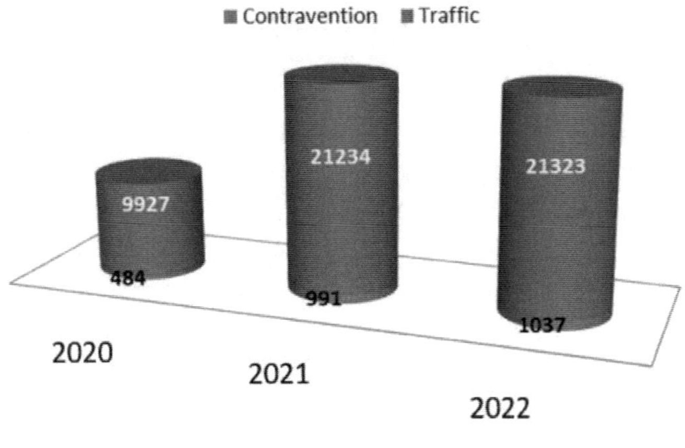

Figure 3. Graph of TSS Traffic and Violation Comparison

From Figure 3. which shows a comparison of the amount of traffic with the number of violations on the TSS, it can be seen that the least quantity of violations was in 2020, in the number 9927, while in 2021 it experienced a significant increase of 21234 and in 2022 it has increased again with a number of achievements of 21323.

Table 2. Table of Comparison of Traffic and TSS Violations

Year		Jan	Feb	Mar	Apr	May	June	July	Aug	Sept	Okt	Nov	Dec
2020	aoc							1598	1711	1708	1731	1616	1563
	vio							110	106	59	96	60	53
2021	aoc	1492	1261	1577	1600	1804	1987	2019	1964	1831	1910	1851	1942
	vio	53	43	67	110	102	81	113	90	70	97	89	76
2022	aoc	2050	1754	1958	1961	1846	1928	1860	2067	1880	2071	1948	
	vio	107	76	78	121	100	126	97	72	42	59	159	

**aoc: amount of*
vio: volation

From the Table 2. above it can be seen from Table 2. that in 2020 from January to November there was a decrease in the number of violations. As well as in 2021 and 2022 from January to September there was a tendency for violations to decrease, but the number of violations in November 2022 increased sharply. Graph of Increased Violations in the Sunda Strait Traffic Separation Scheme (TSS) Since July 1, 2020 Implementation

Figure 4. Graph of The Number of Offending Vessels

Data on the number of ships recorded is the data of ships monitored on VTS Merak, Ships recorded as vessels that commit violations (Contravention) are ships that do not report and are called via VHF Radio communication CH 22 but do not respond to VTS Merak calls, some ships can respond summons and was advised by the VTS Operator to follow traffic procedures in the Sunda Strait TSS but did not heed the advice given. Types of violations in the TSS

Every ship that sails through the TSS must follow the rules of the Collision Regulation, especially rule 10 rules in the TSS, including:

a) Sailing in the direction of an Appropriate traffic flow;
b) Stay away from traffic separation lines/zones;
c) Entering/leaving the traffic at the end of the traffic;
d) Avoid cutting traffic traffics;
e) Avoid anchoring;
f) Avoid catching fish;
g) Vessels of less than 20 m in length or sailing vessels shall not obstruct the passage of the TSS;
h) Prohibition of ships to overtake other ships at a certain LOA size in accordance with the provisions of the route system;
i) The ship is anchored in an unspecified area;
j) Dispose of garbage, waste and other materials from the operation of the ship.

VTS Merak recorded 3 categories of types of violations that were often committed by ships, namely:

a) The ship is sailing not in the direction of traffic / the ship is on the wrong track (false track);
b) The ship crosses the traffic traffics (Crossing Separation Zone/ line);
c) Ships whose LOA exceeds 20 M enter the Inshore Traffic (ITZ) area

IOP Conf. Series: Earth and Environmental Science 1294 (2024) 012028 doi:10.1088/1755-1315/1294/1/012028

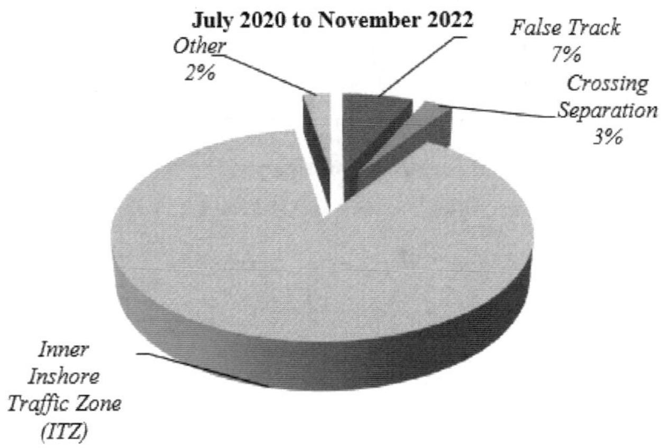

Figure 5. Graph of violation type

From the Figure 5. above it is explained that 3 violations were occupied by False Track with percentage of 7%, crossing separation with percentage of 3%, other with percentage of 2%, and the biggest influence is occupied by Inner Inshore Traffict Zone (ITZ) with 88%.

Table 3. Table of violation type

False Track	175
Crossing Separation	70
Inner Inshore Traffic Zone (ITZ)	2276
Other	66

From the types of violations committed by ships, the most were violations by ships entering the Inshore Traffic Zone (ITZ) with a percentage of 88%, then types of violations by ships on the wrong track 7% and ships that cut the separation zone and separation line 3%, the remaining 2 % is a ship that violates other rules in TSS such as anchoring, drifting and others.The large number of vessels that commit violations by entering the Inshore Traffic Zone, with a total of 2,276 violations committed by the same ship or by different ships. It was recorded that several ships committed the same violation more than once for the same type of violation. VTS Merak also records the type of ship that often commits violations, dominated by tugs with barges. Types of ships that often carry out can be seen can be seen in the graphic data below:

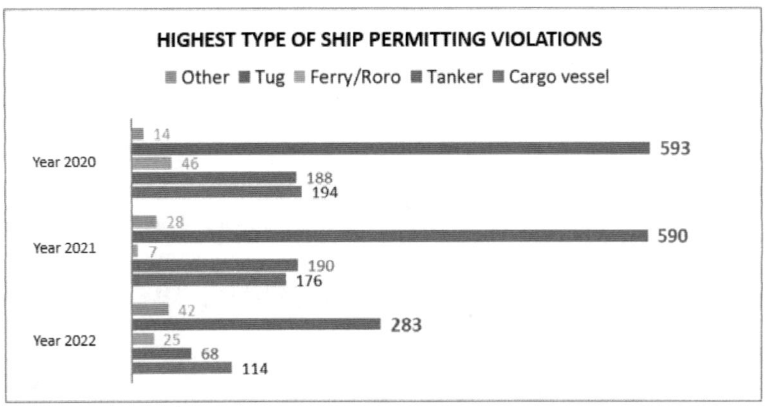

Figure 6. Graph of The highest violation ship types

VTS Merak also records foreign or local ship categories that do not comply with regulations in the TSS, the data can be seen in the graph below:

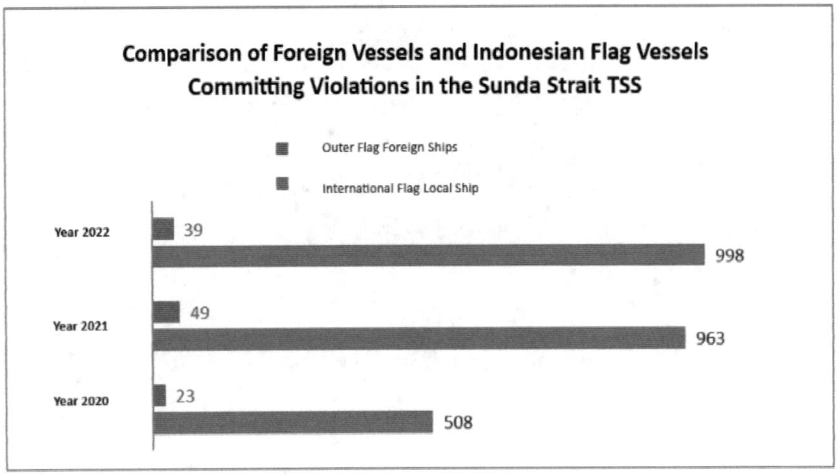

Figure 7. Comparison chart of foreign and Indonesian Violater Vessels

From Figure 7. above, it can be seen that from 2020 to 2022 there has been an increase in violations in the Sunda Strait TSS, where the most significant of these occurred in the vulnerable years 2020 to 2021.VTS Merak also recorded data on the ports of destination and origin of ships that committed violations in the TSS. Panjang Port is the port of origin and destination for Tug ships which often commit violations by entering the Inshore Traffic Zone.

In accordance with the VTS Merak SOP and existing regulations, if a ship in traffic commits a violation, such as sailing in the dividing channel or does not turn on AIS, the VTS operator will inform the Tanjung Priok PLP Base patrol boat via VHF Radio CH 67 for follow-up.

4.3. Findings Research with The Existing Literature on Violations in Ferries

VTS Merak has made a major contribution to the Sunda Strait region in increasing the level of safety by reducing collisions and grounding accidents, more effective time with a fast response in the event of a fire or explosion, and reducing fatalities such as people at sea. or death. at sea.

In line with research conducted [16] where the results obtained from the basic study show that there is still room for improvement considering the still high number of violations. Technical and technological developments in the field of shore-based monitoring, including e-Navigation and rapid digitalization, as well as better traffic management, are expected to soon enable or even require more proactive traffic control in terms of organization and coordination of vessels. traffic. Supported by research conducted by [17] that the implementation of the Traffic Separation Scheme in the Sunda Strait will guarantee the safety of ships sailing both on the Indonesian Archipelagic Sea Route 1 and the crossing route between Marak Harbor and Bakauheni Harbor, as well as the occurrence of collisions between vessels. ship can be avoided. Furthermore, the government urges the implementation of the Traffic Separation Scheme in the Sunda Strait, including a traffic monitoring system, as a solution to traffic congestion on the Merak and Bakauheni crossings and its intersection with the Indonesian Archipelago Sea Route 1, in order to maintain traffic safety in the strait.Therefore, operational procedures need to be further developed, and operators on land must be sufficiently trained to be ready to carry out their complex tasks to ensure the safety and efficiency of maritime transport.

The presence of VTS can provide certain information to ships and prevent unwanted physical contact with navigational landmarks, such as offshore objects or objects at sea. Like the Merak VTS, the existence of the Warnemünde VTS Center is very much in line with the VTS function described by the IMO. For example, there is another VTS in Germany that reports 1,000 deliveries to VTS in a month. Fire or explosion accidents occurred one and eight years ago, and two cases of grounding were reported. Recently, almost no collisions have occurred in the water area. There is a strong relationship between the technology used on the VTS and the ease of operation by personnel to ensure navigation safety. Technology that can help avoid undesirable situations such as wider ground radar coverage to reach long-range ships as well as in bad weather and emergency conditions, more Navigational Aids as navigation markers. Following specific area requirements and advanced VTS software systems to provide alarm or warning management. However, 24-hour internet connection access, CCTV along the water and complete communication equipment are also needed to support and facilitate VTS services.

5. Conclusion

There are 4 (four) main things which are the most significant influencing factors in traffic safety in the Sunda Strait TSS, consist of all ships sailing in the TSS of both straits must comply with the provisions of rule 10 COLREGs 1972 related to the TSS, Communication between ships on voyages in the Sunda Strait is carried out through easy-to-understand and short conversations, the Reporting and Navigational Information System in the Sunda Strait is mandatory for all Indonesian-flagged ships passing, crossing or crossing the TSS in the alert area. and last about report format, when in the course of implementing the TSS in the Sunda Strait, it has been observed that there has been an evaluation of its implementation because violations have been found. Even though it can be said that the level of violations that occur is quite fluctuating every year and with a variety of different causes, but the number cannot be said to be small, this clearly represents that there is still a large enough gap in the possibility of violations so that further efforts are needed to be able to suppress the occurrence of violations in the Sunda Strait TSS in particular.

References

[1] IALA. 2012. IALA Vessel Traffic Manual 2012 (5 ed.): International Association of Marine Aids to Navigation and Lighthouse Authorities.

[2] IMO Circullar number COLREG.2/Circ.74 dated 14 Juni 2019 about New Traffic Separation Scheme and number SN.1/Circ.337 Routeing Measures Other Than Traffic Separation Schemes IMO

[3] Seignette, R. 2012. Vessel traffic management in the Port of Rotterdam. Port Technology International, 105- 107

[4] Porathe, T., de Vries, L., & Prison, J. 2013. *Ship voyage plan coordination in the MONALISA project: user tests of a prototype ship traffic management system*. Paper presented at the Human Factors and Ergonomics Society Europe Chapter 2013 Annual Conference, Torino

[5] Weick, K. E., & Sutcliffe, K. M. 2007. Managing the Unexpected: Resilient Performance in an Age of Uncertainty. San Francisco: Jossey-Bass.

[6] Hollnagel, E. 2011. Prologue: The Scope of Resilience Engineering. In E. Hollnagel, J. Pariès, D. Woods & J. Wreathall (Eds.), Resilience Engineering in Practice. A Guidebook. Farnham, Surrey, UK.: Ashgate Publishing.

[7] Achmadi, A., dan Narbuko. 2015. Metodologi Penelitian. Jakarta: Bumi Aksara

[8] Sugiyono, 2012. Memahami Penelitian Kuantitatif. Bandung: Alfabeta

[9] Undang-Undang Nomor 17 Tahun 1985 tentang Pengesahan United Nations Convention on the Law of the Sea (Konvensi Perserikatan Bangsa-Bangsa Tentang Hukum Laut) (Lembaran Negara Republik Indonesia Tahun 1985 Nomor 76);

[10] Undang-Undang Nomor 6 Tahun 1996 tentang Perairan Indonesia (Lembaran Negara Republik Indonesia Tahun 1960 Nomor 22, Tambahan Lembaraan Negara Republik Indonesia Nomor 1942);
[11] Peraturan Pemerintah Nomor 36 Tahun 2002 tentang Hak dan Kewajiban Kapal Asing Dalam Melaksanakan Lintas Damai Melalui Perairan Indonesia (Lembaran Negara Republik Indonesia Tahun 2002 Nomor 70, Tambahan Lembaran Negara Republik Indonesia Nomor 4209);
[12] Peraturan Pemerintah Nomor 37 Tahun 2002 tentang Hak dan Kewajiban Kapal dan Pesawat Udara Asing Dalam Melaksanakan Hak Lintas Alur Laut Kepulauan Melalui Alur Laut Kepulauan Yang Ditetapkan (Lembaran Negara Republik Indonesia Tahun 2002 Nomor 71, Tambahan Lembaran Negara Republik Indonesia Nomor 4210);
[13] Peraturan Pemerintah Nomor 20 Tahun 2010 tentang Angkutan di Perairan (Lembaran Negara Republik Indonesia Tahun 2010 Nomor 26, Tambahan Lembaran Negara Republik Indonesia Nomor 5108) sebagaimana telah diubah dengan Peraturan Pemerintah Nomor 22 Tahun 2011 tentang Perubahan atas Peraturan Pemerintah Nomor 20 tahun 2010 tentang Angkutan di Perairan (Lembaran Negara Republik Indonesia Tahun 2011 Nomor 43, Tambahan Lembaran Negara Republik Indonesia Nomor 5208);
[14] Peraturan Pemerintah Nomor 21 Tahun 2010 tentang Perlindungan Lingkungan Maritim (Lembaran Negara Republik Indonesia Tahun 2010 Nomor 27, Tambahan Lembaran Negara Republik Indonesia Nomor 5109);
[15] Peraturan Pemerintah Nomor 5 Tahun 2010 tentang Kenavigasian, Menteri Perhubungan wajib menetapkan alur-pelayaran, sistem rute, tata cara berlalu lintas, dan daerah labuh kapal sesuai dengan kepentingannya;
[16] Baldauf M, Fischer S, Kitada M, Mehdi R A, Al-Quhali M A and Fiorini M 2019 Merging conventionally navigating ships and MASS – merging VTS, FOC, and SCC? TransNav 13 pp 495-501
[17] Sunaryo S, Priadi A A and Tjahjono T 2015 Implementation of traffic separation scheme for preventing accidents on the Sunda Strait International Journal of Technology 6 6

IOP Conf. Series: Earth and Environmental Science 1294 (2024) 012029 doi:10.1088/1755-1315/1294/1/012029

Performance of asphaltic concreate binder course (AC-BC) using gilsonite modified asphalt

Hanafi [1], **F Rusgiyarto** [1], **A Juhara** [1]

[1] Lecturer, Departement Of Civil Engineering, Jenderal Achmad Yani University, Indonesia

hanafi@lecture.unjani.ac.id

Abstract. The road pavement structure is the most important part of the design and manufacture of highways, so there is a need for innovation in designing pavement to get a type of pavement that has good quality and is efficient in its design. At this time, many road pavements have suffered damage caused by poor quality. Gilsonite is a natural asphalt that has a relatively high asphalt content compared to other additives, which is 70.9%. In addition, Gilsonite has a maltene content of 27% and 2% oil, then has a nitrogen content of 3.2%. After that, to carry out this test, a test object was made with the addition of Gilsonite Resin using 3 variations of the addition, namely 6%, 7%, and 8%, and compared to the mixture without Gilsonite. The mixture using the addition of Gilsonite (6%, 7%, 8%) an increase in the stability value of the mixture using the addition of Gilsonite, but the flow becomes increased so the properties of the mixture have large deformations. From all the tests that have been carried out, the optimum Gilsonite content is 8%

1. Introduction

Bitumen is defined as a tarlike mixture of hydrocarbons derived from petroleum naturally or by distillation and used for road surfacing and roofing [1] . Asphalt is a solid or semi-solid material that is black to dark brown has a cementious nature and will soften and melt when heated. According to Sukiman, asphalt is thermoplastic, that is, it melts when heated and re-freezes when the temperature drops. This property of asphalt is used in the road construction process and is known as hot mix asphalt.

Current conditions indicate an increasing traffic load, so the quality improvement of pavement materials must be able to meet these conditions. To improve pavement quality, pavement engineers often add additives to bitumen to improve asphalt properties and use alternative materials (substitutes) in asphalt mixtures [2]. Based on where it is obtained, asphalt is divided into natural asphalt and oil asphalt. Natural asphalt is asphalt that is found in nature and can be used as it is obtained or with a little processing, while oil asphalt is asphalt which is the residue of petroleum refining. Asphalt produced from the crude oil refinery industry is known as residual bitumen, which is produced from crude oil through a distillation process.

Asphalt additive is a material used to be added to the asphalt. The use of asphalt additives ranges from types of synthetic rubber and also from rubber that has been processed (for example: used tires), and also from plastic materials. The use of rubber additives can change the characteristics of asphalt. Research on rubber shows that the increasing levels of natural rubber sap, the harder the asphalt and not sensitive to temperature. This is caused by the content of natural rubber latex which is non-bituminous and elastic[3]

Gilsonite has a high asphaltene content (70.9%) when compared to other additive materials, in addition to other mineral content, namely 27% maltene, 2% oil, and 3.2% nitrogen, it is expected that this material can improve aggregate adhesion and water stripping [4]. Research conducted on the addition of gilsonite to 60/70 penetration bitumen turned out to be able to change the properties of asphalt, namely penetration, softening point flash point, and burning point so that asphalt with gilsonite-added material becomes hard [5]

Asphalt mixture testing with variations of Gilsonite additives was carried out. the results of testing the use of gilsonite in asphalt mixtures can increase the stability of the mixture with an asphalt content of 5.5% with an optimum content of 6.8% Gilsonite [6]. Tests on the ac-wc mixture obtained a gilsonite content of 8% which in the conditions of the Marshall test parameters were met [7]. Research on the use of Gilsonite in other types of mixtures was also carried out, one of which was porous asphalt. This mixture is an asphalt mixture with a uniform gradation type so this mixture forms larger cavities compared to other types. The addition of gilsonite to the modified porous asphalt mixture lowers the permeability coefficient where the low permeability value is caused by the small cavity in the mix (VIM) value. [8]

Based on the issue above, it is necessary to develop research into the latest gradation mixtures, one of which is AC-BC. This layer is part of the surface layer between the top bearing (base course) and the wear layer (wearing course) which aggregate grades continuously, generally used for roads with heavy traffic loads [9]. By analyzing the Marshall characteristics of the Gilsonite-modified AC-BC mixture, and reviewing the durability of the AC-BC mixture using Marshall immersion.

2. Material and Research Methodology

Gilsonite Resin is a hydrocarbon mineral with a high content of resin, asphalten, and nitrogen, and mixes easily with asphalt. Gilsonite Resin is a kind of mineral bitumen that resembles asphalt known as asphaltites. Gilsonite Resin when mixed with aggregate and asphalt mixture is expected to increase the stability of flexible pavement.

The content of Gilsonite Resin has 70.9% higher asphaltene compared to other additives as well. contains 27% maltene and 2% oil. The materials used in this study were asphalt pen 60/70, Gilsonite resin, and coarse and fine aggregates.

Mixing Gilsonite resin with asphalt pen 60/70 and testing the properties of asphalt. In testing 3 variations of the use of Gilsonite Resin, namely 6%, 7%, and 8% by weight of asphalt. Tests for these mixes were carried out by the standard tests for hot mix asphalt. The design of the hot mix pavement uses the intermediate layer Laston gradation (AC-BC). To obtain the optimum asphalt content (OAC) value using the Marshall method to measure stability, melting, and cavities in the mixture and test residual strength using the Marshall immersion method. After the Marshall test, an analysis was carried out to determine the optimal gilsonite content. The research flow process can be seen in Figure 1.

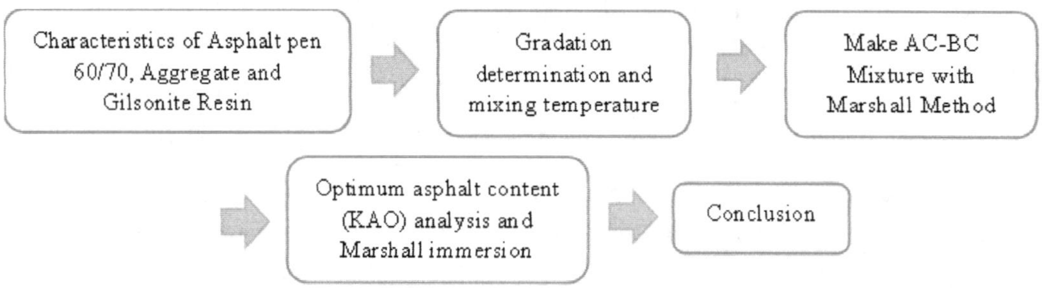

Figure 1 Research Methodology Mixture of Asphalt Modified Gilsonite

3. Result and Discussion

3.1. Gilsonite Modified Material

Analysis of pen 60/70 asphalt modified with Gilsonite Resin added to the penetration test of 6%, 7%, and 8% Gilsonite asphalt obtained penetration values of 47.5, 46.5, and 42.5. This means that more Gilsonite additions make asphalt brittle. on testing the flash and burning point of asphalt added to Gilsonite resin 6%, 7%, and 8%, high flash points, and burning points were obtained, approximately, 305°C to 335°C. where Gilsonite resin-modified bitumen has non-flammable properties because the temperature obtained to get the flash point and burning point is very high

Testing the softening point modified bitumen with Gilsonite resin added 6%, 7%, and 8% obtained values of 50.2°C, 52.6°C, and 54.3°C, where it makes the asphalt properties harder, meaning the higher the temperature, the lower the penetration because it requires a high temperature to reach its softening point. Testing the ductility of modified Gilsonite asphalt 6%, 7%, and 8%, the results were 697.9, 597.9, and 497.9 cm. From the results obtained the ductility of asphalt added to Gilsonite becomes inelastic or brittle, meaning that Gilsonite has the property of hardening asphalt.

Table 1 Characteristics of Gilsonite Resin Modified Bitumen

No	Type of Test	Test Result			Unit
		6%	7%	8%	
1	Penetrate	47.5	46.5	42.5	0.1 mm
2	Flash Point	305	315	335	°C
3	Burning Point	315	330	340	°C
4	Softening Point	50.2	52.6	54.3	°C
5	Spesific Gravity	1.04	1.03	1.03	gr/cc
6	Ductility	697.9	597.9	497.9	cm
7	Viscosity on 135°C	687	704	719	cSt

Viscosity is an important component in mix planning because viscosity can show asphalt behavior due to temperature changes. To check the viscosity of asphalt added with Gilsonite additives, the viscosity of asphalt was obtained, namely, for Gilsonite additives of 6%, 7%, and 8%, the results were obtained for mixing temperatures of 161-167°C, 162-169°C, and 166-172°C . The solidification temperature is 149-154°C, 150-155°C, and 153-159°C. From the results obtained, the mixing temperature is greater than the compaction temperature, meaning that the mixing temperature is higher because the asphalt will unite with the aggregate at high temperatures. The effect of Gilsonite on the viscosity test is that the higher the Gilsonite content, the higher the temperature obtained to obtain the mixing temperature and compaction temperature.

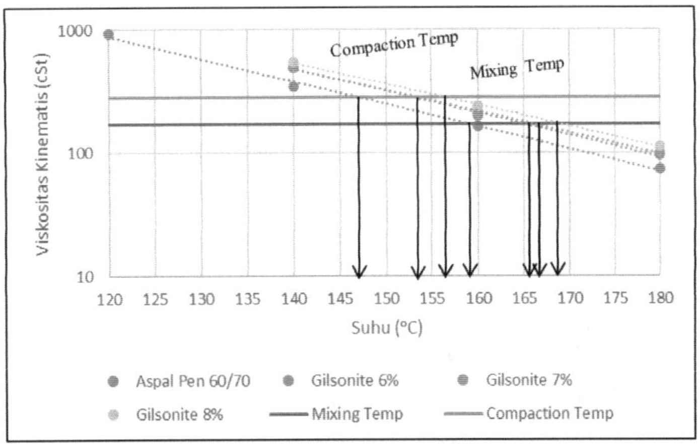

Figure 2 Viscosity Asphalt Modified Gilsonite

3.2. Testing of Mixed AC-BC Marshall Modified with Gilsonite

Testing the Marshall test object aimed to determine the value of voids in the mixture (VIM), voids filled with asphalt (VFB), voids in aggregate (VMA), density (density), stability, Flow, to determined the optimum asphalt content (OAC) value. This OAC value will later be used for the Marshall immersion test to obtain the residual stability index value

The addition of Gilsonite can cause the density value of the test object to decrease. This is due to the decrease in aggregates in the mixture as the asphalt content increases, the properties of the mixture cannot be compacted anymore, due to the occurrence of bonds between the aggregates with the asphalt content. For Gilsonite content of 6%, 7%, there is an increase in density value but at 8% Gilsonite content the density value is the same as that given by Gilsonite with 7% content.

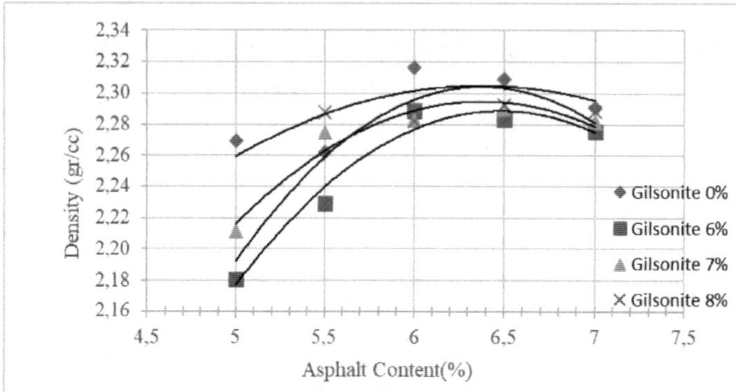

Figure 3 Density of the Modified Gilsonite Mixture

The addition of 6% Gilsonite content, the VMA value increases, this proves that increasing the percentage of Gilsonite content in the mixed composition will increase the void value in the aggregate. The addition of Gilsonite will cause the inner cavity to enlarge.

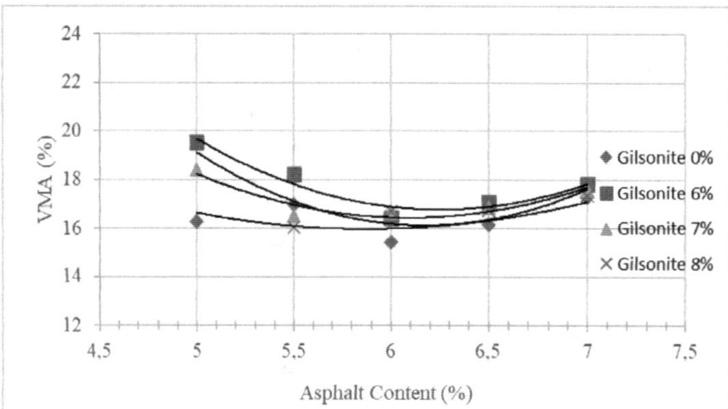

Figure 4 VMA of the Modified Gilsonite Mixture

Testing the VFB value for the asphalt mixture with the addition of 6%, 7%, and 8% Gilsonite decreased compared to the asphalt mixture without the addition of Gilsonite or 0% Gilsonite content, this is also because the voids between the aggregates are getting smaller. The lower the percentage of asphalt content in the mixture, the more surfaces and voids in the mixture that are not covered with asphalt so the binding between aggregate and asphalt is not optimal. For the Gilsonite content of 8%, a decrease will cause the mixture to be less impermeable to water and air because the asphalt layer will become thin and will crack easily.

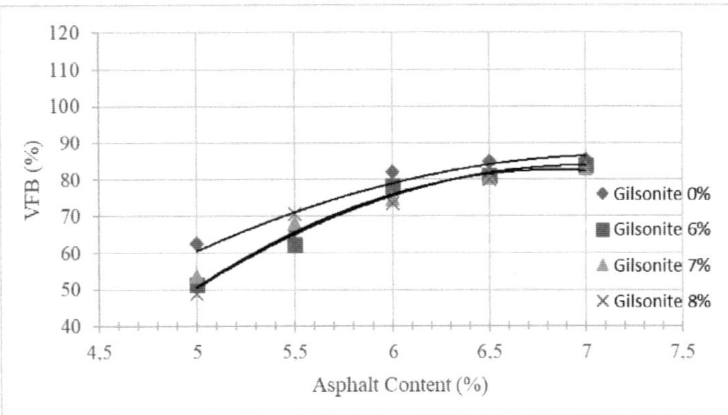

Figure 5 VFB of the Modified Gilsonite Mixture

For a mixture with 0% Gilsonite content, a value of VIM 3.82% was obtained, while for a value VIM with the addition of 6%, 7%, and 8% Gilsonite each had a value of 3.74%, 3.62%, and 3.96 %. Based on the results obtained, the highest number was obtained at the Gilsonite content of 8%, this means that the voids in the mixture are getting larger so that the mixture is porous or absorbs easily, which causes the mixture to become less dense so that water and air easily enter the cavities in the mixture which will cause asphalt easily oxidized, causing the adhesion between aggregate grains to decrease resulting in the release of granules and peeling off the surface of the pavement layer.

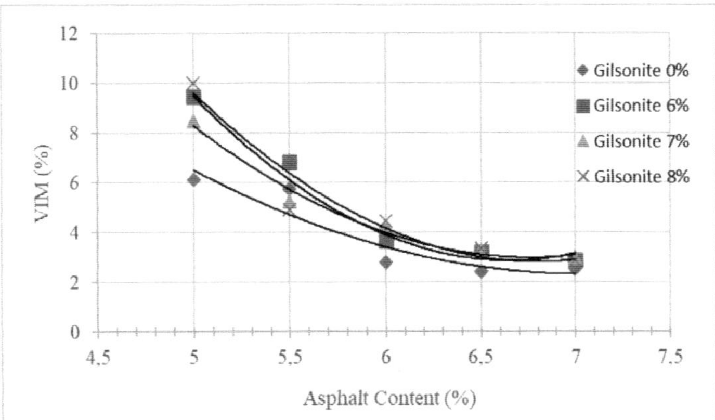

Figure 6 VIM of the Modified Gilsonite Mixture

Improving Gilsonite to the mixture with the addition of 6% and 7% will increase the stability value of the mixture compared to the mixture without the addition of Gilsonite. These results also conclude that the mixture with the addition of Gilsonite is more durable and strong in accepting large amounts of vehicle loads. the value obtained for a mixture with a Gilsonite content of 0% is 1357.10 kg, then for a mixture with a Gilsonite content of 6%, 7%, and 8% respectively has a value of 1314.58 kg, 1598.42 kg, and 1658.67 kg.

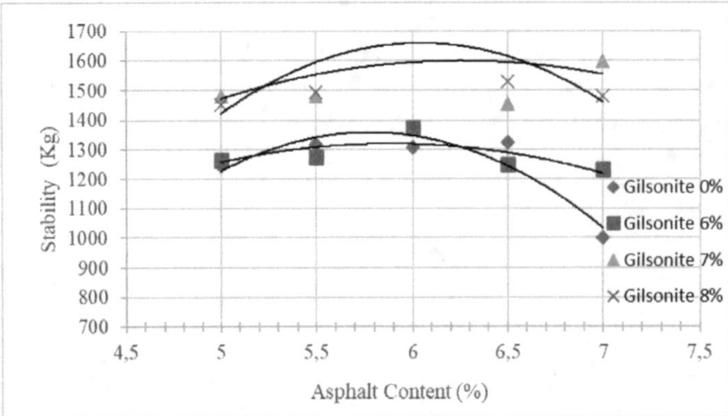

Figure 7 Stability of the Modified Gilsonite Mixture

Flow is affected by asphalt content, viscosity, aggregate gradation, as well as the amount and temperature of compaction of the mixture. The values obtained from the test were 3.8mm for mixtures with 0% Gilsonite content, then 3.32mm, 4.24mm, and 5.61mm for mixtures with 6%, 7%, and 8% Gilsonite content. This also shows that the addition of Gilsonite can increase Flow of the asphalt mixture so that it can cause large deformations when given a load.

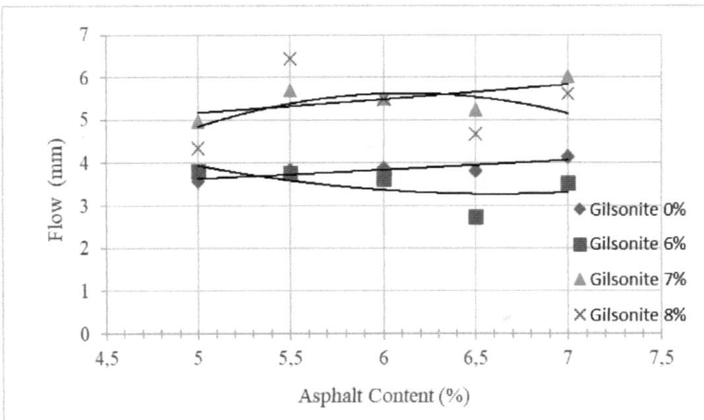

Figure 8 Flow of the Modified Gilsonite Mixture

The Optimum Asphalt Content (OAC) is 5.8% for asphalt without added Gilsonite, while the OAC value for mixture with the addition of Gilsonite Resin from 6%, 7%, and 8% content respectively is 6.1%, 6.2%, and 6.0%, from tests that have been carried out that the addition of Gilsonite can increase the Optimum Asphalt Content of the mixture

The Marshall immersion test aims to see the mixture resistance to temperature. The Marshall immersion test results for the mixture without the addition of Gilsonite obtained a value of 90.78%, while for the Marshall immersion value in the asphalt mixture with the addition of Gilsonite 6%, 7%, and 8% obtained a value of 91.51%, 93.08 %, and 93.44%. It can be concluded that the mixture with the addition of Gilsonite can increase the Marshall immersion value of the asphalt mixture when compared to the mixture without the addition of Gilsonite

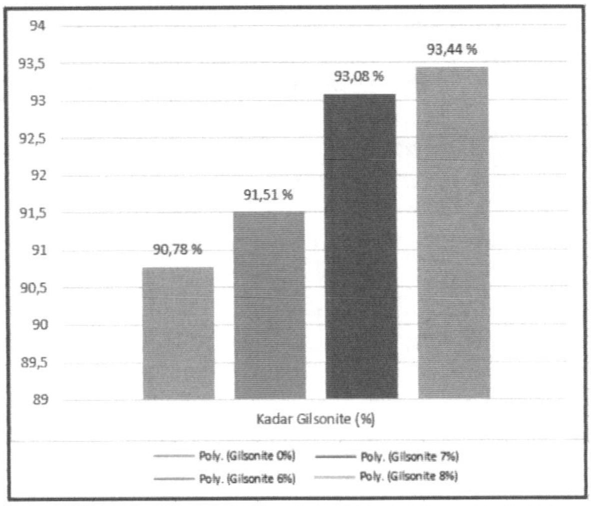

Figure 9 Immersion of the Modified Gilsonite Mixture

3.3. Gilsonite Content Analysis

The determination of the optimum Gilsonite content was carried out using 2 methods. The first was to review the properties of adding Gilsonite resin to asphalt. This analysis uses a relationship approach between the asphalt test results, namely Penetration, Ductility, and Softening Point.

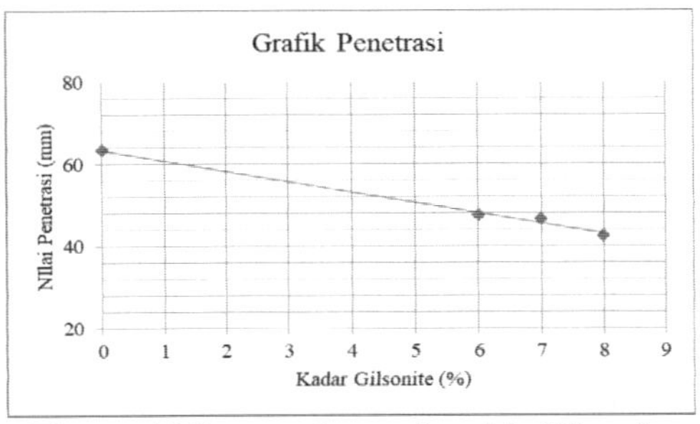

Figure 10 Penetration ratio with Gilsonite Content

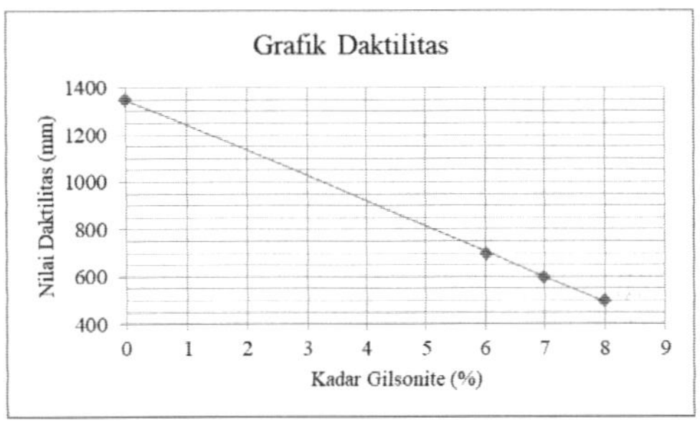

Figure 11 Ductility ratio with Gilsonite Content

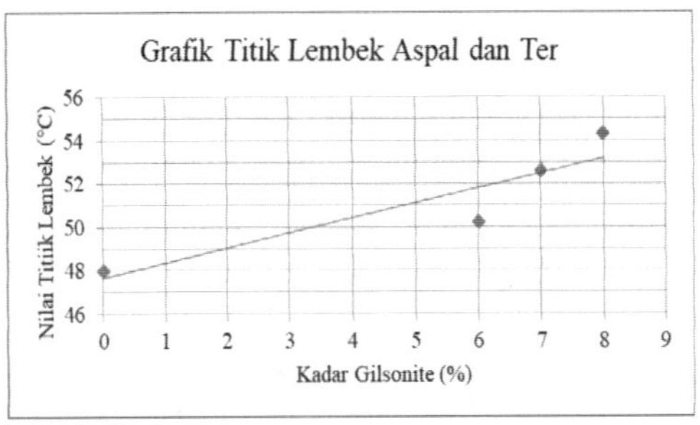

Figure 12 Softening Point with Gilsonite Content

From the analysis of adding Gilsonite to asphalt, it can be concluded that with increasing levels of Gilsonite added to asphalt, the penetration value and the ductility decreases, then the softening point value of asphalt increases. It can also be said that the addition of Gilsonite can make the properties of asphalt brittle or stiff when compared to asphalt without the addition of Gilsonite.

The other method is looking at the characteristics of the mixture added to Gilsonite and reviewing the Marshall test results. Analysis of the characteristics of the mixture of the added Gilsonite examined the stability and melting values of the mixture. Based on Table 2, the Gilsonite modified asphalt mixture uses 6% Optimum Asphalt Content and the results show an increase in the stability value of up to 22% and a melting point of 10%. Analysis of the addition of Gilsonite to asphalt pen 60/70 as well as in mixed testing, it can be said that adding Gilsonite to asphalt and to the mixture can increase the stiffness and strength values of the asphalt mixture itself, along with the many additions of Gilsonite, the stability and melting values also increase. But in this case, if the addition of Gilsonite eats too much it will behave unfavorably towards the asphalt, this can cause the pavement to crack or break easily when loaded. With the addition of Gilsonite the properties of the AC-BC mixture are able to withstand high traffic loads and have high strength. high on the pavement. However, with this addition, the flow value for adding Gilsonite tends to be high for 7% and 8% additions, so a good grade for this addition is 6%.

Table 2 Effect of asphalt mixture on gilsonite additives

Gilsonite Content %	Stability(kg)	Flow (mm)
0.0	1357,10	3.83
6.0	1314.58	3.36
7.0	1598.42	4.22
8.0	1658.67	5.61

4. Conclusion

Based on the analysis and discussion of the addition of Gilsonite to the Characteristics of Asphalt Concrete Intermediate Coating (AC - Binder Course) mixture, the following conclusions are obtained:

1. The use of Gilsonite resin on asphalt pen 60/70 makes asphalt properties harder, this can be seen from the decreased penetration value, increased softening point, and decreased ductility value, this is due to the mineral content in Gilsonite resin which affects the properties of asphalt pen 60/70.
2. The results of penetration testing of 6% modified Gilsonite resin asphalt obtained a value of 47.5 mm, the addition of 7% Gilsonite obtained a value of 46.5 mm, and the addition of 8% Gilsonite obtained a value of 42.5 mm. The softening point value with the addition of 6% Gilsonite resin was 50.2 °C, 7% Gilsonite resin was 52.6 °C, and 8% Gilsonite resin was 54.3 °C. The ductility value of the modified asphalt with the addition of 6% gilsonite resin was 697.9 mm, 7% was 597.9 mm, and 8% was 497.9 cm.
3. The Marshall test showed that the addition of Gilsonite to the AC-BC mixture increased the stability and melting characteristics of the AC-BC mixture. The Marshall stability value increased with the addition of 6% Gilsonite at 1314.58 kg, 7% Gilsonite at 1598.42 kg, and 8% Gilsonite at 1658.67 kg. The melting value of the 6% Gilsonite asphalt mixture was 3.32 mm, 7% Gilsonite asphalt was 4.24 mm, and 8% Gilsonite asphalt was 5.61 mm. Based on the results of the analysis, it can be concluded that adding Gilsonite to asphalt and the mixture can increase the stiffness and strength of the asphalt mixture itself, Along with the many additions of Gilsonite, the stability and melting values also increase. However, if you add too much Gilsonite, it will behave less well with the asphalt, this can cause the pavement to crack or break easily when loaded.
4. The characteristics of the intermediate layer mixture (AC-BC), for the provisions of the properties of the final mixture (AC) used can be said to be good, and suitable to be an intermediate layer (AC-BC) if you look at the quality that has been obtained in the mixed test. If you look at the provisions for the properties of the AC-BC mixture, the effect of Gilsonite on the AC-BC layer can be said to

be good. If you look at the table above, the AC-BC layer with the addition of Gilsonite. With the addition of Gilsonite the properties of the AC-BC mixture can withstand high traffic loads and have high strength on the pavement. However, with this addition, the flow value for adding Gilsonite tends to be high for 7% and 8% additions, so a good grade for this addition is 6%.

5. References

[1] John and D. Whiteoak, The Shell bitumen handbook. 2014.
[2] J. F. B. Mashuri, "Pemanfaatan Material Limbah Pada Campuran Beton Aspal Campuran Panas," Mektek, vol. 8, no. 3, pp. 204–212, 2011.
[3] B. W. Widianto and M. I. Faishal, "Perubahan Karakteristik Aspal Pen 60/70 dengan Substitusi Getah Karet Alam Pangkalan Balai, Sumatera Selatan," RekaRacana: Jurnal Teknil Sipil, vol. 6, no. 3, 2021, doi: 10.26760/rekaracana.v6i3.143.
[4] R. Basuki and M. Machsus, "Penambahan Gilsonite Resin Pada Aspal Prima 55 untuk Meningkatkan Kualitas Perkerasan Hot Mix," Jurnal Aplikasi Teknik Sipil, vol. 3, no. 1, 2007, doi: 10.12962/j12345678.v3i1.2564.
[5] C. Kamba, "Tinjauan Teknis Dan Ekonomis Gilsonite Sebagai Bahan Additive Campuran Aspal Beton," Researchgate.Net, no. August, 2020.
[6] Y. Putuhena, A. Wardaningrum, A. Elkarsa, L. Suryatenggara, and E. M. Sebayang, "PENGARUH PENAMBAHAN GILSONITE RESIN PADA CAMPURAN BETON ASPAL EFFECT OF ADDITION OF MIXED RESIN GILSONITE ASPHALT CONCRETE," 2016.
[7] S. D. Y. Putri and K. H. Putra, "MODIFIKASI CAMPURAN PERKERASAN LASTON AC – WC DENGAN PENAMBAHAN KADAR 8% GILSONITE RESIN," Prosiding Seminar Teknologi Perencanaan, Perancangan, Lingkungan dan Infrastruktur, vol. 1, no. 1, 2019.
[8] T. Oktaviani, "Koefisien Permeabilitas Campuran Aspal Porus Akibat Penambahan Gilsonite," Jurnal SENDI, vol. 02, no. 01, 2021.
[9] Ismadarni, Risman, and Muh. Kasan, "Karakteritik Beton Aspal Lapis Pengikat (Ac-Bc) Yang Menggunakan Bahan Pengisi Pengisi (Filler) Abu Sekam Padi," Majalah Ilmiah Mektek, no. 2, 2013.

Assessment of road pavement conditions using Surface Distress Index (SDI) and Pavement Condition Index (PCI) methods

Putu Hermawati and N P A S Putri
Corresponding Author : hermawati@pnb.ac.id
Civil Engineering Department, Bali State Polytechnic, Bukit Jimbaran, Bali, Indonesia

Abstract. Road pavement is an important element in transportation infrastructure that affects the comfort and safety of road users. It is important to carry out regular assessments of the condition of the road pavement. This study aims to evaluate the condition of the road pavement using the Surface Distress Index (SDI) and Pavement Condition Index (PCI) methods. It is collecting field data regarding the condition of the road pavement, including the type and extent of existing damage. This data is then analyzed using the formula that has been determined in the SDI and PCI methods. In the research that has been done, the average SDI value for flexible pavement for Jalan Gunung Agung is 18 (Good) and the average PCI value is 63.8 (Good). The technical measures that can be applied to the flexible pavement of Jalan Gunung Agung are P2 (Asphalt), P4 (Filling cracks), P5 (Filling holes), and P6 (Gravelling).
Keywords: Rating, Damage, Road, SDI, PCI.

1. Introduction

1.1 Research background

Tabanan Regency is one of several regencies in the Province of Bali. Located in the southern part of Bali Island, Tabanan Regency has an area of 839.33 km² which consists of mountainous areas and beaches. Geographically, Tabanan Regency is located between 1140 – 54' 52" east longitude and 80 14' 30" – 8030'07" south latitude [1]. The length of National Roads in Tabanan Regency is 65,381 km, Provincial Roads 130,780 km, and roads controlled by the city government 863,218 km, so the total length of roads in Tabanan Regency according to their status is 1,059,379 km[2]. Damage to the road will cause a lot of losses that will be felt by the user directly because it will hamper the comfort of use as well as various problems from road damage if repairs are not immediately carried out [3] . One of the roads in Tabanan Regency that has several damages is Jalan Gunung Agung. This road is located in Dajan Peken Village, Tabanan District. Based on government statistics, Jalan Gunung Agung is categorized as a provincial road [4]. This road is also a primary collector-type road with a length of 1.62 km. Jalan Gunung Agung is a 2/2 UD type road with an average road width of 7.358 meters and the pavement layer on this road is flexible pavement and paving blocks. Jalan Gunung Agung is usually passed by various types of vehicles ranging from light vehicles to heavy vehicles. Several types of heavy vehicles that transport various types of materials and materials also pass through this road [5]. Jalan Gunung Agung is seen to have damaged road pavement such as potholes, cracks, and various other types of damage. Based on this, this research aims to determine the condition of the road surface by conducting a visual observation survey on Jalan Gunung Agung. The methods that will be used are Surface Distress

Index (SDI) and Pavement Condition Index (PCI). This research was conducted to identify type damage percentage, and condition index of flexible pavement at STA 0+000 to STA 1+620 on Jalan Gunung Agung, Dajan Peken Village, Tabanan District, Tabanan Regency.

1.2 Research purposes
The purposes of this research are :
1. To find out the types of damage found in the flexible pavement of Jalan Gunung Agung;
2. To determine the condition of Jalan Gunung Agung using the SDI and PCI methods;
3. To find out the technical handling for the level of damage to the road pavement.

2. Research Methods
The first step in this research was collecting primary data, namely data on road pavement damage in the form of damage type, damage dimensions, and documentation through a visual survey using a road condition survey form. Next, the area of each road segment is measured, and divided into 100 meters. After collecting primary data, the author analyzed the data to obtain SDI and PCI values[6]. The author then analyzes solutions regarding the technical handling of the road. The last step, the author makes conclusions from this research. The research location plan can be seen in Figure 1.

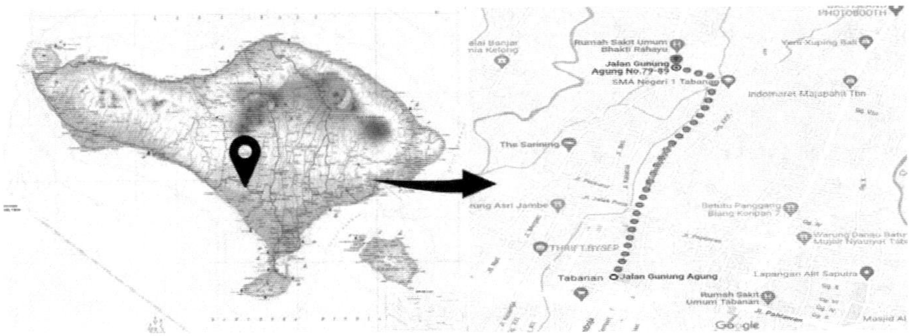

Figure 1. Research location plan.

3. Discussion

3.1 Road Geometric
The geometric of the Gunung Agung road STA 0+000 to STA 1+620 can be seen in Table 1.

Table 1. Geometric of the Gunung Agung road.

Segment	STA	Road shoulder width	Flexible pavement width	Road length	Road area
		(m)	(m)	(m)	(m^2)
I	(STA 0+170 s/d STA 0+270)	1.75	7.75	100	775
II	(STA 0+270 s/d STA 0+370)	1.65	7.65	100	765
III	(STA 0+370 s/d STA 0+470)	1.25	7.55	100	755
IV	(STA 0+470 s/d STA 0+570)	0.9	7.2	100	720
V	(STA 0+570 s/d STA 0+670)	0.97	6.77	100	677
VI	(STA 0+670 s/d STA 0+770)	1.4	6.9	100	690

VII	(STA 0+770 s/d STA 0+870)	1.1	6.7	100	670
VIII	(STA 0+870 s/d STA 0+970)	0.92	7.87	100	787
IX	(STA 0+970 s/d STA 1+070)	1.13	7.63	100	763
X	(STA 1+070 s/d STA 1+170)	1.16	7.96	100	796
XI	(STA 1+170 s/d STA 1+270)	0.97	6.67	100	667
XII	(STA 1+270 s/d STA 1+370)	0.7	6.65	100	665
XIII	(STA 1+370 s/d STA 1+470)	0.95	7.75	100	775
XIV	(STA 1+470 s/d STA 1+570)	0.95	6.85	100	685
XV	(STA 1+570 s/d STA 1+620)	2.97	8.47	50	423.5
	Total			1450	10613.5

3.2 Damage Type

The types of damage found in the flexible pavement of the Gunung Agung road STA 0+000 to STA 1+620 can be seen in Table 2.

Table 2. Types of damage in the flexible pavement of Gunung Agung road.

No	**Type Damage**	**Total area (m^2)**
1	Rutting	7.85
2	Swell	56,432
3	Potholes	0.32
4	Ravelling	71,001
5	Longitudinal Cracking	81.91
6	Alligator Cracking	415.02
7	Patches	58.3695
	Total	690.9025

3.3 Evaluation based on the Surface Distress Index (SDI) method

The steps for calculating road pavement damage using the SDI method are:

1. Count SDia based on the total of cracks area per segment
2. Count SDb based on the width of the average cracks per segment
3. Count SDic based on the number of holes per km
4. Count SDID based on the average rutting depth [7]

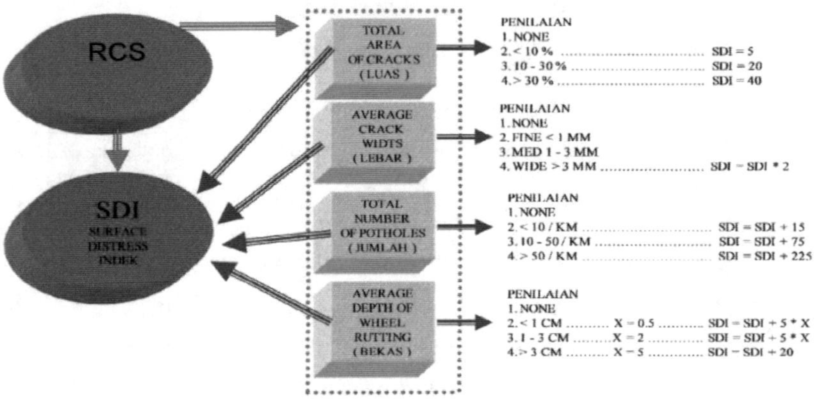

Figure 2. Road damage assessment using the SDI method.

The recapitulation of the Gunung Agung Road assessment index based on the SDI method can be seen in Table 3.

Table 3. Recapitulation of the Gunung Agung Road assessment index based on the SDI method.

Segment	STA	Crack Area (%)	Crack Width (mm)	Amount Hole (/ km)	Rutting Depth (cm)
I	(STA 0+170 to STA 0+270)	2	1-3mm	0	1.5
II	(STA 0+270 to STA 0+370)	0.215	1-3mm	0	0
III	(STA 0+370 to STA 0+470)	0.432	>3mm	1	0
IV	(STA 0+470 to STA 0+570)	2,426	>3mm	1	2.2
V	(STA 0+570 to STA 0+670)	14,206	>3mm	0	0
VI	(STA 0+670 to STA 0+770)	6028	>3mm	0	0
VII	(STA 0+770 to STA 0+870)	2,831	>3mm	0	0
VIII	(STA 0+870 to STA 0+970)	2,160	>3mm	0	2
IX	(STA 0+970 to STA 1+070)	0.406	>3mm	0	0
X	(STA 1+070 to STA 1+170)	2.8517	>3mm	0	0
XI	(STA 1+170 to STA 1+270)	9,520	>3mm	0	0
XII	(STA 1+270 to STA 1+370)	20,496	>3mm	0	0
XIII	(STA 1+370 to STA 1+470)	0.421	>3mm	0	0
XIV	(STA 1+470 to STA 1+570)	5,567	>3mm	0	0
XV	(STA 1+570 to STA 1+620)	2	>3mm	0	1.3

Surface Distress Index (SDI) method per segment by taking a sample unit segment I Jalan Gunung Agung STA 0+170 to STA 270.

Calculation value of SDI for STA 0+170 to STA 0+270

1. Calculation of the total area of the cracks
   total area crack in the segment I = 15.5 m2
   total area road in the segment I = 775 m2
   % Crack area = (Total area cracks / Total area path) x 100%
   So the crack area percentage in the segment I
   = (15.5/775) x 100%
   =2%
   Because the crack area is <10% of the road area, then the SDia value is 5
2. Calculation of the average width cracks
   From the survey results, the average width crack in segment I is 1-3 mm, then the SDIb = SDia = 5
3. Calculation for the number of holes per km
   Number of holes in the segment I = 0
   Because there are no holes in this segment, then SDIc = SDIb = 5
4. Calculation of the average depth of rutting
   Average rutting depth segment I = 1-3 cm
   Due to the average depth of rutting are 1-3 cm, then SDID = SDIc + 5 x X . The value of X is 2, then SDId value = 5 + 5 x 2 = 15.
   From the calculation above, the SDI value of segment I is 15 (Good)

The SDI value of Jalan Gunung Agung can be seen in Table 4.

Table 4. SDI assessment for Jalan Gunung Agung.

Segment	STA	SDI value	SDIb value	SDIc value	SDId value
		Crack Area	Crack Width	Amount of Hole	Rutting Depth
I	(STA 0+170 to STA 0+270)	5	5	5	15
II	(STA 0+270 to STA 0+370)	5	5	5	5
III	(STA 0+370 to STA 0+470)	5	10	25	25
IV	(STA 0+470 to STA 0+570)	5	10	25	35
V	(STA 0+570 to STA 0+670)	20	40	40	40
VI	(STA 0+670 to STA 0+770)	5	10	10	10
VII	(STA 0+770 to STA 0+870)	5	10	10	10
VIII	(STA 0+870 to STA 0+970)	5	10	10	10
IX	(STA 0+970 to STA 1+070)	5	10	10	20
X	(STA 1+070 to STA 1+170)	5	10	10	10
XI	(STA 1+170 to STA 1+270)	5	10	10	10
XII	(STA 1+270 to STA 1+370)	20	40	40	40
XIII	(STA 1+370 to STA 1+470)	5	10	10	10
XIV	(STA 1+470 to STA 1+570)	5	10	10	10
XV	(STA 1+570 to STA 1+620)	5	10	10	20

Recapitulation of the conditions and types of handling of the Jalan Gunung Agung segment is shown in Table 5.

Table 5. Recapitulation of SDI value in Jalan Gunung Agung.

Segment	STA	SDI value	Road Conditions	Type Handling
I	(STA 0+170 to STA 0+270)	15	Good	Routine Maintenance
II	(STA 0+270 to STA 0+370)	5	Good	Routine Maintenance
III	(STA 0+370 to STA 0+470)	25	Good	Routine Maintenance
IV	(STA 0+470 to STA 0+570)	35	Good	Routine Maintenance
V	(STA 0+570 to STA 0+670)	40	Good	Routine Maintenance
VI	(STA 0+670 to STA 0+770)	10	Good	Routine Maintenance
VII	(STA 0+770 to STA 0+870)	10	Good	Routine Maintenance
VIII	(STA 0+870 to STA 0+970)	10	Good	Routine Maintenance
IX	(STA 0+970 to STA 1+070)	20	Good	Routine Maintenance
X	(STA 1+070 to STA 1+170)	10	Good	Routine Maintenance
XI	(STA 1+170 to STA 1+270)	10	Good	Routine Maintenance
XII	(STA 1+270 to STA 1+370)	40	Good	Routine Maintenance
XIII	(STA 1+370 to STA 1+470)	10	Good	Routine Maintenance
XIV	(STA 1+470 to STA 1+570)	10	Good	Routine Maintenance
XV	(STA 1+570 to STA 1+620)	20	Good	Routine Maintenance

From the results analysis condition pavement segments I-XV Jalan Gunung Agung with the use method *Surface Distress Index* (SDI) the values range from 5-40, with the lowest SDI value located in segment II (5) and the highest SDI value found in segments V and XII (40), where values in the range of 0-50 are indicated Good condition and appropriate care is Routine Maintenance[8].

3.4 Evaluation based on the method of Pavement Condition Index (PCI)

The steps in determining road pavement conditions using the *Pavement Condition Index* (PCI) method are as follows:

1. Determine the severity level
2. Calculates the density value
3. Determine the deduct value of each type of damage
4. Count the allowable maximum deduct value (m)
5. Count the Total Deduct Value
6. Determine the Corrected Deduct Value
7. Calculating PCI value [9]

Calculation of PCI segment I value

1. Severity Level

   The results of the classification of the level of damage for each type of damage at STA 0+ 170 to STA 0 + 270 can be seen in Table 6.

Table 6. Severity level in Segment I.

No	Type Damage	Severity Level
1	Patching	M
2	Swell	H
3	Longitudinal Cracking	L
4	Rutting	L
5	Ravelling	M

2. Density (Damage Percentage)

$$Density = \frac{Ad}{As} x100\% \quad (1)$$

Where,
Ad = total area of the type of damage for each level of damage (m^2)
As = total segment unit area (m^2)[9]

Density calculations at STA 0+ 170 to STA 0+ 270 can be seen in Table 7.

Table 7. Density in Segment I.

No	Type Damage	*Severity Level*	Area (m^2)	*Density (%)*
1	Patches	M	9.4325	1.217
2	Swell	H	14,952	1929
3	Longitudinal Cracking	L	15.5	2,000
4	Rutting	L	1.11	0.143
5	Ravelling	M	16.52	2.132

3. Deduct Value

   Deduct Value (reduction value) is a deduction value for each type of damage obtained from the relationship curve of the density (density) and severity (severity level) of damage . Deduct Value can be seen from the graph contained in ASTMD6433.

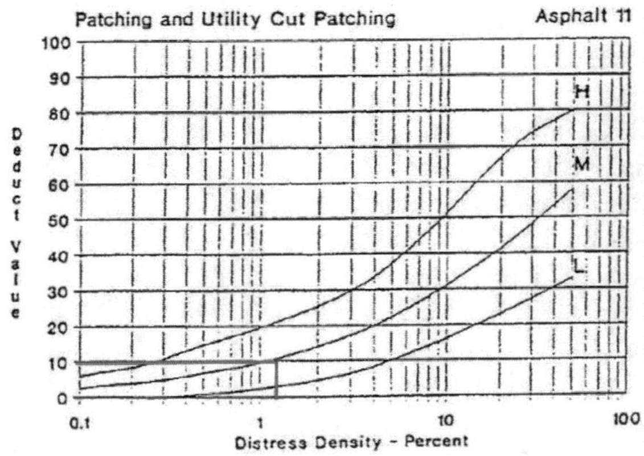

Figure 3. Deduct Value on the Patching.

Results of recapitulation of deduct values at STA 0+ 17 0 to STA 0+ 27 0 can be seen in Table 8.

Table 8. Deduct Value in segment I.

No	Type Damage	Severity Level	Density	Deduct Value
1	Patches	M	1.217	10
2	Swell	H	1.929	38
3	Longitudinal Cracking	L	2.000	6
4	Rutting	L	0.143	1
5	Ravelling	M	2.132	10

4. Allowable maximum deduct value (m) and q

$$m = 1 + \left(\frac{9}{98}\right) x\ (100 - HDV) \quad (2)$$

where,

m = Allowable maximum deduct value
HDV = Highest value deduct value

From the equation above, m = 6.693. Because the m value is 6.693 which is larger than the number of the unit deduct value which is 5, then all values deduct value can used for the next stage. However because condition q value on the CDV graph is marked deduct value which is > 5, then deduct value used only 4 units.

5. Total Deduct Value (TDV)
The results of the total deduct value calculation at STA 0+ 170 to STA 0+ 270 that iterated can be seen in Table 8.
6. Corrected Deduct Value (CDV)
From the analysis results in Table 8, the highest corrected deduct value (CDV) is 53, and this value will be used to calculate the PCI value.

Table 9. TDV values in segment I.

Iteration	*Deduct Value*				TDV	q	CDV
1	38	10	10	6	64	4	36
2	38	10	10	5	63	3	40
3	38	10	5	5	58	2	43
4	38	5	5	5	53	1	53

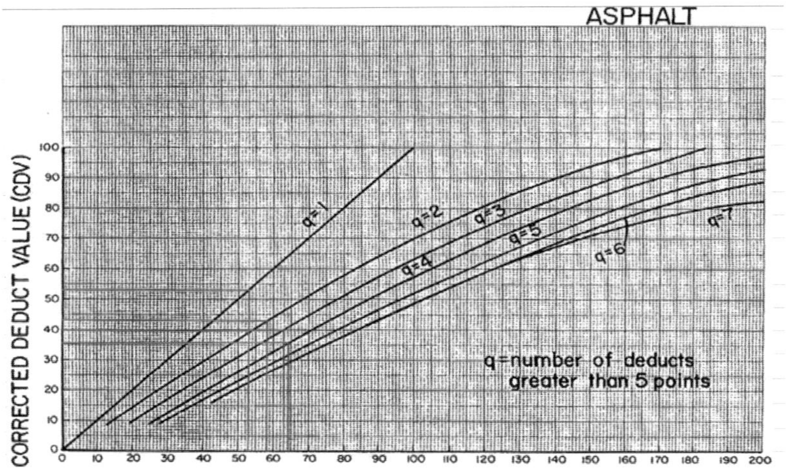

Figure 4. Corrected Deduct Value in Segment I.

7. PCI value
   PCI = 100-CDV
   PCI = 100-53
   PCI = 47

Based on the results of the PCI value calculation at STA 0+ 17 0 to STA 0+ 27 0 is 47, the pavement conditions in this segment are classified as Fair. The results of the PCI value calculation for all segments of the Gunung Agung road can be seen in Table 10.

Table 10. Recapitulation of PCI values for Jalan Gunung Agung.

Segment	STA	PCI Values	*Rating*
I	(STA 0+170 to STA 0+270)	47	*Fair*
II	(STA 0+270 to STA 0+370)	90	*Excellent*
III	(STA 0+370 to STA 0+470)	94	*Excellent*
IV	(STA 0+470 to STA 0+570)	67	*good*
V	(STA 0+570 to STA 0+670)	24	*Very Poor*
VI	(STA 0+670 to STA 0+770)	36	*Poor*
VII	(STA 0+770 to STA 0+870)	62	*good*
VIII	(STA 0+870 to STA 0+970)	66	*good*

IX	(STA 0+970 to STA 1+070)	90	*Excellent*
X	(STA 1+070 to STA 1+170)	71	*Very Good*
XI	(STA 1+170 to STA 1+270)	52	*Fair*
XII	(STA 1+270 to STA 1+370)	42	*Fair*
XIII	(STA 1+370 to STA 1+470)	85	*Excellent*
XIV	(STA 1+470 to STA 1+570)	56	*Good*
XV	(STA 1+570 to STA 1+620)	70	*Good*
	Average	**63.4**	***Good***

3.5 Comparison based on SDI and PCI methods

Based on the average of SDI and PCI values, the Gunung Agung road is in good condition. Although the rating of the Gunung Agung road pavement using the PCI method varies in each segment, the average value is 18 which shows that the road is in good condition. The comparison of the road values can be seen in table 11.

Table 11. Gunung Agung road values based on SDI and PCI methods.

Segment	STA	SDI Values	Rating	PCI Values	Rating
I	(STA 0+170 s/d STA 0+270)	15	Good	47	Fair
II	(STA 0+270 s/d STA 0+370)	5	Good	90	Excellent
III	(STA 0+370 s/d STA 0+470)	25	Good	94	excellent
IV	(STA 0+470 s/d STA 0+570)	35	Good	67	Good
V	(STA 0+570 s/d STA 0+670)	40	Good	24	Very Poor
VI	(STA 0+670 s/d STA 0+770)	10	Good	36	Poor
VII	(STA 0+770 s/d STA 0+870)	10	Good	62	Good
VIII	(STA 0+870 s/d STA 0+970)	20	Good	66	Good
IX	(STA 0+970 s/d STA 1+070)	10	Good	90	Excellent
X	(STA 1+070 s/d STA 1+170)	10	Good	71	Very Good
XI	(STA 1+170 s/d STA 1+270)	10	Good	52	Fair
XII	(STA 1+270 s/d STA 1+370)	40	Good	42	Fair
XIII	(STA 1+370 s/d STA 1+470)	10	Good	85	Excellent
XIV	(STA 1+470 s/d STA 1+570)	10	Good	56	Good
XV	(STA 1+570 s/d STA 1+620)	20	Good	70	Very Good
	Average	**18**	**Good**	**63,4**	**Good**

3.6 Technical Handling

The technical handlings based on the Standard Improvement for Road Repair of Bina Marga 2011 are shown in table 12.

Table 12. Technical handlings based on Bina Marga 2011.

Handling		Type of damages
P1	a	Asphalt bleeding on road pavement
	b	Asphalt bleeding on shoulder
P2	a	Cracking, width < 2mm

P3	a	Shoulder drop off
	b	Alligator Cracking, width < 2mm
	c	Longitudinal Cracking, width < 2mm
	d	Ravelling
P4	a	Longitudinal Cracking, width < 2mm
P5	a	Longitudinal Cracking, width > 2mm
P6	a	Potholes, depth >50 mm
	a	Bump, depth >30 mm
	b	Rutting, depth >50 mm
	c	Depression, depth >50 mm
	d	Swell, depth >50 mm
	e	Shoulder drop off
	f	Alligator cracking, width > 2 mm
	g	Potholes, depth >50 mm on shoulder
	h	Depression, depth >50 mm on shoulder
	i	Swell, width >50 mm on shoulder
	j	Alligator cracking, width >2 mm on shoulder

Based on the type of damages listed in the Standard Improvement for Road Repair of Bina Marga 2011, the technical handling that can be applied to Jalan Gunung Agung is shown in Table 13.

Table 13. Recapitulation of technical handling of damage to Jalan Gunung Agung.

Segment	STA	Handling
I	(STA 0+170 to STA 0+270)	P2, P4, P5, P6
II	(STA 0+270 to STA 0+370)	P2, P4, P5 , P6
III	(STA 0+370 to STA 0+470)	P2, P4, P5, P6
IV	(STA 0+470 to STA 0+570)	P2, P4, P5, P6
V	(STA 0+570 to STA 0+670)	P2, P4, P5 , P6
VI	(STA 0+670 to STA 0+770)	P2, P4, P5 , P6
VII	(STA 0+770 to STA 0+870)	P4, P5 , P6
VIII	(STA 0+870 to STA 0+970)	P2, P5, P6
IX	(STA 0+970 to STA 1+070)	P2, P4, P5
X	(STA 1+070 to STA 1+170)	P2, P4, P5 , P6
XI	(STA 1+170 to STA 1+270)	P5
XII	(STA 1+270 to STA 1+370)	P5
XIII	(STA 1+370 to STA 1+470)	P5
XIV	(STA 1+470 to STA 1+570)	P2, P4, P5 , P6
XV	(STA 1+570 to STA 1+620)	P2, P4, P5, P6

Description :
P1 : Sand spreading
P2 : Paving
P3 : Crack closure
P4 : Crack filling
P5 : Hole patching
P6 : Flattening [10]

The technical handling that can be applied to the flexible pavement of Jalan Gunung Agung are P2 (Asphalt), P4 (Filling cracks), P5 (Filling holes), and P6 (Gravelling).

4. Conclusion

Based on the results of research conducted by the author on the Gunung Agung road from STA 0+170 to STA 1+620, several things can be concluded as follows:

1. There are 7 (seven) types of damage along the flexible pavement of the road, namely Rutting, Swell, Potholes, Ravelling, Longitudinal Cracking, Alligator Cracking, and Patching. The damage that occurred to the flexible pavement of the Gunung Agung road was 690,903 m2 or 6,510% of the total area of the flexible pavement. The biggest damage was found in alligator Cracking, which is 415.02 m2 of the total road area.
2. The SDI value of segment I to segment XV is in the range of 5-40, and the values are included in the category <50 with a description of Good condition. The average SDI value of all segments is 18, with the type of handling Routine Maintenance. The PCI value of segment I value up to segment XV is variated, the segments with the smallest PCI values are segment V (STA 0+570 to 0+670) is 24 (*Very Poor)*, segment VI (STA 0+670 to STA 0+770) is 36 (*Poor*), and segment XII (STA 1+270 up to STA 1+370) is 42 (*Fair*). The average PCI value of the whole segment is 63.4 where the rating value the categorized as Good.
3. Technical handling that can be applied on Jalan Gunung Agung pavement is P2 (Asphalt), P4 (Filling crack), P5 (Patching hole), and P6 (Flattening).

Bibliography

[1] Pemerintah Kabupaten Tabanan 2023 *Topografi* https://tabanankab.go.id/home/mengenal-tabanan/topografi (accessed Mar. 21, 2023)

[2] Naraputra A 2022 *Koordinasi Simpang Bersinyal di Kabupaten Tabanan (Studi Kasus Simpang Gerokgak, Simpang Kasih Ibu, dan Simpang Dukuh*

[3] Fatikasari A 2021 *Analisa Tingkat Kerusakan Jalan Menggunakan Metode PCI Untuk Mengevaluasi Kondisi Jalan di Raya Cangkring, Kecamatan Krembung, Kabupaten Sidoarjo* **vol. 6** no. 2 pp. 1–6

[4] Direktorat Jenderal Bina Marga 1997 *Tata Cara Perencanaan Geometrik Jalan Antar Kota No. 038/TBM/1997 Direktorat Jenderal Bina Marga* no. 038

[5] Kementerian Pekerjaan Umum 2010 *Perbaikan Standar Untuk Pemeliharaan Rutin Jalan* no. 001

[6] Sugiyono 2012 *Metode Penelitian Kuantitatif Kualitatif dan R&D*

[7] Rachman D N and Sari P I 2021 A*nalisis Kerusakan Jalan Dengan Menggunakan Metode Pci Dan Strategi Penanganannya (Studi Kasus Jalan Nasional Srijaya Raya Palembang Km 8+149 Sd Km9+149*) *J. Tek. Sipil* **vol. 10** no. 1 pp. 13–24 doi: 10.36546/tekniksipil.v10i1.456

[8] Aptarila G, Lubis F, and Saleh A 2020 *Analisis Kerusakan Jalan Metode SDI Taluk Kuantan - Batas Provinsi Sumatera Barat* Siklus *J. Tek. Sipil* **vol. 6** no. 2 pp. 195–203 doi: 10.31849/siklus.v6i2.4647

[9] ASTM International 1995 *Standard Practice for Shotcrete*

[10] Kementerian Pekerjaan Umum dan Perumahan Rakyat 2006 *Peraturan Pemerintah Nomor 34/2006 Tahun 2006 tentang Jalan* **vol. 44** no. 2

IOP Conf. Series: Earth and Environmental Science 1294 (2024) 012031 doi:10.1088/1755-1315/1294/1/012031

Road Surface Identification, Roads (Tanah Tinggi – Jati Perumnas) Ternate City Using The PCI Method

Sabaruddin[1]**, Rizky KS**[1] **and Nurul I**[2]

[1]Khairun University, Faculty of Engineering, civil engineering study program, postal code 97716, Indonesia
[2]Khairun University, Civil engineering master's degree program, postal code 97716, Indonesia
sabaruddin.new@gmail.com

Abstract. The section (Highway of Tanah Tinggi – Jati Perumnas) is a primary local road as a road that connects the sub-districts of Central Ternate city and South Ternate city. the Tanah Tinggi – Jati Perumnas highway is a road that has a high traffic density on weekdays and weekends. As a main road, it is important to know the surface condition of this road. To find out the condition of the road surface, identification was carried out using the Pavement Condition Index (PCI). The identification results show that the Jalan Raya Tanah Tinggi – Perumnas Jati section in segment 5 (sta 200 to sta 250) obtained a total score of 23 with a reduction correction value of 24 resulting in a PCI score of 76 (very good), where the damage that occurred was: Bleeding , Polished Aggregate, Patching & util Cut Patching, Block Cracking, Long & Trans Cracking.

1. Introduction

Highways are a unit of roads that connect and bind each other between growth centers so that roads can be interpreted as the most important part in the framework of realizing the development of a region and as a means to improve people's lives for a better life, as defined in the definition of roads (UU No. 22 of 2009) concerning road traffic and transportation. [1]
applying the above understanding and observing the condition of the Tanah Tinggi - Jalan Jati Perumnas road which is very congested in segment 5 (sta 200 to sta 250), it is considered important to carry out identification..[2]
This research aims to determine the surface condition of Jalan Raya Tanah Tinggi – Perumnas Jati in segment 5, which is 50 meters long and 6.9 meters wide.

2. Literature review

Please follow these instructions as carefully as possible so all articles within a conference have the same style to the title page. This paragraph follows a section title so it should not be indented.

2.1. pavement distress

Pavement distress is an external element of the pavement in the form of damage to the pavement caused by several things, for example loading, environmental factors, construction weaknesses, or a combination of both. Common damage that occurs is cracks, wheel marks, and weathering on the pavement surface.[3].There are three levels of severity used in PCI calculations, namely low severity (L), medium severity (M) and high severity (H).[3].

2.2. *Pavement Condition Index (PCI)*

a numerical rating of the pavement condition that ranges from 0 to 100 with 0 being the worst possible condition and 100 being the best possible condition Pavement condition rating- a verbal description of pavement condition as a function of the PCI value. Fig. 1 show an example of the PCI rating scale

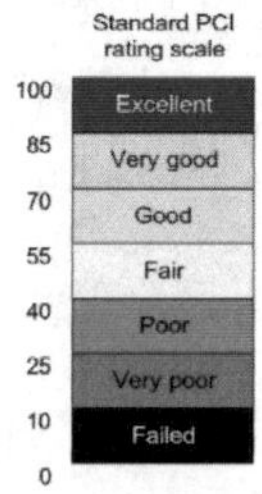

Figure 1. PCI, rating scale, and suggested colors. [4],[5]

3. Figures and tables

3.1. *Overview of risearch location*

This research was carried out in Ternate City by collecting data on roads as shown in Figure 2

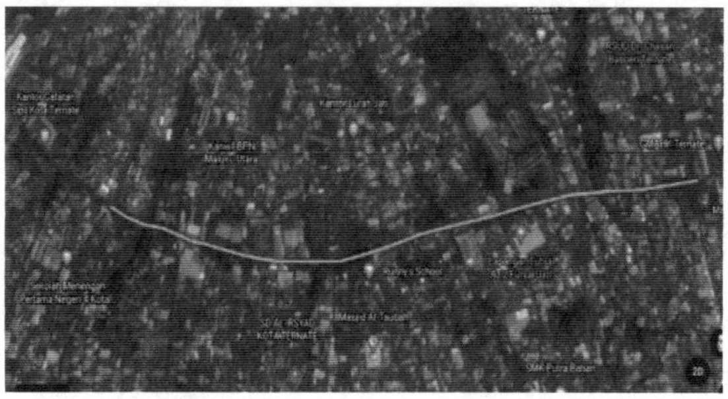

Figure 2. Map of research location.

3.2. *Research flowchart*

The flow chart in this study can be seen in the following figure 3

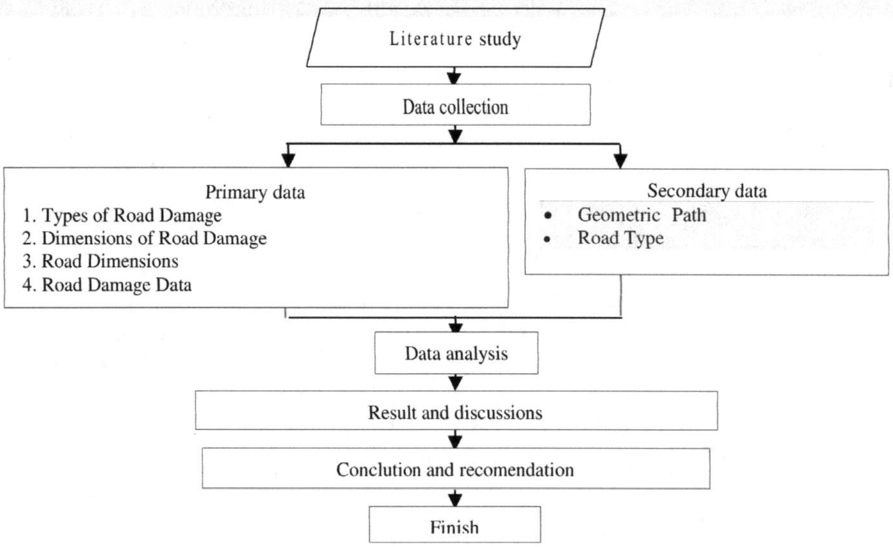

Figure 3. Research flowchart. [2]

3.3. *Data collection technique*

The data collected consists of primary data and secondary data

3.3.1. *Primary data* [2]

- Types of damage and dimensions of road damage were obtained by conducting surveys in the field.
- The equipment used during the survey was a tape measure, paper, stationery, survey forms and a camera.
- The required data is : Measurement of the type of damage, road damage dimensions,data results from field surveys. recording of the location of the damage

3.3.2. *Secondary data* [2]

- Data obtained from related agencies.
- The data required are as follows: road geometric data existing pavement structure data, street type.

Figure 4. roads (Tanah Tinggi – Jati Perumnas). Ternate City

3.3.3. *Data processing* [6]

The identification and determination of the PCI score and ranking are carried out as follows

- Check the sample unit, determine the type and degree of damage, then measure density;
- The reduction value is determined from the reduction value curve for each type of difficulty and severity level;
- The total deductible amount (TDV) is calculated by adding up all the individual deductible amounts.;
- After the TDV is calculated, the corrected correction value (CDV) can be determined correction curve. When determining the CDV, if there is an individual reduction value higher than CDV, CDV is set equal to the highest individual withholding value;
- PCI is calculated using the relation PCI = 100 – CDV.

4. Result and discussion

Identification results, processed and calculated to obtain a PCI value. In this article show segments 5 (200-250) m

Table 4.1 Result Data on the PCI Survey Form.

ASPHALT SURFACED ROAD :	**SKETCH : 50 M**
CONDITION SURVEY DATA SHEET FOR	6.9 **M**
SAMPLE UNIT	Luasan 345 m2

1. Alligator cracking	6. Depression	11. Patching & util cut patching	16. Shoving
2. Bleeding	7. Edge cracking	12. Polished aggregate	17. Slippage cracking
3. Block cracking	8. Jt reflection cracking	13. Potholes	18. Swell
4. Bumps and sags	9. Lane/shoulder drop off	14. Rairoad crossing	19. Weathering/raveling
5. Corrugation	10. Long & trans cracking	15. Rutting	

SEGMEN	DISTRESS SEVERITY	QUANTITY		TOTAL	DENSITY (%)	DEDUCT VALUE	TOTAL DV
5(200 – 250) m	2 L	10.56		10.56	3.06	1	23
	12 L	3.80	9.00	12.80	3.71	1	
	11 L	1.24		1.24	0.36	0	
	3 M	15.75	43.20	58,95	17.09	20	
	10 L	2.10	1.70	3.80	1.10	1	

In the 200 - 250 m segment there is damage as follows:

- Bleeding = 10.56 m^2
- Polished Aggregat = 12.80 m^2
- Patching & util cut patching = 1.24 m^2
- Block cracking = 58.95 m^2
- Long & trans cracking = 3.80 m^2

Density = [Ad/As] x 100 % : [7]

- Bleeding $= \frac{10.56}{/6.9 \times 50}$ X 100% = 3.06%
- Polished Aggregat $= \frac{12.80}{/6.9 \times 50}$ X 100% = 3.71%
- Patching & util cut patching $= \frac{1.24}{/6.9 \times 50}$ X 100% = 0.36%
- Block cracking $= \frac{58.95}{/6.9 \times 50}$ X 100% = 17.09%
- Long & trans cracking $= \frac{3.08}{/6.9 \times 50}$ X 100% = 1.10%

Deduct Value

- Bleeding

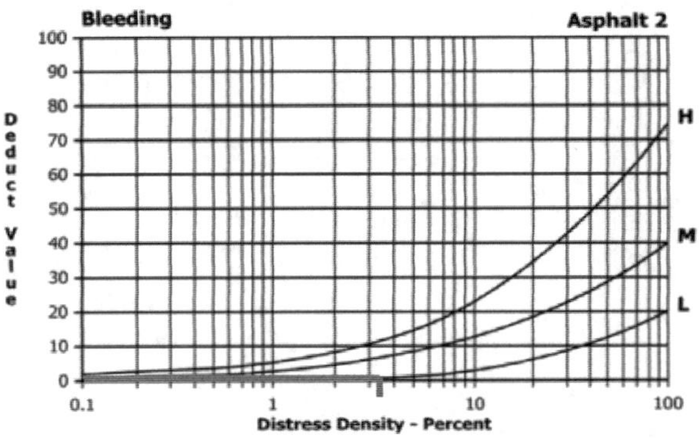

Figure 5. Deduct value bleeding.[3]

- Polished Aggregat

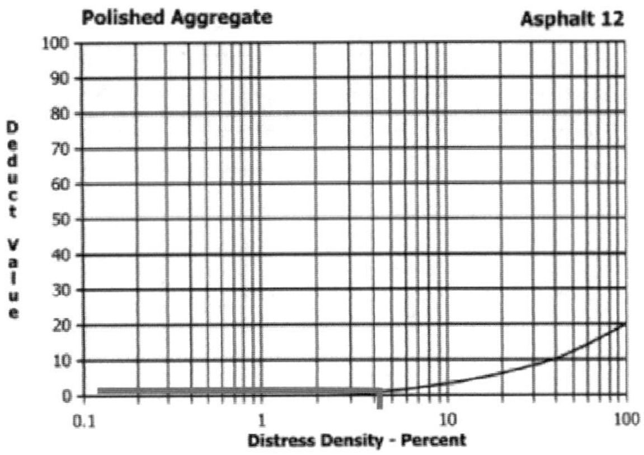

Figure 6. Deduct value polished aggregat. [3]

- Patching & util cut patching

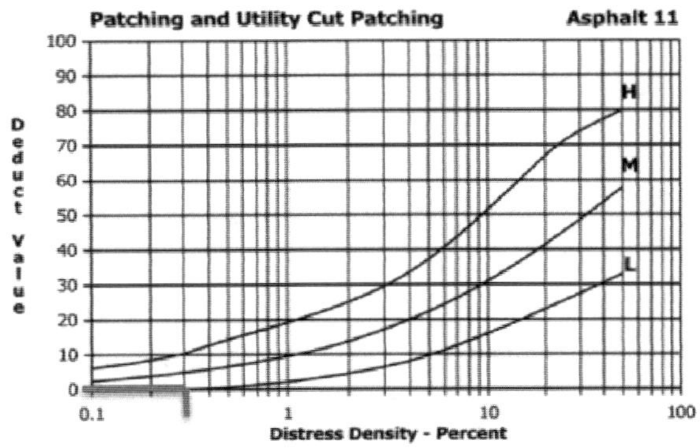

Figure 7. Deduct value patching & util cut patching. [3]

- Block cracking

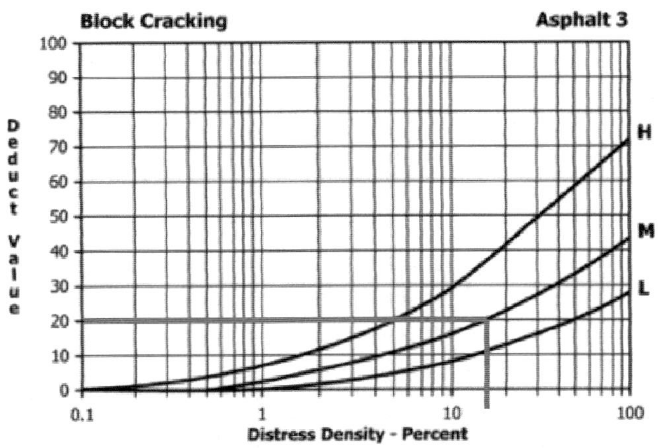

Figure 8. Deduct value block cracking. [3]

- Long & trans cracking

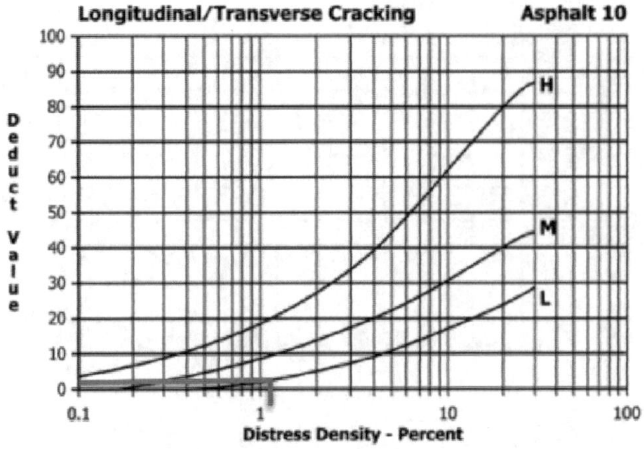

Figure 9. Deduct value Long & trans cracking. [3]

Corrected Deduct Value

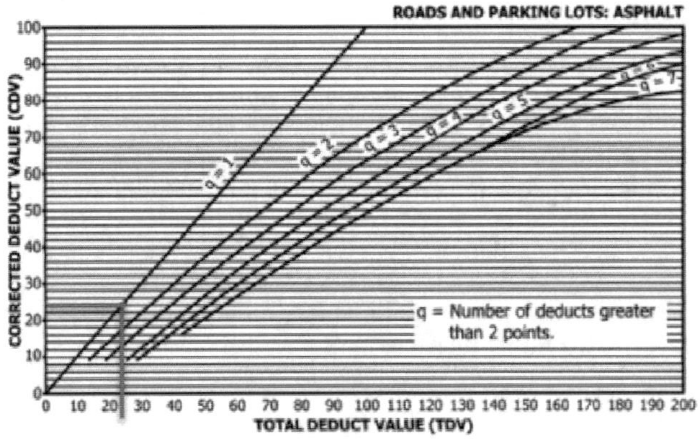

Figure 10. Total Deduct value. [3]

5. Pavement Condition Index (PCI) [8]

PCI = 100 – CDV Max 9
PCI = 100 – 24
PCI = 76 (Very good)

6. Conclusions

The Jalan Raya Tanah Tinggi – Perumnas Jati section in segment 5 (sta 200 to sta 250) obtained a total score of 23 with a corrected reduction value of 24, so a PCI score of 76 (very good) was obtained, where the damage that occurred was • bleeding, polished aggregate, Patching & util cut patching,block cracking, Long & trans cracking.

7. References

[1] "Pemerintah Republik Indonesia.. *Undang-undang Nomor 22 Tahun 2009 Tentang Lalu Lintas dan Angkutan Jalan*. Jakarta, 2009"

[2] "Aldikri Soemawikarta, "*Analisa Kerusakan Jalan Pada Lapis Permukaan Menggunakan Metode Pavement Condition Index (PCI)*, 2021" .

[3] "ASTM D6433-09 / Designation: D6433 – 11, *American Society for Testing and Materials, 'Standard Practice for Roads and Parking Lots Pavement Condition Index Surveys'*, United States, 2009"

[4] "ASTM, Designation: D5340-12. *Standard test method for airport pavement condition index surveys (Reapproved 2018."*

[5] "Kan Wu, *'Development of PCI-based Pavement Performance Model for Management of Road Infrastructure System*, 2015"

[6] "Bubaker M.B. Mohammed, *Application of Pavement Maintenance Management System in Benghazi City Using Micro PAVER Program* (The Higher Institute of Engineering Technology – Benghazi, 2017)."

[7] "Sabaruddin, A Deni, *Application of pavement condition index (PCI) on the assessment of the Kalumata highway section of the City of South Ternate.* , 2020"

[8] "Sabaruddin, Arbain, *Assessment of the Jati Besar Highway of the City of South Ternate Uses Pavement Condition Index (PCI)*, Journal of Physics: Conference Series, 2021."

Acknowledgments

The author thanks the ISTSDC Committee for their willingness to accept this article. The author hopes that the ISTSDC activities will go well.
Finally always successful.

IOP Conf. Series: Earth and Environmental Science 1294 (2024) 012032 doi:10.1088/1755-1315/1294/1/012032

Sustainable aviation fuel development: case study in Indonesia

D A Nugroho, M R Sitompul, and N Widadi

Transport Policy Agency, Ministry of Transportation Republic of Indonesia, Medan Merdeka Timur 5, Central Jakarta, Jakarta, 10110 Indonesia

dio.agronugroho@gmail.com, rafiqi.riqi@gmail.com

Abstract. The usage of fossil fuel in aviation contributes to 2% of CO_2 emission globally. Though the industry keeps improving the efficiency of the aircraft, the increase in activity offset this improvement. The usage of sustainable aviation fuel (SAF) may become a potential option in order to reduce emissions. SAF immediate compatibility and low emission become the main attraction for its usage. Indonesia in order to fulfill its NZE commitment is currently developing its own biofuel. This study's objective is to obtain information on SAF or biofuel development in Indonesia by discussion with experts and stakeholders related to aviation and fuel development. By conducting this study it is known that the development of biofuel in Indonesia has successfully resulted in two products, J2.0 and J2.4, biofuel developed based on palm oil. Though these fuels have been certified domestically, their implementation still faces several issues, such as the sustainability aspect related to the CORSIA scheme, the production volume and availability, and potential price surge. Therefore through this study several recommendations are presented in the hope to accelerate the implementation. Although the direction of SAF development in Indonesia is through the usage of Palm oil as its main feedstock, Indonesia also has used cooking oil (UCO) and Sugarcane as potential feedstock to be explored. The development and implementation for these feedstock into SAF is necessary as a possible solution for reducing the greenhouse gas emission, therefore a support through financial, technological, and regulation means are necessary to be provided by all stakeholders.

1. Introduction

Heavy usage of fossil fuel nowadays such as in the transportation sector contributes heavily to global warming [1]. The usage of this fuel increases the contain of several gasses, such as Carbon dioxide (CO_2), Methane (NH_4), Sulfur dioxide (SO_2), and Nitrogen dioxide (NO_2) [2]. Human activity contributes to 94 percent of carbon emission through the usage of natural resources. It is also estimated that the emissions produced in aviation affect the health of people around the globe [3]. The CO_2 emission becomes the major contributor due to its quantity and the duration its resides in the atmosphere [4]. In 2018, the transportation sector was the second largest contributor to greenhouse gas emissions in Indonesia, namely 26.4 percent or around 157,325 Gigagrams CO2 equivalent. In the past decade, there has been a notable annual increment of 7.88 percent on average [5]. The emission is reduced for several years due to the COVID-19, but the number will soon bounce back to its pre-pandemic condition [6].

Indonesia, along with many countries, try to address the emission produced by the aviation Industry. At COP26 in Glasgow, Scotland (Oct-Nov 2021), it was emphasized by Indonesia that emissions reduction should be achieved through increased utilization of new and renewable energy sources, including the use of biofuels. This is also aligned with the Republic of Indonesia Minister of

Transportation's Decree No. KP. 201 of 2013 on the Determination of the National Action Plan for Greenhouse Gas Emission Reduction in the Transportation Sector and the greenhouse gas inventory in the transportation sector from 2010 to 2020, which states that the initiative to use biofuels in the transport category of air transportation is in accordance with the ICAO Guideline, with a 3% implementation pilot conducted from 2018 to 2020. In November 2022, G20 Summit in Bali was held and one of the outcomes of the summit is a declaration document called the G20 Bali Leaders Declaration, in which G20 member countries committed to reducing environmental impacts by changing unsustainable consumption of energy [7].

Biofuels, in this case the Sustainable Aviation Fuel (SAF), become one of the potential candidates for reducing the aviation sector's climate impact [8]. As an agricultural country, Indonesia has many potential feedstock as the ingredient for its SAF. However, for the producer to provide the necessary quantity, the production may compromise with the sustainability aspect such as the land use change, such as the case in Indonesia's choice of Palm Oil as the SAF feedstock. Therefore in order to obtain insight for the further development of an ICAO Acceptable SAF, the understanding for the development of SAF in Indonesia is necessary. SAF in Indonesia is a complex study with aspects consisting of regulation, technological limitation, availability of feed stock, political and economical aspects.

2. Method

The method of this research is a descriptive method, which elaborates the condition supported by the data obtained. The data was obtained by deep interview and discussion with experts in the environmental aspect in Air Transport in the Ministry of Transportation of Republic Indonesia, experts in the Ministry of Energy and Mineral Resources of Republic of Indonesia. Other than the discussion and interview, data is also obtained through regulation, article, research paper, and other literature. Regulations pertaining to SAF are identified, and the potential development of SAF feedstock is expounded. Suggestions and recommendations are formulated by establishing connections between the current SAF development, local regulations, and the available resources for SAF production in Indonesia.

3. Result and discussion

3.1. SAF development

Indonesia aims to achieve Net Zero Emission by 2060 or earlier. This is reinforced by the President's statement at the 21st Conference of the Parties (COP) of the United Nations Framework Convention on Climate Change (UNFCCC) in Paris (Nov-Dec 2015), where Indonesia targeted a 29% reduction in greenhouse gas (GHG) emissions (with its own capabilities) or a 41% reduction (with international assistance) by 2030.

3.1.1. Sustainability of SAF. SAF became the main option in facing the emission issue due to its versatility. One advantage of SAF is its similarity to conventional fuel, so if a processed feedstock meets the ICAO criteria based on conversion methods following the standards of the American Standard Testing and Material (ASTM International), the resulting fuel can be used in aircraft directly without any modifications to the aircraft structure or engine.

ICAO, as the international aviation governing body, states that in order for a biofuel or bio-aviation fuel to be classified as sustainable aviation fuel (SAF), a Life Cycle Assessment (LCA) is necessary to determine the emissions of the product throughout its life cycle. ICAO has approved nine conversion methods for producing SAF, including examples of feedstocks such as palm oil, animal fat, grease, and used cooking oil.

3.1.2. SAF development in Indonesia. Development of SAF in Indonesia is pioneered by the Ministry of Transportation, Ministry of Energy and Mineral Resources, Pertamina and Institut Teknologi Bandung. This effort is enforced by the release of Ministerial Regulation of Ministry of Energy and Mineral Resources No 12 Year 2015, which states that the mixture target of SAF or Bioavtur, produced by Indonesia minimum should be 2% by 2016, 3% by 2020, and 5% by 2025 [9]. Despite the target in the Regulation, the progress of development of SAF in Indonesia faces some challenges. Biofuel mixture developed in Indonesia currently only resulted in a 2.0 % and 2.4% mixture even though the mixture target is 3% by 2020. These fuels are produced in the form of J2.0 and J2.4 fuel respectively, produced by Pertamina in PT. Kilang Pertamina Internasional in Cilacap Refinery.

3.2. *Regulation in Indonesia related to SAF*

Indonesia's Government has issued several regulations to emphasize its commitment for the NZE target. The regulations are issued through several levels of government organization, from Presidential Regulation, Governmental Regulation, Ministerial Regulation, and many others. Table 1 shows several regulations regarding the usage of renewable energy and biofuel.

Table 1. Regulation sample concerning renewable energy and biofuel in Indonesia

No.	Regulation	Description
1	Presidential Regulation No 5 Year 2006 about National Energy Policy	The use of biofuel as stated in Article 2 which states that the realization of an optimal mixed energy by 2025, namely the role of each type of energy in national energy consumption of biofuel becomes more than 5%. In line with this, the use of petroleum has become less than 20%, and regarding New Renewable Energy especially biomass, nuclear, small-scale hydropower, solar power and wind power, it has become more than 5%.
2	Ministerial Regulation of Ministry of Energy and Mineral Resources No 12 Year 2015	Regulate regarding minimum mixture requirement of biofuel usage. The biofuels regulated are biodiesel (B100), bioethanol (E100), and Pure vegetation oil (E100). The stages of minimum obligations for the utilization of pure bio-oil as a blend in fossil fuels is shown in Table 2.
3	Presidential Regulation No 98 Year 2021	The regulation acts as the basis of carbon economical value and guidance in reducing greenhouse gas emission (GHG) through policy and action in order to reach the NDC target and control the greenhouse gas emission in national development. The NDC targets are: 1. Establish the policy and action including implementation of activities according to the Government commitment to reduce GHG emission 29% - 41% in the year 2030 compared to GHG emission baseline 2. Develop national, territorial and societal resilience from various risks due to climate change.
4	Governmental Regulation No 79 Year 2014 regarding National Energy Policy.	Mixture target of Primary Energy Source: 1. In 2025 renewable energy will contribute at least 23% and 31% in 2050. 2. In 2025 the contribution of crude oil will be less than 25% and less than 20% in 2050. 3. In 2025, the contribution of coal will be less than 30% and in 2050 will be less than 25%. 4. In 2025, the contribution of natural gas will be less than 22% and in 2050 will be less than 24%.

Tabel 2. The stages of minimum obligations for the utilization of pure bio oil as a blend in fossil fuels[a]

Sectors	April 2015	January 2016	January 2020	January 2025	Note
Industrial (Low and Medium speed engine)	10%	20%	20%	20%	Against total needs
Marine Transport	10%	20%	20%	20%	Against total needs
Air Transport	-	2%	3%	5%	Against total needs
Power plans	15%	20%	20%	20%	Against total needs

[a] Ministerial Regulation of Ministry of Energy and Mineral Resources Number 12 Year 2015

3.3. Potential feedstock in Indonesia

3.3.1. Palm Oil. As the world's largest palm oil producer, Indonesia possesses significant potential for the processing and marketing of palm oil-derived products, including palm kernel oil. Palm kernel oil can serve as a readily available feedstock for biofuels and sustainable aviation fuels. Understanding the potential production of palm kernel oil as SAF necessitates an examination of its production trends over years, as depicted in Figure 1.

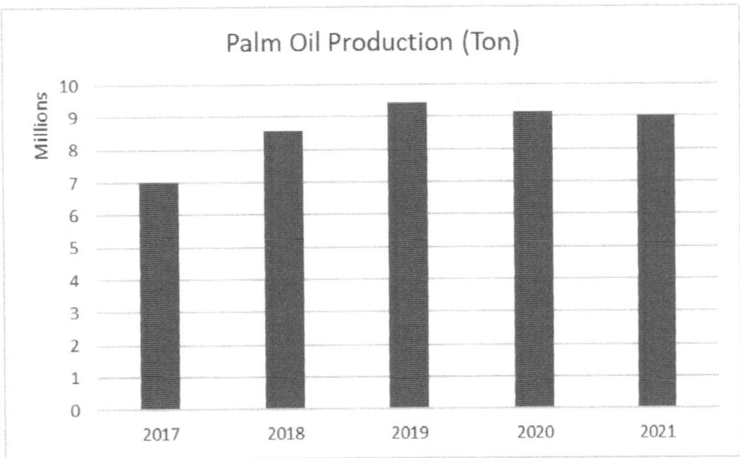

Figure 1. Indonesian Palm Oil Production 2017-2021
Source: BPS-Statistics Indonesia [10]

Due to the large quantity of palm oil produced in Indonesia, The biofuel developed by Pertamina is made from palm kernel oil as previously mentioned. This J2.4 biofuel has been tested in 2021 by using CN-235 aircraft owned by the Indonesian Air Force. The decision to use palm oil is based on the abundance of the palm oil in Indonesia as one of the leading producers in the world [11]. Although both of the biofuels have been certified by Indonesia's own Oil and Gas Testing Agency (LEMIGAS), their acceptance into the CORSIA scheme in ICAO may still be an issue. These biofuels have yet to get the SAF certification recognized by ICAO in order to be eligible for the CORSIA scheme, mainly due to the sustainability aspect of the fuel itself [12]. The loss of the rainforest in Indonesia due to the land use change is significant, around 40 million ha of land [13]. This issue became the main obstacle regarding the sustainability aspect in LCA for the recognition of J2.4 and J2.0. Conducting a more comprehensive review regarding LCA calculations may become one of the solutions, by considering that the previous LCA calculations by ICAO were based on default values. ICAO allows the producer or country to calculate the actual LCA by providing proper and necessary technical information by scientific procedure and tests [14]. The authorized body responsible for calculating Actual LCA should be appointed to formulate emission reduction solutions for the product life cycle. This can be learned from Brazil, which also develops SAF based on corn, soy, and sugarcane, but has low emissions. The development of SAF from other sources such as used cooking oil should also be considered.

The development of bio-aviation fuel by Pertamina is also still constrained by the current limitations of the refinery, where the Cilacap refinery used for biofuel production is not dedicated solely to biofuel. The new Cilacap Phase 2 refinery is planned to be built to meet the 5% blend target by 2025, which has become a national strategic program coordinated by the Ministry of Energy and Mineral Resources. Though the Indonesian Government, Pertamina, Institut Teknologi Bandung, and Garuda Indonesia plan to conduct the next phase of the test by using commercial aircraft in 2023, the production volume capability still remains. The construction of the refinery requires support in the form of investment funds and cooperation.

Based on sources from Indonesia National Aircraft Carriers Association (INACA), it is predicted that there will be a 10% increase in fuel prices with a 5% biofuel blend. This prediction may hinder the usage of the biofuels/SAF even after the successful development and production. The Ministry of Energy and Mineral Resources and the Ministry of Transportation, as regulators, need to develop temporary incentive schemes to assist airlines in implementing SAF in the future. These incentive schemes can be provided not only to airlines but also to attract other producers or developers of SAF to further progress SAF development in Indonesia.

3.3.2. Used Cooking Oil. Heavy usage and production of Palm oil Indonesia for Food usage resulted in the abundance of Used Cooking Oil (UCO). The usage of UCO as feedstock of SAF is already known among SAF Producers. NESTE, one of SAF Producer from Finland is currently producing an ICAO approved SAF which also complies with CORSIA Scheme using UCO as feedstock. Currently one of NESTE's refineries is the newly finished refinery located in Singapore with production capacity of 1.3 million tonnes annually [14]. This refinery is producing SAF with UCO as one of its feedstock.

Press release of the Ministry of Energy and Mineral Resources in 2020 mentioned that in the year of 2019, Indonesia consumed 16.2 million kilo liters (KL) [15]. Despite the potential of UCO as the more sustainable alternative for SAF feedstock, it has issues that need to be addressed before it becomes feasible. The two main issues of UCO as feedstock is the collecting mechanism and the export attractiveness. Out of 16.2 million KL, 6.46-9.72 million KL were produced and only 3 million KL of UCO were successfully collected [14]. From those quantities, 2.4 million KL were recycled to be reused for cooking and exported. Therefore in order for the UCO to be considered, the availability of UCO needs to be solved first.

Regulation or policy that can boost the appeal for UCO to be sold or used domestically need to be introduced. By introducing this policy, the distributor of UCO then will shift their customer from export to domestic companies instead. However, this policy will need to be accompanied by incentive and numerous benefits and mandates, which will also result in mixed reaction.

3.3.3. Sugarcane. Sugarcane is a potential feedstock for SAF through its derivative of sugar and alcohol. The two derivatives are enlisted in the feedstock and conversion process of SAF production based on ASTM D7566 and D1655. The development of SAF is not yet commercialized and requires an investment for its development for production in industrial scale. The potential of sugarcane as the feedstock is shown from the production quantity, as shown in Figure 2.

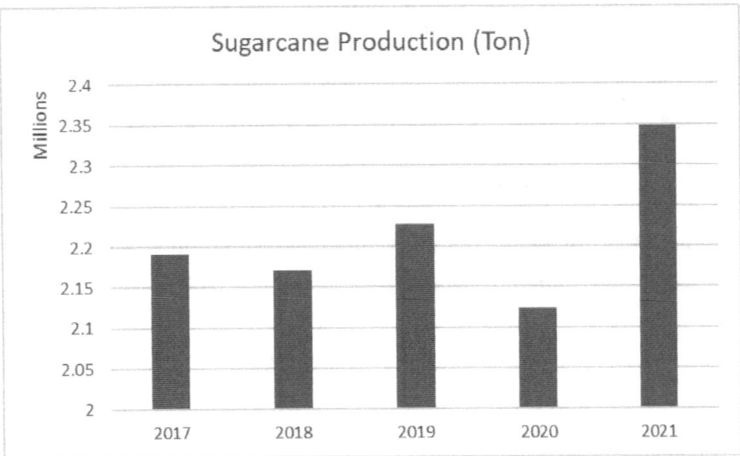

Figure 2. Indonesian Sugarcane Production 2017-2021
Source: BPS-Statistics Indonesia [16]

4. Conclusion

Development of SAF Indonesia, in order to fulfill the Commitment for NZE in 2060, Indonesia through the Ministries, State owned Companies, and University, has been started. Currently Indonesia through Pertamina and Institut Teknologi Bandung successfully developed Indonesia's own biofuel, the 2.0 percent mixture J2.0 and 2.4 percent mixture J2.4. Though currently these fuels face several obstacles in the form of the sustainability aspect, the volume production capability, and the inescapable threat of price surge compared to regular jet fuel, both of the fuels have been successfully certified by Lemigas. Moreover, a flight test has also been conducted through the usage of military aircraft owned by Indonesia Air Force in 2021, with the commercial aircraft flight test scheduled in 2023. Although currently the direction of SAF development in Indonesia is through the usage of Palm oil as its main feedstock, Indonesia also has UCO and Sugarcane as potential feedstock to be explored, though there are some challenges that need to be addressed first for them to be considered. The development and implementation for these feedstocks into biofuels or even SAF in Indonesia is necessary as a possible solution for reducing the greenhouse gas emission, therefore a support through financial, technological, and regulation means are necessary to be provided by all stakeholders.

5. Suggestion

In order for the fuel to be successfully implemented, several actions may be required:

a) Mapping of stakeholders involved in SAF development according to their respective focuses, such as Implementation in transportation, Research for alternative feedstocks, Biofuel industry towards SAF, potential scheme and participant in Investment, research and innovation for the further development.
b) Encouraging the formulation of cooperation through MoUs and the establishment of task forces involving multi-stakeholders.
c) Establishing an authorized institution or agency to calculate the Actual Life Cycle Emissions (LCA) at the national level. Therefore, Indonesia isn't necessarily required to use the default value provided by ICAO.
d) Further exploration is needed regarding other potential sources for biofuel, such as used cooking oil since its abundance in Indonesia due to the extensive consumption of frying food. A proper regulation concerning the limitation of UCO export may be necessary to reinforce the availability of UCO domestically.
e) Pertamina, with support from the Ministry of Energy and Mineral Resources, needs to develop a new refinery dedicated to producing SAF. Focus for the palm oil industry is also needed in order to provide the ample amount of material for the biofuel production.

IOP Conf. Series: Earth and Environmental Science 1294 (2024) 012032 doi:10.1088/1755-1315/1294/1/012032

f) As the prices of the SAF are possibly higher than the regular jet fuel, correct policies need to be formulated. This policy will address the increase in operating costs for the commercial implementation of SAF, in order to assist air transport operators in maintaining profitability. Studies and research need to be conducted to determine suitable incentives. Currently, only biodiesel made from palm oil as feedstock has incentives.
g) In addition to domestic bio-aviation fuel development, stakeholders need to coordinate and consider establishing cooperation with foreign SAF producers who have succeeded in producing and developing SAF. It is necessary to conduct benchmarking studies in countries that have succeeded and advanced in SAF development.

References

[1] Chandel S S, Shrivastva R, Sharma V, Ramasamy P 2016 Overview of the initiatives in renewable energy sector under the national action plan on climate change in India (Renew Sustain Energy Rev 54)
[2] Mawardi I, Winanti W S, Sudinda T W, Alimin A 2022 Analysis Of Achievements Of Net Zero Carbon Emissions As Indonesia'S Reference In Fulfillment Of Climate Change Mitigation Commitments (Journal of Science and Disaster Mitigation Technology 16)
[3] Yim S H L, Lee G L, Lee I W, Allroggen F, Ashok A, Caiazzo F, Eastham S D, Malina R and Barrett S R H 2015 Global, regional and local health impacts of civil aviation emissions (Environ. Res. Lett. 3 034001)
[4] Terrenoire E, Hauglustaine D A, Gasser T, Penanhoat O 2019 The contribution of carbon dioxide emissions from the aviation sector to future climate change (Environ Res Lett 14(8):084019)
[5] Pusat Data dan Teknologi Informasi Energi dan Sumber Daya Mineral 2019 Inventarisasi Emisi GRK Sektor Energi. Jakarta. Available:https://www.esdm.go.id/assets/media/content/content-inventarisasi-emisi-gas-rumah-kaca-sektor-energi-tahun-2019.pdf.
[6] Cabrera E, Sousa J M M d 2022 Use of Sustainable Fuels in Aviation—A Review (Energies 15) p 2440
[7] G20 Bali Leaders' Declaration 2022. https://www.g20.org/content/dam/gtwenty/gtwenty_new/about_g20/previous-summit-documents/2022-bali/G20%20Bali%20Leaders%27%20Declaration,%2015-16%20November%202022.pdf access on 14th of July 2023
[8] Grimme W 2023 The Introduction of Sustainable Aviation Fuels—A Discussion of Challenges, Options and Alternatives (Aerospace 10) p 218
[9] Ministerial Regulation of Ministry of Energy and Natural Resources No 12 Year 2015.
[10] BPS-Statistics Indonesia 2022 Indonesian Oil Palm Statistics 2021. (Jakarta. ISSN: 1978-9947)
[11] Hinandar D, Sari D W 2020 Output Of Indonesian Crude Palm Oil (Cpo) Industry: a Review - Running Head : Output Of Cpo In Indonesia (International Journal of Management 11) p 6
[12] ICAO. Annex 16 - Environmental Protection, Volume IV, Carbon Offsetting and Reduction Scheme for International Aviation (CORSIA)
[13] Wicke B, Sikkema R, Dornburg V, Faaij A 2011 Exploring land use changes and the role of palm oil production in Indonesia and Malaysia (Land Use Policy 28)
[14] ICAO 2021. ICAO Document - CORSIA Methodology for Calculating Actual Life Cycle Emissions Values.
[15] Ministry of Energy and Mineral Resources 2020 Press Release. NUMBER: 388.Pers/04/SJI/2020 Potential Energy Business from Used Cooking Oil
[16] BPS-Statistics Indonesia 2022 Indonesian Sugar Cane Statistics 2021 (Jakarta. ISSN: 2338-6991)

AUTHOR INDEX